Topics in Applied Physics Volume 2

Topics in Applied Physics Founded by Helmut K.V. Lotsch

Volume 1	**Dye Lasers** Editor: F. P. Schäfer
Volume 2	**Laser Spectroscopy** of Atoms and Molecules Editor: H. Walther
Volume 3	**Numerical and Asymptotic Techniques in Electromagnetics** Editor: R. Mittra
Volume 4	**Interactions on Metal Surfaces** Editor: R. Gomer
Volume 5	**Mössbauer Spectroscopy** Editor: U. Gonser
Volume 6	**Picture Processing** and **Digital Filtering** Editor: T. S. Huang
Volume 7	**Integrated Optics** Editor: T. Tamir
Volume 8	**Light Scattering in Solids** Editor: M. Cardona
Volume 9	**Laser Speckle** and Related Phenomena Editor: J. C. Dainty
Volume 10	**Transient Electromagnetic Fields** Editor: L. B. Felsen
Volume 11	**Digital Picture Analysis** Editor: A. Rosenfeld
Volume 12	**Turbulence** Editor: P. Bradshaw
Volume 13	**High-Resolution Laser Spectroscopy** Editor: K. Shimoda
Volume 14	**Laser Monitoring of the Atmosphere** Editor: E. D. Hinkley
Volume 15	**Radiationless Processes** in Molecules and Crystals Editor: F. K. Fong

---

Laser Spectroscopy

of Atoms and Molecules

Edited by H. Walther

With Contributions by
F. A. Blum J. M. Cherlow B. Decomps
M. Ducloy M. Dumont K. M. Evenson
E. D. Hinkley K. W. Nill F. R. Petersen
S. P. S. Porto K. Shimoda H. Walther

With 137 Figures

Springer-Verlag Berlin Heidelberg New York 1976

Professor Dr. HERBERT WALTHER

Sektion Physik der Universität München
D-8046 Garching, Am Coulombwall 1, Fed. Rep. of Germany

ISBN 3-540-07324-8 Springer-Verlag Berlin Heidelberg New York
ISBN 0-387-07324-8 Springer-Verlag New York Heidelberg Berlin

Library of Congress Cataloging in Publication Data. Main entry under title: Laser spectroscopy of atoms and molecules. (Topics in applied physics; v. 2). Includes bibliographical references and index. 1. Laser spectroscopy. 2. Atomic spectra. 3. Molecular spectra. I. Walther, Herbert, 1935—. II. Blum, Fred A., 1939—. QC454.L3L37 535.5'8 75-37819

Monophoto typesetting, offset printing, and bookbinding: Brühlsche Universitätsdruckerei, Giessen

Preface

When the laser was invented one and a half decades ago it became obvious at once that this new device would be a unique light source for many spectroscopic applications. Spectroscopists were initially engaged in looking for new laser materials having transitions suitable for laser action. Many new optical transitions were found in this way and a large amount of new spectroscopic data especially in the infrared spectral region was gathered.

Concurrently the interaction of intense laser radiation with matter was investigated, and many nonlinear effects explored. At the same time useful spectroscopic information was obtained by studying Raman spectra in solids, liquids and gases, using the laser as source of intense light.

For the spectroscopy of atoms and molecules at this time the biggest step forward was the discovery of the "Lamb-dip". It opened up the possibility of eliminating the Doppler width and achieving a resolution corresponding to the natural line width. The applications of this phenomenon were more or less restricted to the investigation of the laser transitions themselves or to transitions in molecules having an accidental coincidence with laser lines. However, it was recognized soon that the phenomenon could play a significant role in the frequency stabilization of lasers. The corresponding techniques were developed rather rapidly; they have now reached such a state of development that the frequency stabilization of lasers has become extremely important for setting metrological standards.

The steadily increasing number of available lasers and laser lines made it possible to use the accidental coincidences with electronic transitions of molecules in order to study molecular fluorescence. With pulsed lasers the relaxation of molecular vibrations has been studied. Meanwhile also techniques for the generation of pulses of picosecond duration were developed so as to open up relaxation studies in liquids and solids. Furthermore, the coherent optical field provided by the laser was used to investigate a variety of optical analogues of phenomena which previously have only been observed in magnetic resonances.

A large widening of the application of lasers to spectroscopy was initiated in the late sixties when methods to frequency-tune laser radiation were found. Tunable lasers are now available from the near ultraviolet to the far infrared, as illustrated by the following diagram, so that the spectroscopic means for investigating atoms and molecules are almost unlimited.

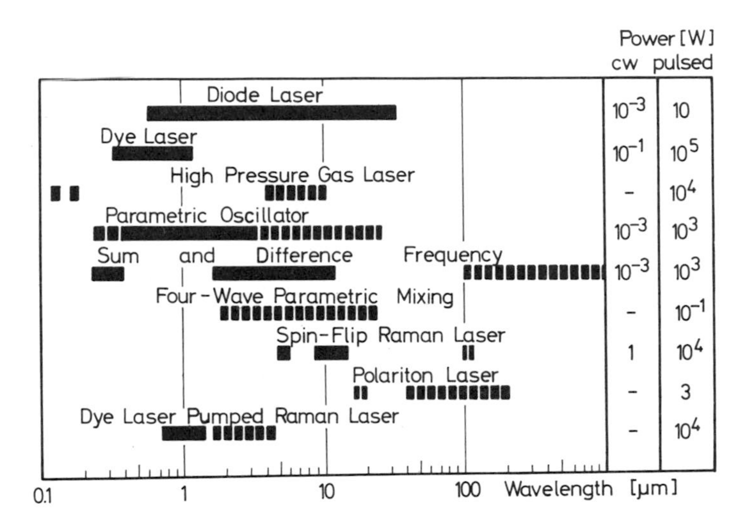

Available tunable laser sources [taken from the paper of J. KUHL and W. SCHMIDT: Appl. Phys. **3**, 251 (1974)]. Details on the infrared tunable lasers are given in Chapter 2 by HINKLEY, NILL, and BLUM. For details on dye lasers see the first volume of this series. In the wavelength regions where the black bars are interrupted tunable lasers are still under development or proposed

In an other direction, a very important development presently under way is the direct measurement of the frequency of laser light. Apart from its metrological significance, this means that in the future spectroscopists may be able to replace the relatively inaccurate wavelength measurement by a frequency measurement, providing a precision several orders of magnitude higher than that of conventional methods.

The development of new experimental possibilities during recent years makes the field of laser spectroscopy more exciting than it ever has been. This is a good time to review the main achievements during the past few years so as to bring to light where more research is possible and needed.

The first two chapters of this volume give a survey on laser spectroscopy in the visible and infrared spectral regions. The following two deal with the application of important methods to molecular spectroscopy. The fifth and sixth chapters finally discuss the linear and nonlinear effects in optical pumping, and the direct measurement of the light frequency. No attempt was made by the editor to suppress the individuality of the different contributions. Therefore different notations may have been used for the same physical quantities; in addition, there is some overlap in the subjects discussed in different articles. This has the great advantage that the reader is confronted with problems from a different point of view.

I would like to thank the contributors to the book for writing the articles which were prepared much quicker than that of the editor. I would like to thank Dr. LOTSCH of Springer-Verlag for help and advice, and for eliminating some of the typical "German" sentences in my contribution. In this connection also the corrections of Dr. W. RASMUSSEN have been quite valuable.

Occasionally my coworkers Dr. K. W. ROTHE and Dr. J. HÄGER helped to collect literature. I would like to thank them for their valuable assistance.

Köln, September 1975 HERBERT WALTHER

Contents

3. Double-Resonance Spectroscopy of Molecules by Means of Lasers.
By K. SHIMODA (With 35 Figures)

Contributors

BLUM, F. A.

Science Center, Rockwell International, Thousand Oaks, CA 91360, USA

CHERLOW, JOEL M.

Departments of Physics and Electrical Engineering, University of Southern California, University Park, Los Angeles, CA 90007, USA

DECOMPS, BERNARD

Université Paris-Nord, Laboratoire de Physique des Lasers, Place du 8 Mai 1945, F-93206, St. Denis-Cedex 01, France

DUCLOY, MARTIAL

Laboratoire de Spectroscopie Hertzienne de l'E.N.S., Associé au C.N.R.S., 24 rue Lhomond, F-75231 Paris-Cedex 05, France

DUMONT, MICHEL

Université Paris-Nord, Laboratoire de Physique des Lasers, Place du 8 Mai 1945, F-93206, St. Denis-Cedex 01, France

EVENSON, KENNETH M.

Precision Laser Metrology Section, National Bureau of Standards, Boulder, CO 80302, USA

HINKLEY, E. DAVID

MIT, Lincoln Laboratory, Lexington, MA 02173, USA

NILL, K. W.

Laser Analytics, Inc., Lexington, MA 02173, USA

PETERSEN, F. RUSSEL

Precision Laser Metrology Section, National Bureau of Standards, Boulder, CO 80302, USA

PORTO, Sergio P. S.

Departments of Physics and Electrical Engineering, University of Southern California, University Park, Los Angeles, CA 90007, USA

SHIMODA, KOICHI

Dept. of Physics, University of Tokyo, 7-3-1 Hongo Bunkyo-ku, Tokyo 113, Japan

WALTHER, HERBERT

Sektion Physik dcr Universität München, D-8046 Garching, Am Coulombwall 1, Fed. Rep. of Germany

(The book was prepared during H. W. was still at the Universität Köln, Fed. Rep. of Germany)

1. Atomic and Molecular Spectroscopy with Lasers

H. WALTHER

With 32 Figures

In this volume the spectroscopic methods as applied to free atoms and molecules are discussed. The first chapter is intended to give a broad overview with an extensive collection of reference material. It is divided into four parts and starts out with a discussion of spectroscopic applications which do not require single-mode lasers. The second and the third part deal with the applications of lasers to chemistry and isotope separation, respectively; these two fields have gained great significance in recent years and are closely related to spectroscopy. The fourth and final part reviews briefly the different methods used in high-resolution spectroscopy where the observed linewidth is limited by the natural width of the transitions.

With the continuous development of the experimental techniques and the theoretical understanding of the observed phenomena, the subject of main interest in a research field is continuously changing. This is, of course, strongly reflected in the selection of topics in a review which, as the present one, is limited in length. To give as much information as possible the papers in which spectroscopic data have been presented are compiled in tables, whereas in the main text the methods used and the results obtained are discussed in some detail. The present survey is essentially restricted to the literature since 1969; for older work the reader is referred to [1.1].

1.1. Doppler-Limited Spectroscopy

We consider first laser spectroscopy characterized by spectral resolution limited by the Doppler width of the vapors under investigation. The lasers are used mostly in multimode operation so that the width of their spectral distribution is determined by the gain profile of the laser transition. For gas lasers at low pressure this profile is identical to the Doppler width of the transition. The actual laser types used in the experiments will not be discussed here; for details the reader is referred to the original literature or recent reviews [1.2].

1.1.1. Absorption and Emission Measurements

Absorption Spectroscopy

A laser beam has a very low divergence and can thus be passed through a sample many times to enhance its absorption. Most absorption measurements have been performed in the infrared spectral region in order to investigate rotational or vibrational-rotational transitions. Since the lifetimes of the corresponding excited states are generally on the order of $10^{-6}-10^{-3}$ sec, the observation of fluorescence emission would be very ineffective.

The rotational and vibrational-rotational transitions are so numerous that they frequently coincide with the molecular-laser lines used in the measurements. Sometimes the laser transition is Zeeman tuned to scan the laser line over the absorption structure [1.3], or conversely the absorption line is shifted by either the Zeeman or the Stark effect [1.4, 5]. When single-mode lasers are used to scan the absorption profile, one obtains a resolution of up to two orders of magnitude better than with conventional infrared spectrometers [1.6].

EVENSON and HOWARD [1.7] have recently made some very significant measurements using magnetic tuning. In their measurements the absorption cell was placed inside the cavity of a H_2O or D_2O laser, having lines at 78, 79 and 119; or at 72, 84, and 108 μm, respectively. They investigated the free radicals OH, CH, HO_2, and HCO. A detection sensitivity of about 10^8 radicals per cm^3 was achieved with a one-second integration time. This high sensitivity makes it possible to study free radicals in a flow reaction system so that useful information on their reaction kinetics can be obtained. These investigations are crucial for a complete understanding of the chemistry of discharges, flames, and other combustion processes.

Most of the recent absorption work in the infrared has been done with tunable lasers; it is reviewed in Chapter 2 by HINKLEY et al.

For detecting trace absorptions, a broad-banded dye laser [1.8] is superior to other laser systems. When the absorption cell is placed inside the laser cavity, the absorption observed in the broad-band laser spectrum is several orders of magnitudes larger than that in the single-pass case [1.9]. This sensitivity gain is not simply due to an increase in the number of passes through the sample cell. It has been shown that the enhancement factor corresponds to the total number of modes excited simultaneously in the laser [1.10]. Theory and experiment give a linear relationship between the absorption for samples inside and outside the laser cavity so that intra-cavity absorption in dye lasers can be used for quantitative analysis [1.11]. The enhancement factors observed range from 100 with a flashlamp-pumped dye laser, e.g. [1.12], to 10^5 with a cw dye laser [1.10]. Interestingly, in the latter case the absorption of I_2 in the intracavity cell was detected by monitoring fluorescence from an I_2 sample placed outside the cavity. This eliminated problems associated with detector resolution. Measurements with flashlamp-pumped dye lasers can be made in 100 nsec or less; therefore, this technique provides a powerful tool for detecting short lived transients in very small concentrations. The species investigated by this technique in the vapor phase were Na [1.13], Cs [1.12], I_2 [1.10, 11, 13], Ar^+ [1.14], HCO [1.15], NH_2 [1.15]; in solution, $Eu(NO_3)_2$ dissolved in methyl alcohol [1.9]; Ba^+, Sr, and the transient molecules BaO, HCO, CuH [1.16] were investigated in flames.

The high spectral brightness achievable with dye lasers means that selected excited levels can be substantially populated, so that the absorption starting from these levels, in addition to the ground state absorption, can be observed against a continuous background. This naturally extends the possibilities which are conventionally used for the detection of the higher members of a series and of autoionizing levels by the observation of absorption spectra. The gain is tremendous, since levels can be reached in absorption which cannot be populated from the ground state directly. There is also a further practical advantage as the autoionizing levels can be observed without having to work in the experimentally difficult vacuum ultraviolet spectral region, since the absorption takes place from levels well above the ground state.

Experiments of this type have been performed on Ca [1.17, 18] Rb [1.19], and Ba and Mg [1.20]. In the case of measurements on the Ba spectrum, the series $6s6p\,^1P_1^0 - 6snd\,^1D_2$ with up to $n=41$ was observed. Furthermore, terms of the series $6s6p\,^1P_1^0 - 5dns\,^1D_2$ have been measured for $n=8$ to $n=11$ (in this series, the terms with $n=9$ to $n=11$ are autoionizing levels). In the Mg spectrum the series $3s3p\,^1P_1^0 — 3snd\,^1D_2$ has been recorded to $n=24$. The autoionizing resonance corresponding to the transition $3s3p\,^1P_1^0 - 3p^2\,^1S_0$ has also been observed. These first measurements demonstrated clearly that the method may be very useful for investigating even more complex atomic spectra, too.

Another very interesting application of absorption measurements is in the field of atmospheric probing by means of lasers. Most of these absorption measurements were performed in the infrared spectral region. For these investigations the use of frequency-tunable lasers is generally required. The absorption lines of some atmospheric compounds coincide with the lines of infrared lasers such as, e.g. the CO, CO_2, HF, and N_2O lasers. They can therefore be used directly for absorption measurements in the atmosphere. To demonstrate how powerful this technique can be, some absorption measurements performed by means

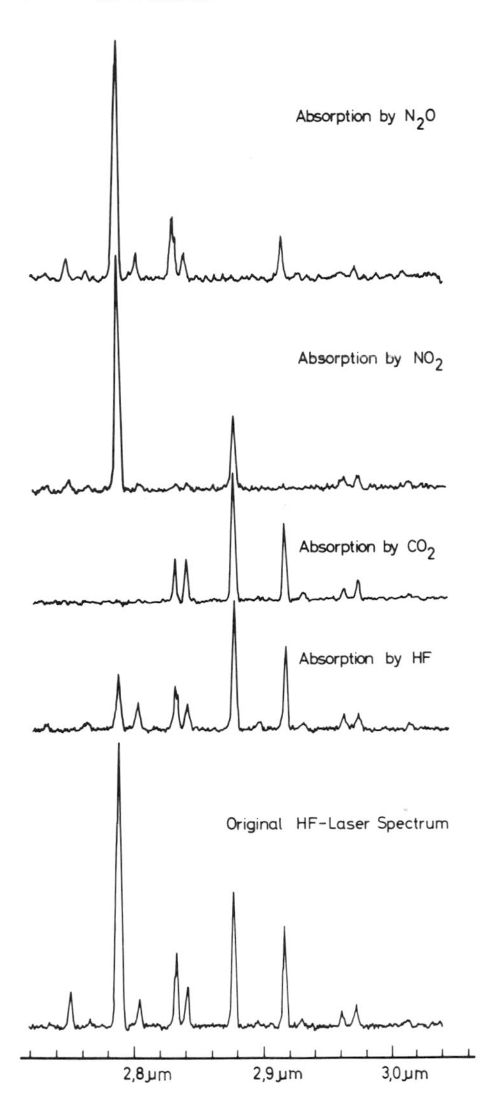

Fig. 1.1 Absorption of the HF-laser lines by different gases [1.21]. For the measurements the laser beam was passed through an absorption cell which was filled with the respective gas with a partial pressure of about 100 Torr. The total pressure was 760 Torr; the buffer gas was air. The lower spectrum is the laser spectrum without absorption

of a transversely-excited HF-laser [1.21] are shown in Fig. 1.1. Different gases show different absorption for the HF-laser lines. It is obvious that the characteristic change of the HF-laser spectrum by the absorption may be used to partially analyse the composition of the absorbing gases.

Using pulsed lasers, the atmosphere can be probed in a *Radar*-like fashion by means of various kinds of scattering processes as, e.g. Raman [1.22], Rayleigh [1.23], Mie [1.24], and fluorescence scattering [1.25].

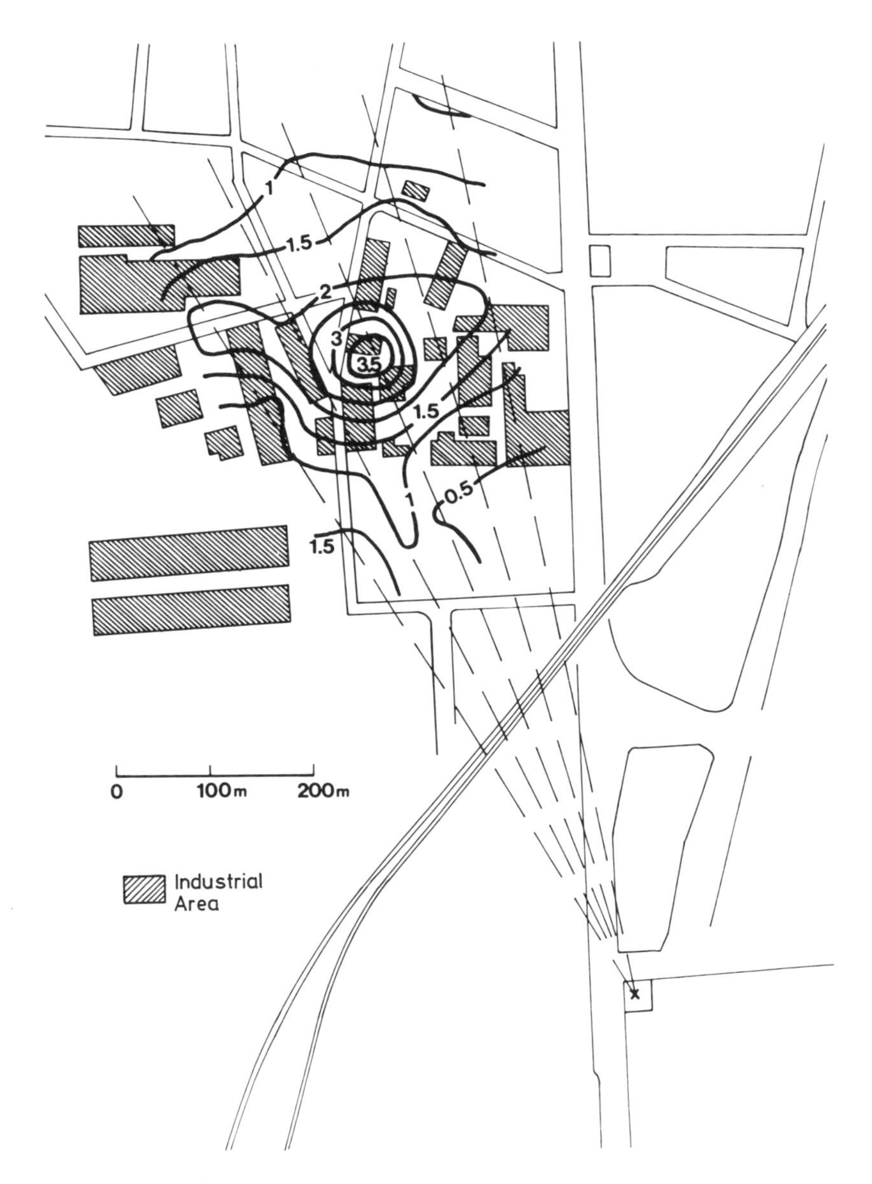

Fig. 1.2. NO_2 distribution over a chemical factory as derived from measurements using the differential absorption method. The laser used was a tunable dye laser. The area of the factory was scanned repetitively by turning the Lidar set-up horizontally in 5 directions at an altitude of 45 m. The concentration is given in ppm [1. 34]

These methods are commonly called *Lidar* because light is used instead of radio waves. *Lidar* may be utilized to get information on atmospheric properties, as e.g. clear air turbulence [1.26], wind [1.27], dust and smog [1.28] molecular species [1.29] and especially pollutants (see also Chapt. 2 and a forthcoming Topics volume).

The observation of absorption seems to preclude the possibility of a *Radar*-like measurement since either the detector or a mirror must be positioned at a distant point. The result of such a measurement would provide only an average value for the density of the pollutant and gives no information on the distribution as a function of distance. This difficulty can be overcome by using the ubiquitous Mie and Rayleigh scattering in the atmosphere as a reflector to obtain spatial resolution of atmospheric constituents. Mie and Rayleigh scattering have a monotonic wavelength dependence. The selective absorption in the radiation backscattered by atmospheric constituents thus permits an analysis of the pollutants. The method, called the "differential absorption method", has been discussed in feasibility studies. Computer simulation comparing this and other laser techniques show differential absorption to be the most sensitive one [1.30–32].

Meanwhile the first experimental application has been carried out [1.33]. Using this method the distribution of NO_2 over a chemical factory could be measured [1.34] (Fig. 1.2).

Emission Spectroscopy

The simplest way to do spectroscopy with a laser is to study its own emission spectrum. In this way many investigations have been performed, and useful information on hitherto unknown levels and splittings has been obtained. Especially with infrared lasers many rotational and vibrational splittings have been determined [1.1]. A very nice example for such a measurement is the determination of the frequencies of $^{12}C^{18}O_2$, $^{13}C^{16}O_2$ and $^{13}C^{18}O_2$ isotope lasers by comparison with a $^{12}C^{16}O_2$ reference laser [1.35]. The frequency differences between the laser lines were measured to better than 3 MHz using heterodyne techniques. The results have been used to calculate new values for the band centers and the rotational constants of the rare isotopes.

Isotope shift measurements have also been performed for atoms by means of lasers filled with different isotopes for which the wavelength differences have been measured interferometrically. For investigations on Xe see e.g. [1.36]. For the Xe isotope shift also heterodyne measurements between different lasers have been performed [1.37].

Other interesting results have been obtained in the investigation of infrared noble gas laser lines originating from transitions between

Rydberg states with high principle quantum numbers [1.38]. The wavelength determination of these lines and the measurement of their relative intensity are quite important for the test of approximations introduced in the calculation of term values and transition probabilites. The accuracy achieved in these experiments is, of course, limited by the Doppler width. When, however, a single-mode laser is used together with frequency stabilization by means of the Lamb dip [1.39] (see Sect. 1.4) a substantial improvement can be obtained. By utilizing this technique, PETERSEN and co-workers have recently determined the rotational constants [1.40] and even absolute frequencies of most of the common $^{12}C^{16}O_2$ laser transitions to within a few kHz [1.41].

A large field of application of lasers is in fluorescence spectroscopy performed with laser lines which accidentally coincide with electronic transitions of a molecule. The intensity of the laser lines is usually strong enough to give a population in the excited state comparable to that in the lower state. The resulting fluorescence is, therefore, very intense and, since only one or a few levels are excited, the spectrum remains rather simple and easy to identify [1.42, 43].

Difficulties can arise when a multimode laser is used and different rotational-vibrational levels of the excited states are populated, starting from different levels of the lower state. In most cases the problem can be overcome by restricting the laser to single-mode operation and thereby reducing the number of excited transitions. By tuning a single longitudinal mode over the gain profile of the laser which is given by the Doppler width of the corresponding transition, a selective population of different states may be obtained. The laser-induced fluorescence spectrum is much simpler than the absorption spectrum of the same molecule. In the case of a diatomic molecule it consists of a series of single lines (Q lines), doublets (R and P lines), or triplets terminating on different vibrational levels of the lower electronic state.

The use of laser-induced fluorescence allows accurate wavelength measurements to be performed, from which the molecular ground state constants can be determined. Excitation with different laser lines also makes it possible to calculate the energy separation in the excited state [1.43]. Because the admixture of buffer gases causes a transfer of excitation energy to neighbouring rotational-vibrational levels of the upper electronic state, the splitting can be observed in the fluorescence spectrum even without laser tuning. It is also possible to change the spectrum steadily by regulating the buffer gas pressure, so that an assignment of the observed lines is easy to perform [1.44–46].

The results coming from rotational-vibrational splittings may be used to derive the potential curves [1.47]. When sufficiently high lying levels of the lower electronic state have been reached due to fluorescence decay the dissociation energy can also be determined by extrapolation [1.44, 48]. The data so obtained can be used to test calculated molecular parameters. This is of interest because the approximation methods that have been developed are much simpler than the conventional self-consistent field calculations, and a comparison can test their reliability. The line intensities measured in the fluorescence spectra can also be used to derive Franck-Condon factors. They provide a good test of the vibrational wavefunctions of the upper and lower states.

Measurements performed by means of laser-induced fluorescence are given in Table 1.1. Only the work published in or after 1969 is considered; for older literature see [1.1].

The table also includes the studies on collision-induced fluorescence giving a detailed information on inelastic collision processes in excited states.

For the investigation of the collision induced fluorescence the molecules are excited into a particular excited state. Due to inelastic collisions, radiationless transitions to neighbouring vibrational or rotational levels may be induced and their spontaneous decay observed. Wavelength, intensity and polarization of the collision induced fluorescence lines contain much information on the collision processes. For example, from the intensity ratio of the collision-induced fluorescence line to the parent line the absolute cross section for the corresponding inelastic collision can be deduced under the condition that the upper state lifetimes are known.

The experiments performed show that the cross sections for collision induced rotational transitions are very large $\sigma(|\Delta J|=1) \approx 100\,\text{Å}^2$ [1.80] whereas the probability for vibrational energy transfer is about one order of magnitude smaller.

The studies on the collision induced rotational transitions in Na_2 [1.79, 80] and Li_2 [1.46] showed that the cross section $\sigma(+\Delta J)$ is remarkably different from $\sigma(-\Delta J)$ for an odd change in the rotational quantum number but equal for an even change. The observed asymmetry $\sigma(+\Delta J)/\sigma(-\Delta J)$ is due to a symmetry change of the electronic wave functions for an odd ΔJ. Calculations have shown that the asymmetry is very sensitive to the interaction potentials between the colliding particles [1.82] and is therefore useful for testing model potentials.

The development of tunable lasers, of course, is of considerable significance to the study of laser-induced fluorescence since the excited-state splitting can now be scanned easily over a large range. First remarkable results have already been obtained on NO_2 [1.75] with a N_2-laser pumped dye laser. In these experiments dye lasers continuously tunable over a broad spectral range are of great importance. Recently a widely tunable set-up was described by WALLENSTEIN and HÄNSCH

[1.83] who used a laser arrangement of the Hänsch type [1.84] with an intra-cavity Fabry-Perot and an extra-cavity confocal resonator, which were both pressure-tuned. Since the reflection grating is also tuned when the pressure of the surrounding gas is changed, a rather large wavelength range can be covered (about 150 GHz). A still larger tuning range is obtained with a solid intra-cavity etalon which is turned in correlation with the diffraction grating. This may be achieved by two computer-controlled step motors. The linewidth obtainable with such a set-up, though being comparable to the Fourier-limited linewidth of the dye laser, is larger than that achieved with the above mentioned arrangement. The tuning range may be of the order of 40 Å or more. A part of a fluorescence spectrum measured with such a set-up [1.85] is shown in the lower part of Fig. 1.3. For comparison see also the set-up described by KLAUMINZER [1.86].

The quasi continuous tuning of cw dye lasers over a spectral range of several Å has also been realized. A set-up was described by GREEN et al. [1.87]. The laser incorporates pressure tuned intra-cavity etalons enabling the single-mode output laser frequency to be tuned over more than 10 Å in a single scan. During the scan the laser frequency jumps from one cavity mode to the next in about 50 MHz steps which are negligible in comparison to molecular Doppler widths.

Another set-up was described by MAROWSKI et al. [1.88]. They used a novel scheme of four intra-cavity, highly dispersive Abbe prisms for wavelength selection and tuning. With that arrangement it was possible to achieve single frequency operation and semi-continuous tuning over a range of about 50 Å with one single micrometer-driven precision translation assembly. Other widely tunable configurations are possible in connection with ring laser arrangements [1.89]. The latter set-up can also be used in connection with flashlamp or N_2-laser pumping.

A very interesting application of dye lasers was opened up some time ago, with the measurement of the $2p_{3/2}-2s_{1/2}$ energy difference in muonic helium [1.90]. The fine-structure splitting of these exotic atoms is large enough to be induced with visible laser radiation. These experiments which are described in [1.91] are very important for the investigation of vacuum polarization effects.
A very interesting application of dye lasers was opened up some
of the Lamb shift in hydrogen-like atoms with a high Z-number. For hydrogen-like sulphur and phosphorus, for example, the $2p_{3/2}-2s_{1/2}$ transition also lies in the visible spectral range and can therefore be induced by dye lasers [1.92]. There are presently several experiments under way. The highly ionized sulphur or phosphorus can be produced by stripping high energy ion beams by foils [1.93].

Table 1.1. Measurements performed by means of laser-induced fluorescence (Doppler linewidth limited). Only the investigations performed in or after 1969 are included; for older literature see [1.1]. This table also contains the measurements on the collision-induced fluorescence. (Further relaxation studies will be discussed in Subsect. 1.1.2.) Abbreviations see end of table on next page.

System	Laser	Wavelength region (excitation) [Å]	Measurements	Ref.
Li_2	Ar^+	4579–5145	$(B^1\Pi_u - X^1\Sigma_g^+)$ ES, VS, RS, SC, DE, ET ($^6Li_2$, $^6Li^7Li$, $^7Li_2$)	[1.44]
	Ar^+		$(B^1\Pi_u)$ ET	[1.49]
	Ar^+	4579–5145	$(B^1\Pi_u - X^1\Sigma_g^+)$ ET	[1.46, 81]
Na_2	Ar^+	4658–5145	$(B^1\Pi_u - X^1\Sigma^+)$ ES, VS, RS, SC, PC, FCF	[1.43]
	HeNe	6118–6401	$(A^1\Sigma_u^+ - X^1\Sigma_u^+)$ ES, VS, RS	[1.50]
	Ar^+	4765–4880	$(B^1\Pi_u - X^1\Sigma^+)$ ET	[1.79, 80]
NaK	Ar^+	4880–5145	$(D^1\Pi - X^1\Sigma)$ ES, VS, ET	[1.51]
	Kr^+		$(B^1\Pi - X^1\Sigma)$	
NaLi	Ar^+	4965	$(^1\Pi - {}^1\Sigma)$ ES, VS, RS, SC, ET	[1.52]
	Ar^+		$(B^1\Pi)$ ET	[1.53]
Br_2	Ar^+	5145	$(B\,^3\Pi_{0u}^+ - X^1\Sigma_g^+)$ ES, VS	[1.54]
Cl_2	Ar^+	4880	$(B\,^3\Pi_{0u}^+ - X^1\Sigma_g^+)$ ES, VS	[1.54] [1.55]
ClO_2	Ar^+	4579–4880	$(A(^2A_2) - X(^2B_1))$ ES, VS, SC	[1.56]
	Ar^+	4765	ES, VS, SC	[1.57]
I_2	Kr^+	5682	$(B\,^3\Pi_{0u}^+ - X\,^1\Sigma_g^+)$ ES, VS, RS	[1.58]
	Ar^+	5145		
	Xe^+	5971		
	Ar^+	5017, 5145	$(B\,^3\Pi_{0u}^+ - X\,^1\Sigma_g^+)$	[1.55] [1.54]
	HeNe	6328	$(B\,^3\Pi_{0u}^+ - X\,^1\Sigma_g^+)$ ES, VS, RS	[1.59] [1.48]
	HeNe	6328	$(B\,^3\Pi_{0u}^+ - X\,^1\Sigma_g^+)$ ES, VS, RS	[1.60]
	Ar^+	5017, 5145		

I_2	Ar^+	5145	$B\,^3\Pi^+_{0u} - X\,^1\Sigma^+_g$) ET	[1.78]
	Kr^+	5208–6471	($B\,^3\Pi^+_{0u} - X\,^1\Sigma^+_g$) ES, VS, RS	[1.61]
	HeNe	6328	($B\,^3\Pi^+_{0u} - X\,^1\Sigma^+_g$) ES, VS, RS, SC, PC, FCF Magnetic quenching	[1.62]
	Ar^+	5017, 5145	($B\,^3\Pi^+_{0u} - X\,^1\Sigma^+_g$)	[1.63]
	Kr^+	5208–5682	Orientation of the optically excited molecule due to predissociation	
	HeNe	6328	($B\,^3\Pi^+_{0u} - X\,^1\Sigma^+_g$) ET	[1.64]
AlO	Ar^+	4658, 4880	($B\,^2\Sigma^+ - X\,^2\Sigma^+$) ES, VS	[1.65]
BaO	Ar^+	4545–5145	($A\,^1\Sigma - X\,^1\Sigma$) ES, VS, RS, SC	[1.66]
CH	Dye laser	4315	($A\,^2\Delta - X\,^2\Pi$) RS Molecule generated in a flame	[1.67]
CN	Dye laser	~3800	($B\,^2\Sigma^+ - X\,^2\Sigma$) ES, VS	[1.68]
CuO	Ar^+	4880	CuO isolated in various matrices ($B\,^2\Sigma - X\,^2\Pi$, $B\,^2\Sigma - A\,^2\Pi$) ES, VS	[1.69]
NO	Dye laser	4545–4525	($A\,^2\Sigma^+ - X\,^2\Pi$) RS, SC Two photon excitation*	[1.70]
OH	Dye laser	2822	($A\,^2\Sigma^+ - X\,^2\Pi$) ET	[1.71]
BO_2	Ar^+	5145, 4880	($A\,^2\Pi_u - X\,^2\Pi_g$) ES, VS, RS, SC	[1.72]
NO_2	Ar^+	4579–5145	$^2A_1$ VS, RS, SC, ET	[1.73]
	Kr^+	4619–5208		[1.74]
	Dye laser	5934–5940	$^2B_2 - {}^2A_1$	[1.75]
CO_2	CO_2	10600	ET	[1.76]
SO_2	Dye laser (frequency doubled)	~3000	ES, VS, ET	[1.77]

Abbreviations: ES: Electronic excitation. VS: Vibrational splitting. RS: Rotational splitting. SC: Spectroscopic constants. PC: Potential curves. DE: Dissociation energy. FCF: Franck-Condon factors. ET: Energy transfer, quenching by buffer gases.

* For further two-photon studies see Subsect. 1.1.5.

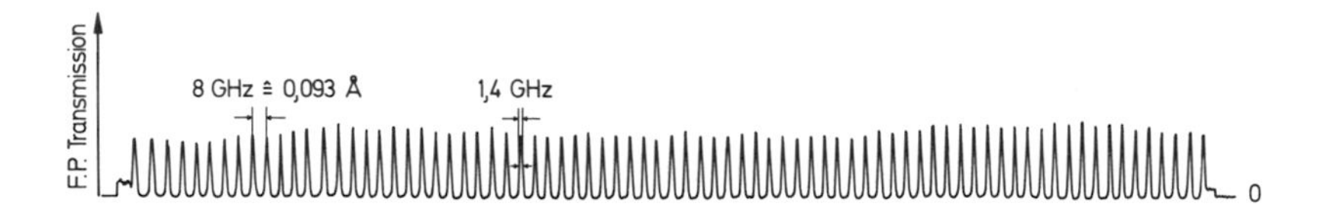

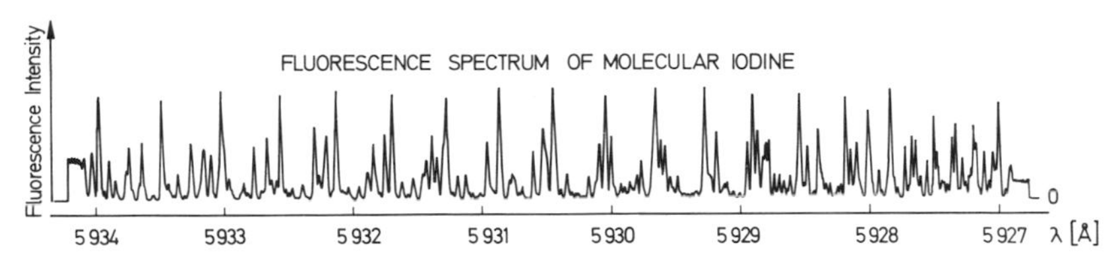

Fig. 1.3. Fluorescence spectrum of I_2 obtained by means of a broadly tunable N_2-laser pumped dye laser. The diffraction grating and the intra-cavity etalon are turned by computer-controlled step motors [1.85]. The upper part of the spectrum shows the reference spectrum obtained by means of a Fabry-Perot interferometer with a free spectral range of 8 GHz

At the end of this sub-section it should still be mentioned that the fluorescence induced by lasers and especially by tunable lasers has gained large significance for analytical purposes. Several examples will be mentioned in the following. Using the 4880 Å line of an argon-ion laser, TUCKER et al. [1.325] have monitored NO_2 in a concentration of 3 ppb in a sample cell. Formaldehyde was detected in a concentration of 50 ppb in air at atmospheric pressure with a frequency doubled, flashlamp-pumped dye laser in the 3200–3450 Å range [1.326]. The highest detection sensitivity in a fluorescent experiment was so far achieved by FAIRBANK et al. [1.327]. Using a cw dye laser sodium vapor was still detected in a density as low as 10^2 atoms/cm^3. In this paper a scheme is proposed which can be used to detect sodiumquark atoms, if they exist, with the sensitivity of one quark atom per $4 . 10^{13}$ normal atoms.

Laser fluorescence may also be useful in connection with classical analytical methods. This was demonstrated by ZARE and co-workers [1.328] who used a nitrogen laser as the excitation source to probe the fluorescence of thin-layer chromatography plates in order to detect 0.2 ng samples of aflatoxins which are cancerogen and present in bread.

The ultimate sensitivity limit for detection of samples of low concentration by various methods has theoretically been discussed by SHIMODA [1.329].

1.1.2. Lifetime Measurements and Relaxation Studies

Lifetime Measurements

There are essentially three different experimental methods which are used for lifetime studies of electronically excited states of atoms and molecules in connection with laser excitation. These are: the phase shift method (see, e.g. [1.43]), pulsed excitation and subsequent observation of the radiation decay (see, e.g. [1.94]), and finally the zero-field level crossing effect (Hanle effect) (see e.g. [1.95]).

In the phase shift method modulated light is used for the excitation of the atoms or molecules. The fluorescent light is also modulated with the same frequency, but phase-shifted with respect to the modulation phase of the incoming light. The phase shift $\Delta\varrho$ and the lifetime τ are connected by the relation $-\omega\tau = \tan\Delta\varrho$, where ω is the modulation frequency. Recently it was pointed out by ARMSTRONG and FENEUILLE [1.96] that under the experimental condition of a single-mode excitation, e.g. by a tunable dye laser, and the scattering of light on atoms or molecules of a well collimated atomic beam without Doppler broadened absorption a phase shift should be observed, which depends on the intensity of the laser light. In the limit of low intensities they derived the relation $-2\omega\tau = \tan\Delta\varrho$, which differs by a factor of two from the result obtained for broad-banded light excitation. Furthermore they found that the strength of the modulated component has a behaviour related to the Autler-Townes effect [1.97] (see Subsect. 1.1.5).

A straightforward method for lifetime measurements with lasers is to use pulsed excitation. Most of the lasers can be pulsed easily by electro-optical or acousto-optical methods, so that pulses which are short compared to the lifetimes of most of the electronically excited states are obtained. Other lasers as, e.g. the N_2-laser pumped dye laser, provide pulses which are also short enough for this purpose. Since the intensities of the laser radiation are rather high, the decay of the induced fluorescence can be monitored directly on an oscilloscope. For measurements on lines which are difficult to excite or which belong to complex spectra, transient recorders with averaging capability can be used. Another method for the detection is time-to-pulse-height conversion with subsequent averaging by a multichannel analyser. In the latter method the intensity of the laser must generally be reduced so that there is less than one signal photon for each laser pulse; otherwise the first signal pulse will always stop the time-to-pulse height converter so that an incorrect time distribution is measured. It is advantageous to employ lasers with a high pulse repetition rate in this scheme.

An elegant method for lifetime measurements on ions is represented by the application of selective laser excitation to fast ion beams. The first experiment of this type was performed by ANDRÄ et al. [1.98]. A fast beam of Ba^+ ions was irradiated by the 4545 Å line of an Ar^+ laser. The $6p\ ^2P_{3/2} - 6s\ ^2S_{1/2}$ transition at 4554 Å does not coincide with the laser line, however, by irradiating the fast beam under a small angle with respect to the beam axis, the light frequency seen by the ions could be Doppler tuned so that a matching could be achieved. The decay constant of the excited state was obtained by measuring the reemitted intensity as a function of the distance from the excitation region. This method can be considered as a modification of the beam-foil spectroscopy technique [1.93], where the foil is replaced by selective laser excitation. The method becomes still more versatile when a tunable laser is used instead of the fixed-frequency laser. This was realized by HARDE et al. [1.99] in experiments with a tunable dye laser. A similar experiment has been performed by ARNESEN et al. [1.100].

The application of lasers to lifetime measurements has greatly been increased when tunable dye lasers became available. The wavelengths of the available radiation are suitable to investigate many electronic transitions in atoms or molecules. Due to the narrow spectral width obtainable with these lasers single fine-structure levels of atoms or individual vibrational or rotational levels in molecules can be excited.

The lifetimes obtained in this way for atomic spectra may give useful information on configuration interaction. In addition, they can be used to test calculations of the radial part of eigenfunctions.

Especially important are lifetime measurements for the spectra of the transition elements (e.g. [1.101]) or of the rare earths, since these values are of considerable astrophysical interest. For these spectra the relative oscillator strengths can frequently be measured quite accurately and the lifetime values are then used to determine the absolute oscillator strengths. They are essential for the determination of element abundances in solar or stellar atmospheres.

For the evaluation of the absolute oscillator strengths it is especially important to investigate the lifetimes of very highly excited levels. Therefore it is quite essential that the dye laser can be used to perform stepwise excitation [1.102, 103].

A typical experimental set-up which was used for lifetime measurements by means of stepwise excitation in the iron spectrum is shown on Fig. 1.4 [1.102]. The two dye lasers were similar to that described by HÄNSCH [1.84]. They were both pumped simultaneously by splitting the beam of the nitrogen laser, thus problems in obtaining a temporal overlap of the two laser pulses were avoided. The radiation of "dye laser I" was frequency doubled by means of a KDP crystal. For the lifetime measurements the exponential decay of the fluorescent radiation

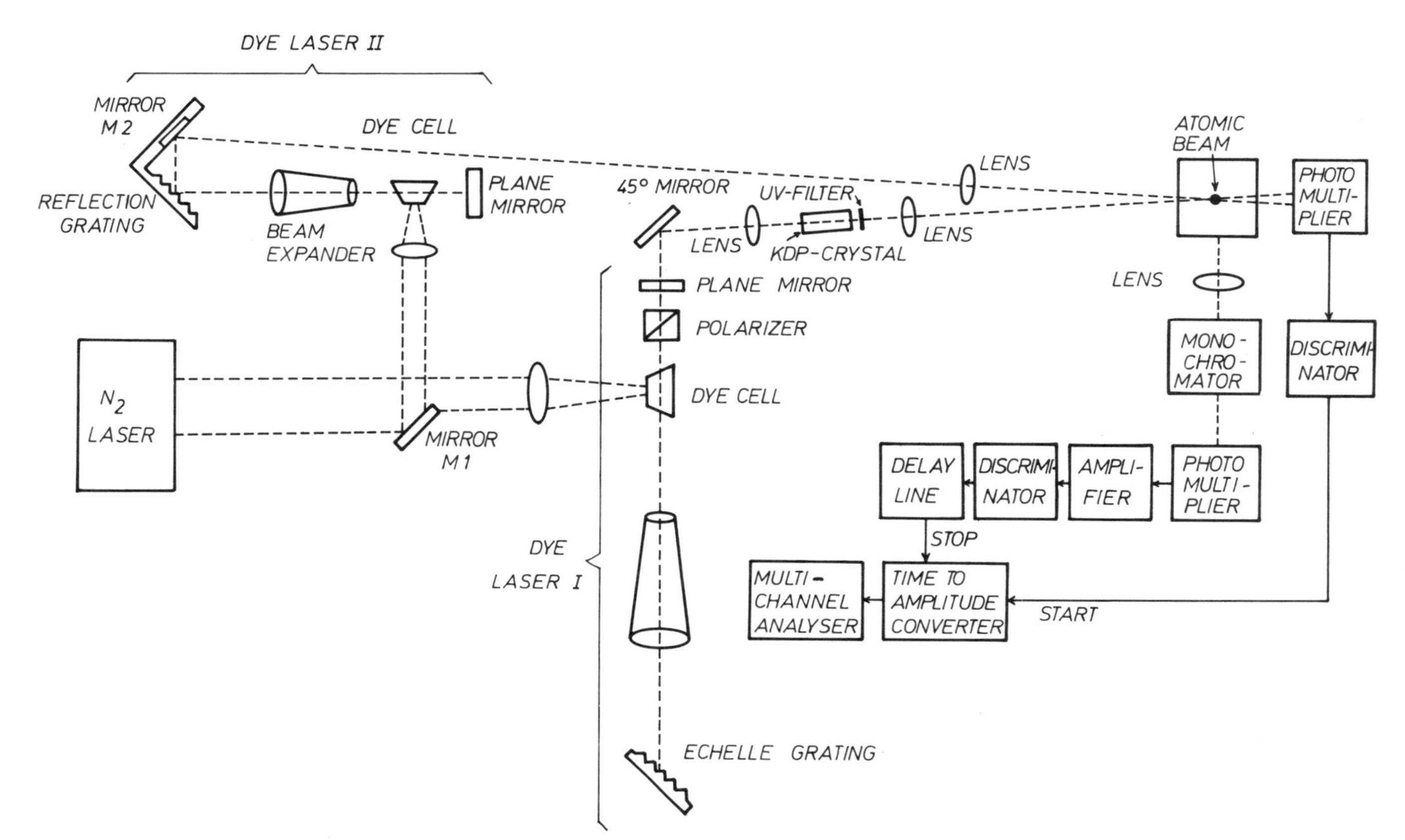

Fig. 1.4. Typical set-up for lifetime measurements by means of stepwise excitation by nitrogen-laser pumped dye lasers. (From [1.102])

Table 1.2. Lifetime and relaxation measurements. Only the investigations performed in or after 1969 are included; for older literature see [1.1]. (Note that 10000 Å = 1 μm.) Abbreviations see end of table.

System	Method	Laser	Wavelength region (excitation) [Å]	Measurements	Ref.
Na	SP	Dye laser	5895	$(3p\,^2P_{1/2})$ LT	[1.109]
	SP	Dye laser	5890, 5896; 5682, 5688 (two step)	$(4d\,^2D)$ LT	[1.103]
Ca	SP, LC	Dye laser		$(4s\,5s\,^3S_1)$ LT; $(4s\,4d\,^3D_1)$ LT	[1.110]
	SP, LC	Dye laser	4227	$(4s\,4p\,^1P_1)$ LT	[1.111]
Yb	SP	Dye laser	5556	$(6s\,6p\,^3P_1)$ LT	[1.112]
Cd^+	SP	HeCd	4416	$(5p\,^2P^0_{3/2})$ LT	[1.113]
Fe	SP	Dye laser	2912–2967; 5269–5406	$(3d^7\,4p\,y^5\,F^0)$ LT; $(3d^6\,4s\,4p\,z^5\,F^0)$ LT	[1.101]
	SP	Dye laser	2967; 5941 (two step)	$(3d^6\,4s\,4d\,e^5\,G_6)$ LT	[1.102]
Ni	SP	Dye laser	2981–3031	$(3d^8\,4s\,4p\,y^3\,D^0$; $3d^8\,4s\,4p\,y^3\,F^0$; $3d^8\,4s\,4p\,z^1\,G^0)$ LT	[1.114]
H_2	SP	ruby	6934	$(v=1,\ J=1)$ RL, VT, SQ, FQ	[1.115]
K_2	PS	HeNe	6328	$(B\,^1\Pi_u)$ RL, SQ	[1.116]
	PS	HeNe	6328	$(B\,^1\Pi_u)$ LT	[1.117]
Na_2	LC	Ar^+	4765	$(B\,^1\Pi_u)$ RL	[1.95]
	PS	Ar^+	4579–4879	$(B\,^1\Pi_u)$ RL, SQ	[1.116]
Rb_2	PS	Ar^+	4579–4879	(B and C states, respectively) LT, SQ	[1.116]
Cs_2	PS	Ar^+	4579–7879	LT, SQ	[1.116]
AlO	SP	Dye laser	4840–4470	$(B\,^2\Sigma^+ - X\,^2\Sigma^+)$ VL	[1.65]
BaO	SP	Dye laser		$(A\,^1\Sigma - X\,^1\Sigma)$ VL, FQ	[1.118]

CO	SP	CO_2 (frequency doubled)	~47.000	(v_1 mode) matrix isolated VL	[1.119]
	SP	CO_2 (frequency doubled)	~48.000	(v_1 mode) VL, VT, FQ	[1.120]
I_2	LC	Ar^+	5017, 5145	$B\,^3\Pi^+_{0u}$ RL	[1.121]
	SP	Dye laser	4990–6400	$B\,^3\Pi^+_{0u}$ VL, FQ	[1.105]
					[1.94]
	SP	Ar^+	5145	$B\,^3\Pi^+_{0u}$ RL determined from line center	[1.122]
	SP	Dye laser	5017–5682	$B\,^3\Pi^+_{0u}$ RL, SQ	[1.104]
ICl	SP	Dye laser	5750–6100	$(^3\Pi_1)$ LT	[1.123]
HF, DF	SP	HF, DF (chemical laser)	28.000 and 40.000	$(v=1, 2)$ VL, VT, FQ, SQ	[1.124–129]
HCl, DCl	SP	HCl, DCl (chemical laser)	38.000 and 53.000	$(v=1, 2)$ VL, VT, FQ, SQ	[1.130–135]
HBr, DBr	SP	HBr, DBr (chemical laser)	43.000 and 60.000	$(v=1, 2)$ VL, VT, FQ, SQ	[1.136–138]
OH	SP	Dye laser (frequency doubled)	~3000	$(A\,^2\Sigma^+)$ VL, FQ	[1.139]
CaF				$(A\,^2\Pi, B\,^2\Sigma^+)$ LT	
CaCl	SP	Dye laser	3900–6400	$(A\,^2\Pi, B\,^2\Sigma^+, C\,^2\Pi)$ LT	[1.140]
CaBr				$(A\,^2\Pi, B\,^2\Sigma^+, C\,^2\Pi)$ LT	
CaI				$(A\,^2\Pi, B\,^2\Sigma^+)$ LT	
SrF				$(A\,^2\Pi)$ LT	
SrCl	SP	Dye laser	4000–6840	$(A\,^2\Pi, B\,^2\Sigma^+, C\,^2\Pi)$ LT	[1.140]
SrBr				$(A\,^2\Pi, B\,^2\Sigma^+, C\,^2\Pi)$ LT	
SrI				$(A\,^2\Pi, B\,^2\Sigma^+, C\,^2\Pi)$ LT	
BaF					
BaCl	SP	Dye laser	4950–5570	$(C\,^2\Pi)$ LT	[1.140]
BaBr					
BaI					
CO_2	SP	HF	28.000	(00°1) VL, VT, FQ	[1.127]
	SP	CO_2	106.000	(00°1) VL, VT, FQ, SQ,	[1.135, 1.141–153]

Table 1.2 (continued)

System	Method	Laser	Wavelength region (excitation) [Å]	Measurements	Ref.
CS_2	SP	N_2	3371	LT, SQ, FQ	[1.154]
N_2O	SP	N_2O	107.000	(00°1) VT, SQ	[1.135] [1.151]
NO_2	SP	Dye laser	4220–5940	$(^2B_1, {}^2B_2)$ LT, SQ	[1.75, 1.106–108]
H_2CO	SP	N_2	3371	$(A\,^1A_2)$ LT, SQ	[1.155]
CH_4	NA	HeNe (magnetically tuned)	33.900	$(v=1, J=7)$ VL, VT	[1.156]
	PS	HeNe	33.900	(ν_3, ν_4) VL, VT, FQ, SQ	[1.157]
CH_3Cl	SP	CO_2	100.000	$(\nu_6=1)$ VT, SQ, FQ	[1.158]
CH_3F	SP	CO_2	96.000	$(\nu_1, \nu_3, \nu_4, \nu_6)$ VL, VT, FQ, SQ	[1.159–161]
C_2H_4	SP	CO_2	106.000	$(\nu_7+\nu_8, \nu_6+\nu_{10})$ VL, VT, FQ	[1.162]
C_3H_2O	SP	Dye laser	3821	$(^1A'', {}^3A'')$ LT, SQ, FQ	[1.163]
SF_6	SP	CO_2	100.000, 160.000	$(\nu_3, \nu_4, \nu_3+\nu_6)$ VL, FT, SQ, FQ	[1.164–169]

The following abbreviations are used:

For the methods:
- PS: Phase shift.
- LC: Level crossing (Hanle effect).
- SP: Short pulse excitation.
- NA: Nonlinear absorption (see Subsect 1.4.2)

For the experimental results:
- LT: Lifetime of atomic state or for molecules level not specified.
- VL: Lifetime measurement with resolution of vibrational levels.
- RL: Lifetime measurement with resolution of rotational levels.
- VT: Vibrational-vibrational, rotational, translational energy transfer.
- SQ: Self quenching.
- FQ: Foreign quenching.

For more details see text.

was analyzed by means of standard time-to-pulse height conversion techniques.

For molecules the study of the lifetimes of individual vibrational levels of an excited electronic state may give useful information on the change of the molecular wavefunction with vibrational excitation. Especially the non-radiative depopulation of an excited state by pre-dissociation through a dissociative state can be investigated. This mixing with a near repulsive state can be increased by self-quenching [1.104] or by collisions with other gases [1.105].

Lifetime measurements in molecules may also be useful for the classification of complex molecular spectra. Levels belonging to the same electronic state should not have very different lifetime values [1.106–108]. Therefore additional information for the classification of the levels becomes available.

A survey of recent lifetime measurements on electronically excited levels of atoms and molecules is compiled in Table 1.2. This table also includes the measurements on the quenching of excited molecular states by the same or other gases. Furthermore the measurements on the energy transfer from vibration to vibration, and vibration to translation and rotation of molecules are included. These measurements will be discussed in the following.

Molecular Vibrational Relaxation Studies

The redistribution of energy between the vibrational, rotational, and translational degrees of freedom of molecules is a problem of great importance. The determination of rate constants for processes connecting individual quantum states and their variation with electronic, vibrational and rotational quantum numbers can greatly increase the understanding of chemical kinetics. The investigations give detailed information on the dynamics and coupling forces which are effective in isolated molecules and during collisions between molecules.

The laser has proved to be a useful tool to investigate such energy transfer processes. Especially in the field of vibrational energy relaxation a large amount of measurements have been performed. Many processes involving energy transfer from vibration to vibration ($V \rightarrow V$) and vibration to translation and rotation ($V \rightarrow T, R$) have been studied.

Beside their significance for the understanding of chemical kinetics the vibrational energy transfer rates also provide insight into the mechanism of chemical and molecular lasers. Furthermore they are important for any non-equilibrium system as, for example, when laser pumping is used in order to induce chemical reactions (see Sect. 1.2), for the understanding of gas dynamic processes in shock waves, in ultrasonic waves, in expansion flows, in combustion, and in electrical discharges. For recent reviews on the subject see [1.169–171].

Vibrational relaxation in solids and liquids is also an important and interesting problem. Here the times are on a picosecond scale so that mode-locked lasers become a useful tool for these investigations (see, e.g., [1.172]). Since this review only considers the processes in free molecules this interesting group of experiments will not be discussed here.

The vibrational relaxation processes have usually been measured by flash kinetic spectroscopy, by infrared emission from shock waves, and by ultrasonic measurements [1.173]. Lasers have been applied for the first time in 1966 [1.174, 175]. Many vibrational bands are precisely matched by laser frequencies. This is especially the case for molecules such as, e.g. CO_2, N_2O, HCl, HF, and others where molecular lasers are available. Therefore, measurements in these or similar systems are easily possible.

Relaxation studies can also be performed by means of the stimulated Raman effect [1.176]. Here no frequency matching is required and the technique can be applied to the excitation of vibrations exhibiting sufficient stimulated Raman gain. With the advent of the tunable laser sources in the infrared the experimental possibilities have now been increased enormously. (See also Chapter 2.)

In order to study vibrational relaxation the molecules are vibrationally excited by a pulsed laser with an intensity in the kilowatt range and a duration of several microseconds. Usually many rotational energy levels of the excited vibrational state are excited and in most cases collisions serve to establish an equilibrium rotational distribution in a time short to the laser-pulse duration. The population in the vibrational state is then investigated by observing the infrared fluorescence [1.130, 138]. In mixtures with other molecules fluorescence may be observed from any vibrations active in the infrared, which were excited during the relaxation process [1.130, 177]. With vibrational excitation of polyatomic molecules energy transfer is observed among the vibrational modes of the molecule [1.175, 178], as well as from molecule to molecule.

The population of the excited vibrational level can also be measured by absorption of the light from a probe laser which has a frequency close to the vibrational transition. The absorbed intensity is then proportional to the population difference between the upper and the lower vibration-rotation levels [1.167, 179]. In the case where excited electronic states can easily be reached by flashlamp or laser excitation, absorption [1.180], and fluorescence resulting from electronic excitation, can also be utilized to determine the population of the excited vibrational levels. The release of energy to translation can be measured by phase contrast detection of the change in the refractive index, resulting from the pressure change, [1.115] and by sensitive microphones [1.168]. Furthermore, the translational temperature changes are detectable through infrared emission [1.175, 181].

The recent results obtained on the vibrational relaxation of free molecules by means of lasers are compiled in Table 1.2. Some conclusions which can be drawn on the vibrational relaxation are discussed in the following.

a) Distributions for Relaxing unharmonic Oscillators. The vibrational energy exchange in a pure gas with unharmonic levels results in the release of energy ΔE which is used for rotation or translation. The same is the case for the energy exchange in a gas which consists of a mixture of two kinds of molecules with different vibrational energy spacings. The equilibrium constant for the energy exchange process is given by $\exp(\Delta E_{vv'}/kT)$. Therefore the quanta of vibrational excitation are shifted to highly excited levels of an unharmonic oscillator ($\Delta E_{vv'}$ is smaller for this energy transfer) or in the gas mixture to the molecules with the lower vibrational frequency. Thus if $V \rightarrow V$ processes come to equilibrium, the vibrational levels with rather high quantum numbers may still be populated and even a population inversion may be obtained [1.182, 183]. For the mixture of two kinds of molecules the one with the lower vibrational frequency will have a higher vibrational temperature.

These effects were experimentally observed in the distribution of vibrational energy in CO lasers [1.184] and in CO–N_2 mixtures [1.185]. They also play an important role in the dissociation of molecules which are vibrationally heated by laser excitation [1.186] (see also Sect. 1.2).

b) Intramolecular Transfer. For the collision of a vibrationally excited molecule with atoms or diatomic molecules whose vibrational degrees of freedom do not enter into the process, the only possible $V \rightarrow V$ processes are transitions from one vibrational level to another. The resulting energy difference ΔE is transferred into translation and rotation. Experiments reveal that achievable values of ΔE for collisions with rare gases may not be larger than kT [1.187]. However, for molecular collision partners, as e.g. H_2, ΔE may be larger. The reason for this is due to the influence of rotation [1.188] of both the vibrating molecule and its collision partner. The intra-molecular $V \rightarrow V$ energy transfer probabilities are on the order of 10^{-4}. In general, the $V \rightarrow V$ energy transfer are much faster than the $V \rightarrow T, R$ rates. This is again a consequence of the fact that $V \rightarrow V$ equilibrium can be reached through processes with a fairly small energy excess ΔE whereas for $V \rightarrow T, R$ energy transfer the total vibrational energy must be transferred into translation and rotation. However, there are some examples where the $V \rightarrow V$ process may be slower than the $V \rightarrow T$ transfer. This is especially the case when a small vibrational energy and a large quantum-number change is involved. These cases are quite interesting since it is often possible to create vibrational population inversions which may be suitable for laser action.

c) Single Quantum Exchange. There is a large amount of data available for the $V \rightarrow V$ single quantum energy transfer between vibrationally excited molecules and collision partners. (For a compilation of the results see Table 1.2 and [1.170]). It is observed that among each group of similar collision partners the probability of energy transfer decreases in general almost linearly as the energy difference ΔE which is lost for vibration increases [1.169]. The different groups, however, differ somewhat in their ΔE dependence indicating that there are different types of inter-molecular forces or collision mechanisms involved.

The $V \rightarrow V$ transfer probability may have a rather strong temperature dependence. The observed decrease in the intermediate temperature region results from the attractive part of the inter-molecular potential, and the repulsive interaction becomes relevant at higher temperatures causing again an increase of the transfer probability [1.189, 190].

The single quantum $V \rightarrow V$ energy transfer probabilities are generally in the range of 10^{-3} to 10^{-1}. The near resonant energy exchange is larger when the vibrations involved are infrared-active. This is due to the large vibrational transition dipole moments and their r^{-3} interaction potential [1.191].

1.1.3. Further Collision Studies

With laser radiation it is, in general, possible to prepare an ensemble of atoms or molecules in a particular excited or fundamental state. Collisions of the so prepared particles with others of the same or different kind have therefore well defined initial conditions. Then the fluorescence of the particles or the laser light itself can be used to probe the change of the initial conditions as a function of time and so makes it possible to study all kinds of relaxation processes, as discussed in Subsections 1.1.1 and 1.1.2.

The main difficulty in the interpretation of these experiments is that multiple collisions influence the results considerably; in addition, no information on the angular and velocity dependence of the collision processes is available, which is quite important for the determination of the inter-atomic or inter-molecular potentials. More refined data are, however, obtainable when the atoms or molecules are studied in crossed particle beams [1.192]. When now, in addition, laser excitation is used to prepare the particles in well defined states details on the scattering dynamics are revealed, which in former experiments have been averaged out.

Presently many groups are preparing scattering experiments with laser-excited crossed particle beams. So far one successful experiment was reported investigating electron scattering on Na atoms in an excited state [1.193, 194].

For the excitation of the atoms a cw dye laser was tuned to the $3\,^2S_{1/2}$, $F^{=1}=2 \rightarrow 3\,^2P_{3/2}$, $F=3$ hyperfine transition, as this choice guarantees that the atom always decays to the same lower hyperfine level so that no optical pumping of the hyperfine levels in the fundamental state can occur. The polarization of the light was either σ^+ or π. In the case of σ^+ excitation an orientation of the atoms in the excited state is produced whereas for π excitation one obtains an alignment.

In the experiment the atoms stay in the identical pumping and scattering regions for about 10^{-6} sec or approximately 100 pumping cycles, as determined essentially by the spontaneous decay time. In spite of the fact that each atom is excited many times stationary conditions are not completely reached when averaged over the excitation region. This has to be considered in the evaluation of the results.

The scattering of electrons on atoms followed by an energy analysis yields the energy loss spectra resulting from inelastic collisions. When the atoms are excited to the $^2P_{3/2}$ state, besides the $3s \rightarrow 3p$ energy loss those losses for $3p \rightarrow 4p$, $3p \rightarrow 3d$, $3p \rightarrow 4s$ are observed, too. Furthermore, the superelastic scattering involving the transition $3p \rightarrow 3s$ can be seen. This process is inverse to the excitation of the resonance line by electron impact. Since the scattering geometry can be varied in the laser experiment, very detailed informations are is obtainable (Fig. 1.5).

The parameters which may be varied are the polarization of the laser light, the angle between incoming light and outgoing electrons, θ, and the angle of the incoming electrons with respect to the outgoing electrons, ϑ_{col}. When linearly polarized light is used the angle ψ between the polarization vector of the light and the scattering plane is also relevant. In Fig. 1.6 an example of the results is shown. The energy of the incoming electrons was $E_{\text{inc.}} = 3$ eV, plotted is $r = I(\psi = 0°)/I(\psi = 90°)$ for different angles θ. The experimental results (crosses) are compared to the theoretical result (solid line) which was calculated using the scattering amplitude of Moores and Norcross [1.196].

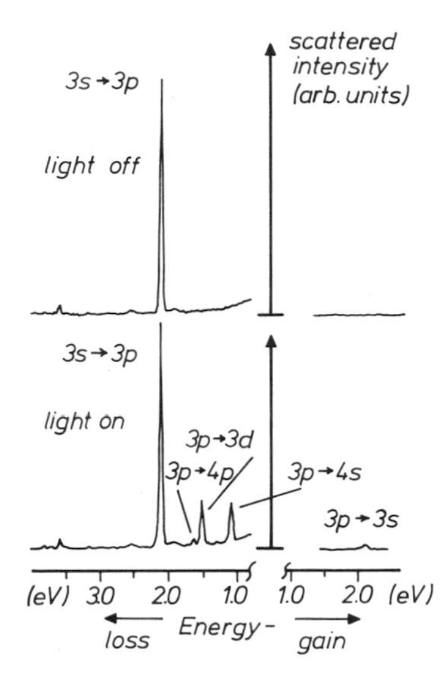

Fig. 1.5. Typical energy loss spectra for electron scattering by sodium atoms with and without laser excitation. (From [1.195])

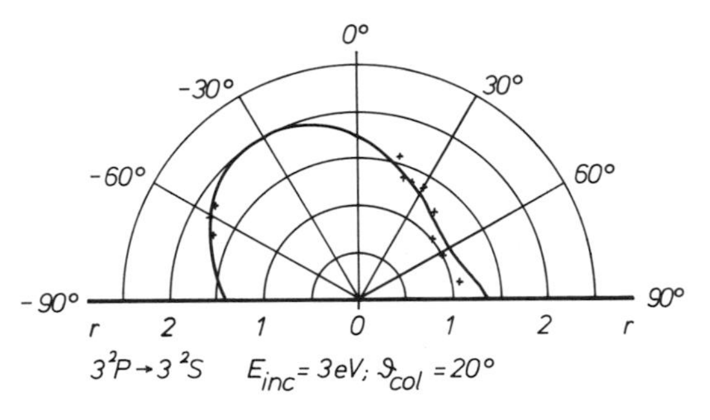

Fig. 1.6. Electron scattering by excited sodium atoms. Comparison between measured and calculated scattering intensities. For details see text. (From [1.195])

These first experiments of scattering at atoms in excited states demonstrated the feasibility, and it is now sure that this technique can also be applied to many more problems, especially to heavy particle collisions. Here, for instance, the influence of the spin-orbit interaction on the scattering process could be investigated in detail. It is apparent that with a similar arrangement experiments on the energy transfer of electronically and vibrationally excited molecules are possible. The well defined initial conditions of a crossed beam experiment would help considerably in the theoretical interpretation of the results.

1.1.4. Photoionization and Photodetachment

Photoionization

The cross sections for photoionization of atoms are important parameters in many fields of science. They are, for instance, used for the determination of the photoionization rates produced in the ionosphere due to the interaction of the extreme ultraviolet radiation from the sun with the earth's atmosphere. Furthermore they are necessary for the understanding of the processes in gaseous discharges; here also molecular ions play an important role. In addition, the investigation of the photoionization cross section as a function of the wavelength gives useful information on autoionizing resonances which are quite important for atomic spectroscopy. The experimental data are necessary to check the validity of the various models for the photoionization process. This is essential since in many cases experimental data may not be obtained or are not available and the theory is the only source for information.

The direct photoionization of atoms from the ground state is only possible by ultraviolet or vacuum-ultraviolet light. This is a wavelength region which is, in principle, accessible with lasers either directly or by frequency doubling. However, very few spectroscopic experiments which use a direct photoionization starting from the fundamental state have been performed. Mostly a stepwise excitation of the atoms is used. In the following several experiments of this kind will be discussed.

A series of measurements on the photoionization of metastable He [1.197, 198], Ar [1.199, 200], and Kr [1.200] atoms has recently been performed by STEBBINGS, DUNNING, and others. For these experiments a beam of metastable atoms was produced by electron impact and then irradiated with the light of a pulsed dye laser. The ions so formed were detected with a particle multiplier. A gating technique was used to distinguish these ions from those liberated at surfaces by scattered laser light and those arising from chemical-ionization in collisions between metastable atoms and any residual gas in the system. In the case of the He measurements, the metastable $2\,^1S$ state, which was also populated by electron excitation, was selectively quenched via transitions of the type $2\,^1S \rightarrow n\,^1P \rightarrow 1\,^1S$ with the radiation from a helium discharge lamp. The $2\,^3S$ state is the lowest level in the triplet system. Data appropriate to each metastable species may be found from observations when the helium lamp is alternatively turned on or off.

In these experiments the metastable atomic beam flux can be determined in a separate experiment by measuring the current of secondary electrons ejected from a surface whose absolute secondary-electron ejection coefficient is known. Furthermore, the absolute cross sections for the photoionization can be deduced. The experimental error for these values may be in the range of about 20%. So far the absolute cross sections for photoionization of metastable He atoms in the $2\,^1S, 2\,^2S, 3\,^1P, 4\,^1P, 5\,^1P, 3\,^3P, 4\,^3P$, and $5\,^3P$ were determined. In the case of the S states the cross section was measured from threshold to 2400 Å. For the P states the population of the levels was performed by laser excitation starting from the S states. Simultaneously some of the atoms in the respective level were ionized. The particle density in the P levels cannot be measured directly, but their values may be inferred when a sufficiently intense laser beam is used, because saturation of the transition will occur almost instantaneously.

The values obtained for the photoionization cross section are essentially in agreement with quantum-defect and close-coupling calculations. The theoretical values obtained by both methods do not differ very much so that within the experimental errors no preference can be given to any one of these calculations. In the case of the photoionization cross section of the He$(2\,^3S)$ state there seems to be a better agreement with the results of NORCROSS [1.196] obtained by close-coupling calculations.

In the argon measurements the autoionization from high-lying $3p^5\,(^2P_{1/2})np^1$ levels has been investigated. These states were populated by laser excitation starting from the metastable $^3P_0$ level. The term values of the autoionizing levels were measured to an accuracy of $\pm 6\,\mathrm{cm}^{-1}$. They are in excellent agreement with values obtained from a quantum-defect extrapolation of the known lower terms of this series. Therefore an unambiguous assignment of principal quantum numbers to the observed levels is possible. From the linewidths observed in the spectra it may be deduced that the lifetime of the levels decaying by autoionization is longer than $2 \cdot 10^{-12}$ sec.

In another paper by DUNNING and STEBBING [1.200] besides $3p^5\,(^2P_{1/2})np^1$ and $3p^5\,(^2P_{1/2})nf^1$ levels of argon also some $4d^5\,(^2P_{1/2})np^1$ and $4d^5\,(^2P_{1/2})nf^1$ levels of krypton have been investigated in a similar way. An example obtained for krypton is shown in Fig. 1.7 with the corresponding level scheme in Fig. 1.8. From the width of the spectral lines it follows that the lifetime of the autoionizing states for krypton is also longer than $2 \cdot 10^{-12}$ sec. This relatively long lifetime suggests the existence of only a weak interaction of all p^1 and f^1 levels with the continuum. The widths of the autoionizing lines decrease as the principal quantum number increases. This comes about since the interaction with the $P_{3/2}$ continuum will be greatest near threshold.

For atoms where excited states can be reached by visible radiation, the photoionization can be performed using dye lasers or other tunable lasers for the first excitation step. This is quite important for applications

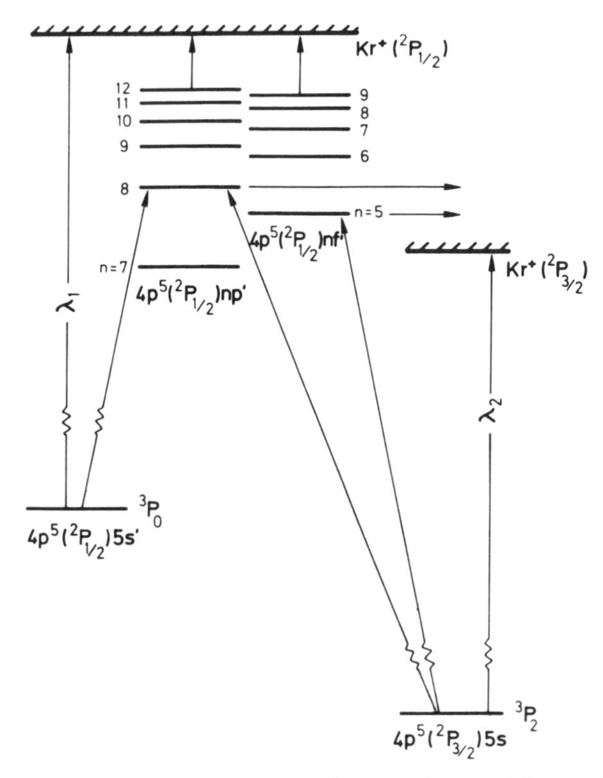

Fig. 1.7. Level scheme of the $4p^5(^2P_{1/2})\,np^1$ and $4p^5(^2P_{1/2})\,nf^1$ autoionizing levels of Kr. Dye laser excitation was performed starting from the $4p^5(^2P_{1/2})\,5s^1$ and $4p^5\,(^2P_{3/2})\,5s$ levels. (From [1.200])

in connection with laser isotope separation. The corresponding experiments will be discussed in Subsection 1.2.2.

A tunable dye laser for the first excitation step has also been used by CARLSTEN et al. [1.201] in an experiment with barium atoms. (In principle this experiment is similar to the one described in Subsection 1.1.1, [1.17–20]). The metastable $6s5d\,^3D$ level of barium was strongly populated by means of a flashlamp pumped dye laser. For this purpose the $6s6p\,^3P$ level was excited which then decayed to the ^3D term. Using this method about 10^{16} atoms/cm^3 in the ^3D metastable term were produced, with the three fine-structure levels equally populated. Several microseconds after the metastable atoms were produced, a flashlamp was fired and the absorption spectrum recorded. Spectra were taken with and without laser excitation to determine the absorption features of transitions starting from the ^3D metastable levels to higher terms.

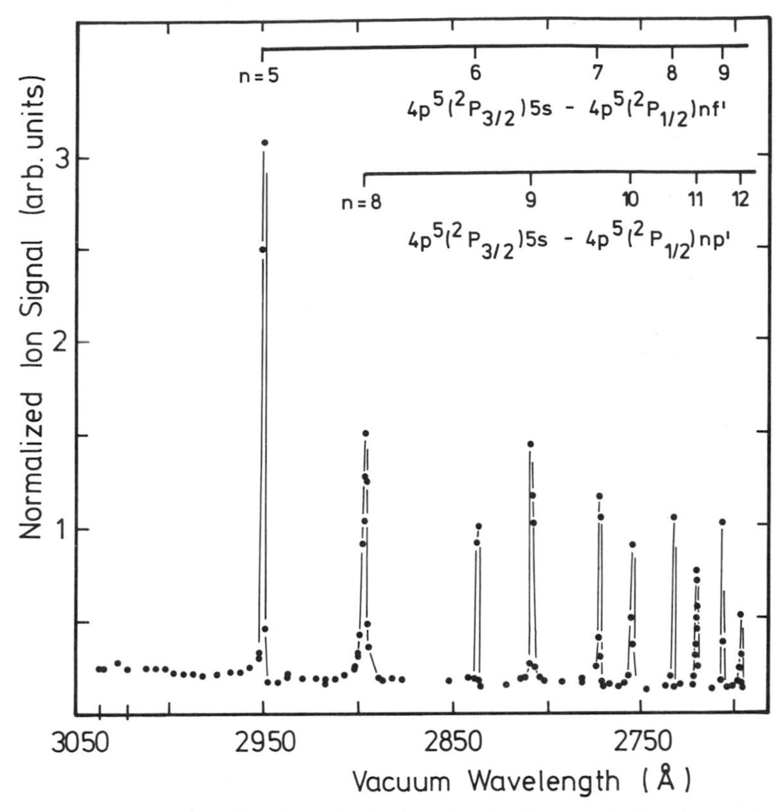

Fig. 1.8. Krypton-ion signal obtained when the dye laser excitation is tuned over the spectral range of 3050–2690 Å. See also Fig. 1.7. (From [1.200])

In the experiment the fundamental triplet series $6s5d\,^3D - 6snf\,^3F$ was identified and observed up to $n = 32$. Furthermore the absolute photoionization cross section was measured at 3030 Å. The relative cross section from 3030 to 2500 Å was also determined and found to be nearly constant. The photoionization cross section for the ^3D metastable level is of importance to the understanding of the photoionization of artifical barium clouds in the upper atmosphere by solar radiation. It has been postulated that the dominant process for the photoionization is a two-step process rather than direct photoionization. The two-step process proceeds via the metastable ^1D and ^3D levels from which the atoms are photoionized. Since the latter process occurs at longer wavelengths than the direct photoionization, more solar photons are available and such a two-step process can dominate. Using the result for the photoionization cross section determined in the above-mentioned measurement it could be shown that the two-step photoionization via the ^3D levels is indeed the predominant mechanism [1.202].

An interesting application of photoionization by means of lasers has been demonstrated by GALLAGHER et al. [1.203]. A beam of metastable barium atoms ($^1D_2$) was photoionized inside the cavity of a He–Cd laser operating at 3250 Å. In this way highly monochromatic photoelectrons ($10^{-12}-10^{-13}$ A) were produced with a kinetic energy of 17 meV and a calculated energy spread of <1 meV. The electron energy was analysed by measuring the 11.08 meV resonance of argon. The observed width of this resonance was about 6 meV, which is larger than the expected energy width. However, there were several perturbing effects as, e.g., the Doppler spreading in the target atomic beam and potential gradients across the collision volume, which may increase the energy spread. There is hope that these perturbing effects may be reduced. An electron beam with an energy spread in the meV region would be an interesting tool for the investigation of resonances in the elastic scattering of electrons on atoms [1.204]. The so far best reported performance for a monochromatic electron source is an energy spread of 3.5 meV [1.205].

Laser-photoionization can be used to produce polarized electrons exploiting the Fano effect [1.206]. This effect can be understood in terms of spin-orbit coupling in the continuum angular momentum states of an alkali atom. For certain wavelengths, the ionization cross-section for circularly polarized light can be very different for the two spin states of the electron. That means that photoionization with certain wavelengths and circularly polarized light results in highly polarized electrons. The effect was experimentally demonstrated by HEINZMANN et al. [1.207]. With a tunable laser in the near ultraviolet a high efficiency for photoionization can be expected. This may be very useful for the construction of sources for polarized electrons. A first successful attempt in this direction was undertaken by VON DRACHENFELS et al. [1.208]. With a frequency-doubled flashlamp-pumped dye laser producing 2 mJ/pulse at 3050 Å the photoionization of an intense cesium atomic beam has generated about $3 \cdot 10^9$ electrons per laser pulse; the polarization of these electrons was 90%.

Photodetachment

A rather interesting field of application for the laser is the negative ion spectroscopy. The electron affinities of atoms, free radicals, and molecules are of fundamental importance to the understanding of the kinetics of chemical reactions. In addition, the precise determination of the electron affinity provides a very good test of quantum calculations.

The *ab initio* theoretical determination of electron affinities of atoms is usually performed by means of a variational method using configuration-interaction trial wave functions [1.209, 210] and gives in some cases a rather good agreement. The empirical method, on the other hand, which uses an extrapolation of known spectral data along an isoelectronic sequence e.g. [1.211, 212] shows considerable scatter.

Since the electron affinities are quite important many experimental attempts have been undertaken to get reliable data. Among the techniques used are, e.g., surface ionization, endothermic charge transfer, pair production in photoionization, and studies of continuous absorption in plasmas [1.213]. In spite of many investigations only few reliable data resulted. The situation improved when the electron affinity was directly determined through photoabsorption studies. However, the negative ion densities necessary for these measurements are difficult to produce; therefore a big step forward was taken when BRANSCOMB and co-workers started to use a crossed beam technique [1.214]. In these experiments a mass selected negative ion beam was crossed by an energy resolved photon beam, and the cross section for the production of electrons was measured as a function of photon energy. A considerable improvement of this technique was then obtained when lasers replaced the conventional light sources since the higher intensity of the laser results in a much better signal-to-noise ratio. Additionally the photon energy resolution is substantially enhanced especially when frequency tunable lasers are used. Furthermore, with lasers, the requirement for intense negative ion beams is removed. Therefore it became possible to study a large number of ions which could not be investigated previously. There are essentially two techniques currently in use in connection with laser-photodetachment:

a) Crossed beam apparatus with tunable dye laser excitation. Here the neutral atoms resulting from laser photodetachment are monitored as function of the photon energy [1.215].
b) Crossed beam apparatus with excitation by a fixed wavelength laser. The energy of the photodetached electrons is analysed in a hemispherical electrostatic condensor (resolution $\approx$ 50 meV) [1.216, 217].

This latter method was also used to investigate the angular distribution of the photodetached electrons. Since the contact potentials in the electron energy analyser are not well known the measurement was performed relative to negative ions whose electron affinities had been determined absolutely [1.218]. A survey on the data obtained by laser photodetachment is given in Table 1.3. For a survey on the experimental techniques and results see also [1.220].

The good signal-to-noise ratio which may be obtained in laser-photodetachment allows a detailed study of variations of the photo-

Table 1.3. Laser photodetachment

Particle	Method	Electron affinity (eV)	Additional remarks	Ref.
He^-	FL	0.080(2)		[1.216]
Li^-	FL	0.620(7)		[1.219]
Na^-	TL	0.543(10)		[1.220]
	FL	0.548(4)		[1.219]
K^-	TL	0.5012(5)		[1.219]
Rb^-	TL	0.4859(15)		[1.219]
	FL	0.486(3)		
Cs^-	TL	0.472(2)		[1.219]
	FL	0.470(3)		
Cu^-	FL	1.229(10)	Angular distribution	[1.221]
Ag^-	FL	$1.303^{+0.007}_{-0.011}$	Angular distribution	[1.221]
Au^-	TL	2.3086(7)		[1.222]
S^-	TL	2.0772(5)		[1.215]
Se^-	TL	2.0206(2)		[1.223]
Pt^-	TL	2.128(2)		[1.222]
OH^-	TL	1.825(2)		[1.224]
	FL	$1.829^{+0.010}_{-0.014}$		[1.225]
OD^-	TL	1.823(2)		[1.224]
O_2^-	FL	0.440(8)	Angular distribution Molecular constants of O_2^-	[1.227]
S_2^-	FL	1.663(40)		[1.225]
SO_2^-	FL	1.097(36)		[1.225]
NH^-	FL	0.38(3)	Angular distribution	[1.225]
NH_2^-	TL	0.74		[1.226]
	FL	0.779(37)		[1‘225]
NO^-	FL	$0.024^{+0.010}_{-0.005}$	Angular distribution Molecular constants of NO^-	[1.217]
PH_2^-	TL	1.26(3)		[1.226]

Abbreviations: FL: Fixed frequency laser, TL: Tunable laser.

detachment cross section. In this way interference effects in the photodetachment can be observed at photon energies corresponding to the opening of a new channel which competes with an already opened channel. In Na^-, for example, a cusp in the photodetachment cross section is observed when the photon energy is sufficient to allow the transition $Na^-(^1S)+h\nu \rightarrow Na(^2P)+e$ in addition to $Na^-(^1S)+h\nu \rightarrow Na(^2S)+e$ [1.220]. The experimental data are in very good agreement with close-coupling calculations by Norcross and Moores [1.228].

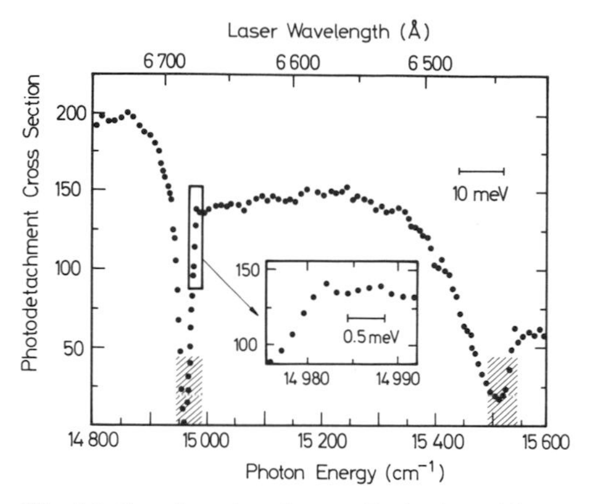

Fig. 1.9. Cs^- photodetachment. Excitation with a tunable dye laser. The signal decrease to zero in the left half of the figure cannot be interpreted as the $^2P_{1/2}$ channel opening. The interpretation is that there is a autodetaching state of the alkali negative ion which lies just below each of the ^2P states. The inset shows a sharp discontinuity above the minimum. This is interpreted as the $^2P_{1/2}$ channel opening. The cross hatched regions indicate the confidence limits for the $Cs(6\,^2P_{1/2})$ and $Cs(6\,^2P_{3/2})$ exit channels. (From [1.219])

In Cs^- and Rb^- it was found that the $^2P_{1/2}$ channel opening is very sharp (see Fig. 1.9), e.g. a width of 150 μeV was observed for Rb^-. Furthermore, the cross sections drop to zero. This strong decrease in the cross section cannot be understood solely on the basis of a new channel opening which competes with an open channel. A possible explanation is that there exists an autodetaching state of the alkali negative ion which lies just below each of the ^2P states of the heavy alkalis.

The interpretation of the molecular photodetachment is, in general, more difficult than for atoms since the manifold of rotational transitions involved complicates the evaluation as the exact form of the threshold law is unknown and must be rotational energy dependent.

If a molecular ion possesses a bound electronically excited state, then two photon photodetachment can be studied using the bound electronically excited state as a real intermediate state. An experiment using this technique has been performed by LINEBERGER et al. [1.229] on C_2^-. The first photon excites the intermediate state, the second photon then induces the photodetachment so that finally the neutral molecule can be detected. The experiment yields spectroscopic information on the intermediate state. The measurements established, for instance, the fact that the Herzberg-Lagerquist band system [1.230] belongs to C_2^-, and that the ground state of C_2^- is an $a\,^2\Sigma_g^+$ state.

This two-photon technique is not limited to intermediate states lying energetically more than halfway up to the continuum. By using two light sources the entire energy range is accessible. Similar measurements can also be performed with molecular positive ions by observing the neutral particles produced by photodissociation from the intermediate state.

1.1.5. Nonlinear Effects

The high spectral brightness obtainable with lasers makes it possible to observe a large number of nonlinear effects in the interaction of radiation with matter. Historically most of these phenomena have first been observed in solids. With the steady improvement of laser technology, however, some of them have now also been demonstrated in gases. This is especially the case for the density dependent effects as, e.g., frequency multiplication and frequency mixing. Here the low density of gases is compensated by the large resonance enhancement which makes the processes competitive with the corresponding effects in solids.

In the following several nonlinear effects observable with lasers will be discussed.

Multiphoton Ionization

The study of multiphoton ionization processes in atoms is a large field of application for the laser. Since the effect can be observed without the laser frequency being in resonance with an atomic transition, high power Nd : YAG lasers or ruby lasers are suitable for the investigations. However, the cross-sections for multiphoton-ionization have resonances at frequencies which coincide with atomic transitions (e.g. [1.231]), therefore the application of tunable lasers is also interesting in order to investigate the near-resonance behaviour.

The theoretical treatment of multiphoton-ionization is essentially performed along two fundamental directions. One of them makes use of the perturbation theory [1.231–238] the other uses non-perturbative methods.

The calculations according to perturbation theory usually consider the lowest-order non-vanishing term relevant for the phenomenon [1.239–240]. In the perturbative calculation the Nth order transition rate consists of a product of N electric dipole matrix elements and $N-1$ sums running over all intermediate states of the atoms. For the calculations, therefore, the atomic wavefunctions must be known so that the dipole matrix elements can be determined and, in addition, a good estimate of the sums over the intermediate state must be available. In the pioneering papers of BEBB and GOLD [1.231, 232] the sums over intermediate states were approximated by averaged energy denominators, and the wavefunctions were determined by the quantum-defect method.

Other calculations of the sum over intermediate states have been performed by means of an implicit technique [1.233, 237, 238] or by using Green's function [1.236, 241].

In general, the lowest-order perturbation theory is not sufficient for a good description of the resonant processes or when the laser intensity exceeds certain critical values. Therefore calculations have also been performed taking into account higher-order processes [1.242, 243]. The result is that the ionization cross-sections are strongly modified near resonance [1.244].

In a recent experiment [1.245] the four-photon ionization of Cs was investigated using light of the frequency near that required to produce three-photon excitation of the $6f$ state. The very rapid variation of the transition probability with frequency and intensity found in this experiment was rather well reproduced by higher-order perturbation theory calculations.

A number of non-perturbation theories for multi-photon ionization of atoms have also been published [1.246–248]. These calculations support the main features derived by perturbation theory. The latest papers published along these lines use unitary transformations of the Schrödinger equation corresponding to either a momentum [1.249, 250] or a space translation [1.251–253]. Furthermore, other theories for ultra-high field intensities have been proposed [1.254, 255]. In this work it has been assumed that the atomic potential can be treated as a perturbation compared to the electromagnetic potential. These calculations predict that the transition rate decreases when the field intensity reaches very large values.

Because the wavelengths of the high power lasers suitable for multiphoton ionization were effectively fixed, the tests of the theoretical calculations have generally been concerned with the change of the ionization cross section with laser intensity rather than with wavelength. In the following some investigations performed during the last years will be discussed.

At first, an experiment will be described which was performed with a Q-switched Nd glass laser on potassium [1.256] and cesium atoms [1.256, 257] with laser powers of more than 10^7 W. It was found that for the cesium measurements the molecular cesium plays an important role in the multiphoton ionization process at low laser intensities. At laser powers of $4 \cdot 10^6$ to $4 \cdot 10^7$ W an ion production rate is found which depends on the third order of the laser power. This is a deviation from theory since four photons are necessary for the ionization process and therefore the order should be 4. But an examination of the scheme of the atomic levels of cesium shows that the difference between the energy of three quanta of the Nd laser and the energy of the $6f$ level is smaller than the bandwidth of the laser radiation. Thus the cesium atoms seem to be ionized via the resonant $6f$ level.

This energy coincidence was later further investigated with a narrow banded (0.4 Å) and tunable Nd-glass laser (tuning range 40 Å) [1.245]. Both, the possibility of accurately tuning the laser wavelength and the better monochromaticity of the radiation now enabled the investigation of the respective roles of the direct multiphoton ionization and the ionization via an intermediate state. In the experiment it was found that a Stark shift takes place in case of a non-resonant excitation of the cesium atoms. This shift explains the strong asymmetry in the change of the multiphoton ionization probability as function of the laser wavelength mentioned above [1.244].

In other experiments the three-photon ionization of cesium was investigated by means of Q-switched ruby lasers [1.258, 259]. In these experiments the relative number of cesium ions produced shows the theoretical power dependence at low laser intensities. With high laser fluxes [1.259], however, the ionization rate decreases. The effect was attributed to the Stark shift of the quasi-resonant $9D$ levels. In the paper by Fox et al. [1.258] it was found that circularly polarized light produces ionization about twice as efficiently as linearly polarized light. This effect will be discussed later.

The dependence of the multi-photon ionization cross section near resonances can be studied extensively when tunable lasers as, e.g., dye lasers are used. The first experiment of this kind was performed by Popescu *et al.* [1.260]. The laser was tuned over a wavelength region from 6550–6950 Å. The resulting dispersion curve for photoionization was interpreted in terms of the two-photon transition from the $6\,^2S_{1/2}$ ground state to resonant n^2D and n^7S intermediate states. The processes occuring are, besides the direct three-photon ionization

$$\mathrm{Cs} + (3h\nu) \rightarrow \mathrm{Cs}^+ + \mathrm{e}$$

through the resonant n^2D and n^2S levels, the collisional ionization of the intermediate state formed by photon absorption

$$\mathrm{Cs} + (2h\nu) \rightarrow \mathrm{Cs}^*(n^2D, n^2S)$$

which proceeds either by associative ionization ($n \leqq 12$)

$$\mathrm{Cs}^* + \mathrm{Cs} \rightarrow \mathrm{Cs}_2^+ + \mathrm{e}$$

or by molecular association followed by dissociative attachment ($n > 12$)

$$Cs^* + Cs \rightarrow Cs_2^* \rightarrow Cs^+ + Cs^- .$$

Two-photon ionization of molecular cesium has also been observed in the wavelength region 6200–6600 Å by means of a tunable dye laser [1.261]. The photoionization as a function of wavelength gives information on the intermediate states. These measurements are therefore analogous to the important technique of resonant multi-photon ionization spectroscopy introduced by LINEBERGER et al. [1.229] in recent studies on the photodetachment of C_2^-.

It was mentioned that multiphoton absorption rates depend on the state of polarization of the photons even if the initial electronic state is totally unpolarized. This is due to the fact that the absorption of the first photon alters the state of the electron, therefore a second photon will see different states of the atom depending whether the first photon was circularly or linearly polarized. This is in contrast to a single-photon transition where the polarization of the light only influences the differential cross section, e.g. the angular distribution of photoelectrons, and not the total transition rate. The dependence of the transition rate of multiphoton processes on the light polarization was first noticed experimentally by FOX et al. [1.258] and later also observed in other experiments [1.330, 331]. Calculations of the transition rates have been performed, e.g. by LAMBROPOULOS [1.332] and KLARSFELD et al. [1.333].

An important result is that for multiphoton ionization linearly polarized light will give an ionization rate which is, in general, higher than that of circularly polarized light. It is also quite important that this ratio may change as a function of the photon frequency. In this connection the observation of the angular distribution of the photoelectrons for light of various polarizations is quite important [1.334], since the results can be used to determine the signs of bound-free matrix elements.

Multi-photon ionization by circularly polarized light in the presence of spin-orbit coupling can lead to photoelectrons with definite spin-polarization. The degree of polarization may reach 100% when the photons are near-resonance with an intermediate atomic state whose fine structure splitting is larger than the laser line-width [1.335–337]. The experiments can be performed with alkali or trivalent atoms.

The electromagnetic field is a vector field which is characterized by polarization and amplitude. The influence of the polarization on the multiphoton process is therefore also related to the influence of photon statistics (correlations of the amplitudes) on the same process. The influence of the radiation statistics on nonlinear processes has been discussed by several authors, e.g. [1.338–343]. Experiments have been performed on the photon correlation enhancement of second-harmonic generation [1.344] and on the second-order photo effect [1.345]. The influence of the temporal coherence effects on multi-photon ionization process was demonstrated recently by LECOMPTE et al. [1.346]. With a single-transverse mode Q-switched Nd-glass laser, which can operate over a variable number of longitudinal modes, the influence of the temporal nature of the laser pulse on the multiphoton ionization probability of Xe atoms has been investigated. For a seven-mode laser pulse, the number of ions formed is increased by several orders of magnitude over that produced by a single-longitudinal mode laser.

Two-Photon Spectroscopy

The theory of multiple quantum transitions was first discussed by GOEPPERT-MAYER [1.262] in 1929. The first experimental observation was made in the microwave region by HUGHES and GRABNER [1.263]. The investigation of double quantum transitions in the optical region became feasible with the advent of lasers. KAISER and GARRET [1.264] performed the first experiment of this kind. $CaF_2:Eu^{2+}$ crystals were illuminated with a ruby laser, and a state of Eu^{2+} with an energy, which corresponds to twice the photon energy of the laser light, was excited; subsequently the fluorescent light at 4250 Å was observed. The experiment was in a very good agreement with calculations by KLEINMANN [1.265] who derived simple formulas for the two-photon absorption.

The first observation of double-quantum transitions in an atomic system was performed by ABELLA [1.266]. In this experiment the $6\,^2S_{1/2}-9\,^2D_{3/2}$ transition in Cs was excited by the intense light of a ruby laser. The $9\,^2D_{3/2}$ levels in Cs are almost at twice the ruby laser frequency. In order to get a good coincidence between the sharp absorption and the laser emission frequency thermal tuning of the ruby laser was required. The absorption was detected by observing the fluorescent decay of the $9\,^2D_{3/2}$ state to $6\,^2P_{3/2}$.

Afterwards double-quantum absorption has been found in a wide variety of materials including: inorganic crystals [1.267], organic crystals [1.268], and organic liquids [1.269].

The observation of saturation in double-quantum absorption was not possible in the early experiments. Since the probability for double-quantum transitions is considerably enhanced by intermediate-state resonances it is possible to saturate the double-quantum transition if the applied frequencies are close enough to intermediate intervals. This was demonstrated by YATSIV et al. [1.270] on potassium atoms for the transition $4S_{1/2}-6S_{1/2}$. The transition $4P_{3/2}-6S_{1/2}$ is rather close to the ruby laser frequency so that temperature tuning can be used. Radiation resulting from the transition $4S_{1/2}-4P_{3/2}$ was obtained by stimulated Raman scattering in 1-bromonaphtalene. The beam of Stokes light was downshifted by the right amount, so that the $4S_{1/2}-4P_{3/2}$ transition was almost matched.

The use of two-photon transitions for spectroscopy became quite exciting recently after it was demonstrated that a Doppler-free observation of the transitions is possible. These experiments will be discussed in Section 1.4 in connection with high-resolution spectroscopy. However, Doppler-limited two-photon spectroscopy also provides useful results since it is possible to investigate transitions to quantum states which are different from those which can be populated with one-photon processes. Furthermore, high lying levels become accessible with visible radiation.

Using tunable dye lasers a series of two-photon experiments has already been performed on free molecules. HOCHSTRASSER et al. [1.271] investigated benzene in the gas phase and as a crystal. Subsequently BRAY et al. [1.70] measured individual rotational levels of the $A^2\Sigma^+$ state of NO. Recently another paper on benzene was published by WUNSCH et al. [1.272, 273]. In this paper even extremely weak hot bands were investigated which have not been seen in other experiments. Further two-photon results on NO and on C_6H_6 and C_6D_6 were published by BRAY et al. [1.274]. In this paper polarization studies were carried out, in particular the anisotropy of the two-photon absorption for incident light either circularly or linearly polarized was investigated. The different absorptions for the two polarization directions gave useful information on the symmetry of the excited states.

Higher-Harmonic Generation and Nonlinear Mixing Processes in Atomic Vapors

Multiple photon absorption which is observed at high intensities is an incoherent process. The mathematical treatment of this interaction must therefore not account for the coherence between incident and scattered photons. Coherent effects in the interaction between light and molecules, of course, also exist and result in macroscopically observable physical properties as, e.g., the refractive index due to the polarization of the medium. At high light intensities the higher-order polarization terms arise which produce the well-known harmonics of the incident wave or frequency mixing when two waves are applied simultaneously. The harmonic generation was initially a domain of solid and liquid systems. Gases and vapors seemed not attractive. Due to the small number density only a low efficiency was to be expected (the nonlinear process is proportional to the square of the number density). The vapors and gases are isotropic media, which would allow only odd-harmonic generation. The first successful experiments have been by WARD and NEW [1.275] using inert gases, $H_2, CO_2, N_2(CH_2)_2, Cl_2, O_2$, and air. The calculations had previously been reported by ARMSTRONG et al. [1.276]. In this paper the phase-matching condition was discussed in detail, and it was shown that it can be realized by mixing gases with proper dispersion behavior.

The expression for the third-order susceptibility responsible for third-harmonic generation is due to ARMSTRONG et al. (see also [1.277])

$$\chi^{(3)}(3\omega) = 1/\hbar^3 \sum_g \sum_{a,b,c} (d_{ga} d_{ab} d_{bc} d_{cg})\, \varrho_{gg} \left[\frac{1}{(\Omega_{ag} - 3\omega)(\Omega_{bg} - 2\omega)(\Omega_{cg} - \omega)} + \frac{1}{(\Omega_{ag} + \omega)(\Omega_{bg} + 2\omega)(\Omega_{cg} + 3\omega)} + \frac{1}{(\Omega_{ag} + \omega)(\Omega_{bg} + 2\omega)(\Omega_{cg} - \omega)} + \frac{1}{(\Omega_{ag} + \omega)(\Omega_{bg} - 2\omega)(\Omega_{cg} - \omega)} \right]. \tag{1.1}$$

The d_{ij} are the electric dipole matrix elements, ϱ_{gg} is the population of the fundamental state, Ω_{ij} denotes the atomic transition frequencies. The summation has to be performed over all states a, b, c of the atom or molecule.

The expression for the third-harmonic nonlinear susceptibility shows a resonant enhancement when the first, second, and third harmonics of the incident frequency ω agree with an atomic frequency. This is the reason that the susceptibilities of the alkalies in the visible or the near-infrared spectral region are several orders of magnitude larger than those which have been observed for the rare gases in the same spectral region. Since the conversion efficiency is proportional to the square of $\chi^{(3)}$ a considerable improvement can be expected when alkali atoms are used. This was pointed out by HARRIS and MILES [1.278]. Phase matching is possible for the alkali metal vapors since they are negatively dispersive for a laser wavelength above and a third harmonic under the wavelength of the first resonance line. The normally or positively dispersive rare gases are therefore suitable to compensate for the negative dispersion of the metal vapor. At a suitable pressure ratio between the alkali vapor and the rare gas it can be achieved that the first and third harmonics have the same velocity in the gas mixture. The third-harmonic generation is therefore cumulative over the whole length of the gas cell. This increases substantially the power conversion because it is proportional to the square of the cell length.

The creation of the third-harmonic can be considered as a step-by-step process. The incident laser wave interacts with the particle in the ground state and produces a macroscopic dipole moment at the driving frequency. Induced dipole moment and incoming field generate a 2ω variation of the excited-state populations. The latter interact again with the incoming field and create a dipole moment at the third harmonic.

The third-harmonic generation in gas mixtures has a number of significant advantages in comparison to third-harmonic generation, or sequential second-harmonic generation, in nonlinear crystals [1.277]. First, the vapors are opaque up to the vacuum ultraviolet region; the absorption cross sections of the alkali vapors above their ionization potentials are so small that no substantial loss is introduced. (The shortest wavelength so far generated by nonlinear crystals is 2128 Å [1.279]). Secondly, the isotropy of the gases does not show the Poynting vector walkoff observed for birefringent phase matching in crystals. Finally the maximum applicable power densities are higher in gases and, in addition, when breakdown occurs the material is virtually not destroyed. A disadvantage, on the other hand, is that for efficient harmonic generation the incident power has to be much higher in the case of gases than of crystals. It was shown [1.277] that the most important limitation in the obtainable conversion efficiency is the violation of the phase-matching condition caused by a small absorption of the incoming and the third-harmonic waves. For example, with a cell of 100 coherence lengths the phase-matching condition is violated when 1% of the alkali atoms are removed from the ground state.

The optimum conversion efficiency is adjusted by increasing the incident power to the value where multiphoton ionization or avalanche breakdown becomes important. Then the density of the metal vapor is increased until a maximum output power is achieved. The metal is mostly evaporated in a heat pipe oven [1.283] which allows it to produce a homogeneous high temperature region of vapor with a cool zone of buffer gas which prevents fogging and corrosion of the optically transparent windows by the metal vapor. To obtain phase matching in vapor-gas or vapor-vapor mixtures usually a double concentric heat pipe according to VIDAL and HESSEL [1.284] is used.

Frequency tripling with an alkali inert gas system has been demonstrated for the first time by YOUNG et al. [1.280]. The medium consisted of a mixture of 1 part rubidium and 412 parts xenon. The laser used was a Nd:YAG laser with a power of about $2 \cdot 10^8$ W, yielding a conversion efficiency of 10^{-4}. Another experiment [1.281] utilized

5 Torr Na, phase matched with 806 Torr Xe and an active optical path length of the medium of 40 cm. With a laser power of about $2 \cdot 10^8$ W a maximum conversion efficiency of 1% was obtained. Other examples for vacuum ultraviolet light generation are the following: Two experiments have been performed with the system Cd-Ar [1.282] which was used to frequency triple 5320 Å and to mix one photon of 1.06 μm radiation with two photons of 3547 Å to yield an output at 1520 Å. In another experiment 3547 Å radiation was frequency tripled in a phase-matched mixture of Xe-Ar [1.285]. This demonstrates the case where the wavelength of the generated harmonic is shorter than the resonance line of one of the inert gases (Xe). Therefore the inert gas is negatively dispersive and plays the role of the metal vapor in the examples above. The conversion efficiency thus obtained was 3% for an incident power density (after focusing) of $6.3 \cdot 10^{12}$ W/cm^2.

The shortest wavelength reported so far was 887 Å due to tripling of 2660 Å radiation in argon [1.281]. The observed efficiency was 10^{-7}. In the experiment the incoming radiation was tightly focussed. It has been shown that in such a case for a negatively dispersive medium a theoretical conversion efficiency of up to 10^{-3} may be obtained when the radiation is focused into a region which is about the coherence length [1.275]. A quasi phase match is obtained since radiation of the fundamental frequency with off- angle k-vectors is mixed.

For the generation of soft X-ray radiation the harmonics of orders higher than the third are interesting. The relative magnitude has therefore been calculated for some examples by HARRIS [1.286]. In this analysis it was shown that if the generated frequency comes close to that of an excited level of the atom, the conversion efficiency is independent of the order of the nonlinear optical polarization. In the case of Li^+ a conversion efficiency of $4 \cdot 10^{-3}$% was determined for the seventh-order process starting with 1182 Å and yielding 169 Å. This result was calculated for a single coherence length, an improvement by means of phase-matching is therefore still possible. It was assumed that the Li ions are generated by the incident laser pulse.

The use of frequency-tunable rather than fixed-frequency lasers allows one to employ the resonance enhancement of the third-order susceptibility to a much larger extent. This was utilized by HODGSON et al. [1.287] to generate tunable, coherent vacuum ultraviolet radiation via third-order sum mixing in Sr vapor. Two dye lasers were used for the experiment. The first frequency ω_1 was tuned to induce the double-quantum transition to a level of the $5p^2$ or $5s5d$ configuration. The second laser frequency ω_2 was tunable in order to get tunable vacuum ultraviolet radiation at $\omega_{vuv} = 2\omega_1 + \omega_2$. In the course of changing ω_2 a strong signal enhancement of the vacuum ultraviolet radiation was observed. The wavelength dependence had the form of resonances and appeared when $2\omega_1 + \omega_2$ was swept over autoionizing states (configurations $6s5p$ and $5d5p$).

The observed resonant enhancement in ω_{vuv} may be easily understood by considering the first term of (1.1)

$$\chi^{(3)}(3\omega) = 1/\hbar^3 \sum_g \sum_{a,b,c} d_{ga} d_{ab} d_{bc} d_{cg} \sigma_{gg} \cdot \frac{1}{(\Omega_{ag} - \omega_{\mathrm{vuv}})(\Omega_{bg} - 2\omega_1)(\Omega_{cg} - \omega_1)} . \tag{1.2}$$

The vacuum ultraviolet radiation generated in the experiment was tunable from 1778 to 1817 Å and from 1833 to 1957 Å. With a laser power of 16 kW at ω_1 and of 1.6 W at ω_2, the vacuum ultraviolet power measured was $5.2 \cdot 10^{-5}$ W. (The ω_2 power was kept small in order to avoid saturation of the detector!). The vacuum ultraviolet power output is linear in the power at ω_2 and quadratic in the power at ω_1. Assuming that this linearity for ω_2 applies also at higher powers, a vacuum ultraviolet output in the range of almost 1 W should be attainable with the two nitrogen-laser pumped dye lasers used in the work. The Sr vapor pressure was 25 Torr, and 460 Torr of Xe was added for phase matching.

The use of a two-photon resonance in the third-order sum mixing process has the advantage that $\chi^{(3)}$ is resonantly enhanced but the input light is not strongly absorbed. It is for the absorption losses that also ω_{vuv} should not precisely coincide with a resonance frequency. In addition, it is difficult to achieve phase matching if ω_1, ω_2 or ω_{vuv} is close to an allowed single-photon transition.

The parametrically generated, tunable, vacuum ultraviolet radiation can also be used to study autoionizing states. This was demonstrated by ARMSTRONG and WYNNE [1.288]. Using the formalism derived by FANO [1.290] an attempt was made to fit the experimental data for the $4d4f$ autoionizing levels of SrI. Later this fit was improved by ARMSTRONG and BEERS [1.289] who showed that the original evaluation was not quite correct.

Using four-wave parametric mixing also tunable coherent infrared radiation has been produced by means of two nitrogen-laser pumped dye lasers [1.291]. The nonlinear medium was potassium vapor in a heat-pipe oven. One of the dye lasers with the frequency ω_1 was tuned to the vicinity of the $4s-5p$ resonance line. With a sufficiently high laser power stimulated electronic Raman scattering was generated and coherent Stokes light was emitted at the frequency $\omega_s = \omega_1 - \omega_{5s-4s}$.

The coherent Stokes light was observed as ω_l was tuned from the low to the high frequency side of the $5p$ resonances: Its intensity shows minima when ω_l coincides with the $5p_{3/2}$ and $5p_{1/2}$ levels. These minima result from the losses induced by the excitation of the $5p_{3/2}$ and $5p_{1/2}$ levels followed by a coherent emission of radiation, which corresponds to the transitions to the $3d$ levels.

The stimulated Stokes radiation mixes with ω_l and ω_p which is the frequency of the second dye laser. The third-order nonlinear response of the vapor then produces a nonlinear polarization at the frequency

$$\omega_{IR} = \omega_l - \omega_s - \omega_p \,. \tag{1.3}$$

This polarization is resonantly enhanced because ω_l is quite close to the frequency of the $5p$ resonance lines and because $\omega_l - \omega_s$ agrees with the $4s-5s$ transition. Phase matching is maintained in the mixing process by tuning ω_p and, in addition, ω_l. An alternative is to alter the linear dispersion characteristic of the metal vapor by adding, e.g., sodium to the potassium and using the above-mentioned concentric heat pipe.

Using pure potassium vapor SOROKIN et al. [1.291] produced infrared radiation from 2 to 4 μm. In rubidium infrared was generated from 2.9 to 5.4 μm [1.292]. In a sodium potassium mixture tunable infrared radiation was obtained in the range of 2 to 25 μm.

The infrared power is proportional to the product of the powers at ω_l, ω_s, and ω_p. Therefore the infrared power is expected to be proportional to the cube of the power of the nitrogen laser which pumps the dye lasers. The use of high-power nitrogen lasers is therefore quite important for the four-wave parametric conversion. The infrared transparency of the alkali vapors makes it quite probable that radiation from 1 to 500 μm may be generated, with a power level which is sufficient for spectroscopic investigations [1.292]. The linewidth of the infrared radiation was in the described experiment determined by that of the dye lasers which was about $0.2\,\text{cm}^{-1}$. However, there seems to be no reason which would prevent a further reduction so that even high-resolution spectroscopy would be possible.

The strong optical excitation of the np levels by tunable dye lasers can result in stimulated emission of the $np-ns$ and/or $np-(n-2)d$ transitions, and in stimulated Raman scattering to the ns states. Therefore spectroscopy on high-lying p, d, and s states can be performed using this radiation. First experiments of this kind have been reported by WYNNE and SOROKIN [1.293] on potassium. In this work, e.g., the transitions $np-(n-2)d$ have been measured from $n=7$ to $n=16$. The $np-ns$ transitions have been investigated from $n=7$ to $n=9$.

Another resonantly pumped system which uses visible and infrared radiation in order to produce vacuum ultraviolet radiation has recently been proposed by HARRIS and BLOOM [1.294]. This scheme uses Mg atoms which are excited from the $3s$ ground state to the $4s$ excited state by means of double quantum transitions with 4597 Å radiation. A second tunable laser generates the sum or difference frequency $2\omega_p \pm \omega_t$, where ω_p and ω_t are the frequencies of the pump and tunable lasers, respectively. The mixing process is especially efficient if the generated frequency lies within a certain range of any of the $np\,^1P$ levels. For $n=4$ thus tunable radiation in the region around 2000 Å is obtained.

As in the case of Sr (see above) the use of the double-quantum transition gives an enhancement of the nonlinear optical susceptibility. The absence of both loss and dispersion at the input and generated frequencies is a big advantage. The proposed system can be used for infrared to visible up-conversion and imaging. The calculations show that up-conversion power efficiencies in excess of 100% should be obtainable with tunable dye lasers having peak powers of hundreds of watts.

Later with a similar scheme using sodium atoms, the infrared up-conversion has been realized by BLOOM et al. [1.295]. In this work radiation at 9.26 μm was converted to 3505 Å with a photon conversion efficiency of 58%. The two-photon transition $3s-3d$ was pumped; the infrared radiation almost agrees with the transition to the $4p$ level. The first experiment using the infrared up-conversion is quite promising. The application of the technique may be quite important for infrared spectroscopy and especially for infrared astronomy.

Distortion of Atomic Structure in High Intensity EM Fields

It is well known that the coupling of an atom or molecule to an electromagnetic wave may give rise to a broadening, shift, or splitting of energy levels. Our fundamental knowledge on these phenomena results from many experiments performed by means of nuclear resonance or radiofrequency spectroscopy. Due to the Doppler width the purely optical observation of the phenomena was not possible until high power laser sources had been developed. In recent years, in addition, the methods of laser spectroscopy became available which allow a Doppler-free observation (see Sect. 1.4), even with low power lasers. Furthermore, tunable lasers now enable the investigation of the frequency dependence of those phenomena.

In the following some laser experiments dealing with the distortion of atomic levels will be reviewed.

a) Power Broadening and Linear Stark Effect. Under the influence of a strong, resonant radiation field the two levels involved in the transition are broadened. One way to observe this broadening is to excite the system with broad-band radiation (frequency width large compared to the inverse of the induced transition probability) and to analyse the absorption or fluorescence by means of a spectrograph. Since the Doppler width is present, a convolution of a Gaussian (Doppler distribution) and a Lorentzian (power broadened line profile) is observed in general. Another way to observe the line profile is the scanning of the resonance curve by means of a laser line with a frequency distribution which is less or comparable to the natural width of the transition. The power-broadened linewidth is deduced from the absorption of the system as a function of the laser frequency. (In this case a shift of the level may be introduced at large detunings of the laser from resonance. This frequency shift is called the quadratic Stark effect and will be discussed under item b) below.)

The power broadening of a two-level system is calculated according to the "Rabi solution" (see, e.g., [1.296]). The broadening is proportional to the power of the incident radiation.

Under the influence of strong narrow-band resonant radiation (frequency width small compared to the induced transition probability and to the natural width of the transition) also a splitting of the levels occurs. This can be observed for one of the levels when it is probed by means of a coupled transition (three level system). The nature of this splitting can be understood if we regard the atom and the electromagnetic field as one quantum system, e.g. [1.297]. For the following we assume a two-level atom with the zero-order energies $E_1^{(0)}$ and $E_2^{(0)}$ ($E_1^{(0)} > E_2^{(0)}$) and a field consisting of N_λ photons in the definite mode λ having frequency ω. If $\omega = (E_1^{(0)} - E_2^{(0)})/\hbar$ the system is degenerate since the energies of the atom and field systems agree and $E_{N_\lambda,1}^{(0)} = E_1^{(0)} + N_\lambda \cdot \hbar\omega = E_{N_\lambda+1,2}^{(0)} = E_2^{(0)} + (N_\lambda + 1)\,\hbar\omega$.

Due to the interaction between the two states the degeneracy is removed. The system splits into two states with energies displaced relative to the original energy by $\Delta E_1^{(1)} = |V_{12}| = \boldsymbol{E}_0\,|d_{12}|/2$ and $\Delta E_2^{(1)} = -\boldsymbol{E}_0\,|d_{12}|/2$. (The applied field is assumed as a plane wave with $\boldsymbol{E} = \mathrm{Re}\,\boldsymbol{E}_0 \exp[i(k_\lambda r - \omega t)]$). The transition to a third level with energy $E_3^{(0)}$ is therefore not a single line with the frequency $\omega_1^{(0)} = (E_3^{(0)} - E_1^{(0)})/\hbar$ but splits into two groups (see also [1.298]): one with frequencies $(E_3^{(0)} - E_1^{(0)} - \Delta E_1^{(1)})/\hbar$, $(E_3^{(0)} - E_1^{(0)} - \Delta E_1^{(1)} \pm 2\hbar\omega)/\hbar, \ldots$ corresponding to one-photon, three-photons, and so on, resonances. Their frequencies are shifted by $\Delta\omega_1 = +|V_{12}|/\hbar$ relative to the position of the undisplaced levels. The other group has the frequencies $(E_3^{(0)} - E_1^{(0)} - \Delta E_2^{(1)})/\hbar$, $(E_3^{(0)} - E_1^{(0)} - \Delta E_2^{(1)} \pm 2\hbar\omega)/\hbar \ldots$. These are shifted by $\Delta\omega_1' = -|V_{12}|/\hbar$. The total splitting is thus $2|V_{12}|/\hbar$. In a field of non-resonant radiation the splitting is given by $[(E_1^{(0)} - E_2^{(0)} - h\omega)^2 + 4|V_{12}|^2]^{1/2}$ which is identical with the Rabi nutation frequency [1.296].

In radio-frequency spectroscopy this Stark splitting was observed for the first time in the spectrum of OCS by AUTLER and TOWNES [1.97]. Therefore it is often called the Autler-Townes effect. The splitting is proportional to the amplitude of the applied electromagnetic field and is thus denominated as the *linear* Stark effect, too.

In a combined radio-frequency optical experiment the linear Stark effect was investigated by HIRSCH [1.299]. In this experiment dilute ruby was placed in an EPR spectrometer, where the four substates ($\pm 3/2$, $\pm 1/2$) of the Cr^{3+} ground state ($^4A_2$) are well separated by the magnetic field. The transition $+3/2\,^4A_2 - +1/2\,^4A_2$ was probed by microwave absorption. A ruby laser was temperature-tuned to the $\bar{E}(^2E) - 3/2(^4A_2)$ transition of the ruby probe. The microwave absorption was measured while the laser was repetitively fired. Under the influence of the strong laser radiation the absorption curves are double-peaked. In a purely optical experiment the linear Stark effect was successfully measured by BONCH-BRUEVICH et al. [1.300] on free potassium atoms. A ruby laser was temperature-tuned to be in resonance with the 6939 Å ($6\,^2S_{1/2} - 4\,^2P_{3/2}$) transition. Under the influence of the strong resonance perturbation the D_2 line ($4\,^2P_{3/2} - 4\,^2S_{1/2}$) splits into two components.

In the optical region the Doppler effect aggravates the direct observation of the linear Stark splitting if convenient powers of CW lasers are used [1.301]. However, utilizing a three-level system where the nonlinear interaction between the two coupled transitions [1.302, 303, 305] allows a partial compensation of the Doppler width, the linear Stark splitting can be observed with rather low laser intensities. Quite recently an experiment of this kind has been published by SCHABERT et al. [1.304, 307]. The result of this measurement is shown in Fig. 1.10. A He–Ne probe discharge was placed inside a single-mode He–Ne laser oscillating at the 0.63 μm wavelength, and close to the $3s_2 - 2p_4$ transition. This field selectively interacts with atoms whose velocities Doppler-shift one of its traveling wave components into resonance. This produces changes in the level population over two narrow intervals symmetrically located about the center of the velocity distribution (see also Sect. 1.4). The discharge was probed with another single-mode He–Ne laser beam at 1.15 μm, which was tunable over the Doppler width of the $2s_2 - 2p_4$ transition. During the frequency of this laser was changed the amplification of the discharge was measured. The part of the signal obtained for the case where the probing and saturating beams travel in the same direction is due to Raman-like two-quantum transitions for which the Doppler effect is partially cancelled. This effect cannot be

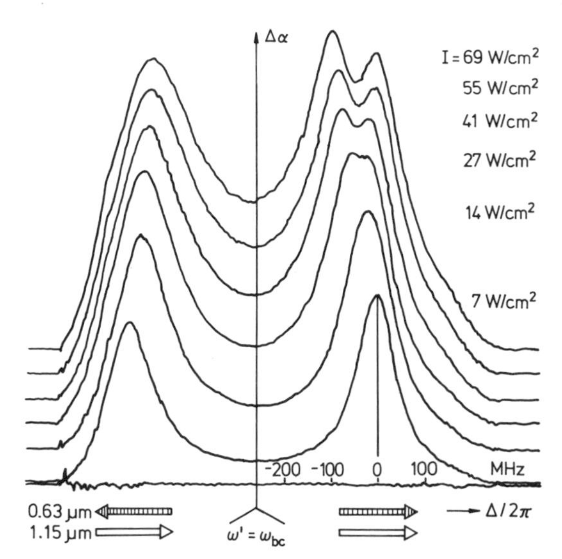

Fig. 1.10. Linear Stark splitting observed in a three-level system. The arrows indicate the direction of the saturating waves (0.63 μm) and the probing wave. The parameters at the different curves give the intensity of the saturating beam. $\Delta\alpha$ is the amplification of the probe beam. $\Delta = \omega' - \omega_{bc} - (k'/k)(\omega - \omega_{ab})$ where ω, ω_1 and k, k' are frequency and wave number of saturating and probe light, respectively. ω_{bc} and ω_{ab} are the resonance frequencies of the coupled transitions. The dynamic Stark splitting is observed when the saturating and probing waves co-propagate. (From [1.307]).

explained by population saturation alone [1.303, 306]. At small intensities of the saturation beam the usual narrowing and peak enhancement is observed, at high intensities the line splitting due to the linear Stark effect is clearly seen.

The linear Stark effect can also be observed in the frequency distribution of the fluorescent light induced by monochromatic resonant radiation. For these experiments the Doppler width of the scattering atoms has to be reduced by the use of atomic beams. The fluorescent spectrum observed under high resolution is a three-peaked structure which results from the splitting of the upper and lower states. The experimental and theoretical work on this subject will be discussed in Section 1.4.

At the end of the discussion of the linear Stark effect it should still be mentioned that this phenomenon is also a powerful tool for the investigation of the statistics of the perturbing radiation. This was first pointed out by ZUSMANN and BURSHTEIN [1.312]. For further hints see also [1.313].

b) Quadratic Stark Effect. In case the applied electromagnetic field is not resonant with the atoms, the system atom plus electromagnetic field is not degenerate. The straightforward time-independent perturbation calculation gives the following level shift of an atomic level (induced by the optical field $\boldsymbol{E} = \mathrm{Re}\,\boldsymbol{E}_0 \exp[i(kz-\omega t)]$ [1.298, 320]

$$\Delta E_n = \frac{1}{4} \sum_{\substack{m \\ n \neq m}} \left\{ \frac{|d_{mn}|^2 \boldsymbol{E}_0^2}{E_n - E_m - \hbar\omega} + \frac{|d_{mn}|^2 \boldsymbol{E}_0^2}{E_n - E_m + \hbar\omega} \right\}. \tag{1.4}$$

The summation has to be taken over all atomic states $|m\rangle$ (bound and continuum) with energy E_m. The shifts are due to virtual transitions between atomic levels caused by non-resonant light. This is analogous to the calculation of transition probabilities of multiphoton transitions. Level shifts are therefore intrinsic to these multiphoton processes. It can be shown that the two-photon transition rate is proportional to $\delta\omega_g \cdot \delta\omega_f$, where $\delta\omega_g$ and $\delta\omega_f$ are the level shifts of the fundamental and excited state, respectively. The level shifts are proportional to the square of the amplitude of the electromagnetic wave. It is therefore termed the *quadratic* Stark effect. (It is interesting that for ω approaching zero the quadratic shifts of the levels become equal to that in a constant field with the same energy density.)

If for the calculation of the level shifts in essence only two neighboring levels (energies E_n and E_m, $E_n > E_m$) have to be considered, it follows from (1.4) that the sign of the shifts is determined by the sign of the difference $E_n - E_m - \hbar\omega$. If $\omega < E_n - E_m$ the levels are displaced so that their distance increases; for $h\omega > E_n - E_m$ their distance decreases. When ω approaches the transition frequency $(E_n - E_m)/\hbar$ the shifts according to (1.4) become infinitely large and are not valid. In this case higher-order perturbations have to be considered. For a two-level atom the solution can be obtained by exact diagonalization of the matrix of the Hamilton operator [1.298]. The result is a dispersion shaped dependence of the level shifts on the energy difference $E_n - E_m$, which agrees, of course, with (1.4) for higher energy differences. The energy shifts for $E_n - E_m = \hbar\omega$ are zero. However, there is the splitting which is due to the linear Stark effect.

The quadratic Stark effect has first been observed by COHEN-TANNOUDJI [1.308] as a small change in the nuclear resonance of optically oriented ^{199}Hg. Similar experiments on other atoms have later been performed by other workers. A survey on this work has been given by HAPPER [1.309].

The first laser experiments on the quadratic Stark effect have been performed by ALEKSANDROV et al. [1.310] and by BONCH-BRUEVICH et al. [1.311]. They observed the level shifts by means of the change of the *D*-line absorption of potassium vapor when the atoms were irradiated with ruby laser light (wavelength 6933 Å) which is nearly resonant with the $4\,^2P_{3/2}-6\,^2S_{1/2}$ transition at 6943 Å. In a later experiment [1.300] the wavelength of the ruby laser was changed in the 6936–6943 Å range by temperature tuning. The quadratic Stark effect was measured as a function of the laser wavelength and intensity. The measurements showed good agreement with theory. Later a similar experiment was performed by MOREY [1.318].

Another system on which the quadratic Stark effect can be studied is the $7\,^3P_2-7\,^3S_1$ transition (11.287 Å) in the Hg I spectrum being rather close to the wavelength of the Nd laser. The level shift of the $7\,^3S_1$ level can be studied in a discharge by observing the strong emission lines at 5461 Å ($7\,^3S_1-6\,^3P_2$), 4358 Å ($7\,^3S_1-6\,^3P_1$), and 4046 Å ($7\,^3S_1-6\,^3P_0$). Measurements of this system have been made by PLATZ [1.314–316]. The line shifts were observed as a function of time during the laser pulse by means of a multi-channel Fabry-Perot interferometer [1.317]. In another more recent experiment the Stark shift of the Balmer H_δ line was studied by DUBREUIL et al. [1.319] in a capillary discharge. Under the influence of cw CO_2 laser radiation with a power density of 3–4 MW/cm^2, the H_δ line was shifted by about 0.5 Å.

The first experiment on the quadratic Stark effect using cw tunable dye lasers was recently published by LIAO and BJORKHOLM [1.321]. In this experiment the Doppler effect was excluded by the use of double quantum transitions (see Sect. 1.4). The experiment was performed with sodium atoms in a cell, which were excited by two dye lasers. The first laser had a wavelength of about 5890 Å which is close to the $3S-4P$ resonance and the other about 5690 Å being close to $4P-4D$. The cell was irradiated by the two lasers in opposite direction so that for the induced double quantum transition $3S-4D$ the Doppler width was determined by the frequency difference $(\omega_1-\omega_2)$ of the two lasers (the residual Doppler broadening was 62 MHz). The resonance denominators in (1.4) assured that the shifts of the $3S$ state where primarily induced by the 5890 Å light while the $4D$ state was shifted primarily by the 5690 Å light. By adjusting independently the intensities of the two lasers, either the $3S$ or the $4D$ level was essentially shifted, and not both. The two-photon transitions were detected by monitoring the $4P-3S$ fluorescence which results from the decay of the $4D$ level.

Figure 1.11 shows the dependence of the shift of the hyperfine level $F=2$ of the $3S$ state versus light intensity and frequency detuning $\Delta\nu$ from the intermediate $3P_{3/2}$ resonance. The latter measurements give good agreement with theory for $|\Delta\nu|>4.6$ GHz. For $|\Delta\nu|<4.6$ GHz no good agreement with (1.4) should be expected due to the possible breakdown of perturbation theory. It is also found that in this region the shifts were no longer linearly dependent on the light intensity.

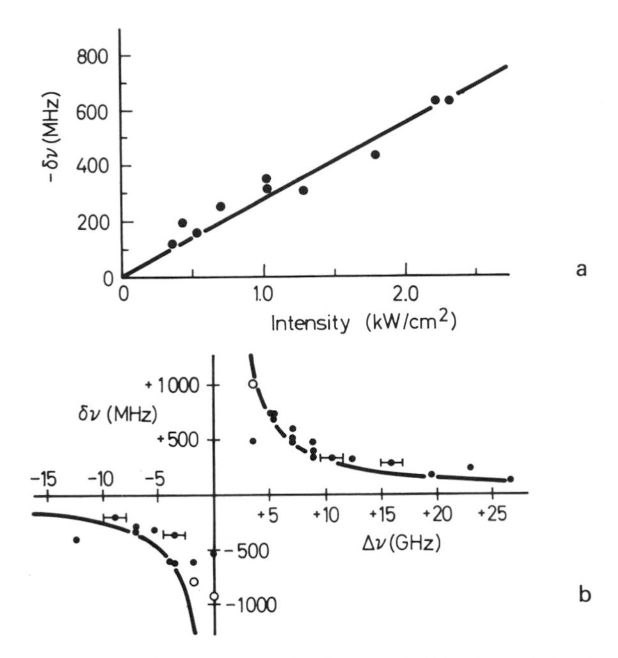

Fig. 1.11a and b. Energy level shift of the $3S$ ($F=2$) level of sodium versus (a) laser intensity with the two 5890 Å laser modes (this laser was not running single mode!) tuned 4.0 and 5.77 GHz from the $3S(F=2)\rightarrow 3P_{3/2}$ intermediate-state resonance and (b) frequency detuning from the $3S(F=2)\rightarrow 3P_{3/2}$ intermediate-state resonance. (The two 5890 Å laser modes have mistunings of $\Delta\nu$ and $\Delta\nu - 1.77$ GHz.) Solid circles give shifts for 5890 Å intensity of 2.15 kW/cm^2. The measurements with open circles are taken with reduced intensity and are then normalized to 2.15 kW/cm^2. The solid lines represent theoretically calculated curves using no adjustable parameters. The two laser modes are responsible for the unsymmetry around $\Delta\nu=0$ of the curves in the lower part of the figure. (From [1.321])

Some interesting features observed in this experiment should still be mentioned. The first is that a broadening in the double quantum transitions is observed which is due to the Doppler effect; different atomic velocity subgroups experience different level shifts since they see the applied light Doppler-shifted in a different way. In agreement with theory it was furthermore realized that the shifts of the $3S$ levels due to near resonance with the $3P_{1/2}$ level are one-half of those due to resonance with $3P_{3/2}$. It was also shown that when the $3P_{1/2}$ level was nearly resonant there were no shifts induced by the 5690 Å light for the $3S-4D_{5/2}$ transitions, whereas the shifts for $3S-4D_{3/2}$ were very strong. This is due to the fact that the dipole matrix element connecting $3P_{1/2}$ and $4D_{3/2}$ vanishes. The same reason should also cause the hyperfine levels of $4D$ to have different shifts. For the observation of this Stark shift, however, the line broadening effect discussed above has to be avoided. This is possible when the light is scattered by atoms of a well collimated atomic beam without Doppler broadening (see Sect. 1.4). In connection with such an experimental set-up the laser induced linear and quadratic Stark effect may utilized as a quite useful tool to collect information on the atom. As discussed in [1.298], the Stark effect induced by ac fields is able to provide the same information as the Stark effect of dc fields but at much smaller field strengths since the resonance enhancement can be utilized. In this way the use of fields of optical frequencies allows measurements at field strengths which are equivalent to values much larger than the breakdown values for dc fields.

In a recent paper FENEUILLE [1.322] has used the "dressed atom" formalism introduced by COHEN-TANNOUDJI and HAROCHE [1.323] to study the structure of an excited state of a two-level system which is distorted by near-resonant cw radiation. The change of the fluorescent light was calculated for the case that a second tunable laser is used to probe the energy distribution. This experiment which has also to be performed with a Doppler-free atomic beam is quite suitable to study the Stark shifts and splittings.

Another distortion of the resonance frequency usually observed in radio-frequency spectroscopy is the BLOCH-SIEGERT shift [1.324]. This phenomenon which will be discussed in Section 1.3 is proportional to $|\boldsymbol{E}^2|/\omega$ and is very small at optical frequencies.

1.2. Application to Chemistry and Isotope Separation

1.2.1. Photochemistry

The laser finds an important application in chemistry to the selective stimulation of chemical reactions. It is obvious to assume that for this purpose tunable lasers will be of a special significance. (For review papers see also [1.347, 348].)

A certain amount of energy is required to start an endothermic chemical reaction, but the initiation of an exothermic reaction usually also requires some activation to overcome the repulsion forces between the reactants. The energy of translational motion may be suitable for this purpose, but the energy of internal degrees of freedom, such as vibrational and rotational energy, also plays an important role. It is well known that in chemical transformations the vibrational energy of the molecules has particular significance; for example, the dissociation probability of an unexcited molecule is exceedingly small, even when the translational energy is larger than the binding energy of the molecule. However, when the vibrational energy exceeds the binding energy, the molecule dissociates [1.349]. Therefore the chemical transformation is mostly started when the vibrational energy acquires the activation energy [1.350]. A heating of the reactant, as it is normally done, leads to a simultaneous increase in translational, vibrational, and rotational energies because all degrees of freedom are usually in thermodynamic equilibrium. Laser excitation, however, can be made to excite a particular bond, and a chemical process involving the molecule may be initiated without restoration of thermodynamic equilibrium. Heating the molecule usually breaks off the weakest bond, whereas laser excitation may affect other bonds as well; this enables the chemist to initiate new reactions and could perhaps play a role in the synthesis of new compounds.

Chemical reactions can also be stimulated by light when atoms are the reaction partners. An electronic excitation changes the potential between the reactants so that the reaction may be accelerated or even becomes possible at all. A classical example for this is the reaction of excited mercury atoms with simple molecules such as H_2O, O_2 or HCl. The interatomic potential change is also important in the formation of molecules such as Xe_2, Kr_2 or related excimers (e.g. [1.351].

A very important application of laser-induced chemical reactions is in the field of isotope separation. The high monochromaticity of the laser can be utilized to excite only the molecules with a particular isotope composition so that they start a reaction, in which the other molecules do not participate. The interesting isotopes may then be isolated by separating the reaction product from the other molecules. (These applications will be discussed in more detail in Subsection 1.2.2.)

Photocatalysis was first applied to chemical reactions many years prior to the advent of lasers. A typical example of the earlier work is the initiation of the reaction between $^{37}Cl_2$ and hydrogen performed by HARTLEY et al. [1.352] using excitation by white light filtered through chlorine gas consisting mainly of $^{35}Cl_2$. Another example is the reaction of "ortho"-iodine molecules with hexane. Here an optical excitation of "ortho"-iodine was used for the initiation of the reaction. The "para" molecules, which were not excited, did not react [1.353]. Other examples of photocatalytical reactions, studied some time ago, are processes investigated in connection with photochemical isotope separation. Some of them are compiled in Subsection 1.2.2. For a more complete survey see [1.354, 355].

The initiation of photochemical reactions with conventional light sources was, of course, rather difficult and generally only possible with strong emission lines of discharge lamps. The high spectral brightness now available with lasers has increased the possibilities tremendously. Although chemical reactions induced by laser radiation have now been studied, the application of tunable lasers in this field is still in its infancy. The problems investigated so far have mostly been influenced by the necessity that the wavelength of laser lines must coincide with transitions of the molecules to be investigated. We discuss some of these studies below.

It has also been demonstrated that reactions with an activation energy in excess of the energy of the laser photon can possibly be carried out. Cascade excitation via collisional transfer of the vibrational energy allows rather high-lying vibrational levels to be reached. Thus, KARLOV et al. [1.349] observed the dissociation of BCl_3 molecules accompanied by recombination radiation when the molecules were irradiated by CO_2 laser light. The energy necessary for the dissociation of the molecule (v_2-oscillation) is $38.8 \times 10^3\,cm^{-1}$. Since the vibrational-translational relaxation time is much longer than the vibrational-vibrational relaxation time, almost no heating of the sample is observed in this experiment. The vibrational cascade effects occuring in the reaction cells at pressures in the Torr region, can complicate the interpretation of the results. Therefore it seems advisable to study laser-induced chemical reactions also in crossed atomic beams. ORDIONE et al. [1.356] reported on such an experiment with the reaction K + HCl. The reaction cross-section was determined to be $0.15\,Å^2$. When HCl was excited to the first vibrational level $v=1$, the reaction cross-section increased by a factor of about 100. Experiments of this kind are quite suitable for determining which degree of freedom of the molecules contributes most to a chemical reaction.

In polyatomic molecules the vibrational levels are quite dense for highly excited levels. However, the uneven level distribution allows in some cases that the lower levels are populated isotope selectively. By the absorption of further laser photons the molecule is then excited to higher levels. This absorption mechanism is complex, it is influenced by the level density, resonantly enhanced multi-photon transitions, ac Stark shifts, and the breakdown of selection rules at high molecular excitations due to distortions. Using this method the isotopes of boron, chlorine, silicon [1.357, 358, 394], and sulphur have been enriched using CO_2 laser pulses. The boron and chlorine were simultaneously enriched using BCl_3, for silicon SiF_4, and for carbon and chlorine CCl_2F_2. The sulphur enrichment was started with SF_6. The fluor atoms produced during the dissociation of the molecules were bound by H_2 or N_2 which were added as scavengers. The big advantage of this method is that a single laser is sufficient to perform the isotope enrichment.

After the excitation of the molecule, a number of relaxation processes may become effective so that the chemical reaction has to be started before the selective excitation of the molecules is lost. The relaxation processes which may occur are: establishment of equilibrium between the intra-molecular degrees of freedom, vibrational-translational relaxation, the population of higher vibrational levels and dissociation. Under real conditions it is usually found that first the equilibrium in each vibrational mode is established; next the equilibrium between the different vibrational modes followed by an intra-molecular equilibrium, and finally the equilibrium between the vibrational and translational degrees of freedom is reached. It is quite obvious that with increasing field intensity the vibrational temperature begins to differ from the translational so that finally in sufficiently strong fields the equilibrium is disturbed and only the bond of the molecule being in resonance with the field becomes selectively heated (see also Subsect. 1.1.2).

The dissociation or a chemical reaction proceeds only when a molecule has a sufficient energy reserve since the probability for both processes is proportional to $\exp(-E/\theta)$, where E is the dissociation or activation energy and θ the average energy of the molecule. When the system is thermally heated the number of active molecules is relatively small and consequently the corresponding processes are slow. However, when the molecules are excited by laser radiation, θ comes quite close to E and the processes of dissociation or chemical reaction become very fast. BASOV et al. [1.359] have shown in calculations, performed under the assumption that the excited vibrational bond of the molecule is almost harmonic up to the activation energy, that the molecules will enter into a chemical reaction before the onset of vibrational-translational relaxation processes and before the system approaches inter-molecular equilibrium. Also the cooling of the excited molecular mode by the interaction with all the degrees of freedom of the molecule does not change the fact that the selected mode can have a temperature exceeding the average temperature of the gas. However, the inter-molecular interaction as, e.g., resonance transfer and the dissipation of the excitation energy between the different modes causes a loss of selectivity; this is especially harmful in connection with isotope separation.

The effect of thermal heating of the molecular system after selective laser excitation was investigated by LETOKHOV and co-workers [1.347]. This experiment was performed on NH_3 molecules, which have been selectively excited by a CO_2 laser. The population of the vibrational levels was probed with a second laser by measuring the intensity decrease of uv absorption lines corresponding to transitions from the vibrational levels of the ground electronic state to an excited electronic state. The laser excitation was discriminated from thermal population by modulating the laser light with an audio-frequency. The modulation amplitude of the uv light due to the population of the levels by the laser should be frequency independent, whereas the population due to thermal relaxation would cause a decrease of the modulation amplitude when the modulation frequency reaches about 1 kHz (the thermal relaxation time is ≈ 1 msec). This was observed in the experiment for several vibrational levels of NH_3.

Another loss of selectivity may be caused by resonance exchange of excitation. This effect mostly is of no importance in the study of laser induced chemical reactions. In isotope separation, however, the resonance transfer may be crucial. For instance, the excitation exchange between Br_2 molecules with different isotope compositions was the reason for the first experiment of laser isotope separation to fail [1.360]. Therefore it is quite important to investigate these relaxation rates (see Subsect. 1.1.2).

In the process of populating high vibrational levels with strong laser pumping starting from lower vibrational levels exists an important effect which limits the excitation rate. This was first observed by LETOKHOV et al. [1.361] and called the "narrow throat" effect. The monochromatic laser radiation excites a certain fraction of the molecules in the rotational-vibrational sublevels. At large laser intensities the excitation rate of the molecules may become much higher than the rotational relaxation rate. In that case, the lower sublevel quickly becomes depleted since a hole is burned in its population distribution; a further excitation is only possible when the population distribution is restored by rotational relaxation.

In the following some laser induced chemical reactions will be discussed.

Without laser irradiation, no reaction can be observed if the mixture SF_6–NO is heated quickly to 1000 K, and in the mixture N_2F_4–NO the conversions start at temperatures over 600 K; there is no reaction, but N_2F_4 decomposes with the formation of NF_3. At laser powers in excess of 20 W and exposures of less than 0.1 sec, instantaneous reactions occur accompanied by luminescence. In the system N_2F_4–NO the products of the reaction are FNO, NF_3, N_2, F_2, and NO_2, SF_6–NO thionylfluoride (SF_2O) is formed.

Calculations show that if the absorbed laser energy were used to heat the reactants, the temperature of the system SF_6–NO would increase to 1000 K and that of the system SF_4–NO to 500 K. At such temperatures, even with long heating times, thermal reactions do not occur for either system. BASOV et al. [1.362] listed also other systems studied in similar experiments.

Another example for a laser induced chemical reaction was reported by BACHMANN et al. [1.363]. In this experiment diborane B_2H_6 was irradiated resonantly by a CO_2 laser line (power about 20 W). The vibrational excitation results in a dissociation of the molecule. In a subsequent chain reaction isoborane $B_{20}H_{16}$ is formed. The chain reaction must be assumed since otherwise the high product yield cannot by explained with the low power laser input. The reaction was initiated neither by thermal activation nor by excitation of electronic states since the process would then proceed in a completely different way. Later also the laser induced exchange reaction of methyl groups against bromine in $B(CH_3)_3$, $B(CH_3)_2Br$, and $B(CH_3)Br_2$ was investigated [1.364]. Also in this case the non-thermal nature of the reaction could be demonstrated.

Laser fluorescence can be a useful tool to measure the rate constants of chemical reactions. This was demonstrated by HANCOCK et al. [1.365]. In this experiment the reaction $NH_2 + NO$ at 298 K was investigated. The reaction rate was deduced from the change of the NH_2 fluorescence with time. The optical excitation was performed by means of a cw dye laser with a wavelength of about 5703 Å. The NH_2 was produced by pulsed photolysis.

1.2.2. Laser Isotope Separation

A discipline of laser applications much discussed in recent years is isotope separation. The focus of interest is on the enrichment of uranium, because of its enormous commercial importance. It is estimated, for example, that the separative work required to cover the US demand alone will increase from about 4000 t per year at present to about 100000 t per year by the year 2000[1]. The separation facilities available at present will be sufficient only up to the year 1982 [1.366, 367]. The costs of uranium enrichment, totalled over the year 1980–2000, is estimated at about 135 billion dollars, of which over 50% will have to be spent on separation facilities and on the separation process itself. More effective and cheaper separation processes therefore attract very large interest. The very important commercial considerations naturally overshadow many other problems for which simpler and more effective isotope separation methods are also of great importance as, e.g., for the production of radio-nuclides for medical and technological uses. In the following the methods used so far for isotope separation by means of lasers will be briefly discussed.

Survey of Isotope Separation Methods by Means of Lasers

The hitherto known schemes for isotope separation by means of lasers exploit the optical isotope shift in atomic or molecular spectra. The

[1] The production of 1 kg reactor grade uranium with an enrichment of 3% requires approximately 4 kg units of separative work.

atoms or molecules containing the isotope which is to be separated, are selectively excited by laser radiation. A necessary condition is that the isotope shift must be larger than the actual line width, so that a particular isotope can be selected. This means that the particles to be separated must be in vapor form because the lines in liquids are generally broadened into a smooth band contour. In solids the isotope shift can be observed in some crystals. Isotope separation cannot, however, be achieved because the separation processes which have to follow the optical excitation, to be discussed later, are not applicable.

The isotope shift of free atoms is determined, in the case of small atomic mass numbers, mainly by the mass effect [1.368] whereas with heavy atoms the nuclear volume effect plays a decisive role [1.368]. The mass effect diminishes with increasing mass, proportionally to $1/A^2$, and in atoms with $A>100$ it usually amounts to only a fraction of the Doppler width. The volume effect is proportional to the square of the nuclear radius change when additional neutrons are added to the nucleus. The isotope shift between hydrogen and deuterium is approximately $\delta\nu/\nu = 2.7 \cdot 10^{-4}$. In atoms with a larger mass number, e.g. calcium, the isotope shift is only about $\delta\nu/\nu = 0.5 \cdot 10^{-6}$ [1.369]; this corresponds to approximately 25% of the Doppler width. In uranium, where a larger nuclear volume effect is observed, the shift amounts to about $\delta\nu/\nu = 0.6 \cdot 10^{-4}$ [1.370].

As the Doppler width diminishes with increasing mass number, the isotope shift in uranium can become 80 times larger than the Doppler width. Isotope separation by means of lasers should therefore give higher separation factors for heavy atoms than for light ones. This is in contrast to classical separation methods in which the relevant factor is always the relative mass difference.

The isotope shift in the rotational spectra of molecules is caused by the different moments of inertia of the molecules consisting of different isotopes. In the vibrational spectra the effect of different isotopes is observable in the change of the vibration frequency from the change in mass. The isotope shift for diatomic molecules can easily be calculated. It amounts to $\delta\nu/\nu = \mu_2/\mu_1 - 1$ for rotational spectra, μ_1 and μ_2 being the reduced masses of the molecules which are compared to each other. For isotope separation vibrational spectra are more important than rotational ones since the vibrational splitting is at a higher energy and is large compared to kT, so that the thermal occupation of the excited states is of no importance. The thermal occupation, of course, reduces the separation effect since the molecules are excited thermally, regardless of their isotope composition. In the vibrational splitting of a diatomic molecule the isotope shift amounts to $\delta\nu/\nu = (\mu_2/\mu_1)^{1/2} - 1$. This results, for example for HCl molecules containing ^{35}Cl and ^{37}Cl, in a shift of $\delta\nu/\nu = 7 \cdot 10^{-4}$; it is about two orders of magnitude larger than the

Doppler width. In the case where the isotope shift is smaller than the Doppler width, an isotope selective excitation can, in principle, be performed by means of double quantum transitions. For this purpose the particles have to be excited by two suitably polarized light beams travelling in opposite directions [1.371–373]. The efficiency which may be expected with this method is, of course, very low so that this method has no practical significance.

The idea of isotope separation by means of light is, in principle, not new. Attempts were made several years ago with the isotopes of Hg [1.374], C [1.375], O [1.375], and Cl [1.376]. In the case of mercury a resonant lamp filled with ^{202}Hg was used to excite the ^{202}Hg isotope. The excited atoms were captured by reactions with simple molecules such as H_2O, O_2 or HCl. However, as the spectral brightness of the classical discharge lamps is not high enough, the number of the excited atoms was always relatively small, so that these earlier photochemical methods failed to achieve any significant success. Lasers, with their spectral brightness, however, make it possible to induce almost equal populations in the excited state and in the ground state, thus opening the way to an effective separation. A very important advantage, of course, is the availability of frequency-tunable lasers in the relevant frequency ranges, so that the desired excitation wavelength can be preset exactly.

In a paper by KIMEL et al. [1.538] it was shown that the process of stimulated Raman scattering may also be used for photoselective isotope separation. The main advantage of this excitation is that a fixed-wavelength laser can be used with various compounds. The selectivity in the excitation inherent in this process comes from its exponential dependence on the isotope concentration. Therefore, the more abundant species in an isotope mixture will show a higher gain for stimulated Raman scattering. For gases standard Q-switched lasers are adequate to reach significant conversion factors. The rotational and vibrational transitions may be produced by visible or ultraviolet radiation with the same degree of selectivity since for the Raman effect the Doppler width is also determined by the frequency of the material excitation [1.371].

Separation by Radiation Pressure

A simple and direct method for isotope separation is the deflection of the atoms and molecules by radiation pressure, which results from the transfer of the photon's linear momentum to the atom during the absorption process. The added velocity resulting from the momentum transfer is very small but it can be increased considerably by repeated excitation, e.g. [1.377, 378]. A necessary condition for the applicability of this method is that the excited particles must always return to the initial state within a sufficiently short time after each excitation, in order to make multiple excitation possible. This condition cannot be fulfilled for complex atomic or molecular systems. The favorable cases are the isotopes of alkaline earths and of atoms with similar spectra. A recently published paper demonstrates the use of this method for the separation of barium isotopes [1.399]. However, it is difficult to imagine that this method would be less costly than the currently used mass spectrometers. (See also p. 97 and Fig. 1.27.)

Photochemical Separation

Another way of separating excited atoms or molecules from species in the ground state is by means of a chemical reaction. This reaction must be chosen so that it preferentially occurs with atoms and molecules in the excited state, but not with those in the ground state. The isotope of interest can then be separated from the others by isolating the reaction product.

An example of this kind is the enrichment of deuterium with the aid of a HF laser [1.380]. It happens that some lines of the HF laser coincide with strong transitions of methanol but not with deutero-methanol. The excitation activates the reaction

$$H_3COH + Br_2 \rightarrow 2HBr + H_2CO\,.$$

Irradiating with a 90 W HF laser for 60 sec in the presence of bromine, a gas mixture of about 50% H_3COH and 50% D_3COD content almost completely removes the H_3COH so that finally 95% D_3COD remains in the cell.

Separation by Pre-Dissociation

In the case where attractive and repulsive states of molecules overlap, it is possible to observe pre-dissociation. The states in which pre-dissociation can occur are usually sufficiently sharp so that the isotope shift may be resolved. Therefore, an isotope selective excitation with concurrent dissociation can be performed. Thereafter the fragments have to be separated from the starting substance. This method was used for the first time with formaldehyde for separating H from D [1.381]. The method is quite simple, though less general. For further applications performed so far see Table 1.4.

Table 1.4. Survey of laser isotope separation

Separated isotope	Method	Compound investigated	Laser used	Ref.
D	PCR	H_3COH, D_3COD (H_3COH excited by laser and reaction with B_2)	HF chemical laser	[1.380]
D	PD	H_2CO, D_2CO mixture	Frequency doubled ruby laser	[1.381]
^{10}B, ^{11}B	STS	BCl_3; stepwise photodissociation fragments react with O_2	CO_2 laser	[1.390]
^{10}B, ^{11}B	STS	BCl_3; dissociation and reaction with O_2	CO_2 laser	[1.391]
B	STS	BCl_3; dissociation reaction with H_2, N_2		[1.357, 358]
^{10}B, ^{11}B	PCR	BCl_3; selective excitation and reaction with H_2S	CO_2 laser	[1.392]
C	STS	CCl_2F_2; dissociation and reaction with H_2, N_2	CO_2 laser	[1.357, 358]
^{14}N, ^{15}N	STS	NH_4; two step photodissociation and reaction of dissociation products	CO_2 laser and uv-light	[1.393]
Si	STS	SiF_4; dissociation and reaction with H_2, N_2	CO_2 laser	[1.357, 358]
S	STS	SF_6; dissociation and reaction with H_2	CO_2 laser	[1.357, 358]
S	STS	SF_6; dissociation and reaction with H_2	CO_2 laser	[1.394]
Cl	STS	BCl_3, CCl_2F_2; dissociation and reaction with H_2	CO_2 laser	[1.357, 358]
^{35}Cl, ^{37}Cl	STS	HCl; reaction of the dissociation products with NO to NOCl	Raman shifted output and fourth harmonic output of Nd-glass laser	[1.382]
^{35}Cl, ^{37}Cl	PCR	ICl; electronically excited and reaction with C_2H_2Br or C_2H_2Cl	Dye laser	[1.395]

^{35}Cl, ^{37}Cl	PCR	$CSCl_2$; selective excitation and reaction with diethoxy-ethylene	Dye laser	[1.396]
^{40}Ca	STS	Ca atoms; second excitation made use of autoionizing levels. Ions were collected	Dye laser and Ar^+ laser	[1.389]
^{81}Br	PD	Br_2; photopre-dissociation and reaction of the dissociation products with hydrogen iodide	Frequency doubled YAG-laser	[1.397]
Br	PD	Br_2; photodissociation and reaction of dissociation products	Ar^+ laser	[1.398]
Rb	STS	Rb-atoms; ions were monitored	Ruby-laser-pumped dye laser and a frequency doubled ruby laser	[1.382]
Ba	RP	Ba atomic beam; deflected atoms are collected	Dye laser	[1.399]
^{235}U	STS	U atoms; ions were collected	Dye laser uv-lamp	[1.383]
^{235}U	STS	U atoms; ions were collected	Dye laser	[1.400]

STS: Selective two-step or multi-step photoionization or dissociation.
PD: Photopre-dissociation.
PCR: Photochemical reaction.
RP: Radiation pressure.

Note added in proof: For recent work on laser isotope separation see also: *Laser Spectroscopy,* Proceedings of the Second International Conference, Megève, June 23–27, 1975, ed. by S. HAROCHE, J. C. PEBAY-PEYROULA, T. W. HANSCH, S. E. HARRIS (Berlin, Heidelberg, New York: Springer 1975) p. 121, 259–304, and additional references of Chapter 1.

Separation by Photo-Isomerization

Another quite elegant method, which was suggested quite recently [1.379] and which is also based on a simple optical excitation, is photo-isomerization of a molecule. The optical excitation converts the molecule into a stereo-isomer with a structure differing from that of the initial molecule. Stereo-isomers can easily be separated from each other due to their different physical properties. This makes it possible to isolate the molecules with the same isotope composition.

Separation by Two-Step Photoionization or Dissociation

A quite universal method for isotope separation by means of lasers is the two-step photoionization of atoms or the two-step dissociation of molecules [1.382]. In a first step the selective excitation is performed; this is followed by a second excitation in which the atoms or molecules are ionized and dissociated, respectively. In the first case the ions produced by stepwise excitation are collected by means of electric fields. More difficult, in general, is the isolation of the dissociation products; for this purpose mostly chemical reactions involving the fragments of the dissociated molecules are used.

In this connection we should mention the enrichment of ^{235}U which was reported [1.383] (and patent applications [1.384–388]). In the experiment an atomic beam of uranium, generated in a furnace at a temperature of about 2100 °C was excited by radiation from a cw dye laser (isotope selective excitation) and then ionized by light from a high pressure mercury lamp. In this experiment the ^{235}U isotope, which is present in natural uranium in a concentration of 0.7%, was enriched to 60%. The yield achieved is still very small. A considerable improvement can be expected when autoionizing levels are excited during the second excitation step. This has already been demonstrated in an experiment on the isotope separation of calcium atoms [1.389]. The transition probabilities to autoionizing resonances are, in general, almost as high as for resonance transitions; therefore the second excitation step may be quite effective.

In one of the patent disclosures [1.387] for laser isotope separation mention is made, in connection with a two-step excitation process, of a yield of 7 g uranium in 24 hrs. It seems unlikely, however, that such a yield can actually be achieved in the set-up described there.

An application for two-step excitation to molecules was utilized in an experiment on the separation of the ^{10}B and ^{11}B [1.390]. In this experiment BCl_3 was isotope-selectively excited by the radiation of a CO_2 laser which has lines coinciding with vibrational transitions of $^{11}BCl_3$. Then the molecules were dissociated by uv light in the spectral range of 213 and 215 nm. The fragments were bound by a reaction with O_2. It was found that 5 light pulses from the CO_2 laser and a xenon flash lamp, which was used as source for the uv radiation, were sufficient to lower the $^{11}BCl_3$ content of the cell by a factor of 2. A 7% enrichment in the ratio of $^{10}B/^{11}B$ was achieved. For further applications of this method especially with respect to multiple-photon dissociation see Subsection 1.2.1 and Table 1.4.

Economic Aspects of Laser Isotope Separation

A survey on the experiments performed so far for laser isotope separation is given in Table 1.4. Since uranium isotope separation is of great political and economic importance, recent advances in this field are no longer published, so that it is difficult to judge the current status of development in this field.

It is known that large activities are in progress in the US aimed at developing and testing the technical feasibility of uranium isotope separation. The main centers are the AEC laboratories in Livermore and Los Alamos. In addition, the Standard Oil Company and the Avco Corporation have established a joint subsidiary for investigating uranium isotope separation. Obviously some success has already been achieved by this company. It is virtually sure that intensive work is in progress in USSR, too. It is well known that there are considerable activities at the Institute of Spectroscopy of the Academy of Sciences of the USSR in Moscow.

From the methods described above, those based on pre-dissociation and on two-step ionization and dissociation seem to be quite straightforward. Also photo-isomerization is a very elegant method but it is obviously limited to only a few special cases as well as the pre-dissociation method. The use of photochemical processes, without dissociation, entails the problem that the chemical reaction must take place within the lifetime of the excited state. This requires under certain circumstances high particle densities within the reaction chamber. Such densities, in turn, induce various disturbing processes, such as non-radiative transitions to the ground state, which result in thermal heating of the molecular ensemble. As a consequence, especially in conjunction with the use of vibrational excitation, non-selective population of the excited states is observed. Furthermore an excitation transfer may take place during collisions, so that the excitation energy is also transferred to isotopes which are not affected by the selective excitation (see also Subsect. 1.2.1).

We may expect that this important application of lasers will considerably promote the development of lasers, especially of frequency-tunable ones. In addition photochemical processes will be studied extensively. Finally, the study of the spectra of the various compounds interesting for isotope separation will give a further impact to atomic and molecular spectroscopy.

Considering only the separation of uranium isotopes, the enrichment by means of lasers may be expected to present advantages in comparison with classical separation schemes. Firstly, the achievable separation factor should be much higher than that obtained by the currently used diffusion or centrifuge methods. Secondly, it should be possible to utilize natural uranium much more effectively. At present the content of ^{235}U is reduced from 0.7% in natural uranium to 0.2% in the tailings, because this is an economic limit for the process. Methods based on the use of lasers will allow one to reduce the ^{235}U content still further.

An estimate of the economics of uranium separation by means of lasers can be performed by taking into account the number of photons required to excite a given quantity of uranium, the efficiency of the lasers and estimates of cross-sections for the subsequent processes. Such a guess is, of course, burdened with many uncertainties. The result nevertheless is that the production of 1 kg uranium enriched to 3% ^{235}U will require about 50 kWh. This compares to 1500 kWh for the centrifuge method and 9000 kWh for the diffusion method. Comparisons of the required investments for the plants are alike.

There is a big drawback of uranium separation by means of lasers. In the future a set-up can probably be built with relatively simple means and we may not be very far from the enrichment plant of the "amateur scientist", providing fissionable material to unauthorized persons [1.401].

1.2.3. Product State Analysis by Tunable Lasers

Tunable lasers have become a useful tool for analysing the composition of reaction products. In addition, they can be utilized to determine the population of rotational and vibrational states of molecules formed during a chemical reaction (for a review paper see also [1.402]). SCHULTZ et al. [1.403] demonstrated this application for the first time in the reaction $Ba + O_2 \rightarrow BaO + O$. The reactions of barium with hydrogen halides were later investigated by CRUSE et al. [1.404]. In the latter experiments they passed a beam of barium atoms through a scattering chamber filled with the hydrogen halides at a pressure of 10^{-3} Torr. A nitrogen laser-pumped dye laser was used to excite electronic transitions of the molecules. The subsequent fluorescence intensity was recorded as a function of the laser wavelength. The measured intensities were then converted into relative populations of the various product internal states using the appropriate intensity factors. The spectral

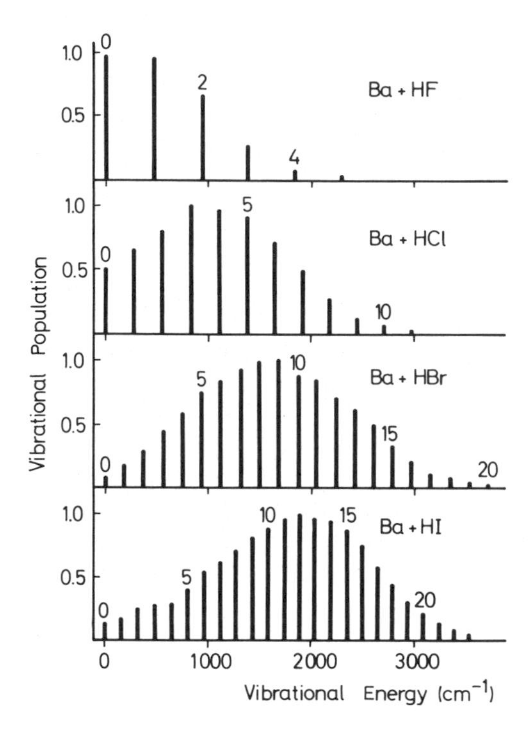

Fig. 1.12. Relative BaX vibrational population for the reactions Ba + HX → BaX + H deduced from laser induced fluorescence. (From [1.404])

resolution necessary in these measurements was provided by the narrow spectral distribution of the laser; the fluorescent light was not spectrally resolved.

From the measurements the vibrational state populations shown in Fig. 1.12 were obtained for the alkaline earth halides. The populations have been normalized for each reaction so that the population of the most probable level equals unity. Only for the reaction with HF a Boltzmann-like population distribution was found. The average percentage of the total available reaction energy that appears as vibration is 12% for BaF and 36% for BaBr. The rotational distributions obtained in a similar fashion indicate that on the average about 15% of the energy is used for rotation. Hence, about 50–75% of the available reaction energy appears as translational energy of the product.

In BaO it is possible to resolve the rotational splitting in the electronic spectrum. Therefore in the reactions $Ba + O_2 \rightarrow BaO + O$ and $Ba + CO_2 \rightarrow BaO + CO$ the rotational state distributions can be investigated, too. Recently the first measurement of SCHULTZ et al. [1.403] has been repeated by DAGDIGIAN et al. [1.405] with an improved apparatus where the collisional relaxation which influenced the results of the first measurements could be reduced. The main result of these

Table 1.5. Internal state distribution

Reaction	Laser	Results	Ref.
$Ba + O_2 \rightarrow Ba + O$	Dye laser	RP, VP ([1.403] corrected)	[1.405]
$Ba + CO_2 \rightarrow BaO + CO$	Dye laser	RP, VP, DE	
$Ba + O_2 \rightarrow Ba + O$	Dye laser	RP, VP	[1.403]
$Ba + HF \rightarrow BaF + H$			
$Ba + HCl \rightarrow BaCl + H$	Dye laser	VP	[1.404]
$Ba + HBr \rightarrow BaBr + H$			
$Ba + HJ \rightarrow BaJ + H$			

Abbreviations: RP: Rotational population distribution.
DE: Dissociation energy.
VP: Vibrational population distribution.

measurements was that the reaction $Ba + O_2$ leads to a small relative population of low rotational states; in addition, a larger number of vibrational states are populated. The striking feature of the reaction is that 30% of the reaction energy appears as product rotation. The average vibrational energy is of the same order of magnitude.

The reaction $Ba + CO_2$ leads to a Boltzmann-like vibrational and rotational distribution. The vibrational temperature T_{vib} is slightly lower than T_{rot}. In contrast to Ba + HX, the reaction $Ba + CO_2$ appears to be characterized by the formation of a collision complex in which the reaction energy is partitioned equally among the available degrees of freedom.

The $Ba + O_2$ reaction has also been studied in a scattering experiment employing a crossed beam arrangement. The most recent angular distribution measurements [1.406] reveal that the BaO products have forward-backward symmetry in the center of mass system, indicating also that the reaction proceeds through a long lived collision complex where the available energy is partitioned among the various modes "statistically".

The vibrational and rotational state distributions can also be investigated using atomic beam deflectors or resonance experiments. The best-studied example of a long-lived collision complex is the reaction $Cs + SF_6$. In this 8-atom complex the translational, vibrational and rotational distributions are found to be well characterized by temperatures which are almost identical for the three degrees of freedom [1.407, 408]. The reactions $Ba + O_2$ and $Ba + CO_2$ studied by means of lasers are so the best examples for a long-lived complex having few degrees of freedom. The essential feature is that the internal state distributions show departures from a Boltzmann distribution. They are therefore especially interesting for testing statistical theories [1.409, 410].

1.3. Coherent Transient Effects

With the advent of the laser coherent optical fields became available so that the optical analogues to many phenomena known from magnetic resonance got accessible. Generalization of magnetic-resonance formulations to electric dipole transitions was first made by FEYNMAN et al. [1.411]. This representation has proved to be quite useful for the understanding of many transient effects. In the following we will introduce this pseudospin vector to explain some of the phenomena observed. (Coherent transient effects are also discussed in Subsect. 3.4.3.)

Feynman-Vernon-Hellwarth (FVH) Representation

A vector quantity $\boldsymbol{r}$ is defined for a two-level system which is analogous to the magnetization vector $\boldsymbol{M}$ of a magnetic spin system. To introduce the "pseudospin" vector a two-level system is assumed, where the time depending wave function is given by $\psi(t)=a(t)\,\psi_a+b(t)\,\psi_b$. ψ_a and ψ_b are the basis states and $E_a=+\hbar\omega_0/2$, $E_b=-\hbar\omega_0/2$ the energies of the upper and lower states, respectively. The components of the vector quantity $\boldsymbol{r}$ are defined by

$$r_1 = a^*b + ab^*$$
$$r_2 = i(a^*b - ab^*)$$
$$r_3 = |a|^2 - |b|^2$$

which are linear combinations of the elements of the one-particle density matrix for the system. The asterisk always indicates the complex conjugate. The third component corresponds to the difference of the populations of the upper and lower states.

The time development of $\boldsymbol{r}$ under the influence of an external electromagnetic field may be obtained from the differential equations for the elements of the density matrix. In the semi classical treatment where the field is assumed to be quasi-monochromatic with a frequency nearly coincident with the transition frequency between the two energy levels the Hamilton operator is assumed to be $\mathscr{H}=\mathscr{H}_0-\mathscr{d}\,\mathscr{E}(t,\boldsymbol{r}_0)$. $\mathscr{d}$ is the particle's dipole moment operator, and $\mathscr{E}(\boldsymbol{r}_0)$ the electric field operator evaluated at the position of the dipole. The result which is obtained for $d\boldsymbol{r}/dt$ is then

$$\frac{d\boldsymbol{r}}{dt} = \boldsymbol{\Omega}\times\boldsymbol{r}\,, \tag{1.5}$$

where $\boldsymbol{\Omega}$ has the components

$$\Omega_1 = (d_{ab}+d_{ba})\,\langle\mathscr{E}\rangle/\hbar$$
$$\Omega_2 = i(d_{ab}-d_{ba})\,\langle\mathscr{E}\rangle/\hbar$$
$$\Omega_3 = \omega_0 = (E_a-E_b)/\hbar\,.$$

The d_{ij} are the dipole matrix elements which are in the general case complex vectors as well as the expectation values $\langle\mathscr{E}\rangle$.

Equation (1.5) is similar to the classical result for the time development of a magnetic moment in a static field. The vector $\boldsymbol{r}$ therefore helps to understand many transient effects. It is also applicable to dipole transitions induced by strong radiation fields when perturba-

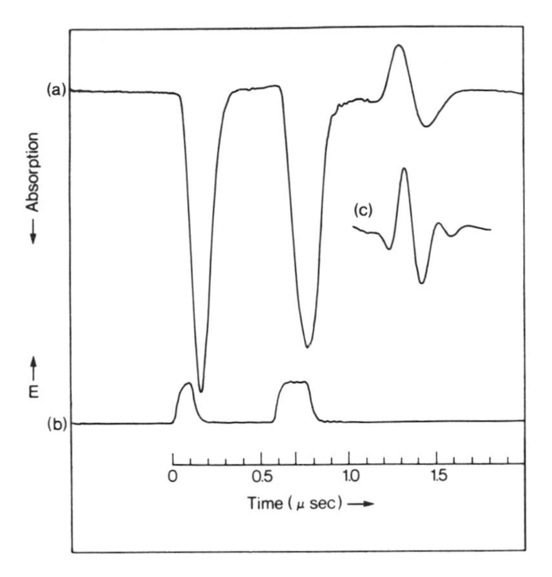

Fig. 1.13. Photon echos in $C^{13}H_3F$ at 9.5 μm (trace *a*). The curve shows the $\pi/2$ and π pulses followed by the echo beat signal. Recorded is the transmission of the laser light through the $C^{13}H_3F$ cell. Trace *b* shows the voltage applied to the Stark plates (resulting field strength 35 V/cm). In Curve *c* the beat frequency of the echo signal has increased over its value in Curve *a* because the amplitude of the Stark pulse has almost doubled (resulting field strength: 60 V/cm). (From [1.418])

tion theory is no longer valid. Relaxation processes can be included in the Feynman-Vernon-Hellwarth representation. As in magnetic resonance treatments the time T_1 describes the decrease of r_3, and T_2 includes the processes which lead to a loss of coherence by dephasing processes.

The vector $\boldsymbol{\Omega}$ is changing rapidly since two components are proportional to the electromagnetic field. In magnetic resonance theory for simplification of such a situation a transformation to a coordinate reference frame is performed which rotates with the light frequency ω.

The vector $\boldsymbol{\Omega}$ can be split into two components rotating in different directions as is done in magnetic resonance descriptions. In the coordinate frame synchronized with $\boldsymbol{r}$ one of these components is constant in time, and the other is counter rotating with frequency 2ω. Usually this part of the field is ignored, which leads to the, so-called, "rotating wave approximation". The influence of the counter rotating part is, however, not negligible, and can give rise to a shift in the true resonance frequency, the Bloch-Siegert shift [1.324]. This shift is proportional to $|\boldsymbol{E}|^2/\omega_0$ and has so far not yet been observed experimentally in the optical case. $\boldsymbol{E}$ is the amplitude of the electromagnetic field.

The energy absorbed by the two-level pseudo-spin system up to the time t is usually expressed by the quantity [2]

$$\theta(t) = \frac{2d_{ab}}{\hbar} \int_{-\infty}^{t} |\boldsymbol{E}(t)| dt$$

which is in the case of resonance $\omega = \omega_0$ the upward tipping angle of $\boldsymbol{r}'$ corresponding to $\boldsymbol{r}$ in the rotating frame.

The transition probabilities for the two-level system with respect to the field strength, time, and detuning of ω is described, in analogy to magnetic resonance theory, by the

[2] For simplicity it is assumed that d_{ab} is real and therefore $d_{ab} = d_{ba}$.

"Rabi solution", e.g. [1.296]. (The variation of population following from this result may be observed in an ensemble of atoms by detecting fluorescence radiation as a function of the input pulse area θ. The experimental result reveals clearly the predicted "Rabi oscillations" (Fig. 1.17).)

1.3.1. Photon-Echoes

A very nice demonstration of the analogy between the magnetic resonance and optical resonance is the photon echoes, which have first been observed by KURNIT et al. [1.412, 413] in solid ruby. To explain the effect for a two-level system it is first assumed that in the FVH representation only r'_3 (in the rotating frame) is different from zero. Applying a laser pulse of the right size a coherent superposition of the lower and upper state is produced, which gives rise to a macroscopic dipole moment. Under the influence of the laser pulse the FVH vector rotates in the $(2', 3')$-plane by

$$\theta = \frac{2d_{ab}}{\hbar} \int_{-\infty}^{+\infty} |\boldsymbol{E}(t)|\, dt = \frac{\pi}{2}. \tag{1.6}$$

Since the atoms have slightly different resonant frequencies the individual dipoles will get out of phase so that the macroscopic dipole disappears. Then a second larger light pulse, a π-pulse, is applied to the system. This causes the probability amplitudes of the upper and lower states to be interchanged. This means in other words that $\boldsymbol{r}'$ rotates 180° about the 1′ axis. In this way the individual dipoles will come in phase again after some time and the original polarization is built up. The macroscopic dipole created in this way can be observed by a light pulse, the echo pulse, emitted by the ensemble. For the generation of the echo pulse it is important that the dipoles retain some memory of their original phase relations, which are in general destroyed in the definite phase memory time T_2. From the size of the echo signal T_2 can thus be deduced. Since the initial demonstration of photon echo in ruby, the study of relaxation times with this effect progressed rather slowly in contrast to the spin echo studies in magnetic resonance [1.414]. Other examples studied so far are SF_6 [1.415, 416] and Cs atoms [1.417]. A useful variation of the method was proposed by BREWER et al. [1.418] who used a continuous-wave laser for the excitation and pulsed the level splitting by means of an electric field bringing the molecules in and out of resonance with the laser. This method has the advantage that it is more convenient than pulsing the laser and allows the experimentalist to study gases which are not exactly resonant with his laser. Measurements of this type have been performed for $C^{13}H_3F$ using a CO_2 laser. The transient optical signals are easily seen in transmission. The $C^{13}H_3F$ was brought into a Stark cell of only 10 cm length. The field is driven by a pulse generator up to a strength of 60 V/cm. For the measurements the ν_3 band transition ($J, K = 4{,}3 \rightarrow 5{,}3$) of $C^{13}H_3F$ was used which coincides within its Doppler width with the P(32) CO_2 laser line. The result for the photon echo is shown in Fig. 1.13.

Initially molecules of a certain velocity are excited in steady state by the laser light. When the first Stark pulse is applied another velocity subgroup of the molecules is brought into resonance with the cw optical field. The same group is also affected when the second Stark pulse is applied and is responsible for the echo signal. Since the second excitation has to be a π-pulse instead of a $\pi/2$-pulse the field is applied twice as long as for the first excitation. After the dipoles are brought in phase again, which happens in the same period of time which elapsed between the first two excitation pulses, they interfere constructively resulting in the photon echo signal. Since the polarization of the medium is performed by the laser beam the emission due to the echo is performed parallel to the laser beam. At the detector therefore a heterodyne beat is produced whose frequency is identical with the Stark shift. This demonstrates the measurements shown in curve *a* and *b* of Fig. 1.13. The echo beat increases from ≈ 6 to 10 MHz when the Stark field-strength is varied from 35 to 60 V/cm. The evaluation of the pressure dependence of the echo signal indicates that the $C^{13}H_3F$ relaxes primarily by rotational energy transfer.

1.3.2. Other Coherent Transient Phenomena

Optical Nutation and Free Induction Decay

Using the Stark pulse technique introduced by BREWER and SHOEMAKER [1.418] it is, of course, quite simple to observe also other coherent transient phenomena. The first which shall be discussed here is *optical nutation.* It is the analogue to the phenomenon observed in pulsed nuclear magnetic resonance [1.419]. Optical measurements have been performed on SF_6 [1.420], CO_2 [1.421] and NH_4 [1.422] using pulsed laser excitation. The Stark pulse technique has been applied to a ν_3 band transition ($J, K = 4{,}3 \rightarrow 5{,}3$) of $C^{13}H_3F$ and to a ν_2 vibrational transition ($J, M = 4{,}4 \rightarrow 5{,}5$) of NH_2D. The good knowledge of the transition assignments for the molecule allows a very good comparison with theories of the transient effects.

The solution for the absorption of energy from the radiation field has been worked out for nuclear magnetic resonance experiments [1.419],

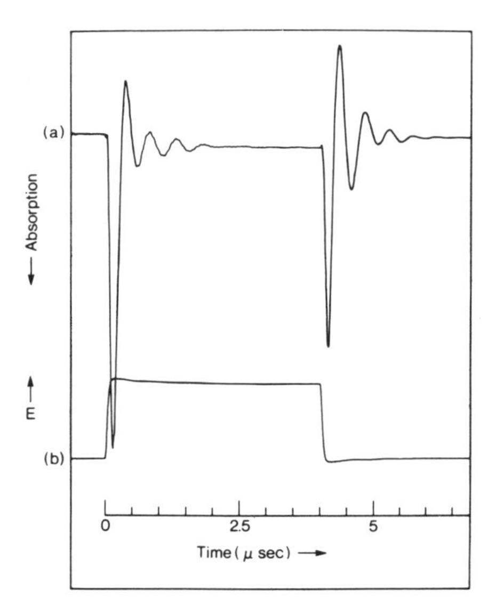

Fig. 1.14. Optical nutation in $C^{13}H_3F$ (Curve *a*). Curve *b* shows the voltage applied to the Stark plates. The nutation period gives the effective saturation parameter directly. The various free induction decay beats which are possible do not show up in this measurement because of their mutual interference and rapid decay

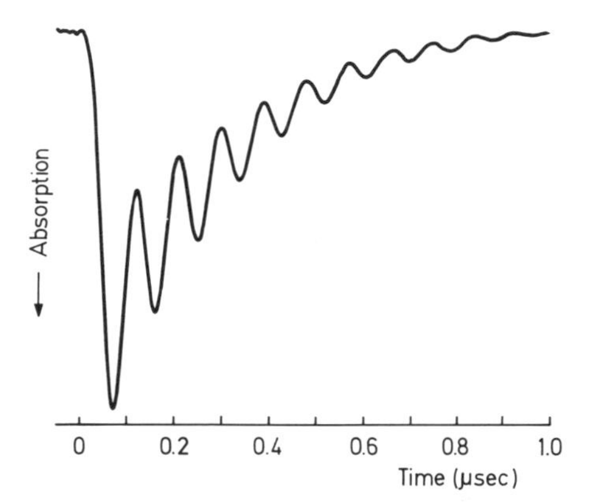

Fig. 1.15. Optical free induction decay in NH_2D. The signal appears when the laser which was used to produce a steady state excitation of the molecules is suddenly switched out of resonance by the Stark effect. The beat frequency represents the Stark shift. It arises because the emission of the ensemble produces a heterodyne signal with the collinear laser beam

and is identical for the optical case when the magnetization is replaced by the vector $\boldsymbol{r}$. When the molecules start to interact with the optical field, the absorption oscillates with a frequency determined primarily by the difference between the resonant frequency of a part of the molecules and the applied frequency. The oscillations are damped out with a time constant which is primarily the relaxation time T_2 (Fig. 1.14). Finally, when the pulse ends another velocity group is excited and it begins to nutate, too, while the group which is now out of resonance emits a free induction signal.

The free *induction decay* following laser excitation is coherent, intense and directional. This behaviour is a consequence of the coherent preparation of the system and of the enhanced spontaneous emission rate due to the cooperative emission process. The intensity is proportional to the square of the number density of the molecules and resembles Dicke's superradiance [1.423]. Also for the free induction decay the analogous effect exists in nuclear magnetic resonance. There it was first predicted by BLOCH [1.424] and experimentally observed by HAHN [1.425], when a group of coherently excited spins was allowed to process freely in a magnetic field, thus inducing a radio frequency current in a pick-up coil. The free induction decay for electric dipole radiation was first demonstrated by BREWER et al. [1.426] using the Stark pulse technique at a Doppler broadened infrared transition of NH_2D (Fig. 1.15).

Coherent Raman Beats

An interesting extension of the investigations described is the observation of coherent Raman beats by SHOEMAKER and BREWER [1.427, 428]. The experiments have been performed using the same experimental set-up as for the other investigations with the Stark pulse technique. The molecule investigated was again $C^{13}H_3F$ excited by the $P(32)$ line of a CO_2 laser.

When by the Stark field the degeneracy of the levels involved in the experiment is removed the transient signal (excitation: $\Delta M = \pm 1$) contains two frequencies which are given by the first-order Stark shift for a $\Delta M = 2$ interval in the upper $(J, K = 5{,}3)$ and the lower $(J, K = 4{,}3)$ levels, respectively. The beat signal decays exponentially with the same relaxation time as that obtained from photon echoes under the same experimental conditions. The transient signal observed in this way is interpreted as a beat signal between the laser wave and the coherent Raman emission of the molecules. The decay of the beat signal is independent of the velocity of the molecules since the Doppler shift for two-photon forward scattering in the coherently prepared sample is $(v/c) \cdot (\nu_1 - \nu_2)$ which is only about 10 Hz in the problem considered. The polarization of the ensemble does not dephase as a result of inhomogeneous broadening and is therefore independent of velocity-changing collisions and to Doppler dephasing. This has the consequence that the two-photon beat signal can be observed still a long time after free induction decay. It is essential to mention that the method opens the possibility to study small Stark splittings of molecules with a precision which is the same as that obtained in double-resonance experiments (see Chapt. 3). Furthermore it allows one to study relaxation phenomena in a selective way.

Two-Pulse Optical Nutation and Multiple-Pulse Spin Echoes

For the study of individual dephasing processes still other coherent transient methods have been developed. The first of these is *two-pulse optical nutation* [1.429] and another is the optical analogue of the *Carr-Purcell* [1.430] *multiple-pulse spin echoes.*

The two-pulse nutation method gives a measurement of the decay time T_1. Here a $\pi/2$ pulse initially produces a coherent superposition of the states involved while a second pulse, which may be a step function, monitors the population in the absorbing level. Molecules which are still in the superposition state at the second pulse have experienced an inhomogeneous dephasing and are not detected. For the *optical Carr-Purcell echoes* a sequence of pulses is applied. The first produce an echo which subsequently dephases but is refocused by the next pulse. This cycle is then repeated thereafter. The multiple pulse echo technique reduces the dephasing caused by velocity-changing collisions to a negligible amount and therefore provides an additional measurement of T_1. However, it should be mentioned that the method still responds to other dephasing processes such as phase-interrupting collisions that exhibit a pure exponential decay. Measurements which have been performed on $C^{13}H_3F$ using two-pulse optical nutation and the multiple-pulse echoes [1.429] show that the results obtained by both methods agree quite well so that it must be concluded that phase-interrupting collisions are unimportant. The relaxation time obtained from the two-pulse echo decay is sensitive to velocity-changing collisions. Therefore the combination of the results obtained with the different methods gives the magnitude of the characteristic velocity jump for molecular collisions, furthermore their cross section can be evaluated.

To be complete still other coherent effects which cannot be discussed in detail in this article should briefly be mentioned. The first one is the *adiabatic following.* The addition of damping terms to the optical Bloch equations which describe the time behaviour of the FVH vector complicates their solution. In the case of a steady external field they remain linear and of first order with constant coefficients and can be solved explicitly [1.419]. Under certain conditions it is also possible to find approximate solutions when the field amplitude is varying slowly enough or adiabatically. The experimental phenomena observed under these conditions are designated accordingly as adiabatic following [1.547–550].

Another coherent effect known for a long time is the *self-induced transparency.* This phenomenon connected with the transport of the radiation through the medium has no analogue in the case of magnetic dipole transitions since there the wavelength is comparable to the external dimensions of the medium. The phenomenon of self-induced transparency manifests itself in the lossless propagation of a wave through an absorbing medium, the breakup of large pulses in smaller ones and by pulse compression, e.g. [1.551–562]. For a comprehensive survey on the theory and the experiments of self-induced transparency see [1.558].

1.3.3. Superradiance

Another interesting coherent effect of an ensemble of particles is the superradiance which was first pointed out by DICKE [1.423]. The effect is due to cooperation of the particles coupled via the common radiation field. Also this effect has first been observed in the microwave region [1.431], where the theoretical treatment is simpler than in the optical region. This is because the particles are confined to a domain small compared to the wavelength of the emitted radiation. First experiments to demonstrate superradiance in the optical region have been performed by SKRIBANOWITZ et al. [1.432] by means of a far-infrared rotational transition in the first excited vibrational state of HF. By optical pumping with a short pulse of a HF-laser operating on a single R or P branch transition starting from the vibrational ground state, population inversion between two adjacent rotational levels of the excited vibrational state was produced. Then the infrared radiation from this transition was studied as function of time. After a delay of 1–2 μsec with respect to the pumping pulse a series of short pulses decreasing in size is emitted (see Fig. 1.16). The short time duration of the pulses (100 nsec) and the angular distribution of the radiation being limited to a small angle along the axis of the gas cell, which had a length between 30 and 100 cm, shows that the emitted radiation is neither an incoherent spontaneous emission, which would cause an exponential decay on the order of several seconds, nor an amplified spontaneous emission in the usual sense, since the transit time through the cell is more than a hundred times shorter than the pulse evolution time. In addition, the peak of the emitted intensity increases with the square of the number N of HF molecules in the cell. It is therefore also different from the intensity behaviour of an ordinary molecular laser system which can oscillate even without mirrors.

The directionality and the N^2 dependence of superradiance can be understood by a simple classical picture: The radiation is assumed to be emitted by an array of dipoles along the z-axis with a macroscopic polarization given by [1.433]

$$\boldsymbol{P} = n d_{ab} \boldsymbol{x} \exp[i(\omega t - kz)] , \tag{1.7}$$

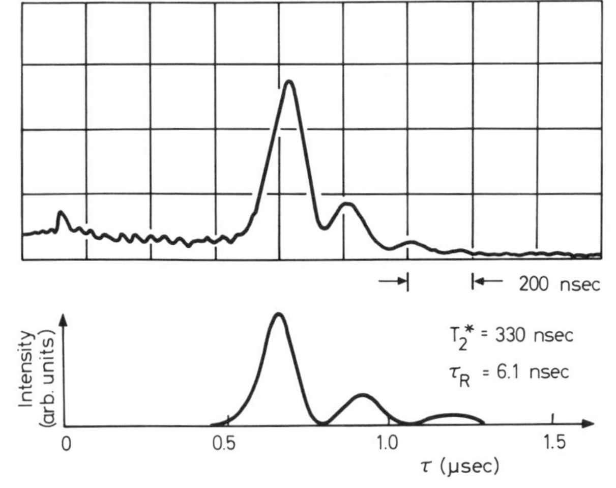

Fig. 1.16. Oscilloscope trace of superradiant pulse at 84 μm ($J=3\rightarrow2$) pumped by the $R_1(2)$ HF laser line. The small peak at the scope trace (upper part of the figure) at $\tau=0$ is the 2.5 μm pump pulse highly attenuated. The lower curve is the theoretical fit to the oscilloscope trace. The pulse evolution can be described simply in terms of the two parameters: τ_R, a characteristic time for the radiation damping of the collective system, and T_2^*, the dephasing time of the rotational transition. (From [1.432])

where n is the number density of the molecules, d_{ab} the dipole moment of a single radiating molecule, k and ω are the wave-number and frequency of the transition. The total far-field intensity of the system of N molecules in direction $\boldsymbol{k}$ is given by ($\boldsymbol{k}=k\boldsymbol{z}$)

$$I(\boldsymbol{k}')=I_0(\boldsymbol{k}')\frac{N^2}{4}|\{\exp[i(\boldsymbol{k}-\boldsymbol{k}')\cdot\boldsymbol{r}]\}_{av}|^2\,. \tag{1.8}$$

$I_0(\boldsymbol{k}')$ is the radiation of a single molecular dipole in the direction $\boldsymbol{k}'$, and the average is taken over the total volume. This result for $I(\boldsymbol{k}')$ is, except for terms of order N, the same as that obtained by DICKE [1.423]. For small $\boldsymbol{k}-\boldsymbol{k}'$, $I(\boldsymbol{k}')$ is proportional to N^2 and the emitted radiation is coherent. This is the result obtained in the experiment of SKRIBANOWITZ et al. [1.432]. In other directions $\exp[i(\boldsymbol{k}-\boldsymbol{k}')\cdot\boldsymbol{r}]|_{av}$ is proportional to N^{-1}. This explains the strong superradiant emission within a small solid angle.

In a quantum mechanical picture the phased array of dipoles can be represented as a coherent mixture of stationary states of molecules.

To explain the time dependence of the emitted radiation one can assume that the emission is initiated by spontaneous emission from one of the excited molecules or by thermal background radiation. This acts as a source for the production of more polarization in the medium and a superradiant state slowly develops over the total sample. The non-uniformity of the polarization over the medium has its reason in propagation effects. The sequence of decreasing pulses in the output pulse is caused by the spatial variation of the polarization. The super-radiant pulses are observed as well in forward as in backward direction. In the FVH formalism the initial state of the medium corresponds to a vector pointing upwards as a pendulum in its unstable position, and a small perturbation can change this initial arrangement. The super-radiant state then corresponds to the FVH vector pointing sideways. After the rapid de-excitation only the lower state is still populated and the FVH vector points downwards.

1.3.4. Spontaneous Decay of Coherently Excited Atoms

From the quantum-electrodynamical discussion of a two-level atom follows (see Subsect. 1.3.1) that the resonance fluorescence induced by coherent radiation is a periodic function of the input pulse area $\theta = 2d_{ab}\hbar^{-1} \int_{-\infty}^{t} |\boldsymbol{E}(t)| dt$ (see p. 69). Thus a maximum in the intensity of the incoherent resonance fluorescence is observed when the excitation produces a pure excited state. This is the case for $\theta = \pi, 3\pi, 5\pi \ldots$. The fluorescence is zero when only the ground state is occupied ($\theta = 0, 2\pi, 4\pi$, etc.). This oscillatory dependence of the fluorescence upon θ has been demonstrated in experiments on Rb atoms [1.539–541] (see also [1.542]). They have been carried out by means of a pulsed Hg II laser which was locked within ± 2 MHz to the center of the 15 MHz residual Doppler-absorption profile of a Rb atomic beam in a 75 k Gauss magnetic field. With a Pockels cell a coherent pulse was extracted from the 1 μsec laser pulse yielding a pulse of area up to about 4π. A result of the measurements is shown in Fig. 1.17. To obtain the signal points the fluorescence intensity was integrated for a certain time interval after each excitation pulse. This demonstration of a simple coherent optical effect can be used for the measurement of dipole moments, relaxation times, for the study of 0π pulses and chirping [1.542]. In addition, the experiment disproves the neoclassical theory of radiation in which the electromagnetic field is not quantized [1.543–546]. This theory assumes that the expectation value of the dipole-moment operator is an actual dipole moment radiating according to classical electrodynamics. The minima in the time behaviour of the fluorescence should occur at input areas of $\pi, 2\pi, 3\pi$, since in that theory a pure state is unable to radiate spontaneously. This is in contradiction to the measurement shown in Fig. 1.17.

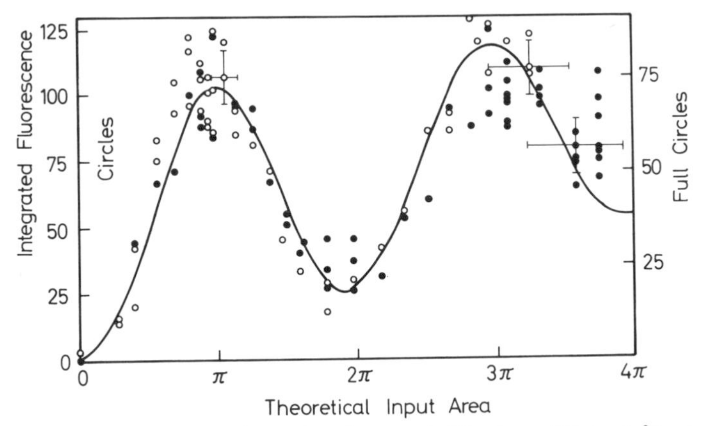

Fig. 1.17. Fluorescence from the Rb state $5p\,^2P_{1/2}(m_J=1/2)$ to $5s\,^2S_{1/2}(m_J=-1/2)$ integrated from 22 to 72 nsec after excitation as a function of input pulse area. The full circles and circles are the counting rates measured in 100- and 80-sec intervals for peak cw absorption of 0.21 and 0.32, respectively. (From [1.540].) The solid curve shows the $(\sin\theta/2)^2$ dependence which follows from quantum mechanics. It is normalized to yield a minimum weighted variance with the circles

1.4. Natural Linewidth-Limited Spectroscopy

In laser spectroscopy basically two groups of methods are available, which allow one to obtain a resolution only limited by the natural linewidth. The first group comprises the methods of radio-frequency spectroscopy, e.g., the optical double-resonance method, and others listed in the right part of Table 1.6. For these applications the laser is only an intense light source providing a continuous wave, a modulated excitation or a pulsed one. All the methods have already been demonstrated years ago with classical light sources. The laser, however, brought a considerable improvement due to its high spectral brightness, which allows a very efficient population of the excited states. The width of the spectral distribution required for the experiments can be comparable with, or is greater than, the Doppler width of the investigated transitions.

The methods compiled on the left side of Table 1.6 require a narrow-band excitation. They became applicable only after single-mode lasers have become available. Experiments of both groups will be discussed in the following.

1.4.1. High Resolution Spectroscopy with Broad-Band Excitation

In this subsection the classical methods of high resolution radio-frequency spectroscopy applicable to the investigation of electronically excited

Table 1.6. Survey of methods used for high-resolution spectroscopy. (Resolution limited by the natural line-width)

Narrow-banded excitation (only possible by lasers)		Broad-banded excitation		
Narrow-banded absorption	Broad-banded absorption (Doppler width)	Incoherent population	Coherent popultation of two or more states of an atom or molecule	
Atomic beams	Exclusion of Doppler width by two light beams travelling in opposite directions. Coupling of the interaction of the *two* light beams with atoms by means of	Induction of coherence by radio-frequency	cw excitation and time integral observation	Pulsed excitation and time differential observation
	(a) Nonlinear single-quantum absorption *saturated absorption*	(a) Optical-radio-frequency *double resonance method*	*Level crossing* (Hanle effect)	*Quantum beats* *Periodic excitation* Observation of other transient effects (Sect. 1.3)
	(b) Double-quantum absorption *Two-photon spectroscopy*	(b) *Optical pumping* (orientation of the fundamental state)		
	(c) *Stepwise excitation*	(c) *Anti-crossing*		
	Limitation of observation to a velocity subgroup of atoms *Fluorescence line narrowing*			

atoms and molecules will be briefly discussed and their use in connection with laser-excitation reviewed. Following Table 1.6 we will start with the group consisting of optical radio-frequency double resonance, optical pumping and anti-crossing. These methods have in common that an incoherent non-statistical population of the levels involved has to be produced.

Optical-Radio-Frequency Double Resonance, Optical Pumping, Anti-Crossing

The *optical-radio-frequency double resonance method* [1.434] has served for more than two decades for high-resolution measurements or fine and hyperfine splittings in free atoms. Its broad application to molecules, however, was restricted by the fact that no light sources for an efficient excitation have been available. With the advent of tunable lasers it can be expected that the method will now have a strong renaissance. Also for atoms new measurements can be performed since with the laser a stepwise excitation is possible so that very highly excited levels can be investigated.

The optical excitation used for the double-resonance method produces a non-statistical population in the excited state. This is achieved by using a non-isotropic excitation by polarized light. The non-statistical population is, of course, also obtained even when the levels under investigation overlap within their Doppler width. After this preparation of the states, radio-frequency transitions are induced. They establish a statistical distribution for the levels involved, therefore the resonance can be monitored via the change in the angular distribution of the resonance radiation. The sensitivity of the method is rather high since the absorbed low energetic radio-frequency quanta are detected via optical quanta.

In the double-resonance experiment the natural linewidth is observed because the Doppler effect is proportional to the frequency, and is therefore small for radio-frequency or microwave transitions. This feature is common to all methods of radio-frequency spectroscopy, e.g., the microwave absorption of molecules or the atomic or molecular beam resonance method. A review of the early work performed with the optical-radio-frequency double resonance method has been given by zu Putlitz [1.435] and by Budick [1.436].

Many modifications of the double-resonance method are possible. The optical excitation may be combined with another optical excitation rather than with microwaves. A survey of the various applications is given in the Chapter 3 by Shimoda.

A special case of the optical microwave double resonance is the observation of the *anti-crossing* effect [1.437]. In this case, mixing of the levels is effected by an internal perturbation of the atoms or molecules and not by the radio-frequency field. The effect occurs when levels with the same projection quantum numbers of the total angular momentum repel each other as a function of an applied static field. A change in fluorescence intensity is observed at field values where the levels have their point of closest approach. The effect is basically similar to double resonance with frequency zero.

A non-statistical distribution can also be established in the fundamental state with the use of excitation by circularly polarized light since the non-statistical population (orientation) of the excited state is transferred to the ground state, resulting in an orientation of the atoms or molecules. This process is called *optical pumping* [1.438]. The orientation in the ground state may be destroyed by the radio-frequency transitions, which can then be monitored by the change in the absorbed intensity of the circularly polarized light used for excitation.

The linewidth observed in such an optical pumping experiment of the ground state is, of course, related mainly to the relaxation time during which its orientation is destroyed by collisions with other species or with the wall of the vessel containing the investigated particles. For atoms or molecules in the ground state the linewidth may also be limited by the briefness of the interaction with the radio-frequency field if the transitions are induced in a rather small volume.

The high power density of the laser makes the optical pumping process very effective. It has been demonstrated [1.439] that a single Na atom travelling with thermal velocity through the beam of a cw dye laser (interaction region in the mm range) may be excited as often as 60 times. In another pumping experiment [1.440] on Na vapor, it was demonstrated that the time constant for the polarization of the sodium atoms is about three orders of magnitude smaller than that obtainable with conventional discharge lamps. Meanwhile the dye lasers have been considerably improved so that much better results are now obtainable. The small linewidth achievable for the spectral distribution may, in addition, be useful in selecting particular hyperfine levels for the optical pumping process (hyperfine pumping). It may well be that the application of the dye laser will be especially useful in connection with optical pumping experiments on short lived isotopes [1.441].

With present dye lasers, the atoms of an atomic beam can be polarized completely by optical pumping within a very small interaction region. It has thus been possible to perform an experiment in which the pumping region is separated from the region wherein the polarization of the atoms is probed [1.442]. Such a set-up has many interesting applications. Any

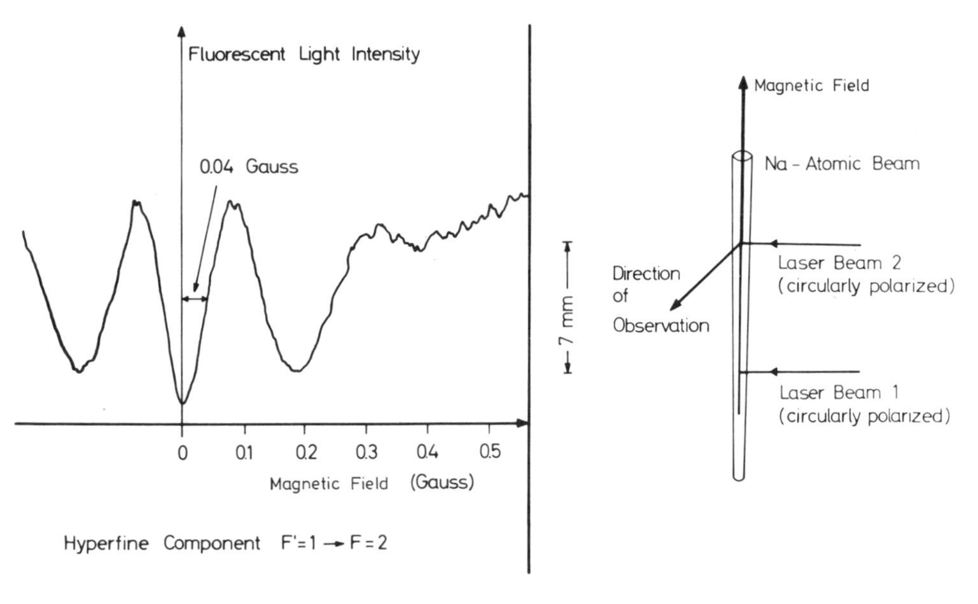

Fig. 1.18. Optical pumping by means of a cw dye laser. Right side: experimental set-up. The direction of observation for the fluorescent light is perpendicular to the laser and atomic beams. Left side: Fluorescence signal induced by the second laser beam as a function of the applied magnetic field. The oscillations of the signal are due to the precession of the macroscopic magnetic dipole moment induced by the first laser beam. (From [1.442])

changes in the phase of the eigenfunctions of the atoms through interaction in an external field (e.g., magnetic field), or through collision or other effects occurring in the light-free region can be sensitively detected with the second probing beam. A result of these measurements is shown in Fig. 1.18. The precession of the macroscopic magnetic dipole moment in the external magnetic field is monitored by the second light beam. In higher magnetic fields the oscillations vanish. The reason is that with a laser beam diameter of 1 mm, the interaction time between atoms and light beams is about 1 μsec; this time is not small compared to the precession period of the dipoles when the magnetic field exceeds about 0.3 Gauss. Furthermore the oscillations are smeared out due to the velocity distribution of the atoms in the beam. (For the optical pumping of the atoms the hyperfine transitions $3\,^2S_{1/2}$, $F=2 \rightarrow 3\,^2P_{1/2}$, $F'=1$ was used.) In principle, the set-up employed for the experiment is similar to a "Rabi apparatus" [1.443] used in atomic beam resonance work. The first light beam plays the role of the A-field and the second one of the B-field. In comparison to the classical "Rabi apparatus" this set-up has the advantage that also ions can be investigated.

Table 1.7. Level-crossing and optical microwave double resonance experiments at electronically excited states. For other optical microwave double resonance experiments see Chapter 3 by SHIMODA

Atom or molecule	Levels	Laser	Direct information	Remarks and references
Na	$3\,^2P_{3/2}$	Pulsed dye laser	Hyperfine splitting	Delayed level crossing [1.452]
	$4\,^2D$	Stepwise excitation with discharge lamp and cw dye laser	Fine structure splitting	[1.470]
K	$7, 8\,^2S_{1/2}$	Stepwise excitation with discharge lamp and cw dye laser	Hyperfine splitting, g-factor	[1.471]
	$6, 7\,^2P_{1/2}, ^2P_{3/2}$	Stepwise excitation with discharge lamp and cw dye laser	Hyperfine splitting, g-factor	[1.471]
	$5, 6\,^2D_{3/2}, ^2D_{5/2}$	Stepwise excitation with discharge lamp and cw dye laser	Hyperfine splitting, g-factor, Tensor polarizability	[1.471]
Rb	$9, 10\,^2P_{3/2}$	Stepwise excitation with discharge lamp and cw dye laser	Hyperfine splitting, g-factor	[1.472, 473]
	$6, 7\,^2D_{3/2}, ^2D_{5/2}$	Stepwise excitation with discharge lamp and cw dye laser	Hyperfine splitting	[1.475, 476]
	$8, 9\,^2D_{3/2}, ^2D_{5/2}$	Stepwise excitation with discharge lamp and cw dye laser	Hyperfine splitting, g-factor	[1.474]
	$6, 7, 8, 9\,^2D_{3/2}, ^2D_{5/2}$	Stepwise excitation with discharge lamp and cw dye laser	Tensor polarizability	[1.474, 481]
	$8\,^2D$	Stepwise excitation with discharge lamp and cw dye laser	Fine structure splitting	[1.483]
Cs	$10, 11\,^2S_{1/2}$	Stepwise excitation with discharge lamp and cw dye laser	Hyperfine splitting	[1.482]

	11, 12, 13 $^2P_{3/2}$	Stepwise excitation with discharge lamp and cw dye laser	Hyperfine splitting, g-factor	[1.472, 478]
	7 $^2D_{3/2}$	Stepwise excitation with discharge lamp and cw dye laser	Hyperfine splitting	[1.478]
	8, 9, 10 $^2D_{3/2}$, $^2D_{5/2}$	Stepwise excitation with discharge lamp and cw dye laser	Hyperfine splitting	[1.475, 476]
	8, 9, 10 $^2D_{3/2}$, $^2D_{5/2}$	Stepwise excitation with discharge lamp and cw dye laser	Tensor polarizability	[1.480, 481]
	11, 12, 13, 14 $^2D_{3/2}$, $^2D_{5/2}$	Stepwise excitation with discharge lamp and cw dye laser	Hyperfine structure	[1.477]
	15, 16, 17, 18 $^2D_{3/2}$	Stepwise excitation with discharge lamp and cw dye laser	Hyperfine structure	[1.478]
	5, 6 $^2F_{5/2}$, $^2F_{7/2}$	Stepwise excitation with discharge lamp and cw dye laser	Hyperfine structure	[1.475, 479]
Ne	$2p_2, 2p_4, 2p_5, 2p_6, 2p_7, 2p_8$	Stepwise excitation with discharge lamp and cw dye laser	Hyperfine structure	[1.566]
BaO	$A\,^1\Sigma^+$	Dye laser	Rotational splitting	[1.484]
I_2	$B\,^3\Pi^+_{0u}$	Gas laser	g-factor times lifetime	[1.485, 486]
Na_2	$B\,^1\Pi_u$	Ar^+-laser	g-factor times lifetime	[1.487]
NaK	$D\,^1\Pi_u$	Ar^+-laser	g-factor times lifetime, Λ doubling, dipole moment	[1.488]

The measurements performed with the optical microwave double resonance method on electronically excited states are compiled in Table 1.7 together with the measurements performed with the level crossing which will be discussed in the next subsection.

Level-Crossing, Quantum Beats, Periodic Excitation

A purely optical technique which allows to observe a lifetime-limited linewidth in the excited state is level-crossing [1.444, 445] and related methods, cited in the last two columns of Table 1.6.

With this method, a coherent mixture $\psi = \alpha|a\rangle + \beta|b\rangle$ of two excited states a, b is produced during the excitation process. When the two states are Zeeman substates differing in their magnetic quantum numbers by ± 2, this coherent mixture is produced by an excitation with light which is linearly polarized perpendicular to the external field. A single photon produces the coherent superposition of the two states whereby the excitation is started from the same lower level. At small magnetic fields where the two coherently excited states have equal energies E_a and E_b within the limit of the natural level width, the condition $|E_a - E_b|/\hbar < 1/\tau$ is fulfilled. In this case the phase relation for the wavefunction established during the continuous excitation process will be conserved when the atoms or molecules decay to the ground state, so that the angular distribution of the re-emitted radiation corresponds to that of the absorbed radiation.

However, when $|E_a - E_b|/\hbar > 1/\tau$, the change of the wavefunction is so rapid that the phase relation produced during excitation is destroyed for the ensemble and the fluorescence is isotropically emitted. Under the condition that the fluorescence is observed perpendicular to both the external magnetic field and the excitation direction the signal shape is an inverted Lorentzian where the minimum is observed for $E_a = E_b$. The halfwidth of the Lorentzian is determined by the natural bandwidth.

The change in angular distribution of the fluorescent radiation observed in the region around $E_a \approx E_b$ is independent of whether the "level crossing" $E_a = E_b$ occurs close to zero of the magnetic field (this is the so-called Hanle signal [1.446]) or at higher fields. Therefore the effect can be used to determine the crossings which belong to Zeeman substates of different fine structure or hyperfine structure states. In addition, it should be mentioned that the effect can also be observed when an external electric field or a combination of magnetic or electric fields is applied.

When level crossing signals at higher fields are observed the method, in principle, compares the magnetic or electric splitting to that at zero field. It therefore provides the parameters of the zero-field splitting in a

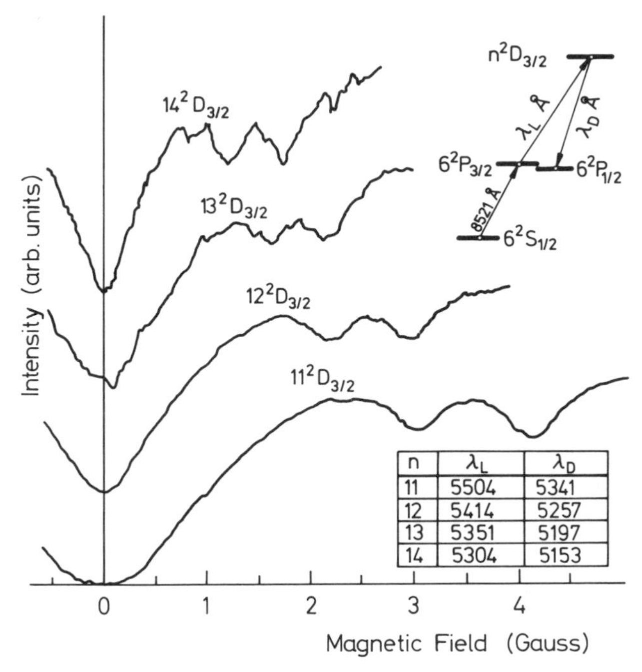

n	λ_L	λ_D
11	5504	5341
12	5414	5257
13	5351	5197
14	5304	5153

Fig. 1.19. Level crossing signals for the $n = 11$, 12, 13, and $14\,^2D_{3/2}$ states in Cs^{133}. The ^2D states were populated in a two-step excitation scheme, involving a conventional radio-frequency lamp and a cw dye laser with wavelength λ_L (see upper right of the figure). The transition designated by λ_D was used for the detection of the signal. (From [1.474])

ratio to the field induced energy change which must be determined by other methods as, e.g., by the optical-radio-frequency double-resonance method in order to obtain the zero splitting absolutely.

A large number of investigations using this method in combination with laser excitation have been performed in alkaline spectra. Here it proved useful that stepwise excitation could be applied in connection with lasers. Since very intense radio-frequency discharge lamps are available for the resonance lines of the alkaline atoms the first excitation step is usually performed by means of such a lamp and the second step by excitation with a dye laser. The recent measurements performed with the level-crossing method and laser excitation are compiled in Table 1.7. An example for a signal curve is shown in Fig. 1.19.

The high population density achievable for the excited states by means of dye lasers makes it quite easy to perform time-delayed level-crossing experiments. For these experiments a pulsed laser is used. The fluorescent light is observed a certain time-lag following excitation. The signal is therefore smaller than the natural width.

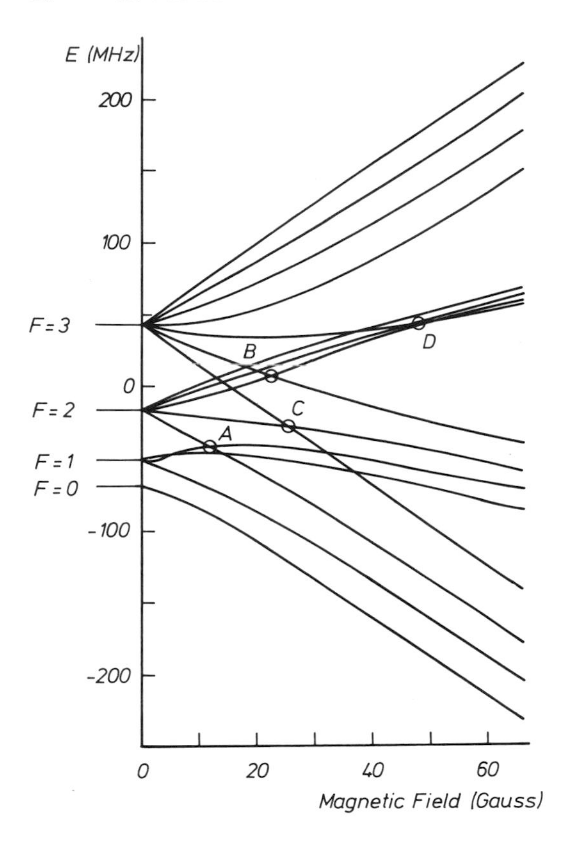

Fig. 1.20. Zeeman splitting of the $3\,^2P_{3/2}$ level of sodium. The crossing points which correspond to levels which differ in their magnetic quantum numbers by ± 2 are indicated by circles. The signal of the zero field crossing (Hanle signal) and of crossing A is displayed in Fig. 1.21. (From [1.452])

Experiments using a delayed observation in order to narrow the observed linewidth below the natural one have been performed many years ago in connection with Mössbauer effect[3] studies (for a survey see [1.447]) as well as in spectroscopy of atoms [1.448, 449] and muonic atoms [1.450].

Using dye lasers the delayed observation of level-crossing signals is strongly facilitated since the excited state can be saturated during the laser pulse; then after eight or ten lifetimes, approximately one in 10^4 atoms is still in the excited state, which would be about the number of atoms initially excited if conventional discharge lamps are used. A zero-field level-crossing experiment (Hanle experiment) of this type has been performed on the lowest lying $^1P_1$ levels of Ca and Ba [1.451]. Higher field crossings have been investigated for the $3\,^2P_{3/2}$ level of Na [1.452]. In this experiment a reduction of the natural linewidth of

[3] See [1.468] for a comprehensive treatment of the Mössbauer effect.

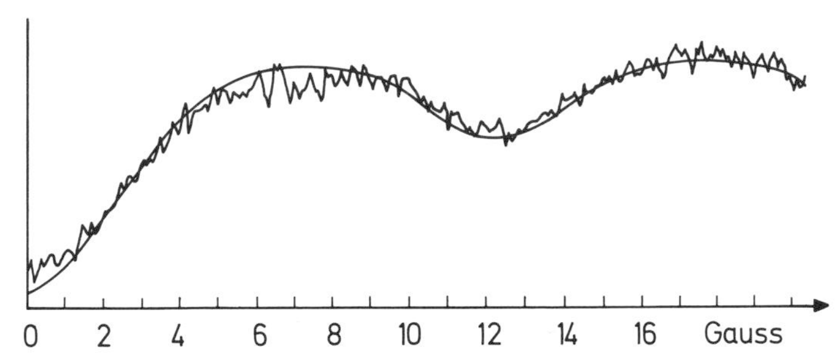

Fig. 1.21. Level-crossing signal for the $3\,^2P_{3/2}$ level of sodium (delayed observation). Shown is a part of the zero-field level-crossing signal and the crossing A (see Fig. 1.20). The measurement was performed with apodisation (see main text for details). The smooth line is a theoretical curve. The shown measurement gives a linewidth which is more than a factor two smaller than the linewidth obtained for a time-integrated observation. In the latter case crossing A is not separated from the zero-field level-crossing signal, see, e.g. [1.563]. (From [1.452])

up to a factor of 6 could be obtained. The hyperfine constants of the $3\,^2P_{3/2}$ level could be determined more accurately than in former experiments. The disadvantage of the method is that for the delayed observation the signal shape is no longer Lorentzian; in addition, undulations are observed at the side of each crossing signal. In Fourier spectroscopy a similar phenomenon occurs which is connected to the finite observation time of the beat signal between the variable and constant branch of the Michelson interferometer [1.453]. Therefore as an analogy to Fourier spectroscopy the side-maxima in the level-crossing experiment may also be suppressed by a suitable apodisation [1.449, 469]. This was done in [1.452] by introducing a time dependent sensitivity of the detection system. The reduction in linewidth obtained with apodisation is, of course, smaller than in the measurements where no apodisation is used (Figs. 1.20 and 1.21).

Another experiment performed with the level-crossing method should also be mentioned here: the observation of the zero-field level-crossing or Hanle effect under monochromatic laser excitation. To treat this latter problem theoretically besides the coherences between Zeeman sublevels of the excited state also the optical coherences between the excited and the fundamental state have to be considered. This has to be done since the coherence time of the laser light used for the experiment is long compared to the lifetime of the excited state [1.454, 455]. The level-crossing experiments have to be performed with a collimated atomic beam to get rid of the Doppler width of the atoms. The experimental results which give a level-crossing signal which is about 45% of the

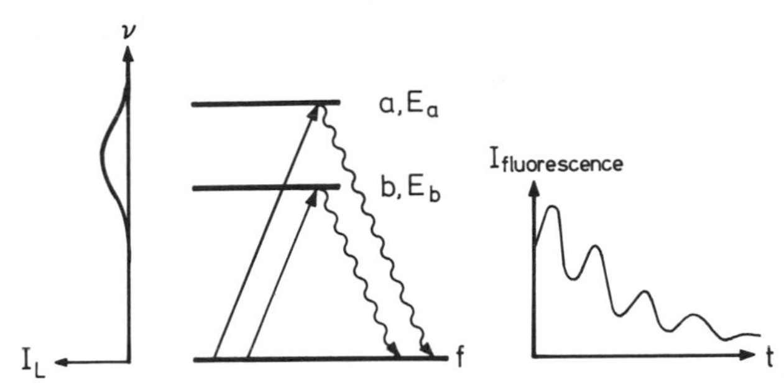

Fig. 1.22. Quantum beat spectroscopy. Left side of the figure: frequency distribution of the laser pulse. Right side: time dependence of the fluorescence signal

width obtained with broad-band excitation [1.456] will also be discussed in Subsection 1.4.2 (p. 93). See also Ref. [1.618].

A method which is very much related to the level crossing method is the observation of *quantum beats*. For this method a pulsed excitation is used; afterwards the fluorescent light is observed as a function of time. When we consider two levels with the energy E_a and E_b the duration of the excitation pulse Δt must be short enough so that the condition $1/\Delta t > |E_a - E_b|/\hbar$ is fulfilled. That means that the energy distribution which follows from the Fourier expansion of the light pulse is larger than the energy splitting between the two levels (Fig. 1.22). Therefore a coherent excitation of the levels is possible. The condition for the duration of the exciting light pulse also guarantees that the atoms or molecules are excited in a time interval which is short compared to the period in which the phase difference between the two levels develops in time. Therefore the contribution of the different atoms and molecules of the ensemble does not average out as in the case of a cw excitation and the "beat frequency" $|E_a - E_b|/\hbar$ is observed to be superimposed on the exponential decay of the fluorescent radiation. The energy difference between the two levels a and b may then be obtained from a Fourier expansion of the signal. For the observation of the quantum beats, it is of no significance which physical effect causes the splitting between a and b; it may be a Zeeman-splitting, hyperfine splitting etc. The essential condition is, however, that the levels between which the energy difference shall be measured, can be populated starting from the same lower level. Compared to the level-crossing method the observation of the quantum beats has the advantage that the splitting between the levels is determined directly; it is not necessary to determine the parameters of a magnetic or electric splitting separately in order to evaluate the data. The laser is switched off when the observation is performed. This gives the advantage that no power broadening of the levels and no intensity shift can occur.

The quantum beat method was applied for the first time in connection with excitation by pulsed light sources by DODD et al. [1.457] and by ALEXANDROV et al. [1.458]. Later the method was also used with pulsed electron beams [1.459] and in beam-foil experiments. In the latter case the excitation of the ions was either performed by the foil [1.460] or by laser light [1.461]. Also in a time differential observation of γ-cascades, the quantum beats have been detected [1.462].

It is quite clear that laser excitation brings many advantages for the use of the quantum beat method. The signal-to-noise ratio is considerably improved compared to the excitation with pulsed discharge lamps or with electrons. In addition, pulses in the nano-second region can be obtained rather easily. Therefore splittings up to almost 1 GHz can be measured. Using mode-locked lasers even shorter pulses are possible, however, the transient recorders available at present do not allow one to investigate larger splittings than about 1 GHz. Employing two synchronously excited dye lasers even a stepwise excitation is possible so that highly excited levels can be investigated.

One of the experiments using stepwise excitation has been performed by HAROCHE et al. [1.503, 504]. In these papers the fine-structure splitting of the Na n^2D levels has been measured from $n=9$ to $n=16$. A first laser pulse, tuned to the wavelength of the D_2 resonance line excites the atoms in a vapor cell into the $3^2P_{3/2}$ level. Then a second laser pulse synchronized with the first one brings the atoms from the intermediate level to the n^2D level of interest. The signal depends strongly on the polarization of the two laser beams and of the detected light. Maximum signal is obtained when the polarization directions of the two laser beams are perpendicular to each other. Figure 1.23 shows as an example two typical spectra obtained by computer Fourier analysis of the beats in level $n=11$ and $n=16$. Even for the very highly excited level still a rather good signal-to-noise ratio has been obtained. The quantum beat method does not give the sign of a splitting; however, the sign can be determined from Stark effect measurements. The calculations of the Stark shift show that sublevels with the same quantum number m_J are shifted by a larger amount in the $^2D_{5/2}$ levels than in the $^2D_{3/2}$ levels. (The shift is due to a repulsion by a very close nF level which lies just above the D levels.) The Stark effect shift of a $\Delta m_J=0$ fine-structure beat should thus be negative if $D_{5/2}$ lies above $D_{3/2}$ (positive fine-structure splitting), and positive when $D_{5/2}$ lies below $D_{3/2}$ (negative fine-structure splitting). Due to the large polarizability of the nD states this check is actually very easy to perform since a field of a few V/cm is sufficient

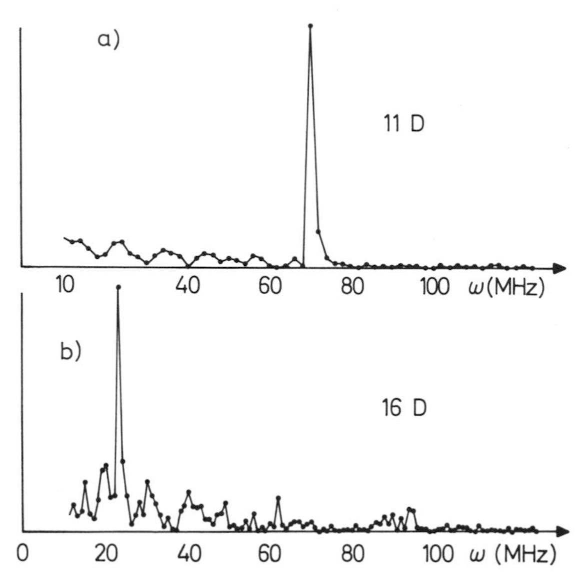

Fig. 1.23a and b. Fourier spectra of the fine-structure quantum beats between the levels $11\,^2D_{3/2,\,5/2}$ and $16\,^2D_{3/2,\,5/2}$. The interval between the points is 2 MHz and 1 MHz for $11\,^2D$ and $16\,^2D$, respectively. (From [1.504])

to induce a shift by several MHz. The result is that the ^2D levels have an inverted fine-structure in contrast to the hydrogen spectrum.

The fine-structure splitting of the Na ^2D levels is compared to the ^2D splitting of hydrogen in Fig. 1.24. Besides the quantum beats experiments also the result obtained by high-resolution two-photon spectroscopy and by classical interferometry [1.463, 464] are included in the Fig. 1.24. It is remarkable that even for very high n the fine-structure is so much different for both atoms. This result is astonishing since the nD levels of sodium behave, as far as their other spectroscopic properties are concerned, as almost perfect hydrogen Rydberg states. The large amount of experimental data now available on the sodium spectrum and the discussed discrepancies will surely inspire the theoreticians to perform more refined calculations. Quite recently it has been shown by LUC-KOENIG [1.465] that the inverted fine-structure splitting in sodium can be explained by a relativistic calculation in the central field approximation.

The quantum beat method in connection with laser excitation has also been applied to molecules. A survey on the experiments performed up to now is given in Table 1.8.

An extension of the method is the use of a *periodic excitation.* In this case, an increase is observed in the degree of modulation of the

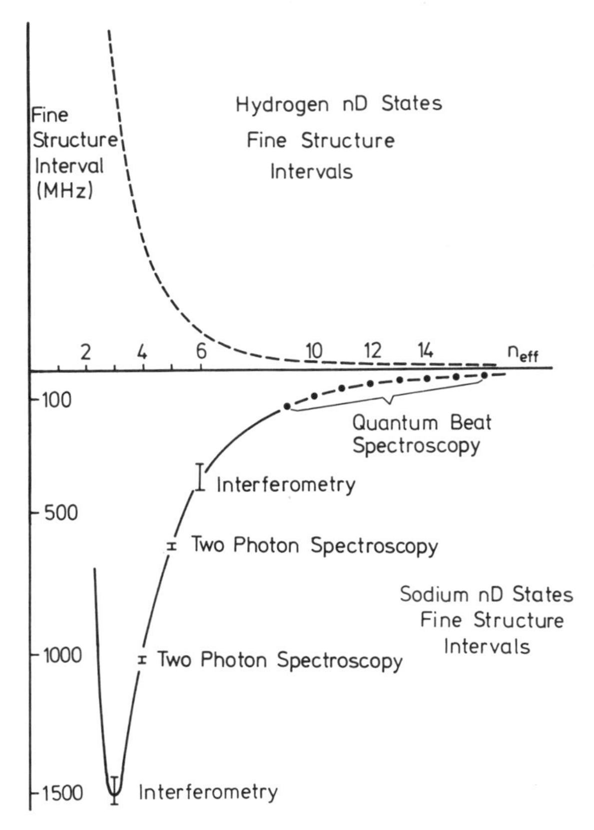

Fig. 1.24. Fine-structure intervals in the levels *nD* of Na as a function of n_{eff}. The full line corresponds to the empirical formula $T = -a/n_{\text{eff}}^3 + b/n_{\text{eff}}^5$ ($n_{\text{eff}} = n - \varepsilon_{\text{d}}$, where $\varepsilon_{\text{d}} = 0.014$, $a = 96\,500$ MHz, $b = 498\,500$ MHz). The dashed curve gives for comparison the variation of the hydrogen fine-structure splitting of the $n\,^2D$ levels. (From [1.504])

fluorescent light when the frequency of the periodic excitation coincides with the energy difference $|E_a - E_b|/\hbar$ of two coherently excited levels. This method was demonstrated by SERIES and CORNEY [1.466], and by ALEXANDROV [1.467]. Its application in conjunction with laser excitation is easy since the laser light can be modulated simply. A restriction is, however, that the laser light modulators are, in general, not very broadband. Therefore the method is mainly applicable in cases where the modulation frequency of the laser can be kept fixed and the energy difference is changed as a function of an external field, as in Zeeman-effect studies. The advantage of the modulation method compared to the optical-radio-frequency double resonance is that it is very easily applicable to electronically excited states with very short lifetimes; those require in

Table 1.8. Quantum-beat spectroscopy

Atom or molecule	Level	Laser	Direct information	Remarks and references
Na	9, 10 ^2D	Stepwise excitation with two dye lasers	Fine structure splitting	[1.503]
Na	11, 12, 13, 14, 15, 16 ^2D	Stepwise excitation with two dye lasers	Fine structure splitting	[1.504]
Na	10, 11, 12 ^2D	Stepwise excitation with two dye lasers	Quadratic polarizability	[1.565]
Cs	7 $^2P_{3/2}$	Dye laser	Hyperfine splitting	[1.505]
Cs	8, 9, 10 $^2D_{3/2}$	Stepwise excitation with dye laser and discharge lamp	Hyperfine splitting, lifetimes	[1.506]
Yb	$6s6p\ ^3P_1$	Dye laser	g-factor	[1.507]
I_2	$^3\Pi^+_{0u}$	Dye laser	Zeeman-splitting	[1.508]

double-resonance experiments a rather high magnetic field strength of the radio-frequency field. The method does not require that the spectral distribution of the laser has a narrow bandwidth; it is, therefore, not identical with the scanning of the structure of the excited state by means of a modulated single-mode laser beam, where the side-bands are used to probe the structure of the state.

1.4.2. High-Resolution Spectroscopy with Narrow-Band Excitation

The second main type of experiments of high-resolution spectroscopy make use of the narrow bandwidth of the spectral distribution of a laser (left side of Table 1.6). Such experiments require a single-mode laser in order to achieve the highest possible resolution. In the first group of these experiments the light is scattered by an atomic or molecular beam where the velocity component of the particles with respect to the laser beam is reduced by collimation.

High-Resolution Spectroscopy by Means of Atomic and Molecular Beams

The absorption width of an atomic molecular beam with the beam axis perpendicular to the laser beam is given by the width

$$\Delta\nu \approx k \cdot \Delta\nu_D\,, \qquad (1.9)$$

where k is the collimation ratio of the beam and $\Delta\nu_D$ the Doppler width of the atoms or molecules at the temperature of the beam source. The collimation ratio can be lowered sufficiently to ensure that $\Delta\nu$ becomes small compared to the natural width. In an experiment where the transitions are scanned by tuning the laser and where the fluorescent light is monitored, a linewidth which is close to the natural width is observed. This is demonstrated in Fig. 1.25. A compilation of the measurements performed with atomic and molecular beams is given in Table 1.9. It can be seen that most of the experiments have been carried out by means of tunable dye lasers. The detection sensitivity for the fluorescent light is high enough so that experiments using rather low beam densities are possible. (Usually the number density of the atoms and molecules is in the range of 10^9 to 10^{12} particles/cm^3. However, much lower densities are still detectable [1.327]).

From the various high-resolution experiments performed with atomic beams in the following those which deal with the frequency distribution of the spontaneously emitted fluorescent light emitted after a monochromatic laser excitation will be briefly discussed. This

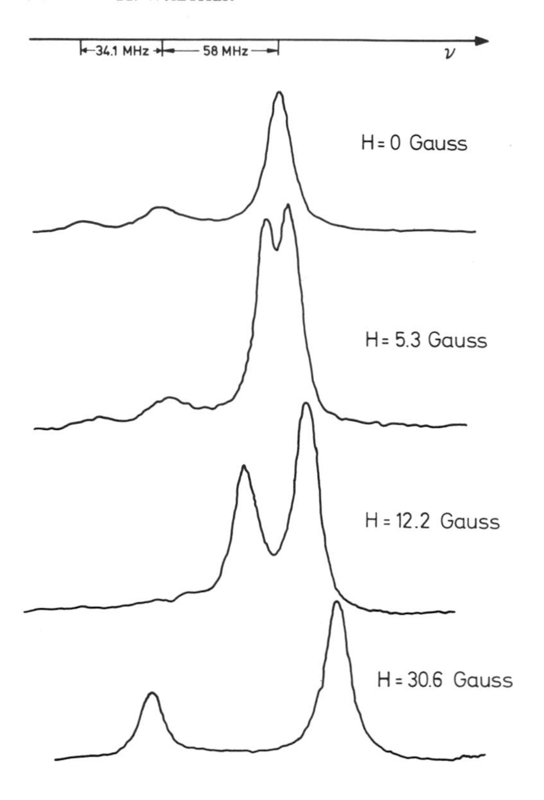

Fig. 1.25. Zeeman splitting of hyperfine components of the sodium D_2 line. The curve for $H=0$ displays the three hyperfine transitions $3\,^2S_{1/2}$, $F=2 \rightarrow 3\,^2P_{3/2}$, $F=1, 2, 3$ (from left to right, see also [1.491]). The observed linewidth is about 10 MHz, which is the natural width of the transition. The fact that the splitting can be observed at rather small fields demonstrates the high resolution obtained in the measurements. (From [1.619])

problem is quite fundamental for the understanding of resonance fluorescence and has therefore received extensive theoretical treatment by many authors [1.576–590] using different methods and approaches. The predicted spectra for high power of the exciting radiation include a Lorentzian with a hole in the middle [1.578, 582], a three-peak distribution [1.577, 579, 581, 584–591], or even more complex structures [1.580, 583].

For the experimental study of the frequency distribution the atomic beam must also be collimated with respect to the observation direction, and the spectrum of the fluorescence has to be studied by means of a highly resolving Fabry-Perot interferometer. In the first experiment [1.592] of this type on the $3\,^2P_{3/2}$, $F=3 \rightarrow 3\,^2S_{1/2}$, $F=2$ hyperfine transition of Na (this transition was chosen to avoid optical pumping of the hyperfine levels of the $^2S_{1/2}$ ground state) a dye laser has been used with a spectral distribution of about 20 MHz. The observed frequency distribution showed an indication of a three-peak structure; but a comparison with theory was not possible. Therefore, recently an im-

Table 1.9. High-resolution spectroscopy by means of atomic and molecular beams

Atom or molecule	Transition	Laser	Direct information	Remarks and references
Na	$3\,^2S_{1/2} - 3\,^2P_{1/2}$ $3\,^2S_{1/2} - 3\,^2P_{3/2}$	cw dye laser	Hyperfine splitting	[1.489, 490] [1.491, 492]
			Hyperfine splitting	Rabi-type atomic beam [1.493]
			Hyperfine splitting	Radiation pressure [1.494]
			Hyperfine splitting	Detection of optical resonance by subsequent photo-ionization [1.495]
Na	$3\,^2P_{1/2} - 5\,^2S_{1/2}$	cw dye laser (stepwise excitation)	Hyperfine splitting	[1.496]
Na	$3\,^2S_{1/2} - 4\,^2D$	Dye laser (double quantum transition)	Fine structure splitting	[1.497]
Li	$2\,^2P - 3\,^2D$	cw dye laser (stepwise excitation)	Fine structure splitting	[1.498]
Na	$3\,^2S_{1/2} - 3\,^2P_{1/2}$	cw dye laser	Isotope shift and hyperfine splitting of radioactive isotopes ^{21,22,24,25}Na	[1.568]
Cs	$10, 11\,^2S_{1/2}$ $9\,^2D_{5/2}$	Stepwise excitation with discharge lamp and cw dye laser	Scalar polarizability	[1.499]
Yb	$^1S_0 - {}^3P_1$	cw dye laser	Isotope shift, hyperfine splitting	[1.500]
Ba	$6\,^1P_1 - 6\,^1S_0$	cw dye laser	Isotope shift, hyperfine splitting	[1.456, 618, 619]
I_2	$^1\Sigma_g^+ - {}^3\Pi_{0u}$	Ar^+-laser	Hyperfine splitting of rotational levels	[1.501]
I_2		cw dye laser	Hyperfine splitting	[1.502]
NO_2	B_2	Ar^+-laser	Hyperfine splitting	[1.564]

proved experiment was carried out [1.593, 594]. The main difference was that the atomic beam was placed inside the Fabry-Perot interferometer which analysed the frequency spectrum of the fluorescence. In such an arrangement the observed signal intensity is enhanced by a factor equal to the "finesse" of the Fabry-Perot [1.595, 596] which was tuned piezo-electrically. The dye laser used in the experiment had a frequency distribution of about 1 MHz and a frequency drift of less than 1 MHz/min. The maximum power density used was about 3500 W/cm^3. The experiment gives a three-peak structure which is in rather good agreement with some of the theories [1.581, 587, 590, 591]. The distance of the side maxima from the central maximum is determined by the Rabi nutation frequency which depends on the matrix element of the transition. The observed structure is therefore influenced by the polarization of the laser readiation. This could also be shown in the experiment [1.594].

For small laser intensities the fluorescence spectrum should exhibit a sharp line [1.576] with a width which is determined by the laser and not by the natural transition. This could not be demonstrated in the Na measurements since the residual Doppler width of the beam and the resolution of the Fabry-Perot gave a resulting width of not better than 10 MHz. However, similar measurements have been performed on the $6s6p\ ^1P_1 \rightarrow 6s^2\ ^1S_0$ transition of ^{138}Ba [1.594]; in this case the linewidth of the fluorescent light observed for low laser intensities was 10 MHz, too. This is smaller than the natural width of the transition being 20 MHz. It supports the prediction of HEITLER [1.576] that at low intensities the spectral width of the fluorescence may be smaller than the natural width.

In connection with high resolution spectroscopy many modifications of the atomic-beam technique have been used. For instance, the optical excitation of the atom changes the population distribution between the Zeeman substates of the fundamental state. This change can then be detected by means of a Stern-Gerlach field or a six-pole magnet which deflects the atoms according to their effective magnetic dipole moments. The atoms may be monitored by means of the standard detectors as, e.g., a hot wire detector. By such a method a non-optical detection of the optical excitation is achieved.

An experiment of this type was recently performed by HUBER et al. [1.568] on the isotope shift of the radioactive ^{21,22,24,25}Na isotopes, which were produced by the spallation reaction ^{27}Al$(p, 3pxn)$ Na with 150-MeV protons. The total beam leaving the oven contained about 10^8 radioactive atoms/sec. The beam was monitored by means of a surface ionizer and a mass spectrometer so that the signal of the different isotopes which were produced simultaneously could be separated. The hyperfine splitting and isotope shift of the radioactive isotopes was measured (see Fig. 1.26). It should be mentioned that the principle

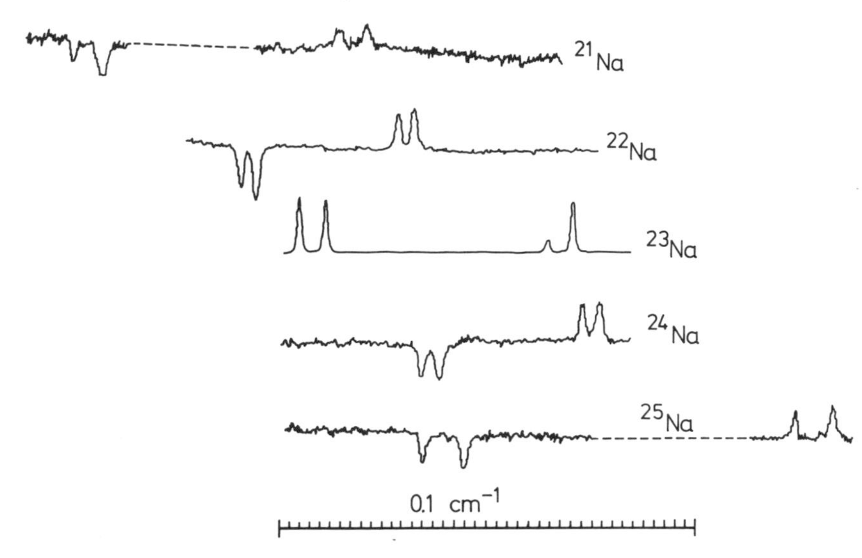

Fig. 1.26. Hyperfine structure splitting and isotope shift of radioactive Na isotopes. The isotopes have been measured individually relative to ^{23}Na. In the figure they are shown together on a common wave-number scale. (From [1.568])

of this method was proposed many years ago by MARRUS et al. [1.569]. At that time no tunable lasers have been available so that the method could not be generally applied. A similar non-optical detection may be obtained with the atomic beam deflection by radiation pressure, which has already been discussed in Subsection 1.2.2. Since a single atom may absorb many photons, the deflection due to radiation pressure may be rather large. In general, detection of atomic absorption via deflection does not give a better sensitivity than detection via fluorescence. However, when the atoms are radioactive and their lifetime is of the order of several seconds to minutes, their radioactivity may be used to monitor the particles. It should, however, be mentioned that not all elements are suitable for this type of investigation since it is based on multiple excitations; many atoms, especially those of the iron group or rare earths are pumped into metastable levels during the excitation process; they therefore get lost for a reexcitation.

The deflection of an atomic beam by a tunable dye laser is demonstrated in Fig. 1.27. The measurement was performed as described in [1.378]; the apparatus, however, was considerably improved [1.570]. High-resolution spectroscopy by means of radiation pressure has also been demonstrated [1.494]. In this experiment, the atoms were excited once during their interaction with the laser beam, therefore only a very small deflection was obtained.

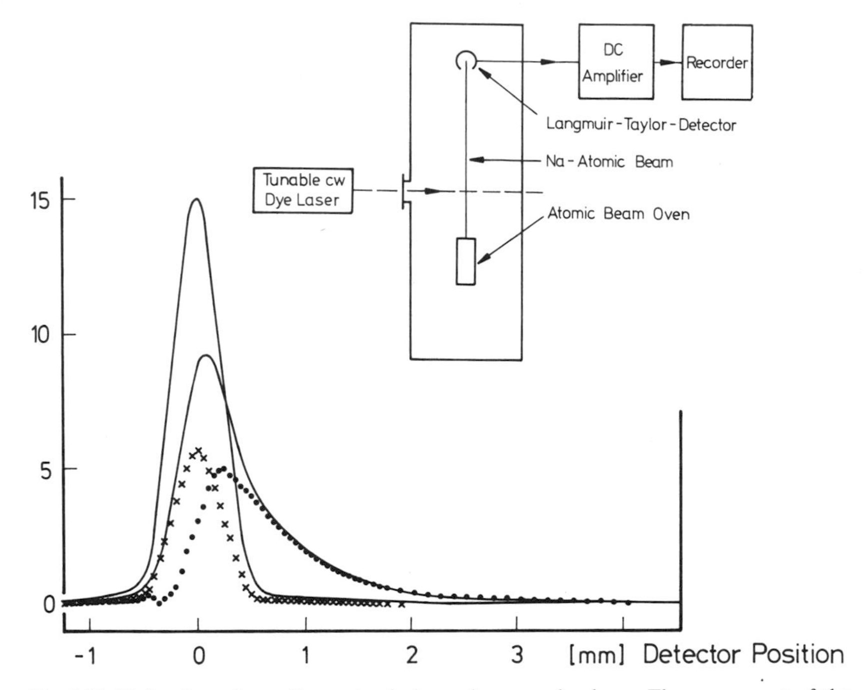

Fig. 1.27. Deflection of a sodium atomic beam by a cw dye laser. The upper part of the figure shows the experimental set-up. The laser was tuned to the hyperfine transition $^2P_{3/2}$, $F'=1 \rightarrow {}^2S_{1/2}$, $F=2$. In the lower part the detector signal is plotted versus the detector position. The symmetric and unsymmetric solid curve is obtained without and with irradiation by the dye laser, respectively. The unsymmetric signal is a superposition of undeflected (crosses) and deflected atoms (points). The undeflected part of the beam consists of atoms in the $F=1$ hyperfine level of the ground state. Thus the light beam performs a state selection of the atoms. It should be mentioned that the deflection induced by the laser beam of roughly 1 mm diameter is much larger than that achievable by hexapole fields (from [1.570])

An interesting possibility which is also based on radiation pressure was pointed out by HÄNSCH and SCHAWLOW [1.571]. If a low-density gas is illuminated with intense monochromatic light confined to the lower-frequency half of the Doppler profile of a resonance line, the velocity of the atoms is reduced since kinetic energy is transferred from the gas to the scattered light. This possibility of cooling by monochromatic light shows that such a radiation is thermodynamically equivalent to mechanical work or electricity rather than heat energy.

The use of atomic beams for high-resolution spectroscopy has many advantages compared to the use of absorption cells, which are mostly employed for the methods discussed below, since highly refractory elements and compounds can also be investigated without particular difficulty. In addition, there is no need to worry about the material reacting with the windows of the cell. However, an investigation by means of a cell, in general, requires less material. This may be important if only small amounts of the substance to be investigated are available.

In connection with the frequency stabilization of lasers, it is noteworthy that the cell can be used for a long time without refilling, and that the systematic errors that can arise with respect to the reproducibility of the absolute frequency are smaller when a cell is used.

The methods that give a natural linewidth limited resolution even when the ensemble shows Doppler broadening are discussed below.

Fluorescence Line Narrowing

The most straightforward of these methods is fluorescent line narrowing, first introduced by SCHWEITZER et al. [1.572] and by FELD et al. [1.573]. SZABO [1.574] demonstrated that the method can also be used for inhomogeneously broadened lines in solids. For the general situation of an absorption-fluorescence sequence, the method is as follows: a velocity subgroup of the atoms or molecules is excited by means of a laser, if the fluorescence is observed in the same or in the opposite direction as the incoming radiation, and a sufficiently small opening angle is used. The spectral distribution of the fluorescent radiation is determined by the natural width of the excited state or by the spectral distribution of the laser. The wavelength of the fluorescent light investigated must be far away enough from that of the laser light to ensure that light of both wavelengths can be separated in a spectrograph. This means in effect that the investigation is limited to intermediate states high enough above the ground state. To obtain the best possible resolution, the observation must be performed by highly resolving spectrographs. This fact and the small opening angle necessary for the experiments means that, in general, intensity problems arise.

The method which will be discussed in the following avoids completely the use of a highly resolving spectrograph. It uses two laser beams which are transmitted through the vapor cell in opposite directions. In principle, it can be said that the first beam selects a certain velocity subgroup and the second beam probes the frequency distribution via a nonlinear interaction. The second beam thus replaces the spectrograph.

Saturated Absorption

The principle of the methods may be explained by means of Fig. 1.28a. Two laser beams coming from the same laser source are transmitted in

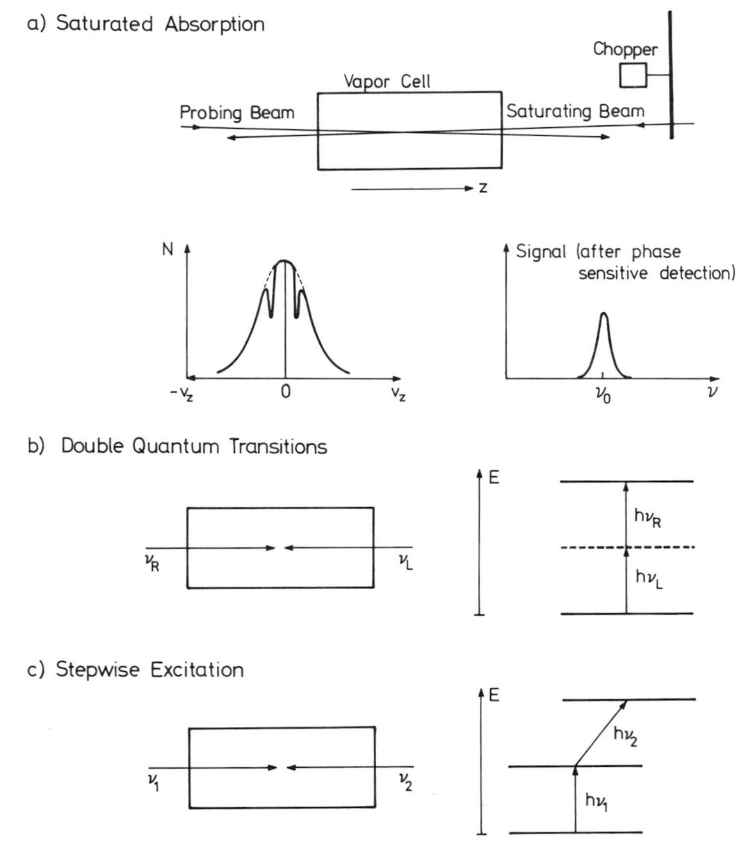

Fig. 1.28a–c. Methods for high-resolution spectroscopy using the simultaneous interaction of the atoms and molecules with two laser beams. For the methods (a) and (b) there is a nonlinear interaction between the two light beams. For method (c) the residual Doppler width is determined by the frequency difference $(\nu_2 - \nu_1)$

opposite directions through the vapor cell. Thus, when the laser frequency does not coincide with the frequency of the transition, the two beams interact with different velocity subgroups of the atoms or molecules in the cell. These are excited resulting in a reduction of their number N in the fundamental state. When the laser frequency coincides with the frequency of the transition, both beams interact with the same subgroup. Due to nonlinear absorption of the laser light by the atoms or molecules, the transmission for each laser beam may be influenced by the other. Therefore, if one beam is modulated, as shown in Fig. 1.28a, the second beam is also slightly modulated. The nonlinear interaction of the two laser beams with the material in the cell can then be traced via phase-sensitive detection of the signal modulation of the second beam. The modulation occurs only when there is a nonlinear interaction with the same subgroup, a linear interaction would, of course, produce no modulation.

The effect was first predicted by LAMB [1.597] in connection with his theory of gas lasers. There the phenomenon is observed as a power dip in the tuning characteristic of a gas laser. The experimental proof was published by two groups [1.598, 599]. The significance of the effect for high-resolution spectroscopy was realized in 1969 when the first successful experiments with absorption cells have been published. The set-up using a chopped saturating beam (Fig. 1.28) was first used by HÄNSCH et al. [1.516] and has proved to be quite useful. It gives a sensitivity which is higher than that obtained by a set-up where the transmission for one of the beams is simply monitored as a function of the laser frequency. The saturated absorption is of great importance for laser stabilization. Here the absorption cell is generally placed inside the laser cavity where a standing wave is present; this wave may be thought of as two waves of the same intensity travelling in opposite directions. The saturation signal is observed from an increase in the intensity of the laser output; the signal can then be used to lock the laser frequency (e.g. [1.603, 606]). See also Section 6.1.

In the set-up shown in Fig. 1.28a the transmission increase $\Delta\alpha_{ab}$ observed as saturation signal may be described by the following formula [1.516]

$$\frac{\Delta\alpha_{ab}}{\alpha_{ab}} = -\frac{1}{2}\frac{I}{I_{\text{sat}}}\frac{\gamma_{ab}^2}{\gamma_{ab}^2+(\Omega_{ab}-\omega)^2}, \tag{1.10}$$

where α_{ab} is the absorption coefficient for the unmodulated beam, I the intensity of the chopped beam, Ω_{ab} the frequency, and γ_{ab} the natural width of the transition. The saturation parameter I_{sat} depends on the matrix element d_{ab} for the transition, on the lifetimes τ_a and τ_b of the levels involved, and on the rate A_{ab} of the direct and cascade spontaneous transitions connecting the upper and lower states

$$I_{\text{sat}}^{-1} = \frac{4\pi}{\hbar^2 c}\frac{|d_{ab}|^2}{3}\gamma_{ab}^{-1}(\tau_a+\tau_b-A_{ab}\tau_a\tau_b)\,. \tag{1.11}$$

The saturated absorption method can be improved by observing fluorescence instead of absorption. When the absorption is reduced by saturation, the fluorescence is, too. This fact was used by FREED and JAVAN [1.604] to detect saturation peaks. This detection method is especially useful at low pressures where the total absorption of the atoms or molecules is small. An experimental set-up first introduced by SOREM and SCHAWLOW [1.517] is especially useful in this connection: the two laser beams, which are transmitted through the cell in exactly opposite directions, are modulated with frequencies ω_1 and ω_2. The atoms or molecules moving perpendicular to the laser beams undergo a nonlinear interaction with both beams; therefore the saturation signal must also show a modulation at the sum frequency $\omega_1+\omega_2$. There are also narrow resonance terms modulated at the frequencies ω_1, ω_2, and zero, but each of these is accompanied by a large background. The fluorescent power at the sum frequency is given by [1.517]

$$F(\omega_1+\omega_2) \propto A \cdot L\Gamma \frac{I_0^2}{I_{\text{sat}}} \cos(\omega_1+\omega_2)\, t \frac{\gamma_{ab}^2}{\gamma_{ab}^2+(\Omega_{ab}-\omega)^2}, \tag{1.12}$$

where I_0 is the intensity, and A the cross section of the laser beams, L is the length of the section of the beams observed in the experiment, and Γ is the absorption coefficient. The formula is valid when the Doppler width is much greater than the homogeneous linewidth: $\gamma_{ab}, I < I_{\text{sat}}$ and $\Gamma L < 1$. The ratio of the power at the frequencies ω_1, ω_2 to the dc fluorescence background is independent of ΓL. The similar ratio in the case of the saturated absorption is proportional to ΓL in the limit $\Gamma L < 1$. Thus, this method which is called *intermodulated fluorescence* has a strong advantage for experimenters working with very weak transitions, very low particle densities, or a poorly populated lower state.

There exist a variety of effects which may be observed in connection with saturated absorption. They cannot be discussed here. For a more complete survey it is referred to [1.605].

The experiments performed by means of saturated absorption are compiled in Table 1.10. As examples, two of these experiments shall briefly be discussed in the following.

The first one deals with the precision measurement of the Rydberg constant by measuring the absolute wavelengths of the Balmer α lines in hydrogen and deuterium. The experiment was performed by HÄNSCH et al. [1.519]. Hydrogen and deuterium was generated electrolytically and pumped through a dc discharge tube, where it was excited to the atomic $n=2$ state. The saturated absorption signal was observed in a certain section of the positive column. The laser was a nitrogen-laser-pumped dye laser. The linewidth was reduced by an external spherical

Table 1.10. Saturation spectroscopy

Atom or molecule	Transitions	Laser	Direct information	Remarks and references
I_2	${}^1\Sigma_g^+ - {}^3\Pi_{0u}^+$	cw dye laser	Hyperfine splitting	[1.509]
	${}^1\Sigma_g^+ - {}^3\Pi_{0u}^+$	Ar^+ Kr^+	Hyperfine splitting	[1.510]
		HeNe	Hyperfine splitting	[1.511]
	${}^1\Sigma_g^+ - {}^3\Pi_{0u}^+$	Ar^+	Hyperfine splitting	Intermod. fluoresc. [1.512]
	$X\,{}^1\Sigma_{0u}^+ - B\,{}^3\Pi_{0g}^+$	HeNe	Hyperfine splitting	[1.513]
	$X\,{}^1\Sigma_{0u}^+ - B\,{}^3\Pi_{0g}^+$	HeNe	Hyperfine splitting	[1.514]
	$X\,{}^1\Sigma_{0g}^+ - B\,{}^3\Pi_{0u}^+$	Ar^+	Hyperfine splitting	[1.515]
	$X\,{}^1\Sigma_{0g}^+ - B\,{}^3\Pi_{0u}^+$	Kr^+	Hyperfine splitting	[1.516]
	$P(13)\ R(15)\ (43-0)$	Ar^+	Hyperfine splitting	Intermod. fluoresc. [1.517]
H	H_α	Pulsed dye laser	Lamb shift	[1.518]
H, D	H_α, D_α	HeNe	Fine structure	Precision measurement of Rydberg constant [1.519]
Na	D_1 and D_2 line	Pulsed dye laser	Hyperfine structure	[1.523]
Ne	$1s_5 - 2s_2$	cw dye laser	Relativistic Doppler shift	[1.615]
CH_4	$v_3=0-v_3=1$	HeNe	Zeeman splitting	Level crossing [1.520]
	$v_3=0-v_3=1$	HeNe	Zero field level crossing	[1.521]
	$v_3=0-v_3=1$	HeNe	Hyperfine structure	[1.522]
BO_2	$A\,{}^2\Pi_u - X\,{}^2\Pi_g$	Ar^+	Hyperfine splitting	Intermod. fluoresc. [1.524]
SF_6	v_3	CO_2	Rot.-vib. splitting	[1.525, 600, 601]
H_2O	$v_2(5_{3,2}-6_{4,3})$	Spin-flip Raman laser	Line width of lamb dip	[1.526]
NH_2D	v_2	CO_2	Stark effect	[1.602]

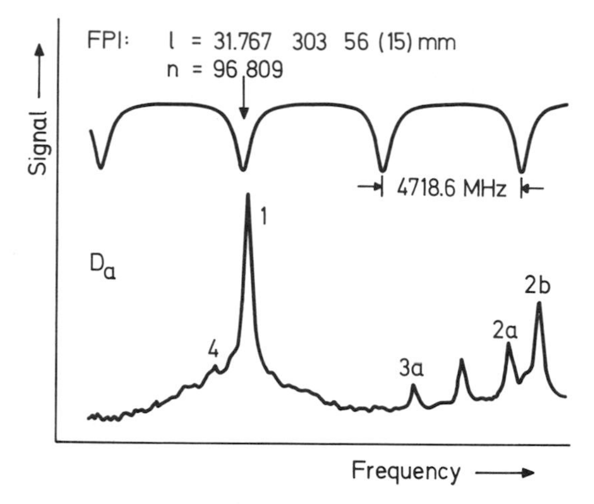

Fig. 1.29. Saturation spectrum of Balmer line D_α with simultaneously recorded transmission maxima of the Fabry-Perot interferometer, whose spacing was accurately known in terms of the HeNe-laser standard wavelength. l is the spacer length and n the order of the interferometer. Line 1 represents the $2\,^2P_{3/2}-3\,^2D_{5/2}$ transition, which was used for the absolute wavelength measurement because it exhibits the smallest hyperfine splitting. (From [1.519])

Fabry-Perot providing a linewidth of 30 MHz. For the absolute wavelength determination the dye laser light was sent through another Fabry-Perot interferometer with an accurately known spacing (Fig. 1.29). Using the experimental line position of the Balmer α line, the Balmer formula which must be corrected for finite nuclear mass, Dirac fine structure, and Lamb shifts [1.607], the Rydberg constant was calculated. The new value $R_\infty = 109\,737.3143(10)\,\mathrm{cm}^{-1}$ is an order of magnitude more accurate than the previous determinations.

The linewidth observed in a saturated absorption experiment for vibrational transitions is mostly determined by the interaction time between the laser beam and the molecules; the natural width of the vibrational transitions is usually much smaller. Therefore a considerable reduction of the observed linewidth can be expected when an expanded laser beam is used. A remarkable resolution was obtained in this way in an experiment on the hyperfine structure of the $F_2^{(2)}$ component in the $P(7)$ line of the ν_3 vibration-rotation band of CH_4. The $F_2^{(2)}$ component which results from Coriolis fine structure of the $P(7)$ line is usually taken to stabilize the 3.39 μm line of a HeNe-laser. By extending the laser beam to 5 cm diameter, in a 13 m long cell containing $^{12}CH_4$ a linewidth of 6 kHz was achieved [1.522]. This corresponds to a resolution of $1.4 \cdot 10^{10}$. To achieve the necessary frequency stability during the measurement, it is essential to control the laser frequency electronically. The concept used for this is the frequency offset locking [1.608, 609] whereby the stability of a reference laser servo-stabilized to methane is transferred to a high power laser through the use of a frequency servo. The high power laser is then used to measure the line structure.

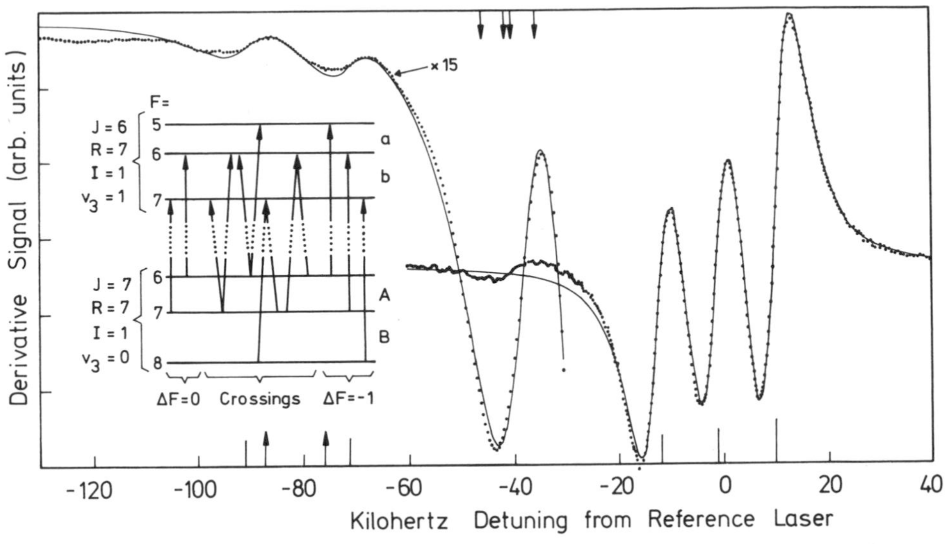

Fig. 1.30. Saturated absorption signal of the hyperfine structure belonging to the $F_2^{(2)}$ component (Coriolis fine-structure) in the $P(7)$ line of the ν_3 vibration-rotation band of CH_4. The optical resolution is above 10^{10}. (From [1.575])

The measured hyperfine structure is shown in Fig. 1.30. From theory three lines for which $\Delta F = -1$ with intensities which increase with F are expected. The 9.7 kHz line is the $8-7$ transition, -1.7 kHz is the $7 \rightarrow 6$, and -12.8 kHz the $6 \rightarrow 5$ transition. Besides these one expects much weaker satellite lines with $\Delta F = 0$. These are the two weak low frequency lines at the left side. In addition to these real transitions, saturated absorption also provides additional signals which occur when two transitions share a common level. They are half way between the real transitions and are called Doppler-generated level crossings [1.605]. Recently, in a similar experiment the laser beam has been expanded to a diameter of 30 cm. With the increased precision it is now possible to see the recoil of the radiation on the molecules [1.612, 613]. This causes lines to split in doublets with a separation of about 2.7 kHz [1.610, 611].

Two-Photon Spectroscopy, Stepwise Excitation

It was pointed out by VASILENKO et al. [1.372] that two-photon transitions may be suitable to eliminate the Doppler width. To understand this effect we assume that the atoms or molecules in a vapor cell (Fig. 1.28b) are excited by two laser beams with frequency ν travelling in opposite direction. An atom moving in the cell with a velocity component v_z sees the frequencies of the two laser beams Doppler shifted by the amount $\nu(1 - v_z/c)$ and $\nu(1 + v_z/c)$, respectively. If the atom performs the two-photon transition (Fig. 1.28b, right side) by absorbing one photon from each of the two beams the influence of the Doppler shift is cancelled since $\hbar\nu(1 + v_z/c) + \hbar\nu(1 - v_z/c) = 2\,\hbar\nu$. [The second order Doppler shift which is proportional to $(v_z/c)^2$ is, of course, neglected.] The essential point is that the Doppler width is compensated for all atoms; therefore the whole ensemble which is illuminated by the laser beams contributes to the signal. The two-photon resonance is usually monitored via the subsequent fluorescence.

If the polarization of the two laser beams is identical, two photons from the same as well as from different beams may be absorbed with the same probability. Since the two-photon transition with photons from the same beam is not Doppler-free a broad background signal is observed together with the sharp transition. Using different polarizations for the two beams (σ^+ and σ^-) eliminates the broad banded background if the two photon transition is only allowed with σ^+ and σ^- photons absorbed simultaneously.

The two-photon spectroscopy complements the saturation spectroscopy since forbidden transitions may be investigated. Since levels can be investigated having an energy twice as large as the photon energy, the energy range accessible by laser excitation is considerably increased.

It was pointed out by CAGNAC et al. [1.373] that the two-photon experiments can be performed even with modest laser powers. The corresponding transition rates may be rather large if an intermediate state of opposite parity exists which has almost half the energy of the two-photon transition [1.616].

The first Doppler-free two-photon experiments have been performed with pulsed dye lasers [1.527, 528] and almost simultaneously with a cw dye laser [1.529], too. Meanwhile the method has been used for many investigations which are compiled in Table 1.11. If the two photon experiments are performed with high power densities it is always necessary to check the influence of the quadratic Stark effect (Subsect. 1.1.5), since this may be well present and affect the result.

A special case of the two-photon transition is the stepwise excitation (Fig. 1.28c). In this type of experiment the two laser beams have different frequencies: the first beam induces the transition to an intermediate level and the second to the final level. In this set-up the atoms in the cell also interact with the two beams simultaneously. However, the Doppler shift is not completely compensated because the frequencies of the two beams are not the same. The residual Doppler shift is deter-

Table 1.11. High resolution measurements using two-photon spectroscopy

Atom or molecule	Transition	Laser	Direct information	Remarks and references
Na	$3\,^2S_{1/2}-5\,^2S_{1/2}$	Pulsed dye laser	Hyperfine splitting	[1.527]
	$3\,^2S_{1/2}-5\,^2S_{1/2}$	Pulsed dye laser	Hyperfine splitting	[1.528]
	$3\,^2S_{1/2}-4\,^2D_{3/2,\,5/2}$	cw dye laser	Fine structure splitting	[1.529]
	$3\,^2S_{1/2}-5\,^2S_{1/2}$	Pulsed dye laser	Zeeman splitting	[1.530]
	$3\,^2S_{1/2}-5\,^2D_{3/2,\,5/2}$	Pulsed dye laser	Fine structure splitting	[1.531]
	$3\,^2S_{1/2}-6\,^2S_{1/2}$	Pulsed dye laser	Hyperfine splitting	
	$3\,^2S_{1/2}-5\,^2S_{1/2}$	cw dye laser	Hyperfine splitting	[1.532]
	$3\,^2S_{1/2}-4\,^2D_{3/2,\,5/2}$	cw dye laser	Fine structure splitting	[1.533]
	$3\,^2S_{1/2}-4\,^2D_{3/2,\,5/2}$	cw dye laser	Fine structure splitting, Paschen-Back effect	[1.534]
	$3\,^2S_{1/2}-4\,^2D_{3/2,\,5/2}$	cw dye laser	Collision broadening and pressure shift	[1.535]
	$3\,^2S_{1/2}-4\,^2D_{3/2,\,5/2}$	cw dye laser	Excitation transfer by collisions	[1.536]
	$3\,^2S_{1/2}-4\,^2D_{3/2}$	cw dye laser	Stark effect	[1.614]
	$3\,^2S_{1/2}-5\,^2D_{3/2}$			
H, D	$1\,^2S_{1/2}-2\,^2S_{1/2}$	Frequency doubled pulsed dye laser	Lamb shift of $1\,^2S_{1/2}$ level	[1.537]
Ne	$3s$, $J=2-4d'(5/2)$, $J=3$	cw dye laser	Isotope shift, hyperfine splitting	[1.567]

mined by the difference in frequency and is given by

$$\delta\nu = (\nu_2 - \nu_1)/(\nu_1 + \nu_2) \cdot \delta\nu_{\nu_1+\nu_2}, \tag{1.13}$$

where $\delta\nu_{\nu_1+\nu_2}$ is the Doppler width that would correspond to that for the direct transition with the frequency $\nu_1 + \nu_2$. If there is splitting of the intermediate level, and ν_1 is in resonance, the structure observed for the splitting of the upper level may be complicated. This, however, is not a theoretical limitation for the method.

In the following two experiments performed with two-photon absorption shall be discussed briefly. The first one was performed by BIRABEN et al. [1.536]. In this measurement the pressure shift and the broadening of the $3\,^2S_{1/2} - 4\,^2D_{3/2}$ transition of Na by neon atoms was investigated. The experiment demonstrates that rather small shifts can be measured. In addition, the two-photon spectroscopy gives the possibility to investigate transitions between levels of the same parity.

In the experiment two cells, one filled with pure sodium and the other with sodium and neon, were irradiated with the light of the same cw dye laser. The laser light is reflected back into the cells by means of two mirrors in order to obtain the second beam necessary for the two-photon experiments. The fluorescence of the two cells was recorded simultaneously during the frequency scan of the laser; in this way a very good accuracy is obtained (Fig. 1.31).

Another remarkable two photon experiment was performed by HÄNSCH et al. [1.537]. This deals with the determination of the $1s-2s$ transition of hydrogen and deuterium. The laser used was a nitrogen-laser-pumped dye laser consisting of a pressure-tuned dye-laser oscillator with optical confocal-filter interferometer and two subsequent dye-laser amplifier stages. The bandwidth of the laser with confocal filter was 120 MHz at an output peak power of 30–50 kW. To induce the double quantum transition radiation at 2430 Å is necessary (Fig. 1.32). This was generated by frequency doubling the laser radiation with a lithium formate monohydrate crystal giving a peak power of 600 W. The hydrogen and deuterium atoms were produced in a Wood-type discharge and then brought into a Pyrex tube where the two photon transitions have been induced by the focused laser light. The two photon transitions have been monitored via the L_α radiation (Fig. 1.32). Simultaneously to the L_α fluorescence also the absorption spectrum of the Balmer-β line at 4860 Å was recorded using the direct dye laser output. In this way a precise comparison between the energy intervals $1S-2S$ and $2S, P-4S, P, D$ could be performed resulting in the Lamb shift of the $1S$ ground state. The results are 8.3 ± 0.3 GHz (deuterium) and 8.6 ± 0.8 GHz (hydrogen). The only previous measurement of the Lamb

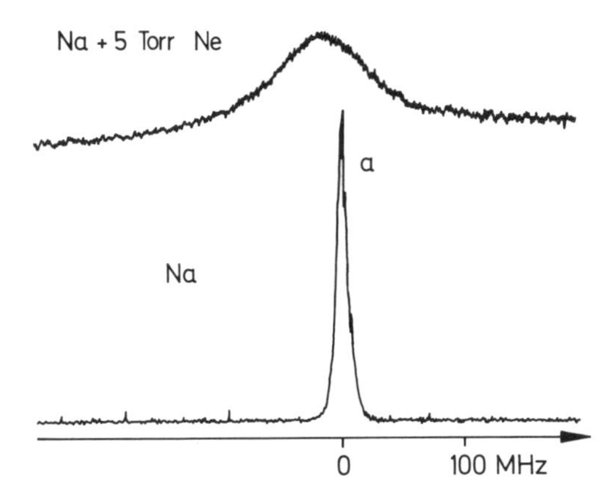

Fig. 1.31. Pressure shift of the Na transition between the $F=2$ hyperfine level of the ground state and the $4\,^2D_{5/2}$ level measured by means of two photon transitions. The upper curve exhibits the pressure shift and pressure broadening due to 5 Torr Ne. The shift is to smaller frequencies and larger wavelengths. (From [1.535])

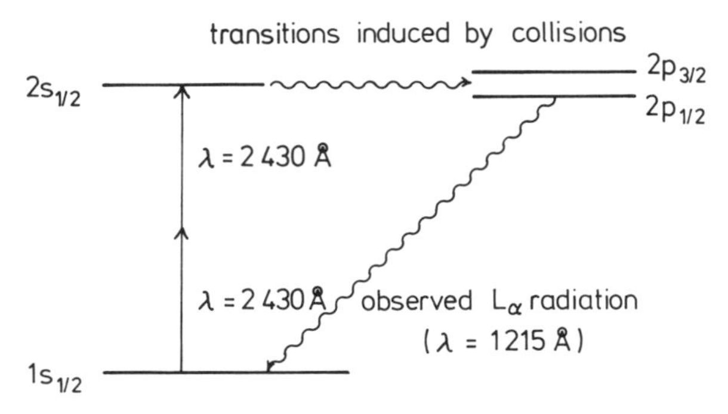

Fig. 1.32. Measurement of the-two photon transition $1s_{1/2}-2s_{1/2}$ in hydrogen. The 2430 Å radiation was generated by frequency doubling dye laser radiation. The two photon transition was minitored via the L_α radiation. (From [1.537])

shift of the $1S$ state of deuterium was performed by HERZBERG [1.617] from an absolute wavelength measurement of the L_α line. The result was 7.9 ± 1.16 GHz.

References

1.1. W. DEMTRÖDER: Phys. Rep. **7**, 223 (1973); *Laser Spectroscopy*, 2nd ed. (Springer Berlin, Heidelberg, New York, 1973).

1.2. *Laser Handbook*, Vol. 1, ed. by F. T. ARECCHI and E. O. SCHULZ-DUBOIS (North-Holland Publ. Co. Amsterdam 1972).

1.3. H. BRUNET: IEEE J. Quant. Electron. QE-**2**, 382 (1966).

1.4. K. M. EVENSON, H. P. BROIDA, J. S. WELLS, R. J. MAHLER, M. MIZUSHIMA: Phys. Rev. Lett. **21**, 1038 (1968).
1.5. T. SHIMIZU, K. SHIMODA: IEEE J. Quant. Electron. QE-**4**, 728 (1968).
1.6. A. D. DEVIR, U. P. OPPENHEIM: Appl. Opt. **8**, 2121 (1969).
1.7. K. M. EVENSON, C. J. HOWARD: In *Laser Spectroscopy*, ed. by R. G. BREWER and A. MOORADIAN (Plenum Press, New York, London, 1974).
1.8. Topics in Applied Physics, Vol. 1: *Dye Lasers*, ed. by F. P. SCHÄFER (Springer, Berlin, Heidelberg, New York, 1973).
1.9. R. A. KELLER, E. F. ZALEWSKI, N. C. PETERSON: J. Opt. Soc. Am. **62**, 319 (1972).
1.10. T. W. HÄNSCH, A. L. SCHAWLOW, P. E. TOSCHEK: IEEE J. Quant. Electron. QE-**8**, 802 (1972).
1.11. R. A. KELLER, J. D. SIMMONS, D. A. JENNINGS: J. Opt. Soc. Am. **63**, 1552 (1973).
1.12. L. S. GOODMAN, W. J. CHILDS: Paper given at 4th Intern. Conf. Atomic Phys., Heidelberg (July, 1974).
1.13. N. C. PETERSON, M. J. KURYLO, W. BRAUN, A. M. BASS, R. A. KELLER: J. Opt. Soc. Am. **61**, 746 (1971).
1.14. M. B. KLEIN: Opt. Commun. **5**, 114 (1972).
1.15. G. H. ATKINSON, A. H. LAUFER, M. J. KURYLO: J. Chem. Phys. **59**, 350 (1973).
1.16. R. J. TRASH, H. VON WEYSSENHOFF, J. S. SHIRK: J. Chem. Phys. **55**, 4659 (1971).
1.17. T. J. MCILRATH: Appl. Phys. Lett. **15**, 41 (1969).
1.18. T. J. MCILRATH: J. Phys. B (Atom. Mol. Phys.) **7**, 393 (1974).
1.19. D. J. BRADLEY, G. M. GALE, P. D. SMITH: J. Phys. B**3**, 11 (1970).
1.20. D. J. BRADLEY, P. EWART, J. V. NICHOLAS, J. R. D. SHAW: In *Laser Spectroscopy*, ed. by R. G. BREWER and A. MOORADIAN (Plenum Press, New York, London, 1974).
1.21. K. W. ROTHE, A. TÖNNISSEN, H. WALTHER: Unpublished results.
1.22. T. HIRSCHFELD, E. R. SCHILDKRAUT, H. TANNENBAUM, D. TANNENBAUM: Appl. Phys. Lett. **22**, 38 (1973).
1.23. G. S. KENT, R. W. H. WRIGHT: J. Atmosph. Terr. Phys. **32**, 917 (1970).
1.24. R. T. H. COLLIS, E. E. UTHE: Opto-Electron. **4**, 87 (1972).
1.25. M. C. W. SANDFORD, A. J. GIBSON: J. Atmosph. Terr. Phys. **32**, 1423 (1970).
1.26. J. CONNEY: Appl. Opt. **11**, 2374 (1972).
1.27. P. A. DAVIS: Appl. Opt. **8**, 2099 (1969).
1.28. CH. S. COOK, G. W. BETHKE, W. D. CONNOR: Appl. Opt. **11**, 1742 (1972).
1.29. H. INABA, T. KOBAYASI: Opto-Electron. **4**, 101 (1972).
1.30. R. M. MEASURES, G. PILON: Opto-Electron. **4**, 141 (1972).
1.31. S. A. AHMED: Appl. Opt. **12**, 901 (1973).
1.32. R. L. BYER, M. GARBUNY: Appl. Opt. **12**, 1496 (1973).
1.33. K. W. ROTHE, U. BRINKMANN, H. WALTHER: Appl. Phys. **3**, 115 (1974).
1.34. K. W. ROTHE, U. BRINKMANN, H. WALTHER: Appl. Phys. **4**, 181 (1974).
1.35. C. FREED, D. L. SPEARS, R. G. O'DONNELL, A. H. M. ROSS: In *Laser Spectroscopy*, ed. by R. G. BREWER and A. MOORADIAN (Plenum Press, New York, London, 1974).
1.36. R. VETTER: Phys. Lett. **31**A, 559 (1970).
1.37. J. H. SHAFER: Phys. Rev. A **3**, 752 (1971).
1.38. W. L. FAUST, R. A. MCFARLANE, C. K. N. PATEL, C. G. B. GARRET: Phys. Rev. **133**A, 1476 (1964).
1.39. W. LAMB: Phys. Rev. **134**A, 1429 (1964).
1.40. F. R. PETERSEN, D. G. MCDONALD, F. D. CUPP, B. L. DANIELSON: Phys. Rev. Lett. **31**, 573 (1973).
1.41. K. M. EVENSON, J. S. WELLS, F. R. PETERSEN, B. L. DANIELSON, G. W. DAY: Appl. Phys. Lett. **22**, 192 (1973).
1.42. W. TANGO, J. K. LINK, R. N. ZARE: J. Chem. Phys. **49**, 4264 (1968).

1.43. W. Demtröder, M. McClintock, R. N. Zare: J. Chem. Phys. **51**, 5495 (1969).
1.44. R. Velasco, Ch. Ottinger, R. N. Zare: J. Chem. Phys. **51**, 5522 (1969).
1.45. K. Bergmann, H. Klar, W. Schlecht: Chem. Phys. Lett. **12**, 522 (1972).
1.46. Ch. Ottinger, D. Poppe: Chem. Phys. Lett. **8**, 513 (1971).
1.47. R. N. Zare: J. Chem. Phys. **40**, 1934 (1964).
1.48. J. I. Steinfeld, J. D. Campbell, N. A. Weiss: J. Mol. Spectrosc. **29**, 204 (1969).
1.49. G. Ennen, Ch. Ottinger: Chem. Phys. **3**, 404 (1974).
1.50. S. E. Johnson, K. Sakurai, H. P. Broida: J. Chem. Phys. **52**, 6441 (1970).
1.51. G. Alzetta, A. Kopystynska, L. Moi: Lett. Nuovo Cim. **6**, 677 (1973).
1.52. M. M. Hessel: Phys. Rev. Lett. **26**, 215 (1971).
1.53. Ch. Ottinger: Chem. Phys. **1**, 161 (1973).
1.54. W. Holzer, W. F. Murphy, H. J. Bernstein: J. Chem. Phys. **52**, 469 (1970).
1.55. W. Holzer, W. F. Murphy, H. J. Bernstein: J. Chem. Phys. **52**, 399 (1969).
1.56. K. Sakurai, H. P. Boida: J. Chem. Phys. **54**, 1217 (1971).
1.57. R. F. Curl, Jr., K. Abe, J. Bissinger, C. Bennet, F. K. Tittel: J. Mol. Spectrosc. **48**, 72 (1973).
1.58. K. Sakurai, H. P. Broida: J. Chem. Phys. **50**, 557 (1969).
1.59. K. Sakurai, H. P. Broida: J. Chem. Phys. **53**, 1615 (1970).
1.60. E. Menke: Z. Naturforsch. **25**a, 442 (1970).
1.61. J. Vigue, J.-C. Lehmann: Chem. Phys. Lett. **16**, 385 (1972).
1.62. G. D. Chapman, P. R. Bunker: J. Chem. Phys. **57**, 2951 (1972).
1.63. M. Broyer, J. Vigue, J. C. Lehmann: Chem. Phys. Lett. **22**, 313 (1973).
1.64. R. B. Kurzel, E. O. Degenkolb, A. J. Steinfeld: J. Chem. Phys. **56**, 1784 (1972).
1.65. S. E. Johnson, G. Capelle, H. P. Broida: J. Chem. Phys. **56**, 663 (1972).
1.66. K. Sakurai, S. E. Johnson, H. P. Broida: J. Chem. Phys. **52**, 1625 (1970).
1.67. R. H. Barnes, C. E. Moeller, J. F. Kircher, C. M. Verber: Appl. Opt. **12**, 2531 (1973).
1.68. W. N. Jackson: J. Chem. Phys. **59**, 960 (1973).
1.69. J. S. Shirk, A. M. Bass: J. Chem. Phys. **52**, 1894 (1970).
1.70. R. G. Bray, R. M. Hochstrasser, J. E. Wessel: Chem. Phys. Lett. **27**, 167 (1974).
1.71. E. L. Baardsen, R. W. Terhune: Appl. Phys. Lett. **21**, 209 (1972).
1.72. D. K. Russel, M. Kroll, D. A. Dows, R. A. Beaudet: Chem. Phys. Lett. **20**, 153 (1973).
1.73. K. Sakurai, H. P. Broida: J. Chem. Phys. **50**, 2404 (1969).
1.74. K. Abe, F. Meyers, T. K. MacCubbin, S. R. Polo: J. Mol. Spectrosc. **38**, 552 (1971).
1.75. C. G. Stevens, M. W. Swagel, R. Wallace, R. N. Zare: Chem. Phys. Lett. **18**, 465 (1972).
1.76. A. A. Offenberger, L. M. Lidsky, D. J. Rose: J. Appl. Phys. **43**, 2257 (1972).
1.77. W. W. Morey, C. M. Penny: J. Opt. Soc. Am. **63**, 1298 (1973).
1.78. R. B. Kurzel, J. I. Steinfeld: J. Chem. Phys. **53**, 3293 (1970).
1.79. K. Bergmann, W. Demtröder: J. Phys. B (Atom. Mol. Phys.) **5**, 1386 (1972).
1.80. K. Bergmann, W. Demtröder: J. Phys. B (Atom. Mol. Phys.) **5**, 2098 (1972).
1.81. Ch. Ottinger, R. Velasco, R. N. Zare: J. Chem. Phys. **52**, 1636 (1970).
1.82. H. Klar: J. Phys. B (Atom. Mol. Phys.) **6**, 2136 (1973).
1.83. R. Wallenstein, T. W. Hänsch: Appl. Opt. **13**, 1625 (1975).
1.84. T. W. Hänsch: Appl. Opt. **11**, 895 (1972).
1.85. H. Figger, F. Jüptner, H. Walther, J. Heldt: Unpublished material.
1.86. G. K. Klauminzer: Opt. Engng. **13**, 528 (1974).
1.87. J. M. Green, J. P. Hohimer, F. K. Tittel: Opt. Commun. **9**, 407 (1973).
1.88. G. Marowski, F. K. Tittel, F. P. Schäfer: Opt. Commun. **13**, 106 (1975).
1.89. G. Marowski: Appl. Phys. Lett. **26**, 647 (1975).

1.90. C.S.WU: In *Atomic Physics*, Vol. 3, ed. by S.J.SMITH and G.K.WALTERS (Plenum Press, New York, London 1973).
1.91. E.ZAVATTINI: In *Atomic Physics*, Vol. 4 (Plenum Press, New York, London, 1975).
1.92. K.D.GARCIA, J.E.MACK: J. Opt. Soc. Am. **55**, 654 (1965).
1.93. S.BASHKIN Ed.): *Topics in Modern Physics*, Vol. 1: Beam Foil Spectroscopy (Springer, Berlin, Heidelberg, New York, 1975).
1.94. K.SAKURAI, G.CAPELLE, H.P.BROIDA: J. Chem. Phys. **54**, 1220 (1971).
1.95. M.MCCLINTOCK, W.DEMTRÖDER, R.N.ZARE: J. Chem. Phys. **51**, 5509 (1969).
1.96. L.ARMSTRONG, JR., S.FENEUILLE: J. Phys. B (Atom. Mol. Phys.) **8**, 546 (1975).
1.97. S.H.AUTLER, C.H.TOWNES: Phys. Rev. **100**, 703 (1955).
1.98. H.J.ANDRÄ, A.GAUP, W.WITTMANN: Phys. Rev. Lett. **31**, 501 (1973).
1.99. H.HARDE, G.GUTHÖHRLEIN: Phys. Rev. A **10**, 1488 (1974).
1.100. A.ARNESEN, A.BENGTSON, R.HALLIN, S.KANDELA, T.NORELAND, R.LINDHOLT: Phys. Lett. **53**A, 459 (1975).
1.101. H.FIGGER, K.SIOMOS, H.WALTHER: Z. Physik **270**, 371 (1974).
1.102. K.SIOMOS, H.FIGGER, H.WALTHER: Z. Physik A **272**, 355 (1975).
1.103. W.GORNIK, D.KAISER, W.LANGE, J. LUTHER, H.H.RADLOFF, H.H.SCHULZ: Appl. Phys. **1**, 285 (1973).
1.104. J.A.PAISNER, R.WALLENSTEIN: J. Chem. Phys. **61**, 4317 (1974).
1.105. G.A.CAPELLE, H.P.BROIDA: J. Chem. Phys. **58**, 4212 (1973).
1.106. K.SAKURAI, G.CAPELLE: J. Chem. Phys. **53**, 3764 (1970).
1.107. P.B.SACKETT, J.T.YARDLEY: Chem. Phys. Lett. **6**, 323 (1970).
1.108. P.B.SACKETT, J.T.YARDLEY: J. Chem. Phys. **57**, 152 (1972).
1.109. T.ERDMANN, H.FIGGER, H.WALTHER: Opt. Commun. **6**, 166 (1972).
1.110. W.GORNIK, D.KAISER, W.LANGE, J.LUTHER, K.MEIER, H.H.RADLOFF, H.H.SCHULZ: Phys. Lett. **45**A, 219 (1973).
1.111. P.SCHENCK, R.C.HILBORN, H.METCALF: Phys. Rev. Lett. **31**, 189 (1973).
1.112. W.GORNIK, D.KAISER, W.LANGE, J.LUTHER, H.H.SCHULZ: Opt. Commun. **6**, 327 (1972).
1.113. M.B.KLEIN, D.MAYDAN: Appl. Phys. Lett. **16**, 509 (1970).
1.114. J.HELDT, H.FIGGER, K.SIOMOS, H.WALTHER: Astron. Astrophys. (in press).
1.115. J.DUCUING, C.JOFFRIN, I.P.COFFINET: Opt. Commun. **2**, 245 (1970).
1.116. G.BAUMGARTNER, W.DEMTRÖDER, M.STOCK: Z. Physik **232**, 462 (1970).
1.117. W.J.TANGO, R.N.ZARE: J. Chem. Phys. **53**, 3094 (1970).
1.118. S.E.JOHNSON: J. Chem. Phys. **56**, 149 (1972).
1.119. H.DUBOST, L.ABOUAF-MARGUIN, F.LEGAY: Phys. Rev. Lett. **29**, 145 (1972).
1.120. W.H.GREEN, J.K.HANCOCK: J. Chem. Phys. **59**, 4326 (1973).
1.121. M.BROYER, J.C.LEHMANN: Phys. Lett. A **40**, 43 (1972).
1.122. P.F.WILLIAMS, D.L.ROUSSEAU, S.H.DWORETSKY: Phys. Rev. Lett. **32**, 196 (1974).
1.123. G.W.HOLLEMAN, J.T.STEINFELD: Chem. Phys. Lett. **12**, 431 (1971).
1.124. J.L.AHL, T.A.COOL: J. Chem. Phys. **58**, 5540 (1973).
1.125. J.K.HANCOCK, W.H.GREEN: J. Chem. Phys. **56**, 2474 (1972).
1.126. J.K.HANCOCK, W.H.GREEN: J. Chem. Phys. **59**, 6350 (1973).
1.127. J.K.HANCOCK, W.H.GREEN: J. Chem. Phys. **57**, 4515 (1972).
1.128. J.R.AIREY, S.F.FRIED: Chem. Phys. Lett. **8**, 23 (1971).
1.129. J.K.HANCOCK, W.H.GREEN: J. Chem. Phys. **56**, 2474 (1972).
1.130. H.L.CHEN, C.B.MOORE: J. Chem. Phys. **54**, 4072, 4080 (1971).
1.131. B.M.HOPKINS, H.L.CHEN, R.D.SHARMA: J. Chem. Phys. **59**, 5758 (1973).
1.132. B.M.HOPKINS, H.L.CHEN: 3rd Conf. on Chem. and Molec. Lasers, St. Louis, Mo., USA (May 1972).
1.133. N.C.CRAIG, C.B.MOORE: J. Phys. Chem. **75**, 1622 (1971).
1.134. P.F.ZITTEL, C.B.MOORE: Appl. Phys. Lett. **21**, 81 (1972).

1.135. M. Margottin Maclou, L. Doyenette, L. Henry: Appl. Opt. **10**, 1768 (1971).
1.136. B. M. Hopkins, H. L. Chen: Chem. Phys. Lett. **17**, 500 (1972).
1.137. P. F. Zittel, C. B. Moore: J. Chem. Phys. **59**, 6636 (1973).
1.138. H. L. Chen: J. Chem. Phys. **55**, 5551, 5557 (1971).
1.139. K. H. Becker, D. Haaks, T. Tatarczyk: Z. Naturforsch. **27**a, 1520 (1972).
1.140. P. J. Dagdigian, H. W. Cruse, R. N. Zare: J. Chem. Phys. **60**, 2330 (1974).
1.141. W. A. Rosser, Jr., A. D. Wood, E. T. Gerry: J. Chem. Phys. **50**, 4996 (1969).
1.142. W. A. Rosser, Jr., R. D. Sharma, E. T. Gerry: J. Chem. Phys. **54**, 1196 (1971).
1.143. J. C. Stephenson, R. E. Wood, C. B. Moore: J. Chem. Phys. **54**, 3097 (1971).
1.144. R. S. Chang, R. A. McFarlane, G. J. Wolga: J. Chem. Phys. **56**, 667 (1972).
1.145. J. C. Stephenson, R. E. Wood, C. B. Moore: J. Chem. Phys. **56**, 4813 (1972).
1.146. M. A. Kovacs, A. Javan: J. Chem. Phys. **50**, 4111 (1969).
1.147. D. F. Heller, C. B. Moore: J. Chem. Phys. **52**, 1005 (1970).
1.148. W. A. Rosser, Jr., A. D. Wood, E. T. Gerry: J. Chem. Phys. **50**, 4996 (1969).
1.149. J. C. Stephenson, C. B. Moore: J. Chem. Phys. **52**, 2333 (1970).
1.150. W. A. Rosser, Jr., E. T. Gerry: J. Chem. Phys. **54**, 4131 (1971).
1.151. M. Gueguen, I. Arditi, M. Margottin-Maclou, L. Doyenette, L. Henry: Compt. Rend. **272** B, 1139 (1972).
1.152. J. C. Stephenson, C. B. Moore: J. Chem. Phys. **56**, 1295 (1972).
1.153. J. C. Stephenson, J. Finzi, C. B. Moore: J. Chem. Phys. **56**, 5214 (1972).
1.154. L. E. Brus: Chem. Phys. Lett. **12**, 116 (1971).
1.155. K. Sakurai, G. Capelle, H. P. Broida: J. Chem. Phys. **54**, 1412 (1971).
1.156. A. Luntz, R. G. Brewer, K. L. Foster, J. D. Swalen: Phys. Rev. Lett. **23**, 951 (1969).
1.157. J. T. Yardley, M. N. Fertig, C. B. Moore: J. Chem. Phys. **52**, 1450 (1970).
1.158. J. T. Knudtson, G. W. Flynn: J. Chem. Phys. **58**, 2684 (1973).
1.159. F. R. Grabiner, G. W. Flynn, A. M. Ronn: J. Chem. Phys. **59**, 2330 (1973).
1.160. E. Weitz, G. W. Flynn, A. M. Ronn: J. Chem. Phys. **56**, 6060 (1972).
1.161. E. Weitz, G. W. Flynn: J. Chem. Phys. **58**, 2679 (1973).
1.162. R. C. L. Yuan, G. W. Flynn: J. Chem. Phys. **57**, 1316 (1972).
1.163. C. A. Thayer, J. T. Yardley: J. Chem. Phys. **57**, 3992 (1972).
1.164. R. D. Bates, Jr., J. T. Knudtson, G. W. Flynn, A. M. Ronn: J. Chem. Phys. **57**, 4174 (1972).
1.165. O. R. Wood, S. E. Schwarz: Appl. Phys. Lett. **16**, 518 (1970).
1.166. R. D. Bates, Jr., J. T. Knudtson, G. W. Flynn, A. M. Ronn: Chem. Phys. Lett. **8**, 103 (1971).
1.167. I. Burak, A. V. Nowak, J. I. Steinfeld, D. G. Sutton: J. Chem. Phys. **51**, 2275; and **52**, 5421 (1970).
1.168. I. Burak, P. Houston, D. G. Sutton, J. I. Steinfeld: J. Chem. Phys. **53**, 3632 (1970).
1.169. C. B. Moore: Accounts Chem. Res. **2**, 103 (1969).
1.170. C. B. Moore: Advanc. Chem. Phys. **23**, 41 (1973).
1.171. C. B. Moore, P. F. Zittel: Science **182**, 541 (1973).
1.172. A. Lauberau, W. Kaiser: Phys. Rev. Lett. **28**, 1162 (1972).
1.173. G. M. Burnett, A. M. North (Eds.): *Transfer and Storage of Energy by Molecules*, Vol. 2: *Vibrational Energy* (Wiley-Interscience, London, 1969).
1.174. L. O. Hocker, M. A. Kovacs, C. K. Rhodes, G. W. Flynn, A. Javan: Phys. Rev. Lett. **17**, 233 (1966).
1.175. J. T. Yardley, C. B. Moore: J. Chem. Phys. **45**, 1066 (1966).
1.176. F. De Martini, J. Ducuing: Phys. Rev. Lett. **17**, 117 (1966).
1.177. P. F. Zittel, C. B. Moore: J. Chem. Phys. **58**, 2922 (1973).
1.178. J. T. Yardley, C. B. Moore: J. Chem. Phys. **49**, 1111 (1968).

1.179. C.K. RHODES, M.J. KELLY, A. JAVAN: J. Chem. Phys. **48**, 5730 (1968).
1.180. R.V. AMBARTZUMIAN, V.S. LETOKHOV, G.N. MARKAROV, A.A. PURETZKI: Chem. Phys. Lett. **16**, 252 (1972).
1.181. R.D. BATES, G.W. FLYNN, J.T. KNUDTSON, A.M. RONN: J. Chem. Phys. **53**, 3621 (1970).
1.182. K.N.C. BRAY: J. Phys. B (Atom. Mol. Phys.) **1**, 705 (1968).
1.183. K.N.C. BRAY, N.H. PRATT: J. Chem. Phys. **53**, 2987 (1970).
1.184. J.W. RICH: J. Appl. Phys. **42**, 2719 (1971).
1.185. K.P. HORN, P.E. OETTINGER: J. Chem. Phys. **54**, 3040 (1971).
1.186. Y.V. AFANAS'EV, F.N. BELENOV, E.P. MARKIN, I.A. POLUENTOV: JETP Lett. **13**, 331 (1971).
1.187. J.T. YARDLEY, C.B. MOORE: J. Chem. Phys. **46**, 4491 (1967).
1.188. C.B. MOORE: J. Chem. Phys. **43**, 2979 (1965).
1.189. R.D. SHARMA, C.A. BRAU: Phys. Rev. Lett. **19**, 1273 (1967).
1.190. R.L. TAYLOR, S. BITTERMAN: J. Chem. Phys. **50**, 1720 (1969).
1.191. B.H. MAHAN: J. Chem. Phys. **46**, 48 (1967).
1.192. H. PAULY, J.P. TOENNIES: In *Advances in Atomic and Molecular Physics*, ed. by D.R. BATES and I. ESTERMANN (Academic Press, New York, London, 1965).
1.193. I.V. HERTEL, W. STOLL: J. Phys. B (Atom. Mol. Phys.) **7**, 570 (1974).
1.194. I.V. HERTEL, W. STOLL: J. Phys. B (Atom. Mol. Phys.) **7**, 583 (1974).
1.195. I.V. HERTEL: In *Atomic Physics*, Vol. 4, ed. by G. ZU PUTLITZ, E.W. WEBER and A. WINNACKER (Plenum Press, New York, London, 1975), p. 381.
1.196. D. MOORES, D. NORCROSS: J. Phys. B (Atom. Mol. Phys.) **5**, 1482 (1972).
1.197. R.F. STEBBINGS, F.B. DUNNING, F.K. TITTEL, R.D. RUNDEL: Phys. Rev. Lett. **30**, 315 (1973).
1.198. F.D. DUNNING, R.F. STEBBINGS: Phys. Rev. Lett. **32**, 1286 (1974).
1.199. R.F. STEBBINGS, F.B. DUNNING: Phys. Rev. A **8**, 665 (1973).
1.200. F.B. DUNNING, R.F. STEBBINGS: Phys. Rev. A **9**, 2378 (1974).
1.201. J.L. CARLSTEN, T.J. MCILRATH, W.H. PARKINSON: J. Phys. B (Atom. Mol. Phys.) **7**, L 244 (1974).
1.202. J.L. CARLSTEN: Planet. Space Sci. (to be published).
1.203. A.C. GALLAGHER, G. YORK: Rev. Sci. Instr. **45**, 662 (1974).
1.204. G.J. SCHULZ: Phys. Rev. Lett. **10**, 104 (1963).
1.205. C.E. KUYATT, J.A. SIMPSON: Rev. Sci. Instr. **38**, 103 (1967).
1.206. M. FANO: Phys. Rev. **178**, 131 (1969).
1.207. M. HEINZMANN, J. KESSLER, J. LORENZ: Z. Physik **240**, 42 (1970).
1.208. W. v. DRACHENFELS, M.T. KOCH, R.D. LEPPER, T.M. MÜLLER, W. PAUL: Z. Physik **269**, 387 (1974).
1.209. A.W. WEISS: Phys. Rev. **166**, 70 (1968).
1.210. A.C. FUNG, J.J. MATESE: Phys. Rev. A **5**, 22 (1972).
1.211. S. GELTMAN: Phys. Rev. **104**, 346 (1956).
1.212. B. EDLEN: J. Chem. Phys. **33**, 98 (1960).
1.213. B. STEINER: In *Case Studies in Atomic Physics*, II, ed. by E.W. MCDANIEL and M.R.C. MCDOWELL (North-Holland Publ. Co., Amsterdam, 1972), p. 483.
1.214. L.M. BRANSCOMB, D.S. BURCH, S.J. SMITH, S. GELTMAN: Phys. Rev. **111**, 504 (1958).
1.215. W.C. LINEBERGER, B.W. WOODWORD: Phys. Rev. Lett. **25**, 424 (1970).
1.216. B. BREHM, M.A. GUSINOW, J.L. HALL: Phys. Rev. Lett. **19**, 737 (1967).
1.217. M.W. SIEGEL, R.J. CELOTTA, J.L. HALL, J. LEVINE, R.A. BENNETT: Phys. Rev. A **6**, 607 (1972).
1.218. J.L. HALL, M.W. SIEGEL: J. Chem. Phys. **48**, 943 (1968).
1.219. T.A. PATTERSON, H. HOTOP, A. KASDAN, D.W. NORCROSS, W.C. LINEBERGER: Phys. Rev. Lett. **32**, 189 (1974).

1.220. W. C. LINEBERGER: In *Laser Spectroscopy*, ed. by R. G. BREWER and A. MOORADIAN (Plenum Press, New York, London, 1974).
1.221. H. HOTOP, R. A. BENNETT, W. C. LINEBERGER: J. Chem. Phys. **58**, 2373 (1973).
1.222. H. HOTOP, W. C. LINEBERGER: J. Chem. Phys. **58**, 2379 (1973).
1.223. H. HOTOP, T. A. PATTERSON, W. C. LINEBERGER: Phys. Rev. A **8**, 762 (1973).
1.224. H. HOTOP, T. A. PATTERSON, W. C. LINEBERGER: J. Chem. Phys. **60**, 1806 (1974).
1.225. R. J. CELOTTA, R. A. BENNETT, J. L. HALL: J. Chem. Phys. **60**, 1740 (1974).
1.226. K. C. SMYTH, R. T. McIVER, JR., J. I. BRAUMANN, R. W. WALLACE: J. Chem. Phys. **54**, 2758 (1971).
1.227. R. J. CELOTTA, R. A. BENNETT, J. L. HALL, M. W. SIEGEL, J. LEVINE: Phys. Rev. A **6**, 631 (1972).
1.228. D. L. NORCROSS, D. L. MOORES: In *Atomic Physics*, Vol. 3, Proc. 3rd Intern. Conf. on Atomic Physics, ed. by S. J. SMITH and G. K. WALTERS (Plenum Press, New York, 1973).
1.229. W. C. LINEBERGER, T. A. PATTERSON: Chem. Phys. Lett. **13**, 40 (1972).
1.230. G. HERZBERG, A. LAGERQUIST: Canad. J. Phys. **4 6**, 2363 (1968).
1.231. H. B. BEBB: Phys. Rev. **149**, 149 (1966).
1.232. H. B. BEBB, A. GOLD: Phys. Rev. **143**, 1 (1966).
1.233. W. ZERNIK: Phys. Rev. A **123**, 51 (1964); **172**, 420 (1968).
1.234. H. B. BEBB: Phys. Rev. **153**, 23 (1967).
1.235. V. M. MORTON: Proc. Phys. Soc. (London) **92**, 301 (1967).
1.236. B. A. ZON, N. MANAKOV, L. P. RAPPAPORT: Sov. Phys. JETP **34**, 515 (1972).
1.237. Y. GONTIER, M. TRAHIN: Phys. Rev. **172**, 83 (1968); A **4**, 1896 (1971).
1.238. Y. GONTIER, M. TRAHIN: Phys. Lett. **36**, 463 (1971).
1.239. N. B. DELONE, G. K. PISKOVA: Sov. Phys. JETP **31**, 403 (1970).
1.240. P. AGOSTINI, G. BARJOT, G. MAINFRAY, C. MAUNS, J. THEBAULT: Phys. Lett. **31** A, 367 (1970).
1.241. L. P. RAPPOPORT, B. A. ZON, L. P. MANAKOV: Sov. Phys. JETP **29**, 220 (1969).
1.242. C. S. CHANG, P. STEHLE: Phys. Rev. A **4**, 641 (1971); A **5**, 1087 (1972).
1.243. Y. GONTIER, M. TRAHIN: Phys. Rev. A **7**, 1899 (1973).
1.244. C. S. CHANG, P. STEHLE: Phys. Rev. Lett. **30**, 1283 (1973).
1.245. B. HELD, G. MAINFRAY, C. MANUS, J. MORELLEC, F. SANCHEZ: Phys. Rev. Lett. **30**, 423 (1973).
1.246. L. V. KELDYSLE: Sov. Phys. JETP **20**, 1307 (1965).
1.247. A. M. PERELOMOV, V. S. POPOV, M. V. TERENT'EV: Sov. Phys. JETP **23**, 924 (1966).
1.248. A. M. PERELOMOV, V. S. POPOV, M. V. TERENT'EV: Sov. Phys. JETP **24**, 207 (1967).
1.249. H. R. REISS: Phys. Rev. A **1**, 803 (1970).
1.250. H. R. REISS: Phys. Rev. D **4**, 3533 (1971).
1.251. W. C. HENNEBERGER: Phys. Rev. Lett. **21**, 838 (1968).
1.252. C. K. CHOI, W. C. HENNEBERGER, F. C. SANDERS: Phys. Rev. A **9**, 1895 (1974).
1.253. F. H. M. FAISAL: J. Phys. B **5**, L 89 (1973).
1.254. S. GELTMAN, M. R. TEAGUE: J. Phys. B (Atom. Mol. Phys.) **7**, L 22 (1974).
1.255. M. H. MITTLEMAN: Phys. Lett. **47** A, 55 (1974).
1.256. B. HELD, G. MAINFRAY, C. MANUS, J. MORELLEC: Phys. Lett. A **35**, 257 (1971).
1.257. B. HELD, G. MAINFRAY, C. MAUNS, J. MORELLEC: Phys. Rev. Lett. **28**, 130 (1972).
1.258. R. A. FOX, R. M. KOGAN, E. J. ROBINSON: Phys. Rev. Lett. **26**, 1416 (1971).
1.259. R. G. EVANS, P. G. THONEMANN: Phys. Lett. A **39**, 133 (1972).
1.260. D. POPESCU, C. B. COLLINS, B. W. JOHNSON, I. POPESCU: Phys. Rev. A **9**, 1182 (1974).
1.261. C. B. COLLINS, B. W. JOHNSON, D. POPESCU. G. MUSA, M. L. PASCU, I. POPESCU: Phys. Rev. A **8**, 2197 (1973).
1.262. M. GOEPPERT-MAYER: Naturw. **17**, 932 (1929); Ann. Phys. **9**, 273 (1931).
1.263. V. W. HUGHES, L. GRABNER: Phys. Rev. **79**, 314 (1950).
1.264. W. KAISER, C. G. B. GARRET: Phys. Rev. Lett. **7**, 229 (1961).

1.265. D. A. Kleinmann: Phys. Rev. **125**, 87 (1962).
1.266. I. D. Abella: Phys. Rev. Lett. **9**, 453 (1962).
1.267. R. Braunstein, W. Ockmann: Phys. Rev. **134**, A 499 (1964).
1.268. W. L. Peticolas, J. P. Goldsborough, K. E. Rieckhoff: Phys. Rev. Lett. **10**, 43 (1963).
1.269. J. A. Giordmaine, J. H. Howe: Phys. Rev. Lett. **11**, 207 (1963).
1.270. S. Yatsiv, W. G. Wagner, G. S. Picus, F. McCluny: Phys. Rev. Lett. **15**, 614 (1965).
1.271. R. M. Hochstrasser, J. E. Wessel, H. N. Sung: J. Chem. Phys. **60**, 317 (1974).
1.272. L. Wunsch, H. J. Neusser, E. W. Schley: Chem. Phys. Lett. **32**, 210 (1975).
1.273. R. G. Wunsch, H. J. Neusser, E. W. Schlag: Chem. Phys. Lett. **31**, 433 (1975).
1.274. R. G. Bray, R. M. Hochstrasser, H. N. Sung: Chem. Phys. Lett. **33**, 1 (1975).
1.275. J. F. Ward, G. H. C. New: Phys. Rev. **185**, 57 (1969).
1.276. J. A. Armstrong, N. Bloembergen, J. Ducuing, P. S. Pershan: Phys. Rev. **127**, 1918 (1962).
1.277. R. B. Miles, S. E. Harris: IEEE J. Quant. Electron. QE-**9**, 470 (1973).
1.278. S. E. Harris, R. B. Miles: Appl. Phys. Lett. **19**, 385 (1971).
1.279. A. G. Akmanov, S. A. Akhmanov, B. V. Zhdanov, A. I. Kovrigin, N. K. Podsotskaya, R. V. Khokhlov: ZhETF Pis. Red. **10**, 244 (1969).
1.280. J. F. Young, G. C. Bjorklund, A. H. Kung, R. B. Miles, S. E. Harris: Phys. Rev. Lett. **27**, 1551 (1971).
1.281. S. E. Harris, J. F. Young, A. H. Kung, D. M. Bloom, G. C. Bjorklund: In *Laser Spectroscopy*, ed. by R. G. Brewer and A. Mooradian (Plenum Press, New York, London, 1974), pp. 59—75.
1.282. A. H. Kung, J. F. Young, G. C. Bjorklund, S. E. Harris: Phys. Rev. Lett. **29**, 985 (1972).
1.283. C. R. Vidal, J. Cooper: J. Appl. Phys. **40**, 3370 (1969).
1.284. C. R. Vidal, M. M. Hessel: J. Appl. Phys. **43**, 2776 (1972).
1.285. A. H. Kung, J. F. Young, S. E. Harris: Appl. Phys. Lett. **22**, 301 (1973).
1.286. S. E. Harris: Phys. Rev. Lett. **31**, 341 (1973).
1.287. R. T. Hodgson, P. P. Sorokin, J. J. Wynne: Phys. Rev. Lett. **32**, 343 (1974).
1.288. J. A. Armstrong, J. J. Wynne: Phys. Rev. Lett. **33**, 1183 (1974).
1.289. L. Armstrong, Jr., B. L. Beers: Phys. Rev. Lett. **34**, 1290 (1975).
1.290. U. Fano: Phys. Rev. **124**, 1866 (1961).
1.291. P. P. Sorokin, J. J. Wynne, J. R. Lankard: Appl. Phys. Lett. **22**, 342 (1973).
1.292. J. J. Wynne, P. P. Sorokin, J. R. Lankard: In *Laser Spectroscopy*, ed. by R. G. Brewer and A. Mooradian (Plenum Press, New York, London, 1974), pp. 103—111.
1.293. J. J. Wynne, P. P. Sorokin: J. Phys. B (Atom. Mol. Phys.) **8**, L 37 (1975).
1.294. S. E. Harris, D. M. Bloom: Appl. Phys. Lett. **24**, 229 (1974).
1.295. D. M. Bloom, J. T. Yardley, J. F. Young, S. E. Harris: Appl. Phys. Lett. **24**, 427 (1974).
1.296. M. Sargent, M. O. Scully, W. E. Lamb, Jr.: *Laser Physics* (Addison Wesley Publishing Company, Reading, Mass., 1975).
1.297. W. Heitler: *The Quantum Theory of Radiation*, 3rd ed. (Oxford University Press, Oxford, 1954).
1.298. A. M. Bonch-Bruevich, V. A. Khodovoi: Sov. Phys. Usp. **10**, 637 (1968).
1.299. H. Hirsch: Ann. Physik **26**, 38 (1971).
1.300. A. M. Bonch-Bruevich, N. N. Kostin, V. A. Khodovoi, V. V. Khromov: Sov. Phys. JETP **29**, 82 (1969).
1.301. J. H. Hertz, K. Hoffmann, W. Brunner, H. Paul, G. Richter, H. Steudel: Phys. Let. **26** A, 156 (1968).

1.302. T. Hänsch, P. Toschek: Z. Physik **236**, 213 (1970).
1.303. B. J. Feldmann, M. S. Feld: Phys. Rev. A **5**, 899 (1972).
1.304. A. Schabert, R. Keil, P. E. Toschek: Appl. Phys. **6**, 181 (1975).
1.305. I. M. Beterov, V. P. Chebotaev: *Progress in Quantum Electronics*, Vol. 3, Part 1, ed. by J. H. Sandars and S. Stenholm (Pergamon Press, Oxford, 1974).
1.306. M. S. Feld: *Fundamental and Applied Laser Physics*, Proceedings of the Esfahan Symposium, ed. by M. S. Feld, A. Javan and N. A. Kurnit (John Wiley, New York, 1973).
1.307. A. Schabert, R. Keil, P. E. Toschek: Opt. Commun. **13**, 265 (1975).
1.308. C. Cohen-Tannoudji: Ann. Physique **7**, 423 and 469 (1962).
1.309. W. Happer: In *Progress in Quantum Electronics*, Vol. 1, Part 1, ed. by J. H. Sandars and S. Stenholm (Pergamon Press, Oxford, 1971).
1.310. E. B. Aleksandrov, A. M. Bonch-Bruevich, N. N. Kosten, V. A. Khodovoi: JETP Lett. **3**, 53 (1966).
1.311. A. M. Bonch-Bruevich, N. N. Kostin, V. A. Khodovoi: JETP Lett. **3**, 279 (1966).
1.312. L. D. Zusmann, A. I. Burshtein: Sov. Phys. JETP **34**, 520 (1972).
1.313. C. Cohen-Tannoudji: Proc. 2nd Laser Spectroscopy Conference, Megève 1975 Springer, Berlin, Heidelberg, New York, 1975).
1.314. P. Platz: Appl. Phys. Lett. **14**, 168 (1969).
1.315. P. Platz: J. Physique **32**, 773 (1971).
1.316. P. Platz: Appl. Phys. Lett. **16**, 70 (1970).
1.317. J. G. Hirschberg, P. Platz: Appl. Opt. **4**, 1375 (1965).
1.318. W. W. Morey: PhD Thesis, Department of Physics and Astrophysics, University of Colorado, Boulder, Colo. (1970).
1.319. B. Dubreuil, P. Ranson, J. Chapelle: Phys. Lett. **42** A, 323 (1972).
1.320. M. Mizushima: Phys. Rev. **133**, 414 (1964).
1.321. P. F. Liao, J. E. Bjorkholm: Phys. Rev. Lett. **34**, 1 (1975).
1.322. S. Feneuille: J. Phys. B (Atom. Mol. Phys.) **7**, 1981 (1974).
1.323. C. Cohen-Tannoudji, S. Haroche: J. Physique **30**, 153 (1969).
1.324. F. Bloch, A. J. F. Siegert: Phys. Rev. **57**, 522 (1940).
1.325. A. W. Tucker, A. B. Petersen, M. Birnbaum: Appl. Opt. **12**, 2036 (1973).
1.326. K. H. Becker, U. Schurath, T. Tatarczyk: Appl. Opt. **14**, 310 (1975).
1.327. W. M. Fairbank, Jr., T. W. Hänsch, A. L. Schawlow: J. Opt. Soc. Am. **65**, 199 (1975).
1.328. R. N. Zare: Proc. 2nd Conference on Laser Spectroscopy, Megève 1975 (Springer Berlin, Heidelberg, New York, 1975).
1.329. K. Shimoda: Appl. Phys. **1**, 77 (1973).
1.330. M. R. Cervenan, N. R. Isenor: Opt. Commun. **10**, 280 (1974).
1.331. P. Agostini, P. Bensoussan: Appl. Phys. Lett. **24**, 216 (1974).
1.332. P. Lambropoulos: Phys. Rev. Lett. **28**, 585 (1972); **29**, 453 (1972).
1.333. S. Klarsfeld, A. Maquet: Phys. Rev. Lett. **29**, 79 (1972).
1.334. M. Lambropoulos, R. S. Berry: Phys. Rev. A **8**, 855 (1973); A **9**, 2459 (1974).
1.335. P. Lambropoulos: Phys. Rev. Lett. **30**, 413 (1973).
1.336. P. S. Farago, D. W. Walker: J. Phys. B (Atom. Mol. Phys.) **6**, L 280 (1973).
1.337. P. Lambropoulos: J. Phys. B (Atom. Mol. Phys.) **7**, L 33 (1974).
1.338. J. Ducuing, N. Bloembergen: Phys. Rev. **133**, A 1493 (1964).
1.339. P. Lambropoulos, C. Kikuchi, R. K. Osborn: Phys. Rev. **144**, 1081 (1966).
1.340. G. S. Agarwal: Phys. Rev. A **1**, 1445 (1970).
1.341. I. V. Tomov, A. S. Chirkin: Sov. J. Quant. Electron. **1**, 79 (1971).
1.342. J. L. Debethune: Nuovo Cimento 12 B, 101 (1972).
1.343. S. Carusotto, C. Strati: Nuovo Cimento **15** B, 159 (1973).
1.344. M. C. Teich, R. L. Abrams, W. B. Gandrud: Opt. Commun. **2**, 206 (1970).

1.345. F. SHIGA, S. IMAMURA: Phys. Lett. **25** A, 706 (1967).
1.346. C. LECOMPTE, G. MAINFRAY, C. MANUS, F. SANCHEZ: Phys. Rev. Lett. **32**, 265 (1974).
1.347. V. S. LETOKHOV: Science **180**, 451 (1973).
1.348. K. L. KOMPA: Z. Naturforsch. **27** b, 89 (1972).
1.349. N. V. KARLOV, YU. N. PETROV, A. M. PROKHOROV, O. N. STELMAKH: JETP Lett. **11**, 135 (1970).
1.350. J. C. POLANYI: Acc. Chem. Res. **5**, 161 (1972).
1.351. C. K. RHODES, P. W. HOFF: In *Laser Spectroscopy*, ed. by R. G. BREWER and A. MOORADIAN (Plenum Press, New York, 1974).
1.352. H. HARTLEY, A. O. PENDER, E. J. BOWEN, T. R. MERTON: Phil. Mag. **43**, 430 (1922).
1.353. R. M. BADGFR, J. W. URMSTON: Proc. Nat. Acad. Sci. (USA) **16**, 808 (1930).
1.354. J. G. CALVERT, J. N. PITTS, JR.: *Photochemistry* (Wiley, New York, 1966).
1.355. W. SPINDEL: *Isotope Effects in Chemical Processes* American Chemical Society, Washington, D.C., 1969).
1.356. T. J. ODIORNE, P. R. BROOKS, J. V. V. KASPER: J. Chem. Phys. **55**, 1980 (1971).
1.357. J. L. LYMAN, R. J. JENSEN, J. RINK, C. P. ROBINSON, S. D. ROCKWOOD: Appl. Phys. Lett. **27**, 87 (1975).
1.358. C. P. ROBINSON: Proc. 2nd Laser Spectroscopy Conference, Megève 1975 (Springer Berlin, Heidelberg, New York, 1975).
1.359. N. G. BASOV, E. M. BELENOV, E. P. MORKIN, A. N. ORAEVSKII, A. V. PANKRATOV: In *Fundamental and Applied Laser Physics*, ed. by M. S. FELD, A. JAVAN and N. A. KURNIT (John Wiley, New York, 1973); Sov. Phys. JETP **37**, 247 (1973).
1.360. W. B. TIFFANY, H. W. MOSS, A. L. SCHALOW: Science **157**, 40 (1967).
1.361. V. S. LETOKHOV, A. A. MARKAROV: Sov. Phys. JETP **36**, 1091 (1973).
1.362. N. G. BASOV, E. P. MARKIN, A. N. ORAEVSKII, A. V. PANKRATOV: Sov. Phys. Dokl. **16**, 445 (1971).
1.363. H. R. BACHMANN, H. NÖTH, R. RINCK, K. L. KOMPA: Chem. Phys. Lett. **29**, 627 (1974).
1.364. H. R. BACHMANN, K. L. KOMPA, H. NÖTH, R. RINCK: Chem. Phys. Lett. **33**, 261 (1975).
1.365. G. HANCOCK, W. LANGE, M. LENZI, K. H. WELGE: Chem. Phys. Lett. **53**, 168 (1975).
1.366. B. B. SNAVELY: Invited Paper, 8th Intern. Conf. on Quantum Electronics, San Francisco, Calif. (1974).
1.367. V. V. ABAJIAN, A. M. FISHMAN: Phys. Today **26**, 23 (1973).
1.368. H. KOPFERMANN: *Nuclear Moments* (Academic Press, New York, 1958).
1.369. R. BRUCH, K. HEILIG, D. KALETTA, A. STEUDEL, D. WENDLAND: J. Physique **30**, C 1 (1969).
1.370. L. J. RADZIEMSKI, JR., D. W. STEINHAUS, R. D. COWAN, J. BLAISE, G. GUELACHVILI, Z. BEN OSMAN, J. VERGES: J. Opt. Soc. Am. **60**, 1556 A (1970).
1.371. P. L. KELLEY, H. KILDAL, H. R. SCHLOSSBERG: Chem. Phys. Lett. **27**, 62 (1974).
1.372. L. S. VASILENKO, V. P. CHEBOTAEV, A. V. SHISHAEV: JETP Lett. **12**, 113 (1970).
1.373. B. CAGNAC, G. GRYNBERG, F. BIRABEN: J. Phys. **34**, 845 (1973).
1.374. B. H. BILLINGS, W. J. HITCHCOCK, M. ZELIKOFF: J. Chem. Phys. **21**, 1762 (1953).
1.375. G. LIUTI, S. DONDES, P. HORTEK: J. Chem. Phys. **44**, 4052 (1966).
1.376. W. KUHN, H. MARTIN: Z. Phys. Chem. B **32**, 93 (1933).
1.377. A. ASHKIN: Phys. Rev. Lett. **25**, 1321 (1970).
1.378. R. SCHIEDER, H. WALTHER, L. WÖSTE: Opt. Commun. **5**, 337 (1972).
1.379. J. I. BRAUMANN, T. J. O'LEARY, A. L. SCHAWLOW: Opt. Commun. **12**, 223 (1974).
1.380. S. W. MAYER, M. A. KWOK, R. W. F. GROSS, D. J. SPENCER: Appl. Phys. Lett. **17**, 516 (1970).
1.381. E. S. YEUNG, C. B. MOORE: Appl. Phys. Lett. **21**, 109 (1972).
1.382. R. V. AMBARTZUMIAN, V. S. LETOKHOV: Appl. Opt. **11**, 354 (1972).
1.383. S. A. TUCCIO, J. W. DUBRIN, O. G. PETERSON, B. B. SNAVELY: Post-deadline paper, 8th Intern. Conf. on Quantum Electronics, San Francisco, Calif. (1974).

1.384. J. ROBIEUX, J. M. AUCLAIR: Patent Brevet d'Invention No. 1391738, Companie Général d'Electricité, application Oct. 12, 1963.
1.385. R. H. LEVY, G. S. JANES: Patent. US Pat. No. 3772519, Jersey Nuclear Avco, Brevet d'Invention No. 2094967, Isotopes Inc., application March 25, 1970.
1.386. K. GÜRS: Patent. Deutsche Offenlegungsschrift No. 1959767, Batelle-Institut, Frankfurt (Main), application June 3, 1971.
1.387. I. NEBENZAHL, M. LEVIN: Patent. Deutsche Offenlegungsschrift No. 2312194, application March 3, 1973.
1.388. G. S. JANES, R. H. LEVY: Patent. Deutsche Offenlegungsschrift No. 2426842, application June 4, 1974.
1.389. U. BRINKMANN, W. HARTIG, H. TELLE, H. WALTHER: Post-deadline paper, 8th Intern. Conf. on Quantum Electronics, San Francisco, Calif. (1974); Appl. Phys. **5**, 109 (1974).
1.390. S. ROCKWOOD, S. W. RABIDEAU: Post-deadline paper, 8th Intern. Conf. on Quantum Electronics, San Francisco, Calif. (1974); Phys. Today (1974), p. 17.
1.391. V. S. LETOKHOV, E. A. RYABOV, N. V. CHEKALIN: To be published; see also Laser Focus, February 1975, p. 16 and R. V. AMBARTZUMIAN et al.: JETP Lett. **20**, 597 (1974).
1.392. S. M. FREUND, J. J. RITTER: Chem. Phys. Lett. **32**, 255 (1975).
1.393. R. V. AMBARTZUMIAN, V. S. LETOKHOV, G. N. MAKAROV, A. A. PURETZKI: JETP Lett. **17**, 91 (1973).
1.394. R. V. AMBARTZUMIAN, Y. A. GOROKHOV, V. S. LETOKHOV, G. N. MAKAROV: JETP Lett. **21**, 375 (1975).
1.395. D. D. S. LIU, S. DATTA, R. N. ZARE: J. Am. Chem. Soc. (in press).
1.396. H. J. DEWEY, R. A. KELLER, J. J. RITTER: Chem. Phys. Lett. **30**, 165 (1975).
1.397. S. R. LEONE, C. B. MOORE: Post-deadline paper, 8th Intern. Conf. on Quantum Electronics, San Francisco, Calif. (1974); Phys. Today (1974), p. 17; Phys. Rev. Lett. **33**, 269 (1974).
1.398. R. M. LUM, K. B. MCAFEE: To be published; see also Laser Focus, March 1975, p. 14.
1.399. A. BERNHARDT, D. DEURRE, J. SIMPSON, L. WOOD: Appl. Phys. Lett. **25**, 617 (1974); Post-deadline paper, 8th Intern. Conf. on Quantum Electronics, San Francisco, Calif. (1974); see also Phys. Today (1974), p. 17.
1.400. R. S. BERRY: Bull. Am. Phys. Soc. **17**, 1129 (1972).
1.401. R. GILETTE: Science **183**, 1172 (1974).
1.402. R. N. ZARE, P. J. DAGDIGIAN: Science **185**, 739 (1974).
1.403. A. SCHULTZ, H. W. CRUSE, R. N. ZARE: J. Chem. Phys. **57**, 1354 (1972).
1.404. H. W. CRUSE, P. J. DAGDIGIAN, R. N. ZARE: Faraday Disc. Chem. Soc. **55**, 1354 (1972), and J. Chem. Phys. **61**, 2464 (1974), and **62**, 1824 (1975).
1.405. P. J. DAGDIGIAN, H. W. CRUSE, A. SCHULTZ, R. N. ZARE: J. Chem. Phys. **61**, 4450 (1974).
1.406. H. J. LOESCH, D. R. HERSCHBACH: To be published.
1.407. S. M. FREUND, G. A. FISK, D. R. HERSCHBACH, W. KLEMPERER: J. Chem. Phys. **54**, 2510 (1971).
1.408. H. G. BENNEWITZ, R. HAERTEN, G. MÜLLER: Chem Phys. Lett. **12**, 335 (1971).
1.409. P. PECHUKAS, J. C. LIGHT, C. RANKIN: J. Chem. Phys. **44**, 794 (1966).
1.410. S. A. SAFRON, N. D. WEINSTEIN, D. R. HERSCHBACH, J. C. TULLY: Chem. Phys. Lett. **12**, 564 (1972).
1.411. R. P. FEYNMAN, F. L. VERNON, JR., R. W. HELLWARTH: J. Appl. Phys. **28**, 49 (1957).
1.412. N. A. KURNIT, I. D. ABELLA, S. R. HARTMANN: Phys. Rev. Lett. **13**, 367 (1964).
1.413. I. D. ABELLA, N. A. KURNIT, S. R. HARTMANN: Phys. Rev. **141**, 392 (1966).
1.414. E. L. HAHN: Phys. Rev. **80**, 580 (1950).
1.415. C. K. N. PATEL, R. E. SLUSHER: Phys. Rev. Lett. **20**, 1087 (1968).

1.416. J.P. GORDON, C.H. WANG, C.K.N. PATEL, R.E. SLUSHER, W.J. TOMLINSON: Phys. Rev. **179**, 294 (1969).
1.417. B. BÖLGER, J.C. DIELS: Phys. Lett. **28** A, 401 (1968).
1.418. R.G. BREWER, R.L. SHOEMAKER: Phys. Rev. Lett. **27**, 631 (1971).
1.419. H.C. TORREY: Phys. Rev. **76**, 1059 (1949).
1.420. G. B. HOCKER, C.L. TANG: Phys. Rev. Lett. **21**, 591 (1968).
1.421. P.W. HOFF, H.A. HAUS, T.J. BRIDGES: Phys. Rev. Lett. **25**, 82 (1970).
1.422. J.M. LEVY, J.H.S. WANG, S.G. KUKOLICH, J.I. STEINFELD; Phys. Rev. Lett. **29**, 395 (1972).
1.423. R.H. DICKE: Phys. Rev. **93**, 99 (1954).
1.424. F. BLOCH: Phys. Rev. **70**, 460 (1946).
1.425. E.L. HAHN: Phys. Rev. **77**, 297 (1950).
1.426. R.B. BREWER, R.L. SHOEMAKER: Phys. Rev. A **6**, 2001 (1972).
1.427. R.L. SHOEMAKER, R.G. BREWER: Phys. Rev. Lett. **28**, 1430 (1972).
1.428. R.G. BREWER, E.L. HAHN: Phys. Rev. A **8**, 464 (1973).
1.429. J. SCHMIDT, P.R. BERMAN, R.G. BREWER: Phys. Rev. Lett. **31**, 1103 (1973).
1.430. H.Y. CARR, E.M. PURCELL: Phys. Rev. **94**, 630 (1954).
1.431. R.M. HILL, D.E. KAPLAN, G.F. HERRMANN, S.K. ICHIKI: Phys. Rev. Lett. **18**, 165 (1967).
1.432. N. SKRIBANOWITZ, I.D. HERMAN, J.C. MCGILLIVRAY, M.S. FELD: Phys. Rev. Lett. **30**, 309 (1973).
1.433. I.P. HERMAN, J.C. MACGILLIVRAY, N. SKRIBANOWITZ, M.S. FELD: In *Laser Spectroscopy*, ed. by R.G. BREWER and A. MOORADIAN (Plenum Press New York, London, 1974).
1.434. A. KASTLER, J. BROSSEL: Compt. Rend. **229**, 1213 (1949).
1.435. G. ZU PUTLITZ: Springer Tracts in Modern Phys. **37**, 105 (1965).
1.436. B. BUDICK: In *Advances in Atomic and Molecular Physics*, Vol. 3, ed. by R.D. BATES and I. ESTERMANN (Academic Press, New York, 1967).
1.437. T.G. ECK, L.L. FOLDY, H. WIEDER: Phys. Rev. Lett. **10**, 239 (1963).
1.438. A. KASTLER: J. Physique **11**, 255 (1950).
1.439. R. SCHIEDER, H. WALTHER, L. WÖSTE: Opt. Commun. **5**, 337 (1972).
1.440. G.A. RUFF, M. HERCHER, H.A. PIKE: Bull. Am. Phys. Soc. **16**, 1340 (1971).
1.441. J. BONN, G. HUBER, H. J. KLUGE, M. KOPF, L. KUGLER, E.W. OTTEN, RODRIGUEZ: In *Atomic Physics*, Vol. 3, ed. by S. SMITH and G.K. WALTERS (Plenum Press, New York, London, 1973).
1.442. R. SCHIEDER, H. WALTHER: Z. Physik **270**, 55 (1974).
1.443. N.F. RAMSEY: *Molecular Beams* (Oxford University Press, Oxford, 1956).
1.444. F.D. COLEGROVE, P.A. FRANKEN, R.R. LEVIS, R.H. SANDS: Phys. Rev. Lett. **3**, 420 (1959).
1.445. P. FRANKEN: Phys. Rev. **121**, 508 (1961).
1.446. A.C.G. MITCHELL, M.W. ZEMANSKY: *Resonance Radiation and Excited Atoms* (Cambridge University Press, Cambridge, 1934).
1.447. G. ZU PUTLITZ: Comments Atom. Mol. Phys. **1**, 74 (1969).
1.448. I.J. MA, J. MERTENS, G. ZU PUTLITZ, G. SCHÜTTE: Z. Physik **208**, 352 (1968).
1.449. G. COPLEY, B.P. KIBBLE, G.W. SERIES: J. Phys. B (Atom. Mol. Phys.) **1**, 724 (1968).
1.450. D. FAVART, P.M. MCINTYRE, D.Y. STOWELL, V.L. TELEGDI, R. DE VOE: Phys. Rev. Lett. **27**, 1336 (1971).
1.451. P. SCHENCK, R.C. HILBORN, H. METCALF: Phys. Rev. Lett. **31**, 189 (1973).
1.452. H. FIGGER, H. WALTHER: Z. Physik **267**, 1 (1974).
1.453. P. JACQUINOT, B. ROIZEN-DOSSIER: In *Progress in Optics*, Vol. 3, ed. by E. WOLF (North-Holland Publ. Co., Amsterdam, 1964).

1.454. C. Cohen-Tannoudji: In *Atomic Physics*, Vol. 4, ed. by G. zu Putlitz, E. W. Weber, and A. Winnacker (Plenum Press, New York, London, 1975).
1.455. P. Avan, C. Cohen-Tannoudji: J. Phys. Lett. **36**, L 85 (1975).
1.456. W. Rasmussen, R. Schieder, H. Walther: Opt. Commun. **12**, 315 (1974).
1.457. J. N. Dodd, R. D. Kaul, D. M. Warrington: Proc. Phys. Soc. **84**, 176 (1964).
1.458. E. B. Alexandrov: Opt. Spectrosc. **17**, 957 (1964).
1.459. T. Hadeishi, W. A. Nievenberg: Phys. Rev. Lett. **14**, 891 (1965).
1.460. H. J. Andrä: Phys. Rev. Lett. **25**, 325 (1970).
1.461. H. J. Andrä: In *Atomic Physics*, Vol. 4, ed. by G. zu Putlitz, E. W. Weber, A. Winnacker (Plenum Press, New York, London, 1975).
1.462. E. Matthias, S. S. Rosenblum, D. A. Shirley: Phys. Rev. Lett. **14**, 46 (1965).
1.463. K. W. Meissner, K. F. Luft: Ann. Phys. **29**, 968 (1937).
1.464. K. W. Meissner, L. G. Mundie, P. Stelson: Phys. Rev. **74**, 932 (1948); **75**, 891 (1949).
1.465. E. Luc-Koenig: Private communication.
1.466. A. Corney, G. W. Series: Proc. Phys. Soc. **83**, 207, 213, and 331 (1964).
1.467. E. B. Alexandrov: Opt. Spectrosc. **14**, 233 (1963); **17**, 522 (1964); **19**, 252 (1965).
1.468. U. Gonser (Ed.): *Topics in Applied Physics*, Vol. 5: Mössbauer Spectroscopy (Springer, Berlin, Heidelberg, New York, 1975).
1.469. J. S. Deech, P. Hannaford, G. W. Series: J. Phys. B (Atom. Mol. Phys.) **7**, 1131 (1974).
1.470. K. Fredriksson, S. Svanberg: Phys. Lett. **53** A, 61 (1975).
1.471. G. Belin, L. Holmgren, I. Lindgren, S. Svanberg: Physica Scripta **12**, 287 (1975).
1.472. G. Belin, S. Svanberg: Phys. Lett. **47** A, 5 (1974).
1.473. G. Belin, L. Holmgren, S. Svanberg: Physica Scripta (to be published).
1.474. S. Svanberg, G. Belin: J. Phys. B (Atom. Mol. Phys.) **7**, L 82 (1974).
1.475. S. Svanberg, P. Tsekeris, W. Happer: Phys. Rev. Lett. **30**, 817 (1973).
1.476. S. Svanberg, P. Tsekeris: Phys. Rev. A **11**, 1125 (1975).
1.477. W. Hogervorst, S. Svanberg: Physica Scripta (to be published).
1.478. G. Belin, L. Holmgren, S. Svanberg: Physica Scripta (to be published).
1.479. S. Svanberg: In *Laser Spectroscopy*, ed. by R. G. Brewer and A. Mooradian (Plenum Press, New York, London, 1974).
1.480. W. Hogervorst, S. Svanberg: Phys. Lett. **48** A, 89 (1974).
1.481. W. Hogervorst, S. Svanberg: Physica Scripta **12**, 67 (1975).
1.482. P. Tsekeris, R. Gupta, W. Happer, G. Belin, S. Svanberg: Phys. Lett. **48** A, 101 (1974).
1.483. G. Belin, L. Nilsson, S. Svanberg: To be published.
1.484. R. W. Field, R. S. Bradford, H. P. Broida, D. O. Harris: J. Chem. Phys. **57**, 2209 (1972).
1.485. M. Broyer, J. Vigue, J.-C. Lehmann: Compt. Rend. **273**, B 289 (1971).
1.486. M. Broyer, J.-C. Lehmann: Phys. Lett. **40** A, 43 (1972).
1.487. M. McClintock, W. Demtröder, R. N. Zare: J. Chem. Phys. **51**, 5509 (1969).
1.488. R. E. Drullinger, M. M. Hessel, E. W. Smith: Bull. Am. Phys. Soc. **18**, 575 (1973).
1.489. F. Schuda, M. Hercher, C. R. Stroud, Jr.: Appl. Phys. Lett. **22**, 360 (1973).
1.490. W. Hartig, H. Walther: Appl. Phys. **1**, 171 (1973).
1.491. H. Walther: Proc. on Doppler-Free Methods of Spectroscopy on Excited Simple Molecular Systems, Aussois (22—25 May 1973), Ed. CNRS No. 217.
1.492. W. Lange, J. Luther, B. Nottbeck, H. W. Schröder: Opt. Commun. **8**, 163 (1973).
1.493. H. T. Duong, P. Jacquinot, S. Liberman, J. L. Pidqué, J. Pinard, J. L. Vialle: Opt. Commun. **7**, 371 (1973).
1.494. P. Jacquinot, S. Liberman, J. L. Picqué, J. Pinard: Opt. Commun. **8**, 163 (1973).

1.495. H.T. DUONG, J.L. VIALLE: EGAS Conference, Lund (1973).
1.496. H.T. DUONG, S. LIBERMAN, J. PINARD, J.L. VIALLE: Phys. Rev. Lett. **33**, 339 (1974).
1.497. D. PRITCHARD, J. APT, T.W. DUCAS: Phys. Rev. Lett. **32**, 641 (1974).
1.498. W. HARTIG, V. WILKE, H. WALTHER: Opt. Commun. **14**, 244 (1975).
1.499. K. FREDRIKSSON, S. SVANBERG: Phys. Lett. A (to be published).
1.500. J.H. BROADHURST, M.E. CAGE, D.L. CLORK, G.W. GREENLEES, J.A.R. GRIFFITH, G.R. ISAAK: J. Phys. B (Atom. Mol. Phys.) **7**, L 513 (1974).
1.501. D.G. YOUMANS, L.A. HACKEL, S. EZEKIEL: J. Appl. Phys. **44**, 2319 (1973).
1.502. R.E. GROVE, I.Y. WU, L.A. HACKEL, D.G. YOUMANS, S. EZEKIEL: Appl. Phys. Lett. **23**, 442 (1973).
1.503. S. HAROCHE, M. GROSS, M.P. SILVERMAN: Phys. Rev. Lett. **33**, 1063 (1974).
1.504. C. FABRE, M. GROSS, S. HAROCHE: Opt. Commun. **13**, 393 (1975).
1.505. S. HAROCHE, Y.A. PAISNER, A.L. SCHAWLOW: Phys. Rev. Lett. **30**, 948 (1973).
1.506. J.S. DEECH, R. LUYPAERT, G.W. SERIES: J. Phys. B (to be published).
1.507. W. GORNIK, D. KAISER, W. LANGE, J. LUTHER, H.-H. SCHULZ: Opt. Commun. **62**, 327 (1972).
1.508. R. WALLENSTEIN, J.A. PAISNER, A.L. SCHAWLOW: Phys. Rev. Lett. **32**, 1333 (1974).
1.509. B. CANILLARD, A. DUCESSE: Opt. Commun. **13**, 398 (1975).
1.510. M.D. LEVENSON, A.L. SCHAWLOW: Phys. Rev. A **6**, 10 (1972).
1.511. I.D. KNOX, YOH-HAN PAO: Appl. Phys. Lett. **18**, 360 (1971).
1.512. M.S. SOREM, T.W. HÄNSCH, A.L. SCHAWLOW: Chem. Phys. Lett. **17**, 300 (1972).
1.513. I.D. KNOX, YOH-HAN PAO: Appl. Phys. Lett. **16**, 129 (1970).
1.514. G.R. HANES, C.E. DAHLSTROM: Appl. Phys. Lett. **14**, 362 (1969).
1.515. M.S. SOREM, M.D. LEVENSON, A.L. SCHAWLOW: Phys. Lett. **37** A, 33 (1971).
1.516. T.W. HÄNSCH, M.D. LEVENSON, A.L. SCHAWLOW: Phys. Rev. Lett. **26**, 946 (1971).
1.517. M.S. SOREM, A.L. SCHAWLOW: Opt. Comm. **5**, 148 (1972).
1.518. T.W. HÄNSCH, I.S. SHAHIN, A.L. SCHAWLOW: Science **235**, 63 (1972).
1.519. T.W. HÄNSCH, M.H. NAYFEH, S.A. LEE, S.M. CURRY, I.S. SHAHIN: Phys. Rev. Lett. **32**, 1336 (1974).
1.520. E.E. UZGIRIS, J.L. HALL, R.L. BARGER: Phys. Rev. Lett. **26**, 289 (1971).
1.521. A.C. LUNTZ, R.G. BREWER, K.L. FOSTER, I.D. SWALEN: Phys. Rev. Lett. **23**, 951 (1969).
1.522. J.L. HALL, C. BORDÉ: Phys. Rev. Lett. **30**, 1101 (1973).
1.523. T.W. HÄNSCH, I.S. SHAHIN, A.L. SCHAWLOW: Phys. Rev. Lett. **27**, 707 (1971).
1.524. A. MUIRHEAD, K.V.L.N. SESTRY, R.F. CURL, J. COOK, F.K. TITTEL: Chem. Phys. Lett. **24**, 208 (1974).
1.525. P. RABINOWITZ, R. KELLER, J.T. LA TOURRETTE: Appl. Phys. Lett. **14**, 376 (1969).
1.526. C.K.N. PATEL: Appl. Phys. Lett. **25**, 112 (1974).
1.527. F. BIRABEN, B. CAGNAC, G. GRYNBERG: Phys. Rev. Lett. **32**, 643 (1974).
1.528. M.D. LEVENSON, N. BLOEMBERGEN: Phys. Rev. Lett. **32**, 645 (1974).
1.529. T.W. HÄNSCH, K.C. HARVEY, G. MEISEL, A.L. SCHAWLOW: Opt. Commun. **11**, 50 (1974).
1.530. N. BLOEMBERGEN, M.D. LEVENSON, M.M. SALOUR: Phys. Rev. Lett. **32**, 867 (1974).
1.531. M.D. LEVENSON, M.M. SALOUR: Phys. Lett. **48** A, 331 (1974).
1.532. F. BIRABEN, B. CAGNAC, G. GRYNBERG: Phys. Lett. **49** A, 71 (1974).
1.533. F. BIRABEN, B. CAGNAC, G. GRYNBERG: C.R. Hebd. Séan. Acad. Sci. **279**, B 51 (1974).
1.534. F. BIRABEN, B. CAGNAC, G. GRYNBERG: Phys. Lett. **84** A, 469 (1974).
1.535. F. BIRABEN, B. CAGNAC, G. GRYNBERG: J. Physique Lett. **36**, 41 (1975).
1.536. F. BIRABEN, B. CAGNAC, G. GRYNBERG: Compt. Rend. Hebd. Séan. Acad. Sci. **280**, B 235 (1975).

1.537. T. W. Hänsch, S. A. Lee, R. Wallenstein, C. Wieman: Phys. Rev. Lett. **34**, 307 (1975).
1.538. S. Kimel, A. Ron, S. Speiser: Chem. Phys. Lett. **28**, 190 (1974).
1.539. H. M. Gibbs: Phys. Rev. Lett. **29**, 459 (1972).
1.540. H. M. Gibbs: Phys. Rev. A **8**, 446 (1973).
1.541. H. M. Gibbs: Phys. Rev. A **8**, 456 (1973).
1.542. H. P. Grieneisen, N. A. Kurnit, A. Szöke: Opt. Commun. **3**, 259 (1971).
1.543. M. D. Crisp, E. T. Jaynes: Phys. Rev. **179**, 1253 (1969).
1.544. C. R. Stroud, Jr., E. T. Jaynes: Phys. Rev. A **1**, 106 (1970).
1.545. D. Leiter: Phys. Rev. A **2**, 259 (1970).
1.546. E. T. Jaynes: Phys. Rev. A **2**, 260 (1970).
1.547. D. Grischkowsky: Phys. Rev. Lett. **24**, 866 (1970).
1.548. D. Grischkowsky, J. A. Armstrong: Phys. Rev. A **6**, 1566 (1972).
1.549. D. Grischkowsky: Phys. Rev. A **7**, 2096 (1973).
1.550. D. Grischkowsky, E. Courtens, J. A. Armstrong: Phys. Rev. Lett. **31**, 422 (1973).
1.551. S. L. McCall, E. L. Hahn: Phys. Rev. Lett. **18**, 908 (1967).
1.552. S. L. McCall, E. L. Hahn: Phys. Rev. **183**, 457 (1969).
1.553. G. L. Lamb, Jr.: Rev. Mod. Phys. **43**, 99 (1971).
1.554. F. A. Hopf: In *Amplifier Theory in High Energy Lasers and Their Applications*, ed. by S. F. Jacobs, M. Sargent III, and M. O. Scully (Addision Wesley, Reading, Mass., 1975).
1.555. H. M. Gibbs, R. E. Slusher: Appl. Phys. Lett. **18**, 505 (1971).
1.556. R. E. Slusher, H. M. Gibbs: Phys. Rev. A **5**, 1634 (1972).
1.557. G. L. Lamb, Jr.: Phys. Rev. Lett. **31**, 196 (1973).
1.558. L. Allen, J. H. Eberly: *Optical Resonance and Two Level Atoms* (Wiley Interscience, New York 1975).
1.559. G. J. Salamo, H. M. Gibbs, G. G. Churchill: Phys. Rev. Lett. **33**, 273 (1974).
1.560. H. M. Gibbs, S. L. McCall, G. J. Salamo: To be published.
1.561. H. M. Gibbs, G. G. Churchill, G. J. Salamo: Opt. Commun. **12**, 396 (1974).
1.562. W. Krieger, P. E. Toschek: Phys. Rev. A**11**, 276 (1975).
1.563. M. Baumann: Z. Naturforsch. **21** a, 1049 (1969).
1.564. W. Demtröder, F. Paech, R. Schmiedl: To be published.
1.565. C. Fabre, S. Haroche: Opt. Commun. **15**, 254 (1975).
1.566. E. Giacobino: J. Physique Lett. **36**, L 65 (1975).
1.567. F. Biraben, E. Giacobino, G. Grynberg: To be published.
1.568. G. Huber, C. Thibault, R. Klapisch, H. T. Duong, J. L. Vialle, J. Pinard, P. Juncar, P. Jacquinot: Phys. Rev. Lett. **34**, 1209 (1975).
1.569. R. Marrus, D. McColm: Phys. Rev. Lett. **15**, 813 (1965).
1.570. H. J. Braun: Diplomarbeit Universität Köln (1973).
1.571. T. W. Hänsch, A. L. Schawlow: Opt. Commun. **13**, 68 (1975).
1.572. W. G. Schweitzer, Jr., M. M. Birky, J. A. White: J. Opt. Soc. Am. **57**, 1226 (1967).
1.573. M. S. Feld, A. Javan: Phys. Rev. **177**, 540 (1969).
1.574. A. Szabo: Phys. Rev. Lett. **25**, 924 (1970); **27**, 323 (1971).
1.575. Ch. Bordé, J. L. Hall: In *Laser Spectroscopy*, ed. by R. G. Brewer and A. Mooradian (Plenum Press, New York, London, 1974).
1.576. W. Heitler: *Quantum Theory of Radiation*, 3rd ed. (Oxford University Press, London, 1964).
1.577. P. A. Apanasevich: Opt. Spectrosc. **16**, 387 (1964).
1.578. S. M. Bergmann: J. Math. Phys. **8**, 159 (1967).
1.579. M. C. Newstein: Phys. Rev. **167**, 89 (1968).
1.580. V. A. Morozov: Opt. Spectrosc. **26**, 62 (1969).

1.581. B.R. MOLLOW: Phys. Rev. **188**, 1969 (1969).
1.582. C.S. CHANG, P. STEHLE: Phys. Rev. A **4**, 641 (1971).
1.583. R. GUSH, H.P. GUSH: Phys. Rev. A **6**, 129 (1972).
1.584. M.L. TERK-MIKAELYAN, A.O. MELIKYAN: Sov. Phys. JETP **31**, 153 (1970).
1.585. C.R. STROUD, JR.: Phys. Rev. A **3**, 1044 (1971); *Coherence and Quantum Optics*, ed. by L. MANDEL and E. WOLF (Plenum Press, New York, London, 1972), p. 537.
1.586. L. HAHN, I.V. HERTEL: J. Phys. B **5**, 1995 (1972).
1.587. G. OLIVER, E. RESSAYRE, A. TALLET: Lettere al Nuovo Cimento **2**, 777 (1971).
1.588. G.S. AGARWAL: Quantum Optics, in *Springer Tracts in Modern Physics*, Vol. 70 (Springer, Berlin, Heidelberg, New York, 1974), p. 108
1.589. M.E. SMITHERS, H.S. FREEDHOFF: J. Phys. B **8**, L 432 (1974).
1.590. H.J. CARMICHAEL, D.F. WALLS: J. Phys. B **8**, L 77 (1975).
1.591. C. COHEN-TANNOUDJI: In Proc. 2nd Intern. Laser Spectroscopy Conference, Megève 1975 (Springer, Berlin, Heidelberg, New York, 1975).
1.592. F. SCHUDA, C.R. STROUD, JR., M. HERCHER: J. Phys. B **7**, L 198 (1974).
1.593. H. WALTHER: In Proc. 2nd Intern. Laser Spectrosc. Conf., Megève 1975 (Springer, Berlin, Heidelberg, New York, 1975).
1.594. R. SCHIEDER, W. RASMUSSEN, W. HARTIG, H. WALTHER: To be published.
1.595. A. KASTLER: Appl. Opt. **1**, 17 (1962).
1.596. L.D. VIL'NER, S.G. RAUTIAN, S.A. KHAIKUI: Opt. Spectrosc. **12**, 240 (1962).
1.597. W.E. LAMB, JR.: Phys. Rev. **134** A, 1429 (1964).
1.598. A. SZOKE, A. JAVAN: Phys. Rev. Lett. **10**, 521 (1963).
1.599. R.A. MCFARLANE, W.R. BENNETT, W.E. LAMB, JR.: Appl. Phys. Lett. **2**, 189 (1963).
1.600. N.G. BASOV, I.N. KOMPANETS, O.N. KOMPANETS, V.S. LETPKHOV, V.V. NIKITIN: Sov. Phys. JETP Lett, **9**, 345 (1969).
1.601. F. SHIMIZU: Appl. Phys. Lett. **14**, 378 (1969).
1.602. R.G. BREWER, M.J. KELLY, A. JAVAN: Phys. Rev. Lett. **23**, 559 (1969).
1.603. J.L. HALL: In *Atomic Physics*, Vol. 3, ed. by S.J. SMITH and G.K. WALTERS (Plenum Press, New York, London, 1973).
1.604. C. FREED, A. JAVAN: Appl. Phys. Lett. **17**, 53 (1970).
1.605. P. TOSCHEK: Colloques Internationaux du CNRS No. 217 (Edition du CNRS, Paris, 1974).
1.606. W.G. SCHWEITZER, JR., E.G. KESSLER, JR., R.D. DESLATTES, H.P. LAYER, J.R. WHETSTONE: Appl. Opt. **12**, 2827 (1973).
1.607. G.W. SERIES: Contemp. Phys. **14**, 49 (1974).
1.608. J.L. HALL: IEEE J. Quant. Electron. QE-**4**, 638 (1968).
1.609. R.L. BERGER, J.L. HALL: Phys. Rev. Lett. **22**, 4 (1969).
1.610. A.P. KOL'CHENKO, S.G. RAUTIAN, R.I. SOKOLOVSKII: Sov. Phys. JETP **28**, 986 (1969).
1.611. E.V. BAKLANOV: Opt. Commun. **13**, 54 (1975).
1.612. J.L. HALL, CH. BORDÉ: To be published.
1.613. V.P. CHEBOTAYEV: Report at the 4th Vavilov Conference, Akademgorodok (June 1975).
1.614. K.C. HARVEY, R.T. HAWKINS, G. MEISEL, A.L. SCHAWLOW: Phys. Rev. Lett. **34**, 1073 (1975).
1.615. J.J. SNYDER, J.L. HALL: In *Proc. 2nd Intern. Laser Spectroc. Conf., Megève 1975* (Springer, Berlin, Heidelberg, New York, 1975).
1.616. J.E. BJORKHOLM, P.F. LIAO: Phys. Rev. Lett. **33**, 128 (1974).
1.617. G. HERZBERG: Proc. Roy. Soc. Ser. A **234**, 516 (1956).
1.618. J. HÄGER: Thesis, University of Köln 1975; to be published.
1.619. R. SCHIEDER: Thesis, University of Köln 1975; to be published.

2. Infrared Spectroscopy with Tunable Lasers

E. D. HINKLEY, K. W. NILL, and F. A. BLUM

With 26 Figures

Since the development of the first lasers in the early 1960's, most applications have relied on their high power and low beam divergence. Wavelength tuning has, by and large, been of secondary importance. Lasers are generally orders of magnitude brighter than thermal sources used with monochromators for conventional spectroscopy, and have inherently better resolution; however, continuous tuning over very wide intervals has been difficult to achieve. Advances in laser spectroscopy to 1970, where continuous tuning was generally restricted to within 0.1 cm^{-1} of discrete gas laser lines, were reviewed by DEMTRÖDER [2.1]. The use of fixed and tunable laser techniques for nonlinear spectroscopy, to 1971, is described in a comprehensive treatise by SHIMODA and SHIMIZU [2.2]. Since that time significant developments have taken place in the field of tunable lasers, and several new and revolutionary spectroscopic observations have been made.

It is our purpose in this chapter to review the state of the art of tunable infrared lasers and their spectroscopic applications. We shall confine our attention primarily to the wavelength region between 1 and 20 μm. This is the so-called infrared "fingerprint" region which is rich in characteristic molecular vibration-rotation lines, and is the most active area for laser spectroscopy at the present time. Excluded, therefore, are near-infrared tunable sources such as GaAs semiconductor diodes and organic dye lasers. We shall also exclude certain applications to chemistry (e.g. chemical reaction inducement, isotope separation, and vibrational energy transfer studies), as most of this work is performed with tunable lasers operating in the visible [2.3].

Perhaps the simplest tunable coherent source is the conventional gas laser (e.g. CO_2, CO, or He–Ne) which can be tuned discretely to any of a number of emission lines by a diffraction-grating cavity mirror. Continuous fine tuning over the Doppler-broadened width of a selected transition can be achieved via a piezoelectric cavity length adjustment. The very limited tuning range of the order of 0.003 cm^{-1} requires nearly perfect coincidence between the lasing transition and the spectral line of interest. Our main attention will be given to more widely tunable infrared lasers, such as semiconductor diodes, spin-flip Raman lasers, devices based on nonlinear effects, and gas lasers (Zeeman-tuned and high-pressure). Most of these can be tuned over at least 1 cm^{-1}, and some over much larger intervals. Tunable light sources in general were recently reviewed by KUHL and SCHMIDT [2.173], and spin-flip Raman lasers in particular by HÄFELE [2.174].

Historically, the first infrared laser spectroscopy was performed by GERRITSEN in 1966 by Zeeman-tuning the 3.39-μm line of a He–Ne gas laser [2.4]. He showed that within a rather limited tuning range of 0.2 cm^{-1}, several absorption lines of the C–H stretch bonds of CH_3F and C_2H_6 could be observed at low pressure. The use of other tunable infrared laser sources for spectroscopy did not begin in earnest until four years later. Doppler-limited absorption spectroscopy using current-tuned semiconductor diode lasers was reported by HINKLEY in 1970 [2.5], following an earlier measurement of the inherently narrow linewidth of this type of tunable coherent source [2.6]. Also in 1970, PATEL et al. published an infrared spectrum of NH_3 taken with a spin-flip Raman laser [2.7]. Using the difference frequency between a tunable dye laser and a ruby laser, DEWEY and HOCKER obtained, in 1971, an absorption spectrum of polyethylene in the 3–4.5 μm region [2.8]. An optical parametric oscillator was used by WALLACE at about the same time [2.9] to scan several isotope-split doublets of HBr in the first overtone region around 2 μm. In 1973 CORCORAN et al. reported the use of two-photon mixing involving a tunable microwave klystron and a fixed-frequency gas laser to reveal the gain lineshape of a molecular transition [2.10]; and BETEROV et al. were the first to use a high-pressure gas laser to perform infrared spectroscopy [2.11]. The characteristics of these lasers, and others with which spectroscopy has yet to be performed, will be discussed in Section 2.1.

Tunable infrared lasers have greatly enhanced the options available for spectroscopic investigations. New modulation techniques can be used, and long-path propagation is possible. Saturation spectroscopy and sensitive emission (heterodyne) spectroscopy are becoming realities. Along with this increased versatility, however, have come new complexities involving not only operation of the tunable laser itself but more fundamental questions related to wavelength calibration and stability. These unique attributes of tunable laser spectroscopy will be covered in Section 2.2.

Section 2.3 is a comprehensive description of the experimental work which has already been carried out using infrared lasers. A wide range of measurements is discussed in detail, encompassing investigations of spectral lineshapes, hyperfine structure, and band analysis. Data taken with several types of tunable lasers will be presented in order to point out their particular usefulness. Section 2.3 concludes with a description of the important applications of tunable infrared lasers to trace gas analysis and pollution monitoring.

Throughout this chapter we shall refer to the spectral position of the laser emission in terms of wavelength (μm), wavenumber (cm^{-1}), or frequency (MHz, GHz). Following conventional usage, ν will, at times, designate wavenumber as well as frequency. Frequency units will usually be used when the tuning increments are small (only a few Doppler widths). For purposes of conversion, recall that $1\,cm^{-1} \approx 30\,GHz$.

2.1. Tunable Infrared Lasers

Most high-resolution infrared spectra have been taken with grating spectrometers which disperse radiation from a thermal source and provide a resolving power given by the expression:

$$\nu/\Delta\nu = qN\,, \tag{2.1}$$

where ν is the infrared frequency, $\Delta\nu$ the instrument bandpass (resolution), N the total number of lines ruled on the grating, and q the order [2.12]. Although spectrometers with resolutions approaching $0.1\,cm^{-1}$ ($\Delta\nu = 3\,GHz$) are commercially available, very large and expensive gratings are necessary in order to reveal spectral lineshapes of gases even at atmospheric pressure where the lines are relatively broad. In addition,

Table 2.1. Properties of tunable infrared lasers

Type	Symbol	Wavelength coverage [μm]	Highest resolution obtained [cm^{-1}]	General references
Semiconductor diode laser	SDL	<1–34	3×10^{-6}	[2.14]
Spin-flip Raman laser	SFR	5–6	3×10^{-5}	[2.15]
		9–14	3×10^{-2}	[2.30]
Nonlinear optical devices				
Optical parametric oscillator	OPO	<1–11	3×10^{-2}	[2.16]
Difference frequency generator	DFG	3–6	2×10^{-3}	[2.17]
Two-photon mixer	TPM	9–11	3×10^{-5}	[2.10]
Four-photon mixer	FPM	2–24	1×10^{-1}	[2.18]
Gas lasers				
Zeeman-tuned	ZTG	3–9	3×10^{-3}	[2.19]
High-pressure CO_2	HPG	9–11	3×10^{-2}	[2.20]

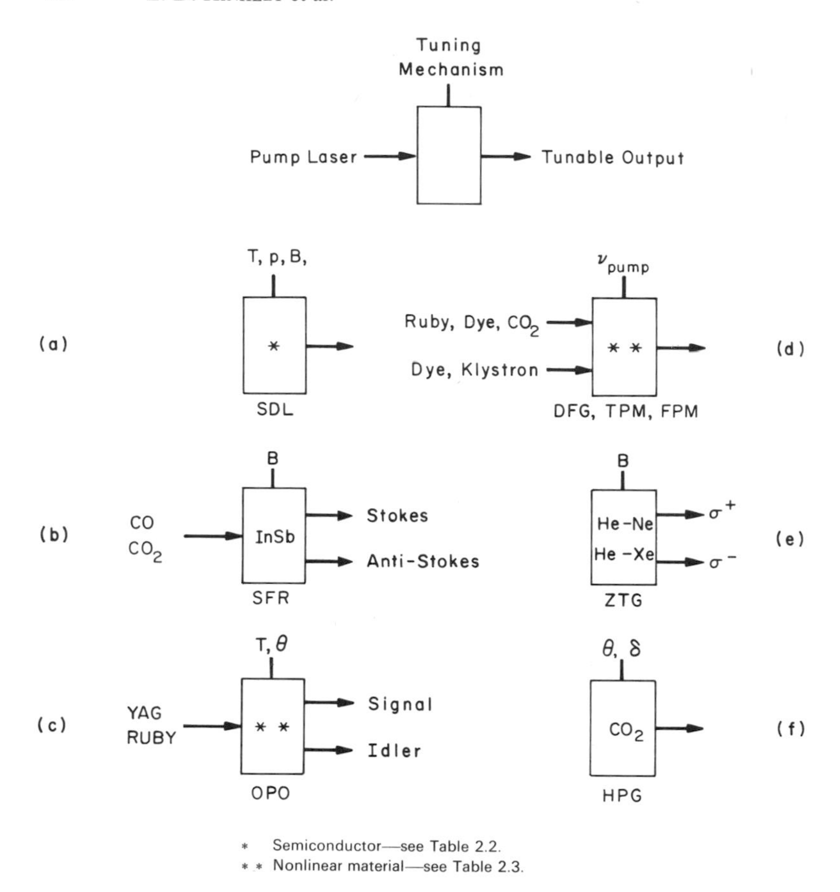

Fig. 2.1a–f. Schematic representations of six basic types of tunable infrared lasers. The semiconductor diode laser (a) and spin-flip Raman laser (b) require cryogenic cooling for normal operation, whereas the others usually operate at room temperature. The optical parametric oscillator (c) consists of a nonlinear crystal (see Table 2.3) pumped by a visible laser. Difference-frequency generation as well as two- and four-photon mixing are shown schematically in (d), where various pairs of pumps are used and tuning is accomplished by changing the frequency of one (or both) of them. In (e) a Zeeman-tuned gas laser is represented, and in (f) a high-pressure gas laser

for a thermal source there is an unavoidable trade-off between improved resolution and useful power. For example, the available power from an 1800 K globar at 10 μm is typically less than 5×10^{-9} W after passing through a spectrometer having a resolution of 0.01 cm^{-1}. Consequently, in order to produce high-resolution spectra having a reasonable signal-to-noise ratio, the wavelength scans must be made extremely slowly—with several days sometimes required.

In 1958 SCHAWLOW and TOWNES derived an expression for the quantum phase noise in a laser oscillator [2.13], which indicated that even for a low-power laser the linewidth would be orders of magnitude narrower than that required for most spectroscopic applications. The first experimental verification of this theory, by HINKLEY and FREED [2.6], revealed a linewidth of 54 kHz (1.8×10^{-6} cm^{-1}) for a 0.24 mW semiconductor diode laser. The spectral power density of 133 W/cm^{-1} is orders of magnitude larger than that available from thermal sources. At the present time, however, no single laser can be tuned continuously over the region from 1 to 20 μm and retain its inherently narrow linewidth. Consequently, several different techniques are being explored. Table 2.1 lists the tunable infrared lasers which have been developed to date, each identified in the second column by a three-letter acronym. Characteristics such as wavelength coverage and highest resolution achieved to the present time are listed. Although the continuous tuning range is also a very important parameter for spectroscopic applications, there is such wide latitude even for similar types of tunable lasers that it must be considered separately. Figure 2.1 is a set of schematic representations of each of the six different basic types of tunable lasers to be discussed in this section. As indicated in the legend, symbols are used to denote the pump laser (if required), the tuning mechanism, and the output beam(s). The information contained in Table 2.1 and Fig. 2.1 will be continually referred to and clarified as each of the tunable lasers is treated in detail in the remainder of this section.

2.1.1. Semiconductor Diode Lasers (SDL)

Extensive use of semiconductor diode lasers for high resolution infrared spectroscopy has been made since the first publication in 1970 of Doppler-limited spectra within the 10.6-μm band of SF_6 [2.5]. These first measurements utilized lasers of $Pb_{0.88}Sn_{0.12}Te$, and subsequent work by NILL et al. on $PbS_{1-x}Se_x$ diodes in 1971 extended the coverage to the shorter wavelengths between 4 and 7 μm [2.21]; and, more recently, to the 3.5 μm region by using $Pb_{1-x}Cd_xS$ [1.22]. SDL's with wavelengths below 3 μm have been fabricated from $Pb_{1-x}Cd_xS$ alloys, although extension to shorter wavelengths is unlikely due to difficulties with crystal growth. A recent review article by CALAWA describes the crystal growth and laser fabrication techniques, as well as some of the spectroscopic uses of SDL's [2.23]. The characteristics of impurity-diffused $Pb_{1-x}Sn_xTe$ diode lasers in the 8–12 μm region are discussed in detail by ANTCLIFFE and PARKER [2.116].

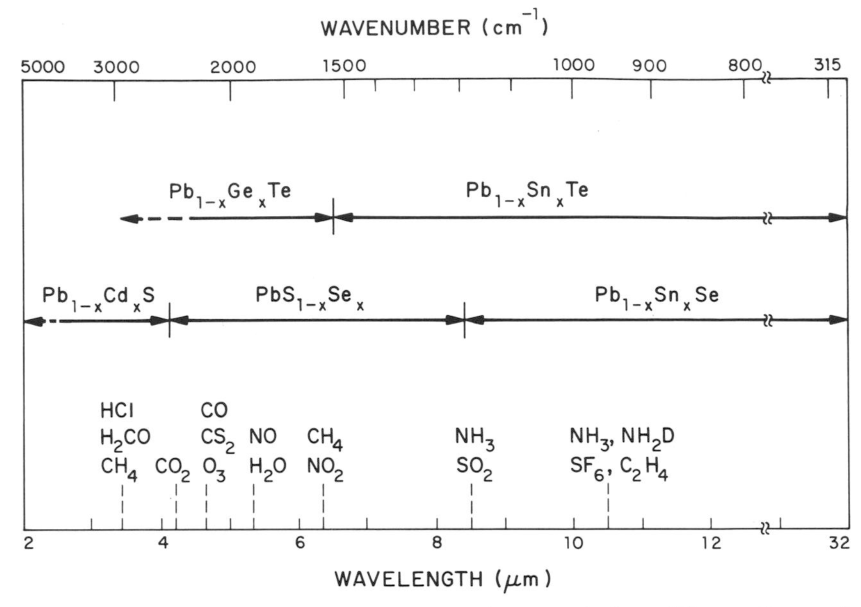

Fig. 2.2. Wavelength ranges for semiconductor lasers made from different alloys and compositions (see also Table 2.2). The dashed lines indicate possible future wavelength coverages. Also shown are regions in which diode laser spectroscopy has been performed on the gases indicated

The Pb-salt compounds which have been used for spectroscopy in the 3–15 μm region are indicated in Fig. 2.2, which shows the compositional tuning ranges of the semiconductors as well as the nominal locations of molecular absorption bands which have been studied to date. The spontaneous emission frequency ν_s varies almost linearly with the alloy composition factor x over most of the range of a given ternary alloy, i.e.,

$$\nu_s \approx \nu_0 + \xi x \,. \tag{2.2}$$

A list of the parameters ν_0 and ξ for the semiconducting compounds is given in Table 2.2, and can be used to select the appropriate crystal composition for the infrared region of interest. A *p-n* junction is formed in the grown crystal, which is then cut into rectangular parallelepipeds typically 1 mm long, and $0.3 \times 0.2\ \text{mm}^2$ in cross section. After cleaving or polishing [2.24], and attaching ohmic contacts, the diode is mounted onto a Cu stud which, in operation, is cooled to cryogenic temperatures.

Table 2.2. Composition parameters for Pb-salt lasers: $\nu_s = \nu_0 + \xi x$

Material	ν_0 [cm^{-1}]	ξ [cm^{-1}]	Composition range	Wavenumber range [cm^{-1}]
$Pb_{1-x}Sn_xTe$	1540	− 3.837	$0 \leqq x \leqq 0.32$	1540– 312
$Pb_{1-x}Ge_xTe$	1540	14.600	$0 \leqq x \leqq 0.05$	1540–2270
$Pb_{1-x}Sn_xSe$	1190	− 8.780	$0 \leqq x \leqq 0.10$	1190– 312
	1190	5.952	$0.19 \leqq x \leqq 0.40$	312–1562
$PbS_{1-x}Se_x$	2295	− 1.105	$0 \leqq x \leqq 1$	2295–1190
$Pb_{1-x}Cd_xS$	2295	29.396	$0 \leqq x \leqq 0.058$	4000–2295

A forward bias applied to the diode produces spontaneous emission at a wavelength determined by the energy gap of the semiconductor, with a bandwidth of approximately 6 cm^{-1}. Tuning of the spontaneous emission frequency can be accomplished by varying the diode temperature [2.25], applying a magnetic field [2.26, 27], or by applying external pressure [2.28] to the device. For example, by increasing the hydrostatic pressure on a PbSe diode laser to 15 kilobars, BESSON et al. tuned the spontaneous emission wavelength from 8.5 to 22 μm [2.28]. Also, with magnetic fields to 100 kG, the wavelength can be tuned over up to 5% of its original value; and 30% tuning with temperature has also been achieved [2.29].

Laser emission within the gain curve of the spontaneous emission is produced by multiple reflections and gain between the parallel end faces of the semiconductor crystal which determine the resonant cavity. Narrow-line emission occurs at a wavelength λ_m determined approximately by the Fabry-Perot equation

$$m\lambda_m = 2nl\,, \tag{2.3}$$

where n is the refractive index, l the cavity length, and n the number of half-waves within the crystal cavity. If, during tuning, the spontaneous emission wavelength closely follows that of the laser emission, then continuous *laser* tuning over the entire range of *spontaneous* tuning can be achieved. However, for the usual tuning mechanisms (T, B, p), the laser frequency does not follow the spontaneous emission, so that continuously tunable regions up to a few cm^{-1} wide are usually separated by discontinuous jumps in wavelength coverage of approximately the same amount. This is discussed in Subsection 2.2.2 and illustrated in Fig. 2.3 for a $Pb_{0.88}Sn_{0.12}Te$ SDL operating nominally around 10.6 μm. As the laser current is increased, there is a heating of the junction which affects both the energy gap and the refractive index. The overall rise in

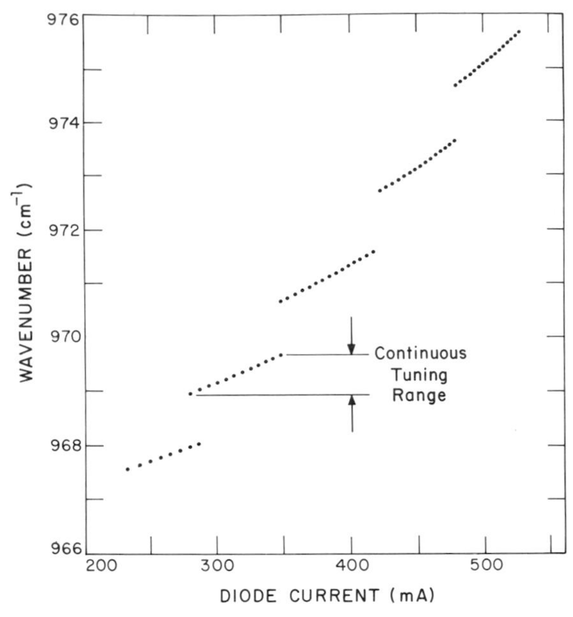

Fig. 2.3. Current-tuning curve for a $Pb_{0.88}Sn_{0.12}Te$ diode laser operating cw in a liquid-helium-cooled dewar. The tuning is continuous within each of the five modes. The data points indicate peaks in transmission of the laser radiation through a Ge etalon having a free spectral range of 1.955 GHz, used for relative frequency calibration [2.144]

laser frequency with current is due to the increase of the energy gap with temperature. Continuous tuning occurred over ranges of 0.5 to 1.5 cm^{-1}, as measured with a Fabry-Perot interferometer having a free spectral range of 1.955 GHz.

Laser linewidth and frequency stability are other important considerations for applications to high-resolution spectroscopy. In 1969 HINKLEY and FREED [2.6] displayed on a spectrum analyzer the beat note between a current-tunable $Pb_{0.88}Sn_{0.12}Te$ SDL and a stable, fixed-frequency CO_2 gas laser. The diode laser linewidth was found to be 54 kHz, and variations in its center frequency over several minutes were on the order of 100 kHz.

SDL's of $PbS_{1-x}Se_x$ with up to 370 mW of multi-mode power (as high as 50 mW in a single mode) have been produced in the 4- to 5-μm region [2.24], and similar improvements at other wavelengths are expected. Significant increases in the operating temperature and tuning range of $Pb_{1-x}Sn_xTe$ SDL's around 10 μm have recently been achieved using double heterostructures [2.29a]. A single cw laser with 1- to 10-mW power was tuned from 10.6 to 8.2 μm (280 cm^{-1}) by varying the heat-sink temperature from 12 to 80 K [2.29b], and Doppler-limited spectroscopy of NH_3 was performed over this range using a variable-temperature, closed-cycle cooler. The double heterostructure holds promise of significant improvements in operating temperature and performance of all Pb-salt SDL's.

2.1.2. Spin-Flip Raman (SFR) Lasers

The SFR laser, illustrated schematically in Fig. 2.1b consists of a fixed-frequency laser whose beam is inelastically (Raman) scattered by electrons in a semiconductor crystal. In the presence of a magnetic field there is a separation in energy levels between those electrons whose spins are aligned along the field and those with opposite spins. When the incident radiation causes an electron to be excited into a higher energy state, the scattered radiation is at a lower frequency and is called the Stokes component. Conversely, when the incident radiation induces an electron to shift to a lower energy state, the scattered radiation at the higher frequency is called the anti-Stokes component. Both of these components, as well as some weaker ones, can be useful for spectroscopy. The output SFR frequency is approximately related to the pump frequency ν_p by the expression

$$\nu \approx \nu_p \pm |g| \mu_B B/h , \tag{2.4}$$

where g is the conduction electron g-factor, μ_B the Bohr magneton, B the applied magnetic field, and h is Planck's constant. For InSb, which has been used as the semiconductor for most of the infrared SFR lasers, $|g| \approx 50$, such that the tuning rate is approximately 2.3 cm^{-1}/kG. The semiconductor must be cryogenically cooled to minimize thermal population of the upper spin level.

The first observation of tunable SFR scattering was reported by PATEL and SHAW in 1970 [2.30]. Using a pulsed CO_2 laser pump, they obtained radiation tunable from approximately 11.7 to 13.0 μm for magnetic fields from 48 to 100 kG. Shortly thereafter, MOORADIAN et al. [2.31a] developed the first cw SFR laser, with a CO laser pump, proving one watt of tunable infrared power in the 5- to 6-μm region. (Recently, ENG et al. [2.31b] achieved tunable SFR emission between 3332 and 3347 cm^{-1} using InAs pumped by an HF laser. The measured g-factor was 15.3, corresponding to a tuning rate of 0.59 cm^{-1}/kG for the spontaneous emission.) The SFR laser linewidth is determined to a large extent by the semiconductor cavity, and heterodyne measurements have shown it to be as low as 1 kHz [2.32a]; although the short-term frequency jitter is usually much higher due to pump laser instabilities, BRUECK has been able to achieve 100 kHz stability using frequency-locking techniques [2.32b]. In order to achieve narrow-linewidth operation, the InSb cavity must have a high Q. Under these circumstances, wide-range continuous tunability is sacrificed and the tuning curves become similar to those for the SDL shown in Fig. 2.3, except that the abscissa is the magnetic field B [2.33].

Some of the properties of the SFR laser which make it attractive for infrared spectroscopy are its potential for wide-range tunability, high output power, and good mode quality. Recently DESILETS and PATEL reported the operation of a low-field ($B < 6$ kG) SFR laser operating cw in a closed-cycle cryogenic cooler [2.34]. Their success suggests that, in addition to yielding adequate power for linear and nonlinear spectroscopic measurements, the SFR laser may find more practical applications, such as the continuous monitoring of atmospheric pollutants. Table 2.1 contains a summary of the SFR properties which have been mentioned above. A further discussion of its tuning characteristics and wavelength coverage will be given in Section 2.2.

2.1.3. Nonlinear Optical Devices

Several important sources of tunable infrared radiation are based on the principle of frequency mixing in a material possessing a nonlinear index of refraction. These devices operate at room temperature, and thus have at least one operational advantage over the SDL and SFR lasers described above. Some general characteristics of the four types to be discussed are given in Table 2.1. More detailed data are shown in Table 2.3.

Optical Parametric Oscillator

The OPO consists of a laser pump source in the visible or infrared, and a nonlinear birefringent crystal. The output frequency is established by the unique conditions required to simultaneously satisfy energy and momentum conservation for the parametric process occurring within the pumped material [2.16]. By rotating the nonlinear crystal, or by varying its refractive index through temperature changes, or by applying pressure or an electric field, a tunable infrared output can be achieved. The first successful operation of an OPO was reported in GIORDMAINE and MILLER [2.47] in 1965. They used a temperature-tuned $LiNbO_3$ crystal and Q-switched Nd : YAG laser pump. Three other types of non-

Table 2.3. Tunable nonlinear optical devices

Type	Coverage [μm]	Nonlinear material	Pump laser [s]	Ref.
OPO	0.4– 4	$LiIO_3$	Ruby	[2.35–37]
	0.4– 3.6	$LiNbO_3$	Nd : Glass, YAG	[2.38, 39]
	1.2– 8.5	Ag_3AsS_3	Nd : $CaWO_4$	[2.17]
	3.4– 7.9	CdSe	Dy : CaF_2, HF	[2.40a, 40b]
	9.8–10.4	CdSe	Nd : YAG, HF	[2.41, 40b]
DFG	2.2– 4.2	$LiNbO_3$	Argon/dye	[2.49]
	3 – 4.5	$LiNbO_3$	Ruby/dye	[2.8]
	3.2– 5.6	Ag_3AsS_3	Ruby/dye	[2.42]
	4.1– 5.2	$LiIO_3$	Ruby/dye	[2.43]
	10.1–12.7	Ag_3AsS_3	Ruby/dye	[2.44]
TPM	9 –11[a]	GaAs	CO_2/klystron	[2.10, 45]
FPM	2 –24	K vapor	Dye/dye	[2.18, 46]

[a] Tuning within 0.2 cm^{-1} of each CO_2 laser line using 6 GHz klystron.

linear crystals have been used in OPO's, as shown in Table 2.3. Recently, HANNA et al. [2.17] obtained radiation tunable from 1.2 to 8.5 μm by rotating a crystal of proustite (Ag_3AsS_3) which was pumped with a *Q*-switched Nd laser. The output power was 100 W (peak), and the linewidth approximately 1 cm^{-1}. By using an intra-cavity etalon the linewidth can be reduced considerably [2.48], but in order to obtain continuous tunability beyond the free spectral range of the cavity, the OPO frequency must be scanned with that of the etalon.

Difference-Frequency Generator (DFG)

Radiation from a fixed-frequency laser can be mixed in a nonlinear crystal with radiation from a tunable laser to obtain an output at the difference frequency. Using a $LiNbO_3$ crystal, DEWEY and HOCKER produced intense infrared radiation in the 3–4 μm region by mixing the output from a tunable dye laser with that from a *Q*-switched ruby laser [2.8]. The infrared power was 1 W (peak), and the effective resolution approximately 10 cm^{-1}, limited by mechanical vibrations.

Tunable cw emission between 2.2 and 4.2 μm has been obtained by PINE [2.49] by mixing radiation from two visible lasers (dye and argon-ion) in a crystal of $LiNbO_3$. The output power was 1 μW and the 15-MHz-wide line was continuously tunable over 1 cm^{-1}. Doppler-limited spectroscopy was performed on H_2O, NH_3, CH_4, and N_2O using this DFG technique.

Two-Photon Mixer (TPM)

Two-photon mixing has been used by CORCORAN et al. to produce precisely-controlled wavelength tuning in the 10 μm region [2.10, 45]. By mixing in a GaAs-loaded waveguide the signal from a millimeter-wave klystron with that from a stabilized CO_2 laser, infrared emission with the spectral purity of the CO_2 laser but tunable beyond its fluorescent linewidth has been produced. Unfortunately, the mixing efficiency is rather low, resulting in output powers of the order of 1 μW. Moreover, the frequency jitter is approximately 10 MHz, caused primarily by mechanical vibrations between the various components. Nevertheless, the high frequency precision of this TPM technique could be important in spectroscopic studies of certain gas laser transitions.

Four-Photon Mixer (FPM)

In 1973, SOROKIN et al. [2.18, 46] reported the development of a four-photon device tunable from 2.0 to 24.5 μm. It is based upon a four-wave mixing process in alkali metal vapor (in this case, potassium), with input beams provided by two tunable dye lasers which are pumped with the same nitrogen laser. The highest observed tunable output power was 100 mW, although the input power was over 1 kW (pulsed). The linewidth of the tunable radiation was estimated to be $0.1\ cm^{-1}$, corresponding to that of the dye lasers. The very broad tuning range of the FPM is attractive for spectroscopy; however, the rather complex operating requirements of this device will probably limit its widespread utilization in the near future.

2.1.4. Gas Lasers

Since conventional gas lasers usually operate at low pressures, the gain profile broadened primarily by the Doppler effect is quite narrow. For example, the Doppler width for CO_2 (10.6 μm, 293 K) is 51 MHz; for CO (5 μm, 293 K) it is 139 MHz; and for He–Ne (3.39 μm, 293 K) it is 222 MHz. Although continuous tuning can be achieved within the gain profile of a transition by, for example, changing the cavity length, the resulting tuning ranges are too restricted for general spectroscopic purposes. Techniques to shift the positions of the atomic transitions have been investigated, as well as attempts to widen them by operating the gas lasers at higher pressures. Two types of tunable gas lasers based upon these approaches are discussed below.

Zeeman-Tuned Gas (ZTG) Laser

Zeeman-tuning of the lasing transitions of atomic gases can result in significant tunability of the output wavelength. The infrared laser lines of He–Ne at 3.39 μm and He–Xe at 3.508 μm have been tuned using axial magnetic fields of several kG. A tuning range of 7 cm^{-1} was obtained by KASUYA [2.50] using a He–Xe laser and fields to 70 kG produced by a superconducting magnet, and spectra of numerous gases (CH_3F, H_2CO, CH_3I, N_2D_4) have been obtained. The tuning was not continuous over the 7-cm^{-1} interval, however. For an inhomogeneously-broadened laser transition, such as those in the He–Ne and He–Xe lasers, several closely spaced (0.005 cm^{-1}) axial cavity modes can oscillate simultaneously, and it is difficult to isolate and utilize a single mode. Furthermore, unless the cavity-mode frequencies are tuned (through mirror displacement) simultaneously with the Zeeman tuning of the gas transition, mode hopping will occur, producing discontinuous frequency increments. Laser tuning characteristics are discussed in detail in Subsection 2.2.2 below.

In order to obtain a broadly-tunable single-frequency output from a ZTG laser, an intra-cavity etalon can be used to select a single axial cavity mode. In addition, a cavity tuning mechanism (such as a mirror translator) for continuous frequency tuning, and a large (probably cryogenically-cooled) magnet capable of producing uniform fields to 100 kG are necessary. These complexities and the scarcity of suitable (high-gain) laser lines suggest that the ZTG laser will have rather restricted use in high-resolution infrared spectroscopy.

High-Pressure Gas (HPG) Lasers

As indicated above, by broadening the gain bandwidth of a gas laser it should be possible to increase its range of continuous tunability, especially if overlap can be created between adjacent vibration-rotation lines [2.20]. As shall be seen in *Lorentzian Profile—Collision Broadening* (in Subsection 2.3.1), pressure-broadening (self-broadening) coefficients for different gases vary from 5.1 MHz/Torr for CO to 25 MHz/Torr for NH_3 and SO_2. Since the adjacent lines of a CO laser are approximately 4 cm^{-1} apart, significant overlap will occur when the gas pressure reaches 30 atmospheres. For the CO_2 laser, overlap occurs at 10–15 atmospheres, and these pressures can be reduced if isotopic mixtures are used [2.51]. As indicated in Table 2.1, continuous tuning in the 9–11 μm region should be possible with a HPG laser of CO_2. It may eventually be possible to cover the 5–8 μm region with a CO HPG laser.

Population inversion and gain in a high-pressure gas are difficult to achieve. The initial development of an atmospheric-pressure CO_2 laser by BEAULIEU utilized electrical discharge between a set of pins situated in the laser cavity [2.52]. Electron-beam-pumping of a 100-atm CO_2 laser was reported by BASOV et al. [2.53], who observed that raising the pressure of the working mixture to that value did not lead to any qualitative changes in the processes of excitation and relaxation of the laser levels. The first spectroscopic experiments using HPG lasers were reported by BETEROV et al. [2.11] in which SF_6 was scanned over 0.007 cm^{-1}, and by BAGRATASHVILI et al. [2.54] in which a weak NH_3 absorption line was displayed at different pressures by 0.15-cm^{-1}-wide scans around the $P(12)$ CO_2 line.

Photoionization by ultraviolet radiation has also been used in an attempt to eliminate the difficulties associated with large-area electron-beam excitation. Recently, ALCOCK et al. [2.55] reported preliminary results of an ultraviolet-initiated transverse-discharge CO_2 laser which operated to a pressure of 15 atm. Tuning was achieved by means of a grating, which served as one of the end mirrors, following pre-dispersion by a prism. In an attempt to further simplify the HPG apparatus, LEVINE and JAVAN [2.56] used a low-ionization-potential organic vapor in the CO_2 gas laser to operate it at 1 atm using an electrically-heated plasma generated via photoionization by xenon flashlamps.

Finally, ABRAMS and BRIDGES recently reported [2.57a] the successful operation of a cw, sealed-off capillary-bore (waveguide) CO_2 laser with an output power of ≈ 1 W at a pressure of a few hundred Torr. By reducing the diameter of the discharge, gas pressure can be increased substantially with little effect on efficiency, yielding gain bandwidths which are large enough to be of spectroscopic interest. ABRAMS has performed opto-acoustic spectroscopy with a waveguide CO_2 laser tunable over 1.3 GHz, and obtained pressure-broadening coefficients for CO_2–CO_2, CO_2–N_2, and CO_2–He [2.57b].

2.2. Spectroscopic Techniques

This section discussed the operating characteristics of the broadly tunable laser sources described in Section 2.1, and the techniques employed to obtain the spectroscopic results which will be presented in Section 2.3. The topics to be covered here include the stability and resolution capabilities of the tunable lasers, their tuning mechanisms, wavelength coverage and calibration, and techniques of modulation and detection. We are primarily concerned with the use of these lasers in infrared *absorption* spectroscopy, since this is the main area of application at the present time. A detailed treatment of the general field of laser spectroscopy can be found in DEMTRÖDER [2.1], which discusses several other potential application areas of the tunable sources as well as the many desirable qualities for use as spectroscopic light sources. The early absorption spectroscopy using He–Ne and He–Xe ZTG lasers is also described in some detail.

In order to gain insight into the resolution requirements for infrared spectroscopy, consider the absorption linewidth of a typical gas at room temperature. The widths (full width at half maximum absorption coefficient) of such lines are typically 30–300 MHz in the middle infrared (2–20 μm) due to the Doppler effect caused by thermal motion of the molecules. These narrow linewidths can be observed only for gas pressures below a few Torr where broadening due to collisions is negligible. The widths of absorption lines of typical gases in the atmosphere are 0.1 to 0.3 cm^{-1}, although several *atypical* atmospheric water lines as narrow as 0.015 cm^{-1} have been observed [2.58]. In order to fully resolve the lineshape of a low-pressure gas absorption line, a resolution at least ten times better than the linewidth is required, namely, 3–30 MHz. Higher resolution is necessary if saturation (nonlinear) spectroscopy is to be performed (see *Natural Linewidth and Saturation Spectroscopy* in Subsection 2.3.1), where linewidths can be 30 kHz or less. In such experiments, the resolution must be on the order of 3 kHz or less in order to fully resolve the lineshapes.

2.2.1. Resolution Capabilities of Tunable Lasers

Figure 2.4 is a graphic comparison between molecular absorption linewidths and the resolution capabilities of infrared spectrometers and tunable lasers. It is apparent that, although a high-quality laboratory spectrometer can provide accurate lineshape data for typical gaseous absorption lines which are pressure-broadened in the atmosphere, its resolution is much too low for Doppler-broadened lines. Even though the inherent linewidths of tunable infrared lasers are orders of magnitude narrower than required for Doppler-limited spectroscopy, there are other characteristics (such as cavity instability) that can greatly reduce their effective resolution. These limiting characteristics are discussed in detail below.

The ultimate resolving power of a tunable laser is limited by its fundamental linewidth, which depends on the laser power level and cavity conditions. It is important to distinguish between this intrinsic linewidth due to quantum phase noise and the laser's frequency instability

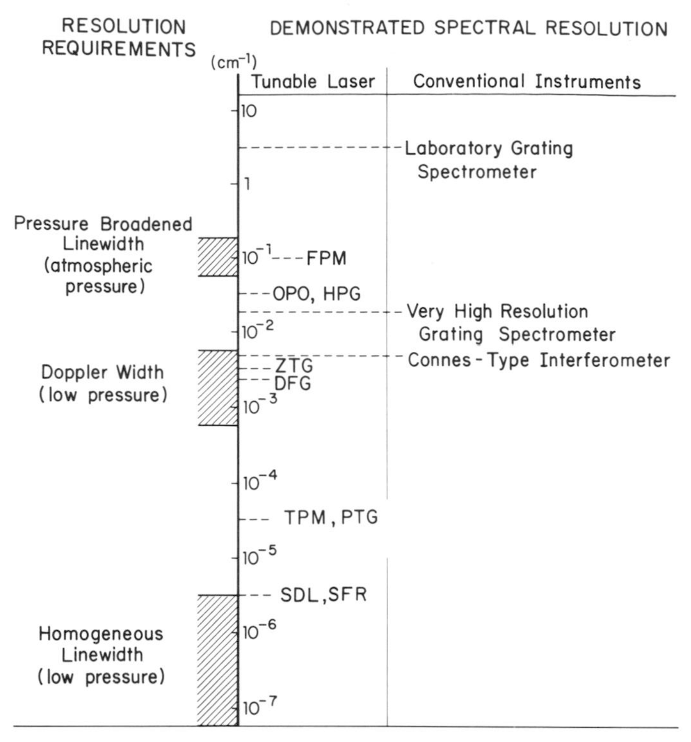

Fig. 2.4. Resolution requirements for infrared spectroscopy, compared with the capabilities of conventional instrumentation and the highest resolution yet achieved for several types of tunable lasers. The indicated resolution for the SFR, TPM, and PTG tunable sources is limited in part by the instability of the associated gas lasers (see Fig. 2.1). A stability of $3 \times 10^{-6}\,\mathrm{cm}^{-1}$ in a 1-sec interval has been achieved by BRUECK [2.71] for a SFR laser phase-locked to a stable CO laser, and the same approximate spectral resolution has been claimed (PATEL [2.74]) on the basis of saturated spectra of H_2O using an unstabilized SFR laser. The Connes-type interferometer was described in [2.59]

due to external factors. In fact, attempts to directly measure the inherent laser linewidth did not succeed for many years because of such instabilities (caused by mechanical vibrations or electromagnetic interactions) [2.60]. The laser linewidth $\Delta\nu_L$ determined by quantum phase noise, is a function of the cavity Q $(=\nu/\Delta\nu_c)$ and mode power P as follows [2.13]

$$\Delta\nu_L = \pi h\nu(\Delta\nu_c)^2/P\,. \tag{2.5}$$

The cold-cavity bandwidth $\Delta\nu_c$ is very small for most lasers, and their fundamental linewidths are usually much narrower than the resolution requirements of Doppler-limited spectroscopy ($\Delta\nu \lesssim 3$ MHz). For SDL's, however, $\Delta\nu_c$ is relatively large and P can be quite small, resulting in linewidths in the vicinity of 30 kHz to 3 MHz. The resolution capability of SDL's was established by direct measurements of the laser linewidth [2.6], which involved heterodyning radiation from a 10.6-μm $Pb_{0.88}Sn_{0.12}Te$ diode laser with that from a highly stable CO_2 laser, as discussed above. Since the CO_2 laser linewidth was extremely narrow, and its frequency was stable to ≈ 1 kHz, the spectral distribution of the beat signal represented the power spectral density of the diode laser emission. A linewidth of 54 kHz was measured for a SDL with 240 μW mode power. A similar measurement has been performed for $PbS_{0.6}Se_{0.4}$ diode lasers operating around 5.2 μm. In this case the tunable SDL emission was heterodyned with that from a stable CO laser [2.61]; and for a diode laser power of 10 μW the linewidth was found to be ≈ 1.5 MHz. These results indicate that for a typical SDL having more than 100 μW of mode power, the linewidth will be less than 100 kHz. Since the long-term frequency instability due to other factors is of the same order as the linewidth, the SDL can easily resolve Doppler-broadened spectra.

The effective resolution of most of the tunable laser sources described in this chapter is limited by frequency jitter (rather than by the fundamental linewidth) which can arise from an instability in the tuning mechanism or from changes in other parameters which affect the output frequency. For example, Zeeman-tuned He–Ne and He–Xe laser cavities must be well-stabilized against temperature drifts and mechanical vibration; for, while it is possible to achieve a stability of 1 kHz with a well-designed gas laser cavity [2.62], drift and fluctuations of the order of 1 to 10 MHz are not uncommon for conventional gas lasers. Also, in order to avoid frequency "chirping" which occurs during pulsed operation of SDL and SFR lasers, cw operation is usually employed in experiments designed to yield high-resolution spectra. (Doppler-limited spectroscopy has been performed with a pulsed SDL operating at 77 K by time-resolution of the frequency scan occurring during the current pulse [2.63].)

Since thermal drifts and similar externally-induced effects can usually be overcome, the frequency stability of most of the broadly tunable lasers usually reflects the stability of the tuning mechanism itself. To illustrate, if a laser can be tuned over 30 cm^{-1} by varying a tuning parameter (magnetic field, temperature, etc.), the frequency stability of the output will be 1 MHz only if that parameter is stable to ≈ 1 part in 10^6 (i.e. ≈ 1 ppm). For example, the tuning rate of the first Stokes line of the SFR laser is 2.3 cm^{-1}/kG. Therefore, for 1 MHz stability of the output frequency, the magnetic field must be stable to 0.014 G—or 2 ppm if the applied field is 7 kG. Similarly, the SDL is current-tunable at rates between 10 and 1000 MHz/mA (depending upon the fabrication process), so that a current stability of 1–100 μA (2–200 ppm at 0.5 A current), is required for 1 MHz stability. These are rather stringent requirements on the tuning control. Finally, some tunable sources, such as the OPO, SFR, and TPM, require a primary fixed-frequency laser pump. The stability of the tunable output frequency then reflects that of the primary laser as well as the stability of the tuning mechanism.

2.2.2. Laser Tuning Characteristics

As indicated in Section 2.1, laser tuning characteristics are complicated by effects such as mode hopping, multimode output, and nonlinearities. This section uses a simple model to indicate the reasons for these effects, and to determine their dependence on laser parameters which can be controlled. In general, there is considerable flexibility in the choice of laser parameters; and wide variations are therefore possible in tuning rate, range, and linearity. In some cases different tuning mechanisms can be selected (e.g., temperature, magnetic field, or pressure for the SDL) which have significantly different tuning characteristics. Some of the compromises involved in these choices are indicated below. A further reason for discussing the tuning mechanisms and characteristics is that *all* potential sources of frequency instability (such as an unrecognized tuning mechanism) must be controlled in order to achieve the full spectral resolution capability of the laser.

The laser emission frequency is determined jointly by the gain spectrum of the lasing medium and the resonances (modes) of the optical cavity [2.64]. Tuning results from changes in either one (or both) of these, and the resulting characteristics can be rather complex. Figure 2.5a illustrates some of the typical tuning characteristics of a simple laser. The frequency tunes continuously at a nearly linear rate over a small range, and then discontinuously shifts to another frequency. This behavior results in incomplete frequency coverage, which presents some difficulty when a source of this type is used for high-resolution spectroscopy. Furthermore, simultaneous oscillation may occur on several axial cavity modes.

In order to illustrate the essential ideas required to understand the tuning characteristics, we will consider the simplest laser model in which oscillation occurs in a single axial mode only. While this model is only appropriate to a homogeneously-broadened gain medium in the absence of spatial inhomogeneity [2.65], it contains features required to explain and understand details such as mode hopping and tuning rates. Figure 2.5b defines the spontaneous emission gain and cavity parameters. Here Γ_s

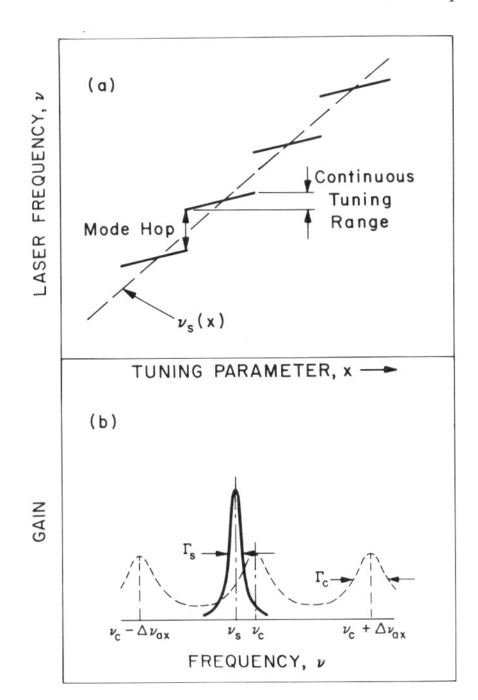

Fig. 2.5a and b. Illustration of the tuning characteristics of a simple laser. (a) shows the continuously tunable segments separated by mode jumps, as the tuning parameter x is increased; (b) defines the spontaneous emission gain and cavity parameters (see text)

and ν_s are the bandwidth and frequency of the gain spectrum, Γ_c is the cavity bandwidth, and ν_c the closest cavity resonance frequency. The relative sizes of Γ_s and Γ_c are typical of the SFR laser (i.e. $\Gamma_s \ll \Gamma_c$). For the SDL, OPO, and HPG lasers, $\Gamma_s > \Gamma_c$. The laser frequency in each continuous portion of Fig. 2.5a is intermediate between ν_s and ν_c in the simplest model, and is given by

$$\nu = (\nu_c \Gamma_s + \nu_s \Gamma_c)/(\Gamma_s + \Gamma_c)\,. \tag{2.6}$$

Tuning of the laser frequency typically results from variations in ν_c and ν_s, with the relative effects of each on ν determined by Γ_s/Γ_c. The frequency of the laser closely follows ν_s if $\Gamma_s \ll \Gamma_c$, and ν_c if $\Gamma_c \ll \Gamma_s$. When $\Gamma_s \ll \Gamma_c$ the laser frequency is "pulled" away from the cavity resonance ν_c toward ν_s. Since ν_c and ν_s typically tune at different rates there is a relative motion between ν_c and ν_s during tuning. As ν_s passes the midpoint between two cavity resonances, the laser frequency ν discontinuously shifts toward the next (adjacent) mode—an effect called a "mode hop". The size of the discontinuity depends on the relative sizes of Γ_s and Γ_c, and it can be a significant fraction of the cavity axial-mode separation $\Delta\nu_{ax}$. The axial mode spacing for a simple, plane-parallel Fabry-Perot resonator can be expressed in terms of the cavity length l and the optical index of refraction n by

$$\Delta\nu_{ax} = \left[2l\left(n+\nu\frac{\partial n}{\partial\nu}\right)\right]^{-1} \mathrm{cm}^{-1}, \tag{2.7}$$

where, for simplicity, ν is given in wavenumber units [cm^{-1}]. From (2.6) the spacing between adjacent longitudinal modes of the laser emission frequency can be derived

$$\Delta\nu_l = \left(\frac{\Gamma_s}{\Gamma_s+\Gamma_c}\right)\left\{2l\left[n+\nu\frac{\partial\nu}{\partial\nu}\left(\frac{\Gamma_s}{\Gamma_s+\Gamma_c}\right)\right]\right\}^{-1}, \tag{2.8}$$

which can be considerably less than the cavity mode spacing $\Delta\nu_{ax}$ if $\Gamma_s \ll \Gamma_c$ (as in the low-concentration InSb SFR laser). The second term in the brackets of (2.7) and (2.8) is typically less than 20% of the first term ($<0.2n$), such that $\Delta\nu_l/\Delta\nu_{ax} \simeq \Gamma_s/(\Gamma_s+\Gamma_c)$. The tuning rate of the laser frequency can be derived by partial differentiation of (2.6) with respect to a tuning parameter "x" (e.g., temperature, magnetic field, mirror spacing) with the result

$$\frac{d\nu}{dx} = \left(\Gamma_c\frac{d\nu_s}{dx} - \Gamma_s\frac{\nu}{n}\frac{\partial n}{\partial x} - \nu_c\frac{\Gamma_s}{l}\frac{dl}{dx}\right)\left[\Gamma_c+\Gamma_s\left(1+\frac{\nu}{n}\frac{\partial n}{\partial\nu}\right)\right]^{-1} \tag{2.9}$$

Using (2.8) and (2.9), we can calculate the tuning rate, tuning range, and frequency shift due to mode hopping. Tuning characteristics of the SDL and the InSb SFR laser are considered here since they exhibit most of the characteristics predicted by this simple model.

Equations (2.6)–(2.9) can be used to estimate the general nature of the tuning characteristics of a laser. There are, however, other features which become important in spectroscopic measurements which will also be briefly considered here. Over the tuning range of a mode, the power will reflect the gain variations at the mode frequency. In a SDL having several simultaneously oscillating modes (multimode), the power in a particular mode rises smoothly from zero, passes through a maximum and gradually falls back to zero. In single-mode SDL's and SFR lasers the power in a mode rises abruptly near a mode hop, and temporal switching between adjacent modes is sometimes observed. Hysteresis often occurs

[2.33, 66] when either of two adjacent modes oscillates, depending on whether the laser frequency is being monotonically increased or decreased.

The difficulties with incomplete wavelength coverage resulting from mode hopping can be largely eliminated if there are two substantially independent tuning mechanisms [x in (2.9)]. For example, simultaneous current (temperature) and magnetic-field tuning of SDL's has been employed in spectroscopic studies using PbSSe and PbTe lasers [2.27]. In these experiments the magnetic field was used to produce laser emission in the spectral region of interest (at some fixed diode current), and then held steady while the mode frequency was tuned continuously by adjusting the diode current. For lasers having more than one convenient tuning mechanism, the method of tuning is chosen from consideration of factors such as 1) the tuning range of a mode [from (2.6) and (2.7)], 2) the required stability and tuning linearity [from (2.9)], and 3) the wavelength overage for the range of usable tuning parameters. For example, current (i.e., temperature) tuning of the laser modes of SDL's is usually chosen over magnetic field tuning because the continuous range for current tuning is 3–5 times larger, and it is more linear. However, a large magnetic field increases the overall wavelength coverage of a device substantially ($\approx 50\,\mathrm{cm}^{-1}$ for 100 kG), and this can be quite useful in spectroscopic studies. Note that the wavelength coverage for a diode laser is determined by the available magnetic field (typically < 100 kG) and the temperature range over which cw operation can be obtained (typically < 50 K). As mentioned above, nearly all high-resolution spectra taken with SDL's have utilized cw operation in order to maintain the required frequency stability.

The laser parameters and tuning characteristics calculated from (2.6)–(2.9) for SFR lasers are shown in Table 2.4. The tuning characteristics of the SFR laser are particularly interesting since the large range of Γ_s (0.01–$1.0\,\mathrm{cm}^{-1}$) results in more than an order of magnitude variations in the mode spacing, tuning rate, and tuning range. The linewidth Γ_s of the spontaneous Raman gain in n-InSb increases rapidly with electron concentration from a value of $\approx 0.01\,\mathrm{cm}^{-1}$ at $n \approx 10^{14}\,\mathrm{cm}^{-3}$ to almost $1\,\mathrm{cm}^{-1}$ at $n \approx 10^{16}\,\mathrm{cm}^{-3}$. Since $\Gamma_s/\Gamma_c \ll 1$ for $n \approx 10^{14}\,\mathrm{cm}^{-3}$, the tuning rate of the laser output with magnetic field, $d\nu/dB$, nearly matches the spontaneous tuning rate $d\nu_s/dB$. In addition, the longitudinal mode spacing $\Delta\nu_l$ is much less than the axial mode spacing $\Delta\nu_{ax}$, so that the tuning range of a mode is nearly the full cavity mode spacing. The tuning range around the pump frequency for low-concentration InSb samples is limited to 10–$20\,\mathrm{cm}^{-1}$ due to the incidence of carrier freezout [2.67] at magnetic fields of around 10 kG. However, any of the numerous laser lines [2.68] available from a CO laser (separated by $< 5\,\mathrm{cm}^{-1}$ in the

Table 2.4. Typical tuning characteristics of SFR lasers and SDL's

	Spin-flip Raman laser $\nu = 1700$–$1900\ \mathrm{cm}^{-1}$, $l = 4.25$ mm, $T = 4.2$ K		Pb-salt diode laser $\nu = 500$–$2500\ \mathrm{cm}^{-1}$, $l = 0.5$ mm, $T = 4.2$ K		
	InSb [$\approx 10^{14}\ \mathrm{cm}^{-3}$] $B \approx 5$ kG	InSb [$\approx 10^{16}\ \mathrm{cm}^{-3}$] $B \approx 30$ kG	B-tuned ($x = B$)	I-tuned ($x = I$)	p-tuned ($x = p$)
Γ_s [cm^{-1}]	0.01	1.0		5–50	
Γ_c [cm^{-1}]	0.15	0.15		0.3–1	
$\Delta\nu_{ax}$ [cm^{-1}]	0.26	0.26		1–2	
$d\nu_s/dx$	$-2.3\ \mathrm{cm}^{-1}$/kG	$-2.3\ \mathrm{cm}^{-1}$/kG	0.5–5 cm^{-1}/kG	10–100 cm^{-1}/A	50–100 cm^{-1}/kB
$\lvert(\nu/n)\partial n/\partial x\rvert$	0.05	0.05	≈ 0.1	2–20	5–20
Continuous tuning rate ($d\nu/dx$)	$-2.2\ \mathrm{cm}^{-1}$/kG	$-0.3\ \mathrm{cm}^{-1}$/kG	$\approx 0.1\ \mathrm{cm}^{-1}$/kG	2–20 cm^{-1}/A	5–20 cm^{-1}/kB
Tuning range [cm^{-1}]	0.25	0.03	0.1–0.4	0.2–2	1–2
Mode hop [cm^{-1}]	0.01	0.23	1–2	0.1–1	1–2
Longitudinal-mode spacing [cm^{-1}]	0.023	0.23		1–2	

5–6 μm region) can be used to pump the InSb SFR laser to provide wide overall wavelength coverage using magnetic fields below 10 kG. An additional advantage of using small magnetic fields is that they can be obtained from a highly stable electromagnet or a permanent magnet. In the latter case, the field at the InSb crystal can be controlled by moving the magnet pole pieces relative to the crystal [2.69].

Although high-resolution spectra have been obtained with a stable, CO-pumped, cw SFR laser operating in the 5-μm region [2.69, 70], spectra obtained with pulsed CO_2-pumped SFR lasers have markedly poorer resolution [2.7, 109, 110]. This suggests either an unstable laser frequency during the pulse (chirping) or the presence of more than one oscillating laser mode (perhaps due to transverse modes in the InSb cavity). The chirping could result from the dependence of spin-flip energy on electron temperature [2.33] which increases during the pulse of pump laser power. Transverse mode structure has been observed [2.71] in a CO-pumped, cw SFR laser even when the pump laser was oscillating in the fundamental mode.

Semiconductor diode lasers can be tuned using any of several different parameters: changes in temperature, applied magnetic field, or pressure affect 1) the energy gap of the semiconductor, hence, the spontaneous gain frequency ν_s, 2) the spontaneous emission linewidth Γ_s, and 3) the index of refraction n. Since Γ_s (5–50 cm^{-1}) usually far exceeds Γ_c ($\leqq 1\ cm^{-1}$), tuning of the laser frequency is primarily the result of changes in the refractive index $n(T, B, p)$ through (2.9). Under these conditions, using our simplified model, the *continuous* tuning range is the change in laser output frequency as the spontaneous line shifts by an amount equal to $\Delta\nu_{ax}$ given in (2.7). One can easily show that the largest continuous tuning range is obtained when the ratio of the *cavity* tuning rate $d\nu_{ax}/dx$ to the *spontaneous* tuning rate $d\nu_s/dx$ is unity. For the SDL this ratio is typically 0.1–0.2 for current (temperature) tuning and 3–5 times smaller for magnetic-field tuning, so that the current tuning range is correspondingly greater. Wide continuous scans are obviously very important for laser spectroscopy, and current tuning of SDL's has been used almost exclusively for this reason.

The tuning characteristics of Pb-salt lasers which have been used in spectroscopic measurements are also listed in Table 2.4. The values are estimates for typical devices and large deviations are possible. Few direct measurements have been made of the parameters listed. The absolute magnitude of the current tuning rates [e.g., $d\nu_s/dI$, $(\nu/n)\partial n/\partial I$] is not of fundamental significance since these are temperature effects, and the variation of the temperature with current is not known *a priori*. More than an order-of-magnitude change of the tuning rate $d\nu/dI$ is

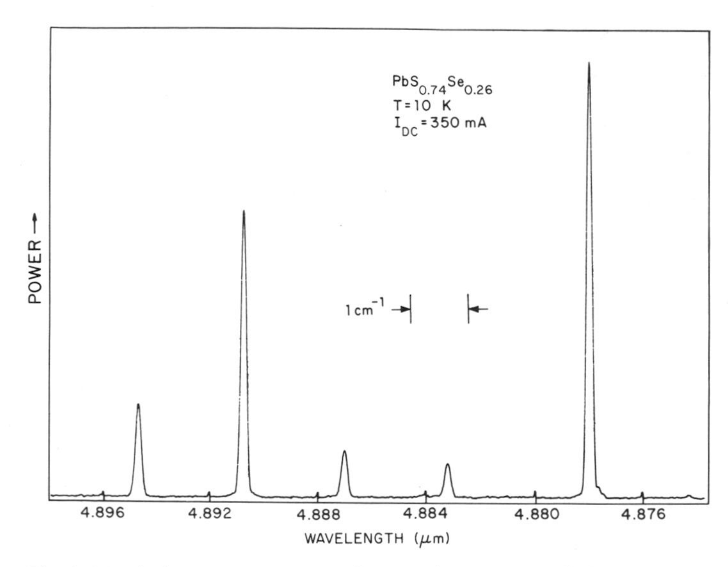

Fig. 2.6. Emission power spectrum from a $PbS_{0.74}Se_{0.26}$ diode laser at a constant current of 350 mA. The laser linewidth is less than 3×10^{-5} cm^{-1}; the spectrometer resolution is approximately 0.1 cm^{-1}. (After NILL and BLUM [2.72])

observed for different lasers due to variations in the heat transfer properties (i.e., dT/dI) of the devices and their dewar mounts.

Figure 2.6 shows the emission power-spectrum from a typical Pb-salt SDL, as recorded with a grating spectrometer having ≈ 0.1 cm^{-1} resolution. (The laser linewidth is much narrower than indicated.) Operating at a constant current of 150 mA at a temperature of around 10 K, the diode laser is emitting radiation in five distinct modes approximately 1.5 cm^{-1} apart. This agrees with a calculation based on a cavity length of 0.4 mm and an effective refractive index (2.7) of ≈ 6. If the current were increased slightly, the mode positions would shift toward higher frequency (to the right). The rate of change of a particular mode frequency with current, $d\nu/dI$, increases smoothly with increasing current, and although usually rather constant, it can vary by as much as 50% over the tuning range (1–2 cm^{-1}). Hence, it is necessary to calibrate the tuning rate at intervals of < 0.1 cm^{-1} in order to obtain an accurate wavenumber scale for transmission spectra.

Finally, the tuning characteristics of the ZTG lasers which have been used for spectroscopy should be mentioned. As typically used, the spectral resolution obtained by magnetic field tuning of this laser is considerably worse than the laser linewidth or frequency stability. It is limited primarily by the axial-mode spacing of the laser cavity, $\Delta\nu_{ax}$ of (2.7). The magnetic

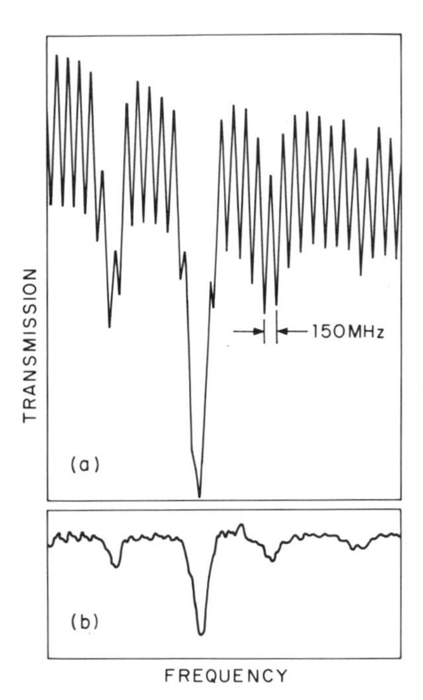

Fig. 2.7a and b. Infrared spectra of H_2CO around 2851.6 cm^{-1} taken with a magnetic-field-tuned He : Xe laser. In (a) the transmitted power through the cell is shown, illustrating the 150-MHz axial cavity-mode resonances of the ZTG laser; and in (b) a ratio technique is used to remove these effects. The 50-cm cell contained 5 Torr of H_2CO. (After KUSUYA [2.50]

field affects the location of the spontaneous emission ν_s of the ZTG laser but has little effect on the laser mode frequency ν [since ν_c is nearly independent of magnetic field and $\Gamma_c/\Gamma_s \ll 1$, cf. (2.6)]. Consequently, the output spectrum of the magnetic-field-tuned ZTG laser consists of a series of discrete cavity-mode frequencies separated by the axial-mode spacing $\Delta\nu_l \approx \Delta\nu_{ax}$. While continuous tuning of the laser frequency by varying the magnetic field produces a negligible continuous tuning range, the laser frequency can be continuously tuned over the spontaneous width $\Gamma_s \approx 0.003\,cm^{-1}$ by scanning one of the laser mirrors. Simultaneous scanning of both the magnetic field and the mirror spacing could increase the continuous range, but this has not been attempted.

The spectral resolution obtained using a magnetic-field-tuned, 1.5-m-long ZTG laser with an axial-mode spacing of 0.003 cm^{-1} is clearly no better than 0.003 cm^{-1}—a resolution far worse than either the laser linewidth (2.5) or its frequency stability. However, a large spectral range of up to $\pm 3.5\,cm^{-1}$ can be continuously scanned *with this resolution.* Figure 2.7a shows the power from a He : Xe laser transmitted through a 50-cm gas cell containing H_2CO, as the frequency is magnetic-field tuned around 2851.6 cm^{-1} [2.50]. The horizontal axis is readily calibrated in terms of laser frequency since pronounced power variations occur when the laser output shifts between cavity modes separated by 150 MHz. The four absorption lines of H_2CO in this region are more apparent in Fig. 2.7b, obtained using a double beam arrangement which will be discussed in *Experimental Arrangement* below.

2.2.3. Experimental Techniques of Laser Spectroscopy

The experimental techniques for high-resolution spectroscopy with tunable lasers are discussed here. For illustrative purposes, emphasis is placed on the use of semiconductor diode lasers—quite similar techniques are used for the other tunable lasers. Particular consideration will be given to experimental procedures which do not arise in conventional spectroscopic measurements.

There are several advantages in using lasers instead of standard incoherent sources, in addition to that of vastly improved spectral resolution. Among these are the ability to propagate the collimated radiation over large distances (such as in a long-path absorption cell). Also, there are several modulation and detection techniques which can be employed with tunable laser sources. These techniques, as well as those used to calibrate the tunable lasers, are discussed below.

Experimental Arrangement

Figure 2.8 illustrates the usual arrangement of equipment for laser absorption spectroscopy. The collimated output of the laser, possibly consisting of several modes, is transmitted through a gas cell and then spectrally filtered by a grating spectrometer which selects the wavelength range ($\gtrsim 1\ cm^{-1}$) of interest. In some cases, a reference beam is used (dashed line) in order to compensate for laser-power variations by a ratio technique. The entrance and exit slits of the spectrometer are relatively wide so that its transmission can be nearly constant over the tuning range of interest. In this configuration the spectrometer serves both to exclude all but one laser mode and to provide an absolute calibration of the laser wavelength to within $\approx 0.1\ cm^{-1}$ (when the slits are narrowed). It does not determine the spectral resolution in this arrangement. For a cw laser, the beam is usually mechanically chopped

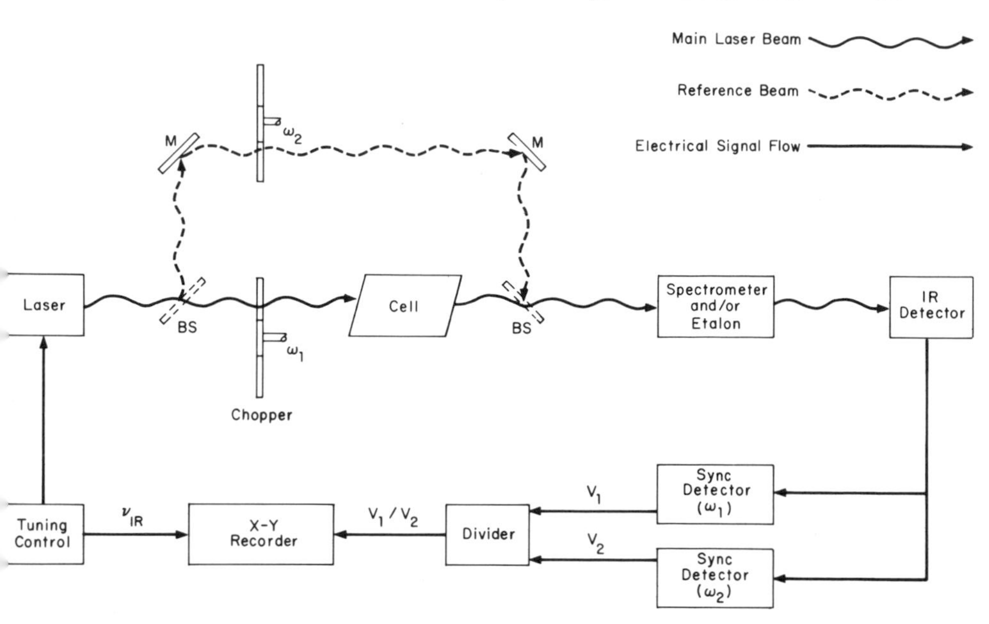

Fig. 2.8. Typical experimental arrangement for tunable laser spectroscopy. The reference beam chopped at ω_2 can be used to ratio out the effects of laser power variations on the recorded spectra (cf. Fig. 2.7)

at 0.1–1 kHz, and then synchronously detected using an appropriate infrared detector [2.73]. Alternate modulation and detection schemes are discussed in *Laser modulation and detection* below.

Calibration of Laser Tuning Rate

In Fig. 2.3 the wavelength of a SDL was tuned by varying the laser current. The resulting recorder display of detector signal vs. current represents the transmission spectrum of the gas in the cell. Conversion of laser current into wavenumber requires an accurate tuning-rate calibration procedure, several of which are discussed below. Absolute calibration of the laser wavelength can be obtained from the spectrometer, although for the typical laboratory-quality instrument, this procedure is accurate to only about 0.1 cm^{-1} at best—far worse than the laser's resolution capability. An alternative procedure, applicable in many instances, is to use a second cell containing a low-pressure gas whose absorption line positions are precisely known (such as NH_3, CO_2, CO). The transmission spectra then contain the absorption lines of the calibration gas as well as those of the gas under study. Calibration

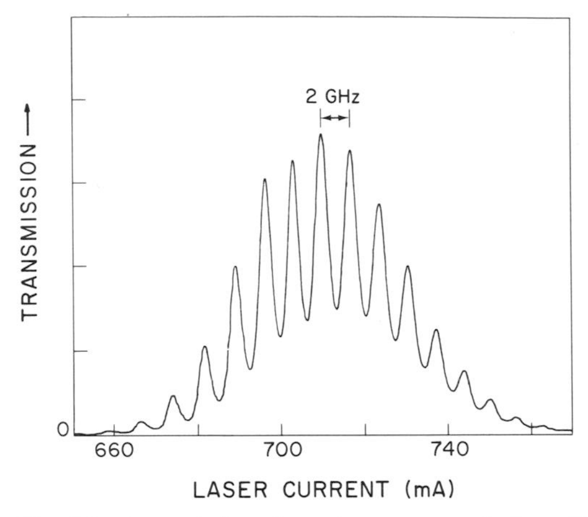

Fig. 2.9. Transmission through an uncoated Ge calibration etalon of radiation from a $PbS_{0.74}Se_{0.26}$ SDL over the full tuning range of one of its modes. Frequency calibration is obtained as the current increment required to sweep the 2-GHz free spectral range of the etalon. (After NILL and BLUM [2.72])

has also been obtained by measuring the heterodyne beat signal between a SDL mode and gas laser lines [2.5]. This technique has not been widely used due to the limited number of gas laser lines in the vicinity of molecular absorption lines and the need for a wide-bandwidth (few GHz) infrared detector.

The most convenient technique for calibrating the frequency shift of a laser mode is to transmit the radiation through a Fabry-Perot interferometer whose free spectral range is a small fraction of the tuning range of the mode. For example, a plane, parallel interferometer with a 5-cm air spacing has a free spectral range of 0.1 cm^{-1}, resulting in ten transmission maxima over a 1 cm^{-1} tuning interval. A convenient interferometer for calibration in the infrared is a solid, uncoated Ge etalon approximately 2 cm long. The high refractive index of Ge ($n \approx 4$) results in a small free spectral range of 0.0625 cm^{-1} (1.9 GHz) even for this relatively short etalon; and the high index of refraction yields a tolerable finesse for the transmission resonances without reflective coatings. Figure 2.9 illustrates the transmission of SDL radiation through a solid Ge etalon with a free spectral range of 1940 MHz (0.0647 cm^{-1}), as obtained by NILL and BLUM [2.72]. Calibration of the tuning rate across the full range of the mode is essential since significant variations may occur.

A second calibration technique is based upon using the Doppler width of the gas as a known tuning increment (see, e.g. [1.27 and 107]). Care must be taken to insure that pressure broadening is negligible

($p \lesssim 1$ Torr), that there is no fine structure of the absorption line which can add to its width, and that the absorption at line center can be accurately determined. It will be clear from some of the Doppler-limited spectra shown in Section 2.3 that the agreement between calculated linewidths and those obtained experimentally is very good.

A third tuning-rate calibration method is based on a heterodyne technique which requires two lasers and a high-speed infrared detector. Photoconductors of Ge : Cu, and photodiodes of HgCdTe with GHz frequency response are available [2.75, 73] and can be used for such applications. With the recent success of SPEARS and FREED [2.75] in detecting a 60-GHz ($2\,cm^{-1}$) beat frequency between two adjacent transitions of a CO_2 laser, it appears that heterodyne techniques may soon find more widespread use in infrared frequency calibration.

Laser Modulation and Detection

The experimental arrangement shown in Fig. 2.8 incorporates the conventional spectroscopic technique of amplitude-modulation of the infrared source with a mechanical chopper, followed by narrow-band, synchronous detection. The ability to rapidly frequency-modulate a tunable laser source allows alternative schemes to be used for obtaining spectroscopic data. These new capabilities offer several advantages, such as the accentuation of weak absorption lines or line structure, and allow for rapid scan of a spectral interval while maintaining a high signal-to-noise ratio. The simplest modification of Fig. 2.8 is to substitute frequency modulation of the laser radiation for the amplitude modulation provided by the chopper. The amplitude of the synchronously-detected signal is then the *derivative* of the transmission spectrum; and weak absorption lines (particularly in the Doppler-broadened regime) are accentuated. The frequency-modulation technique is particularly useful when the laser output occurs in several modes, since only that mode which is absorbed by the gas will produce a derivative signal. Hence, in many instances (such as the specific detection of gases in samples of contaminated air) mode filtering provided by a spectrometer is unnecessary. The optical apparatus thus consists only of the laser, gas cell, detector, and collimating lenses or mirrors. By adding a small (≈ 1 mA) sinusoidal (≈ 200 Hz) modulation current to the slowly-swept tuning (and excitation) current of a $Pb_{0.93}Sn_{0.07}Te$ SDL, the derivative spectra of Fig. 2.10 were obtained. The region is around $1122\,cm^{-1}$ where major lines of SO_2 in the ν_1 band are about 20 times weaker than the strongest ones in this band [2.76]. Using a 7.3-m path, the upper trace was obtained for 0.2 Torr of pure SO_2, and thereby identifies the fundamental line "signature" of this gas. The lower scan was taken for a raw sample of gas from

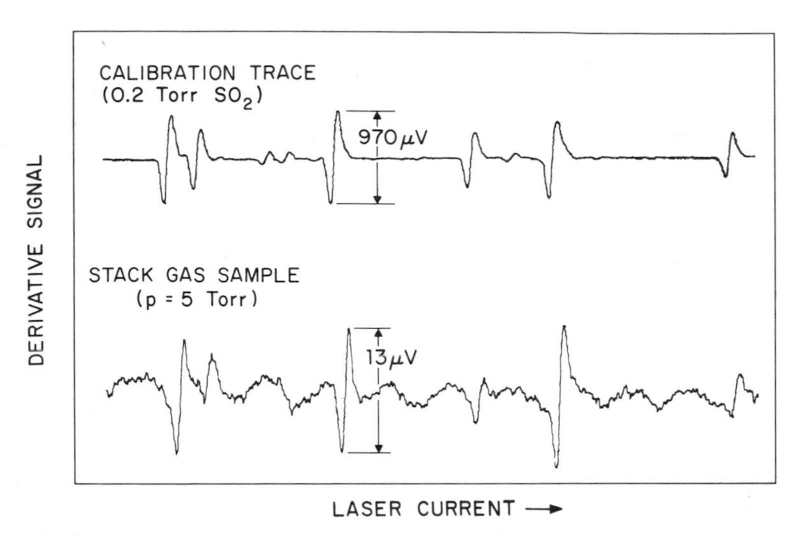

Fig. 2.10. First-derivative spectra in the $1122\,\mathrm{cm}^{-1}$ region of 0.2 Torr SO_2 (upper trace) and a sample of smokestack effluent reduced to 5 Torr pressure (lower trace). These spectra were recorded with a $Pb_{0.93}Sn_{0.07}Te$ SDL whose modulation current was 1 mA peak-to-peak. Cell length was 7.3 m [2.159]

the smokestack of an oil-fired heating plant which was introduced into the cell and partially evacuated to 5 Torr in order to reduce pressure broadening effects. This illustrates the manner in which the derivative technique can be used to identify and quantify gases. The slowly-varying oscillations occurring throughout the lower trace are the result of Fabry-Perot type resonances in the path between some of the optical components, and they can usually be eliminated by reorienting the apparatus slightly.

Because of the high power available from most tunable lasers, it is usually possible to use synchronous-detection bandwidths much larger than 1 Hz and still retain an excellent signal-to-noise ratio. A convenient way to utilize this capability is to sweep the laser mode rapidly (on the order of milliseconds) and display the detector output directly on an oscilloscope. A transmission spectrum of H_2CO (formaldehyde) obtained in this manner by NILL et al. [2.22] is shown in Fig. 2.11.

The absorption lines are evident as dips below the background, which represent the variation of mode power with frequency. In this case, a 1-cm^{-1} region around $2800\,\mathrm{cm}^{-1}$ was scanned in approximately 10 msec. A preamplifier with a 500 kHz bandwidth was used to enhance the signal from an InSb photodiode. The laser wavelength was scanned by adding a small current ramp to the direct (steady) laser current. Wavelength scans of the other types of tunable lasers can also be produced by small modulation of the tuning parameter—for example, a 100-Gauss

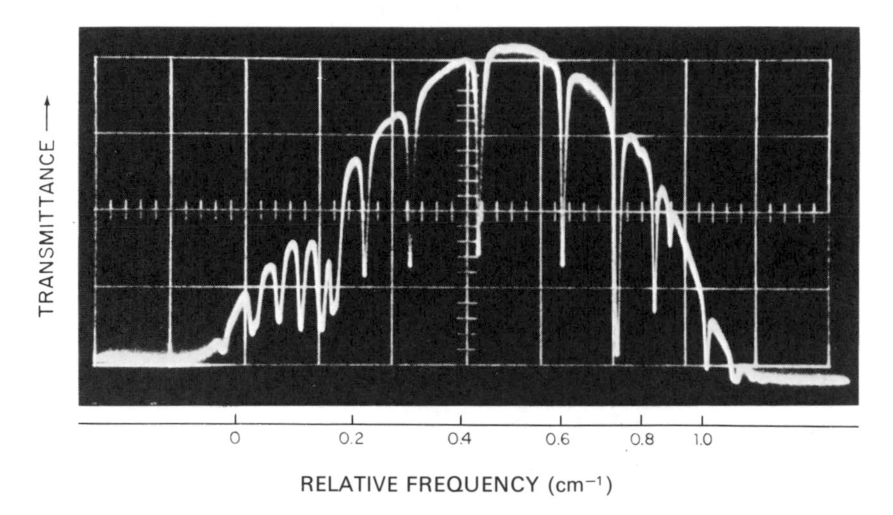

Fig. 2.11. Time-resolved spectra of H_2CO in the 2800 cm^{-1} region, as taken with a current-tuned $Pb_{0.98}Cd_{0.02}S$ SDL. The cm^{-1} interval was scanned in 10 msec. Gas pressure was 5 Torr and cell length 10 cm. (After NILL et al. [2.22])

magnetic-field sweep is sufficient to tune the modes of a SFR laser over their full range. Rapid, continuous tuning of the laser wavelength is readily achieved with tunable lasers which have a simple resonant cavity (e.g. SDL and SFR laser), but the coupled resonances which are present in tunable sources such as the OPO with an intra-cavity etalon preclude rapid continuous tuning without large, undesirable variations of laser power output.

Heterodyne Detection of Infrared Emission Lines

A potential application of tunable infrared lasers is the heterodyne detection of narrow-line emission from thermal sources, particularly molecular gases [2.77–79]. For this technique, laser radiation is mixed in a wideband infrared detector with that from the gas, as shown in Fig. 2.12a. The heterodyne beat signal within the detector bandwidth is measured as the tunable laser scans the region around the infrared frequency of the thermal line source. Heterodyne detection of thermal sources has not been widely studied since there are few coincidences between molecular emission lines and the laser lines of conventional gas lasers. Under ideal conditions, using a photoconductive detector, the minimum detectable power is [2.73]

$$P_{\min} = (2h\nu/\eta)\,(B/\tau)^{1/2}\,, \tag{2.10}$$

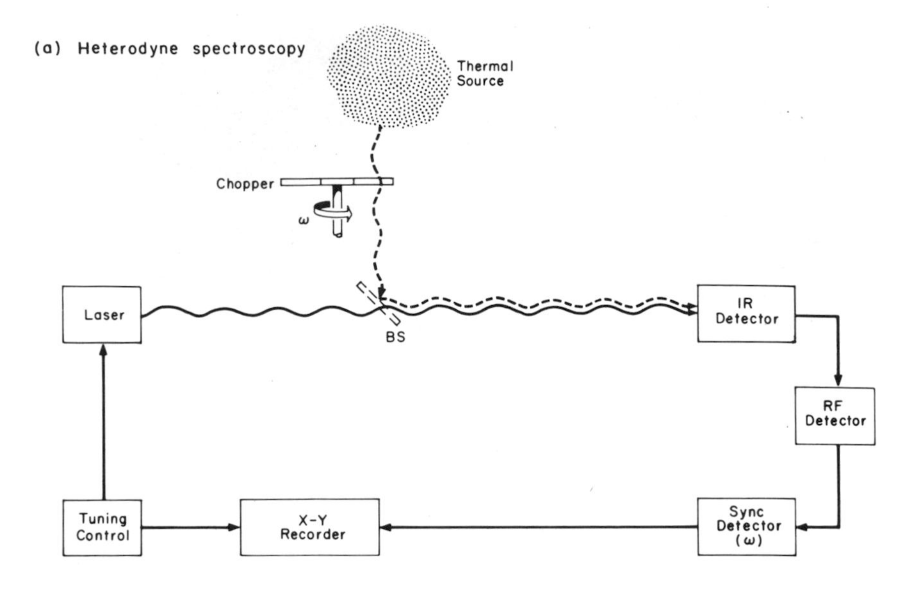

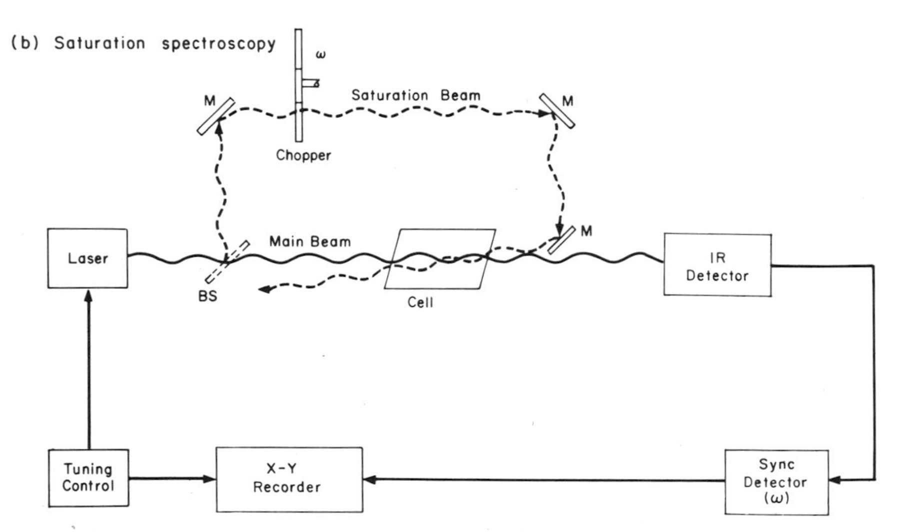

Fig. 2.12a and b. Experimental arrangements for (a) heterodyne detection of infrared line emission from a gaseous source, and (b) "Lamb-dip" absorption spectroscopy

where B denotes the IF bandwidth, η is the quantum efficiency of the infrared detector, and τ is post-detection integration time. For a 1-GHz IF bandwidth, a one-second integration time, and a quantum efficiency of 0.5, $P_m \approx 10^{-15}$ W in the 5–12 μm spectral region. This sensitivity is significantly better than that obtained using typical background-limited detectors for incoherent detection of thermal emission. The highly directional properties of the heterodyne radiometer are also useful in some applications [2.80].

The heterodyne detection sensitivity is dependent on the mode power and mode quality of the laser, whose function is that of a tunable local oscillator. The ultimate sensitivity given by (2.10) is obtained as the laser power is increased until the induced generation-recombination noise in the detector dominates that due to the preamplifier. For a photoconductive detector, the laser power required to overcome amplifier noise is [2.73]

$$P_{\mathrm{LO}}=(h\nu/e)(kT_{\mathrm{A}}/\eta R_{\mathrm{L}}G^2) \tag{2.11}$$

where T_{A} is the noise temperature of the amplifier, G the photoconductor gain, k is Boltzmann's constant, and R_{L} the load resistance. For photovoltaic detectors $G=1$ and there is a factor of two in the denominator of (2.11) since carrier recombination is not a noise source. Using typical parameters for a Ge : Cu liquid-helium-cooled photoconductor ($G=0.1$, $\eta=0.5$, $R_{\mathrm{L}}=50\,\Omega$), and a 500-MHz amplifier with $T_{\mathrm{A}}=240$ K, (2.11) gives $P_{\mathrm{LO}}=12$ mW for optimum heterodyne detection of 10-μm radiation. Similarly, for a photovoltaic PbSnTe or HgCdTe detector with the same quantum efficiency, P_{LO} can be less than 1 mW since $G=1$.

Heterodyne detection of emission lines around 10.6 μm from heated (900 K) C_2H_4 was reported in 1971 by HINKLEY and KINGSTON [2.77]. In later work, MENZIES [2.81] detected radiation from SO_2 and CO_2; and the coverage has been expanded [2.82] to include the gases NO, O_3, and NH_3. In each case the local oscillator power was supplied by a discretely-tunable gas laser. These results, coupled with recent advances in stable, high-power, tunable lasers (e.g. over 200 mW from a SDL [2.24] and 1 W from a SFR laser [2.33]), indicate that heterodyne emission spectroscopy will soon become a useful laboratory technique.

Nonlinear Spectroscopy

The nonlinear response of a molecular gas to an intense optical field can reveal many spectroscopic features beyond those observed in linear absorption experiments [2.83]. Effects such as double resonance, two-photon absorption, and the Lamb dip have been studied in the infrared using He–Ne and CO_2 lasers. These experiments have been limited to the few cases of coincidence between molecular gas absorption lines and the laser output wavelength. Many experiments have been performed in which infrared absorption lines have been tuned into coincidence with laser lines through Stark or Zeeman splitting. The ability to tune the laser into coincidence with an absorption line will greatly increase the scope of nonlinear spectroscopic measurements.

Since the features of a saturated absorption spectrum are typically very narrow (0.1–1 MHz), and the mode power required to achieve a nonlinear response is sizable ($\geqq 10$ mW), the stability and power requirements on the tunable laser source are stringent. The power necessary for a significant nonlinear response depends on the strength of the dipole matrix element μ_D of the transition and the relaxation time τ of the levels involved. An estimate of the power required is given by the expression $\mu_D E\tau/\hbar = 1$, where E is the optical electric field. For a 10-mW laser beam 1 cm in diameter, $E = 3 \times 10^{-2}$ e.s.u., which is adequate for a molecule such as CO with $\mu_D \approx 10^{-19}$ e.s.u. and a relaxation time of 10^{-7} to 10^{-6} sec (determined by intermolecular collisions). In order to fully resolve the saturated absorption structure, the laser linewidth and/or frequency stability must be better than 0.1 MHz (an order of magnitude below that required to fully resolve linear absorption spectra).

The experimental set-up for studying the simplest nonlinear effect, saturated absorption, is shown in Fig. 2.12b. The experiment described here is in the class referred to as inverted "Lamb dip" spectroscopy in which the frequency of a strong saturating beam is tuned across the inhomogeneously broadened absorption line of a molecular gas and the absorption of a weak probe beam is monitored. The saturating beam affects the absorption of the weak radiation through the nonlinear absorption of the gas whenever the two beams are interacting with the same molecules. For two counter-directed beams this occurs when the laser wavelength is tuned to the center of an absorption line, since both beams interact with molecules which are essentially stationary relative to either beam. Under these conditions a narrow resonant increase in the detected signal occurs, and the fine structure otherwise hidden by the inhomogeneous broadening is revealed. Examples of such measurements are given in Section 2.3. (See also Sect. 1.4.)

2.3. Infrared Laser Spectroscopy

Virtually every known molecule has a vibration-rotation absorption band between 2 and 15 μm [2.84]. As mentioned above, the high resolution capabilities of tunable lasers are ideal for studying the characteristic linewidths and fine structure of these bands, where low-pressure Doppler widths range from thirty to several hundred MHz, and molecular energy level fine structure gives line splittings from ten to several hundred MHz.

Absorption lines for excitations in solids and liquids have widths which are typically orders of magnitude larger, and they can usually be resolved using conventional infrared spectroscopic instrumentation. For these reasons tunable infrared lasers have found wide (and virtually exclusive) application in infrared molecular spectroscopic studies of gases, and related areas of research. This section, therefore, will be devoted almost entirely to a review of the important and revolutionary observations in molecular spectroscopy which have been made using tunable infrared lasers. Our approach will be to consider, in turn, each of the fundamental properties of infrared absorption lines (shapes, widths, positions, fine structure) and to give examples of measurements of each. Following this review of fundamental properties, we will discuss the important practical application of tunable lasers to air pollution monitoring. The emphasis throughout this section will be on obtained results.

In accordance with the guidelines mentioned at the beginning of this chapter, we will not consider the following: a) Spectroscopic work performed with tunable GaAs diode lasers [2.85] and tunable dye lasers [2.86]; b) measurements obtained with an untuned laser in conjunction with Stark or Zeeman shifting of molecular lines [2.87, 88] or tunable microwave oscillators for infrared-microwave double-resonance experiments [2.89]; or c) nonlinear level-crossing and saturation (Lamb-dip) spectroscopy performed with the gas species being the active component within a gas laser [2.90, 91]. (While Lamb-dip experiments such as these have important implications for frequency-stabilization of gas lasers and the establishment of infrared frequency standards, they do not constitute a generally applicable technique of tunable laser spectroscopy.)

Thus, we restrict our consideration to situations in which a gas sample is probed by radiation emitted from a separate tunable laser—an arrangement which is clearly the most general, and one which will receive the greatest attention as the availability and utilization of tunable lasers becomes more widespread.

Table 2.5 lists molecules for which high-resolution spectra have been obtained using tunable infrared lasers. The type of measurement is given, as well as the nominal wavelength region, tunable laser used, and references to the original work. The spectra obtained range from the simple diatomic structure of NO and CO to complex spectra associated with molecules such as SF_6 whose line density is so large that adjacent Doppler-broadened lines partially overlap. The experiments listed in Table 2.5 have been carried out over the entire 2–11 μm region, and they could, in principle, be extended beyond 20 μm with some of the tunable lasers of Table 2.1.

Table 2.5. Molecular spectroscopy with tunable infrared lasers

Measurement	Molecule	Wavelength [μm]	Laser	Ref.
Absorption line spectra	HCl	1.2, 3.3	OPO, SDL	[2.92–95]
	HBr	1.9	OPO	[2.9]
	CO	2.4, 4.7	OPO, SDL	[2.21, 92, 96]
	CH_3F	3.4	ZTG	[2.2, 97]
	C_2H_6	3.4	ZTG	[2.2]
	CH_4	3.4, 6.5	ZTG, SDL	[2.2, 27, 98–101]
	H_2CO	3.6	ZTG, SDL	[2.22, 102]
	CO_2	4.2	SDL	[2.29]
	CS_2	4.6	SDL	[2.103]
	O_3	4.7	SDL	[2.29]
	NO	5.4	SDL, SFRL	[2.69, 104–108]
	H_2O	5.3, 6.3	SFRL, SDL	[2.58, 108–110]
	NO_2	6.2	SDL	[2.111]
	SO_2	8.7	SDL	[2.76, 112–114]
	NH_3	8.7, 10.5, 11.8	SDL, SFRL	[2.7, 112, 115, 116]
	C_2H_4	10.5	SDL	[2.79]
	SF_6	10.5	SDL, HPG	[2.5, 11]
	C_2H_3Cl	10.5	TPM	[2.117]
Pressure broadening	CO	4.7	SDL	[2.118]
	H_2O	5.3	SDL	[2.58, 119, 20]
	SO_2	8.7	SDL	[2.76, 112, 113]
	NH_3	10.5	SDL	[2.112]
	CO_2	10.6	TPM	[2.10]
Collisional narrowing	H_2O	5.3	SDL	[2.120]
Nuclear hyperfine structure	CH_4	3.4	PTG	[2.121]
	NO	5.4	SDL	[2.107]
Lambda-doubling	NO	5.4	SDL, SFRL	[2.104, 69, 108]
Zeeman splitting	CH_4	3.4	PTG	[2.98]
	NO	5.4	SDL	[2.104, 122, 123]
	CH_4	3.4	ZTG	[2.99, 102]
	H_2CO	3.4	ZTG	[2.102]
	HDCO	3.4	ZTG	[2.102]
Stark splitting	NH_2D	10.5	SDL	[2.124]
	CH_3Cl	10.6	TPM	[2.117]
	CH_3Br	10.6	TPM	[2.117]
Band analysis	SO_2	8.7	SDL	[2.76]
	SF_6	10.5	HPG, SDL	[2.11, 125]
Laser gain lineshape	CO	5.3	SDL	[2.126]
	CO_2	10.6	TPM	[2.10]
Isotope lines	$C^{12,13}O_2^{16}$	4.2	SDL	[2.29]
	$N^{14,15}O^{16}$	5.4	SDL	[2.122]

Many of the fundamental aspects of molecular spectroscopy are covered in the experiments of Table 2.5. Both Doppler-broadened and pressure-broadened lineshapes have been recorded, absolute and relative line strengths measured, and line identification and band analysis performed. Rotational fine structure arising from the Stark effect, Zeeman effect, nuclear hyperfine coupling, Λ-type doubling, and isotopes has been observed. Several of these observations were the first such measurements in the infrared, and could not have been made without the use of tunable lasers. Most of these effects have been studied in detail for only one or two molecular species; consequently, there is a great deal of interesting and fruitful work yet to be done.

2.3.1. Lineshapes, Linewidths, and Intensities

Three fundamental properties of an isolated spectral line are its location (center wavelength), intensity, and shape. Tunable infrared-laser technology has, to this point, made little impact on absolute wavelength calibration; i.e., we still rely on conventional methods [2.127] to determine line locations. However, recent heterodyne experiments involving fixed-frequency lasers and metal-metal point-contact diodes [2.128] have established very accurate values for certain infrared laser frequencies, and may be precursors to essentially absolute frequency calibration over much of the infrared. With the establishment of a sufficiently large number of secondary frequency standards, one can envision valuable extensions being made using tunable lasers coupled with heterodyne techniques or Fabry-Perot interferometry. Thus, as discussed in Section 2.2, only the *local change* in tunable infrared laser frequency has, in general, been measured precisely, and only *relative* line positions established. On the other hand, the intensity and shape of a spectral line are directly measurable with a tunable laser. The key to such measurements is the laser's narrow linewidth (high resolution) and tunability.

In the rest of this subsection we will discuss measurements of infrared lineshapes and intensities for gases. Fine structure and line splittings will be considered later.

Gaussian Profile—Doppler Broadening

In the regime of low gas pressure the linewidth of a molecular transition is determined by the Doppler effect due to thermal motion of the molecules. The absorption coefficient full width at half-maximum (FWHM) is given approximately by the expression [2.129]

$$\Delta\nu_{\mathrm{D}} \simeq (214/\lambda)\,(T/M)^{1/2}\ \mathrm{MHz}\,, \tag{2.12}$$

where $\lambda[\mu m]$ is the wavelength, $T[K]$ the gas temperature and M its molecular weight. The spectral lineshape is Gaussian with a FWHM determined by (2.12), and the peak intensity related to the molecular density and the size of the transitional dipole moment. For a Doppler-broadened absorption line with integrated intensity S and center frequency ν_0, the linear absorption coefficient is

$$\alpha(\nu) = \frac{2S(\ln 2)^{1/2}}{\pi^{1/2}\Delta\nu_D} \exp\left[-\frac{4(\nu-\nu_0)^2 \ln 2}{(\Delta\nu_D)^2}\right], \tag{2.13}$$

at frequency ν.

An example of a measured Doppler-broadened lineshape is shown in Fig. 2.13a for the $P(7)$ line of CO at 2115.63 cm^{-1}, within the fundamental vibration-rotation band. The data were taken [2.161] with a $PbS_{0.82}Se_{0.18}$ SDL whose radiation was transmitted through a 10-cm cell containing 0.1 Torr CO, using the experimental arrangement of Fig. 2.8. In order to display the fundamental absorption constant as shown, the transmittance data were reduced on a computer to yield the solid line of the experimental scan. The Doppler width for this line is found to be 155 MHz using (2.12), and a theoretical (Gaussian) lineshape having this FWHM is indicated in Fig. 2.13a by the dashed trace. The agreement between the measured and theoretical lineshapes is seen to be quite good.

Doppler-limited absorption spectra have been obtained for most of the gases listed in Table 2.5 using a variety of tunable laser sources. Most extensive measurements have been carried out on CO [2.21], H_2O [2.120], SO_2 [2.76, 112], and NH_3 [2.112]. The agreement between measured and calculated Doppler-broadened linewidths has been excellent, although few detailed lineshape studies have been reported.

Lorentzian Profile—Collision Broadening

As the gas pressure is increased from some arbitrarily small value, the shape and width of an infrared spectral line is affected by intermolecular collisions. Generally, the linewidth increases monotonically due to collisions between molecules of the same species (self broadening) or with different molecular species (foreign-gas broadening). When the collisions are so frequent that collision broadening becomes dominant over Doppler broadening, the lineshape becomes Lorentzian with an absorption coefficient given by [2.129]

$$\alpha(\nu) = \frac{S}{2\pi} \frac{\Delta\nu_p}{(\nu-\nu_0)^2 + (\Delta\nu_p/2)^2}, \tag{2.14}$$

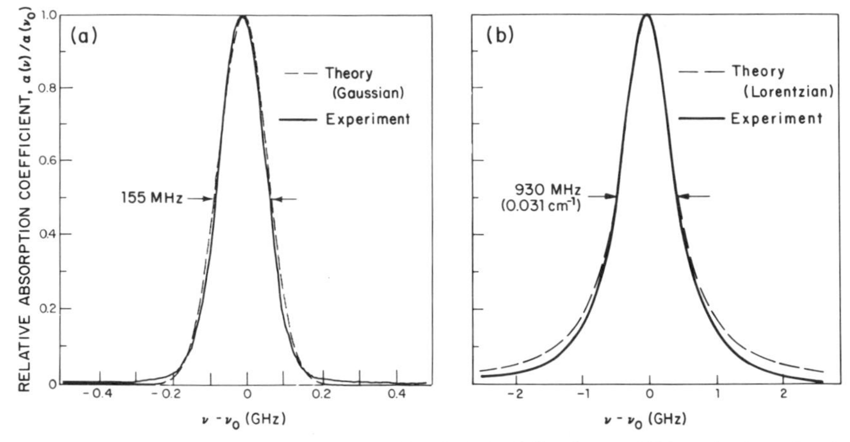

Fig. 2.13a and b. Measured (solid) and calculated (dashed) lineshapes: (a) Doppler-broadened $P(7)$ line of CO at 2115.63 cm^{-1}, 0.1 Torr pressure [2.158]; (b) collision-broadened lineshape of the 13(2, 12)←12(1, 11) transition in the ν_2 water band at 1879.6 cm^{-1}, atmospheric pressure. (After BLUM et al. [2.58])

where $\Delta\nu_\mathrm{p}$ is the absorption coefficient FWHM for the collision (pressure)-broadened line. For a multi-component gas, it has the form

$$\Delta\nu_p = \Sigma p_i \Delta\nu'_{p,i}, \qquad (2.15)$$

where $\Delta\nu'_{p,i}$ is the collision-broadened width per unit pressure of the ith gas species (which has a partial pressure p_i in the sample being analyzed).

Figure 2.13b shows an example of a collision-broadened line in the ν_2 band of water vapor present in a laboratory atmosphere, taken from the work of BLUM et al. [2.58]. The absorption coefficient, plotted as a function of frequency, was deduced from a measurement of SDL transmittance over a 7.4-m atmospheric path. This line has been identified as the 13(2, 12)←12(1, 11) transition [2.130] at 1879.6 cm^{-1}, where the notation $J(K_\mathrm{a}, K_\mathrm{c})$ is used for asymmetric rotor energy level quantum numbers. The line has a measured width of 870 ± 90 MHz, which is less than one-half the value of 2040 MHz calculated by BENEDICT and CALFEE using accepted line-broadening theory [2.130]. A discussion of anomalous widths of this and other ν_2 lines can be found in [2.58, 119, 120]. Since there was no simple way to eliminate the absorbing path for a determination of background intensity variations, extrapolation to zero water vapor had to be made by hand. Consequently, the error in the deduced absorption coefficient is expected to increase rapidly for frequencies well-removed from line center, i.e., in the "wings". The dashed line shows

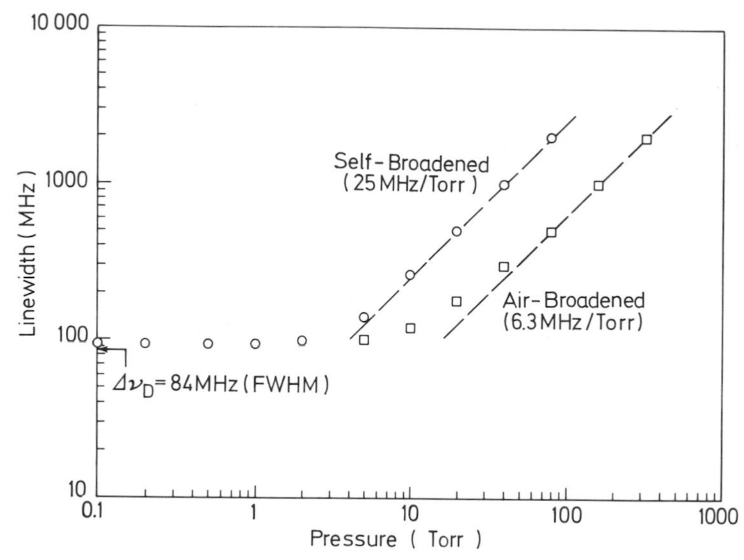

Fig. 2.14. Linewidth (FWHM) of the aQ(9, 3) transition of NH_3 at 939.21 cm^{-1} vs. NH_3 and air pressure [2.112]

the Lorentzian shape calculated using (2.14) with a width of 870 MHz and value of $\alpha(0)$ chosen to match the peak in the experimental data.

As with the Doppler-broadened lineshapes, no detailed studies of collision-broadened spectra have been reported. Deviations from the theoretical Lorentzian shape for isolated spectral lines have traditionally been the object of great interest and theoretical speculation [2.131]. Narrow-linewidth tunable lasers are ideally suited to lineshape measurements, but experiments to accurately measure such effects will require extreme care and the use of ratioing techniques which have not been generally employed in this rather young area of research activity.

The pressure-broadening coefficient is one important parameter which can now be measured for many gases. Self-broadening and foreign-gas broadening have been studied, using SDL's, in H_2O [2.107, 120], SO_2 [2.76], NH_3 [2.112], and CO [2.118]. Broadening rates from 5 to 50 MHz/Torr have been measured, with the self-broadening coefficients always larger than those for foreign-gas broadening; and reasonable agreement has been found with values deduced from more conventional spectroscopic experiments where comparison was possible. Illustrating this type of measurement, Fig. 2.14 shows the FWHM for the $aQ(9, 3)$ line of NH_3 at 939.21 cm^{-1} as a function of NH_3 and air pressure. At low pressures the linewidth asymptotically approaches the Doppler width of 84 MHz. Beyond approximately 4 Torr the linewidth increases linearly with pressure at a rate of 25 MHz/Torr for self-broadening, and beyond 10–20 Torr at a rate of 6.3 MHz/Torr for air broadening. This linear dependence is expected in the collision-dominated regime [2.129].

At intermediate gas pressures the lineshape is represented by a Voigt profile [2.129] formed by the convolution of Gaussian and Lorentzian lineshapes. Although such spectra have been measured, no careful studies using tunable lasers have been made in this regime.

Collisional Narrowing

When the molecules of a gas collide, they undergo largely elastic velocity-changing collisions. It is sometimes possible for these collisions to occur more frequently than the line-broadening collisions; i.e., not every velocity-changing collision is effective in broadening the infrared spectral line through changes in the internal state of the molecule. Even though the collisions are elastic, they can change the effective Doppler speed of a molecule since it is only that velocity component along the direction of propagation of the radiation which is important. That is, the Doppler frequency shift is $\boldsymbol{k} \cdot \boldsymbol{v}$, where $\boldsymbol{k}$ is the wavevector of the radiation and $\boldsymbol{v}$ the molecular thermal velocity. When these changes in instantaneous Doppler shift are very frequent, increasing the gas pressure can actually *reduce* the linewidth below the limiting low-pressure Doppler width [2.132–1.136]. On the average, the probing radiation "sees" molecules with a narrower velocity spread. This collisional narrowing is often called Dicke narrowing after DICKE [2.132], who was the first to discuss this phenomenon for spectral lines. The effect is closely related to the phenomenon of motional narrowing frequently discussed in connection with NMR studies [2.137]. If velocity-changing collisions occur with mean time τ, one expects significant narrowing when

$$\Delta \nu_{\mathrm{D}} \tau \approx k V_{\mathrm{th}} \tau \lesssim 1 , \qquad (2.16)$$

where v_{th} is the mean thermal speed. This expression can be rewritten as

$$l \lesssim \lambda / 2\pi , \qquad (2.17)$$

where l $(= v_{\mathrm{th}} \tau)$ is the molecular mean free path. Thus, we see that collisional narrowing requires that the mean free path for velocity-changing collisions be a) shorter than the mean free path for line broadening, and b) much less than the wavelength of the probing radiation.

Collisional narrowing of infrared spectral lines was first observed in absorption and Raman-scattering studies of H_2 [2.135, 136], and has

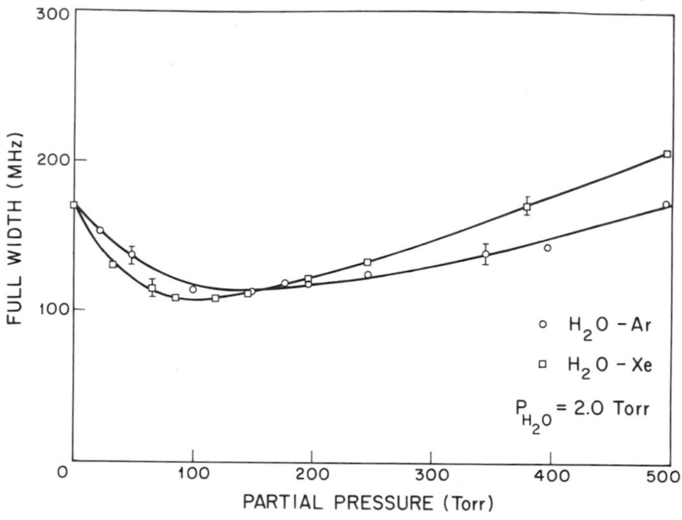

Fig. 2.15. Linewidth (FWHM) of the 16←15 H_2O transition at 1871.01 cm^{-1} vs. Xe and Ar gas pressure, showing collisional narrowing. (After ENG et al. [2.120])

recently been observed in studies of H_2O lines in the 6.3-μm ν_2 band by ENG et al. [2.120] using a tunable SDL. Figure 2.15 illustrates collisional narrowing of the doubly-degenerate 16(0, 16)←15(1, 15), 16(1, 16) ←15(0, 15) H_2O transition at 1879.01 cm^{-1} when subjected to increasing pressures of Xe and Ar buffer gases. This line had previously been found to have an anomalously narrow atmospheric-pressure width of 450 MHz [2.58], less than three times the Doppler width of 170 MHz. The data of Fig. 2.15 were taken with an H_2O partial pressure of 2 Torr, and the observed Doppler width is 170 MHz, as expected. As the buffer gas pressure is increased, the linewidth decreases, reaches a minimum, and then increases as collision broadening becomes dominant. The minimum observed linewidth is 110 MHz for Xe buffer gas, which is 35% smaller than the Doppler width. Collisional narrowing may also be observed in other gases for transitions involving states of high rotational energy where line-broadening collisions are infrequent.

Natural Linewidth and Saturation Spectroscopy

If the molecules in a gas were stationary, there would be no Doppler broadening of its spectral lines; and if the gas pressure were sufficiently low, collision broadening would be negligible and the observed linewidth of a molecular transition would be that characteristic of its radiative lifetime, i.e., its "natural" linewidth [2.129]. These natural widths are on the order of 1 kHz for infrared spectral lines. Thermal broadening is, of course, always present for gases; the temperature required to reduce the Doppler width to 1 kHz would be near absolute zero. Thus, in *linear* absorption spectroscopy, the natural linewidth will always be obscured by Doppler broadening even under the most optimistic conditions.

Since Doppler broadening is inhomogeneous by nature, it is possible, using nonlinear resonance techniques, to measure homogeneous linewidths and fine structure which are normally masked by Doppler broadening. In practice, natural linewidths have not generally been measured [2.83], and the observed linewidths are usually limited by 1) intermolecular collisions or 2) the molecular transit time through the laser beam. Linewidths as narrow as 12 kHz have been measured by this technique [2.121], yielding an effective resolving power three orders of magnitude better than the best Doppler-limited linear absorption spectroscopy. Clearly, the use of saturation spectroscopy techniques with high-power tunable lasers will have a tremendous impact on future measurements of fine structure in infrared molecular spectra.

For a Doppler-broadened line, it can be shown [2.138] that the nonlinear absorption effects produced in a Lamb-dip type configuration (see *Nonlinear spectroscopy* above) change the spectral nature of the absorption constant, so that in lowest order [2.83],

$$\alpha_{NL}(\nu) \approx \alpha(\nu)\left[1 - \frac{2|(\mu\varepsilon/\hbar)|^2}{(\nu-\nu_0)^2 + (\Delta\nu_N/2)^2}\right], \tag{2.18}$$

where $\alpha(\nu)$ is the linear absorption coefficient given by (2.2), μ the dipole moment of the transition, $\Delta\nu_N$ the limiting homogeneous linewidth (the natural linewidth under ideal conditions), and ε is the magnitude of the electric field of the infrared radiation. On the basis of (2.18), we expect a narrow, nonlinear spike to occur at the center of the normally Doppler-broadened line. This spike will have a width $\Delta\nu_N$ and an absorption coefficient of amplitude $8|(\mu\varepsilon/\hbar)|^2/(\Delta\nu_N)^2$. Thus, the size of the effect is directly proportional to the infrared power density. Since typical molecules have vibrational dipole moments of ≈ 0.01 Debye (1 Debye $= 10^{-18}$ in c.g.s. units), and experimental homogeneous linewidths are 0.1–1 MHz, a nonlinear change of a few percent in α requires power densities of only tens of mW per cm^2.

Unfortunately, the use of this elegant technique in tunable infrared laser spectroscopy (in the sense described above) has been very limited. A number of experiments have been performed with lasers in the visible range [2.83, 88]; particularly conspicuous are those of HANSCH et al. [2.86]. In the infrared regime, studies of Lamb-dip effects involving the active species of gas lasers have been made [2.91]. Also, numerous saturation

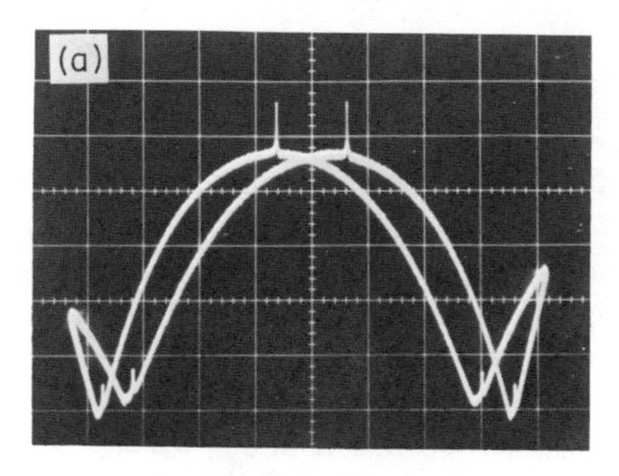

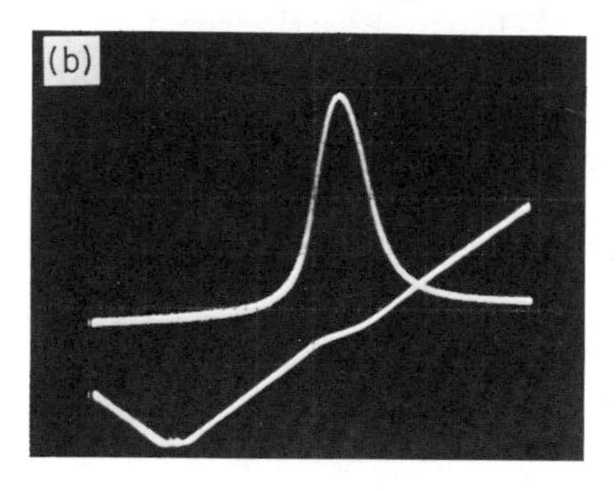

Fig. 2.16 a and b. Inverted Lamb dip spectra of the $P(7)$ line of CH_4 at 3.39 μm: (a) 28 MHz/div; (b) MHz/div. (After BARGER and HALL [2.101])

spectroscopy experiments utilizing fixed-frequency gas lasers and Stark or Zeeman tuning of nearby molecular lines have been performed. BREWER [2.83] has reviewed these and other related measurements involving optical level crossing and transient coherent effects. Most of these measurements lie outside the scope of this chapter (for details see Sects. 1.3 and 1.4).

There is one important class of measurements involving molecular saturation spectroscopy, performed with piezoelectrically-tuned gas (PTG) lasers. The experiments employ a passive, low-pressure absorbing gas cell placed within the resonant optical cavity of an operating gas laser [2.101, 139]. The absorbing cell must contain a gas which has an absorption line in near coincidence with the laser line. In view of the narrow continuous tuning ranges of cavity-tuned gas lasers, this requirement can be a strict one indeed. The nonlinear phenomenon occurs as an inverted Lamb dip, i.e., a narrow, positive spike in the laser emission during tuning. Because the spike is extremely narrow (analogous to the conventional Lamb dip described above), a laser whose frequency is locked to it by a feedback mechanism will be very stable. In addition to providing frequency stabilization for gas lasers, this technique has permitted the establishment of infrared frequency standards, and has yielded the most precise values to date for the speed of light.

Figure 2.16 shows the inverted Lamb dip on the $P(7)$ line in the ν_3 band of CH_4 obtained by BARGER and HALL [2.101] with a He–Ne laser oscillating at 3.39 μm. The saturation peak is centered on the Ne^{20} Doppler gain curve in Fig. 2.16a; the same spectra are shown with an expanded frequency scale in Fig. 2.16b. The peak is 400 kHz wide and has an amplitude of about 2% of the laser's output power of 1 MW. Two He–Ne lasers were *independently* frequency-locked in this manner, and an offset frequency of under 1 kHz was measured with a reproducibility of one part in 10^{11}. Subsequent measurements [2.140–142] using such stabilized lasers and point-contact mixers has led to a new value for the speed of light: $c = 299\,792\,456 \cdot 2 \pm 1.1$ m/sec. This value is *two orders of magnitude* more accurate than that previously determined! Needless to say, infrared frequency measurement and stabilization are important applications of saturation spectroscopy and will receive considerably more attention in the future.

Laser Gain Lineshapes

All of the above discussions have dealt with molecular absorption lines in passive gases. Molecular gas lasers [2.143] emit on many of the same transitions, and the spontaneous gain lineshapes of the laser lines are governed by the same physical processes which determine the absorption lineshapes. However, the laser medium is usually a complex, nonequilibrium mixture of several gases at unknown or varying temperature and pressure. A tunable infrared laser with a narrow linewidth is a powerful tool for the study of such a molecular system. It can be used to determine the lineshapes and peak amplitudes of gain and loss lines in molecular lasers, and to obtain information concerning the molecular kinetics of gas lasers.

The first experiment of this type was reported in 1972 by BLUM et al. [2.126]. A $PbS_{0.6}Se_{0.4}$ SDL was used to probe a CO molecular laser amplifier operating near 5.3 μm. The laser employed a flowing gas mixture of CO, He, N_2, and air, and the tube containing these gases was immersed in a bath of liquid nitrogen. The single-pass peak gain coefficient was measured (with the laser cavity mirrors removed) and found to be in reasonable agreement with measurements made using fixed-frequency matched laser-amplifier arrangements. Spectra of several gain and loss lines were obtained. The effective Doppler width of the $P(9)_{9,8}$ line was found to be 100 MHz, which implied a CO translational temperature of 170 K, as compared with the tube wall temperature of 77 K, and previous indirect estimates of from 140 to 200 K. Measurements were

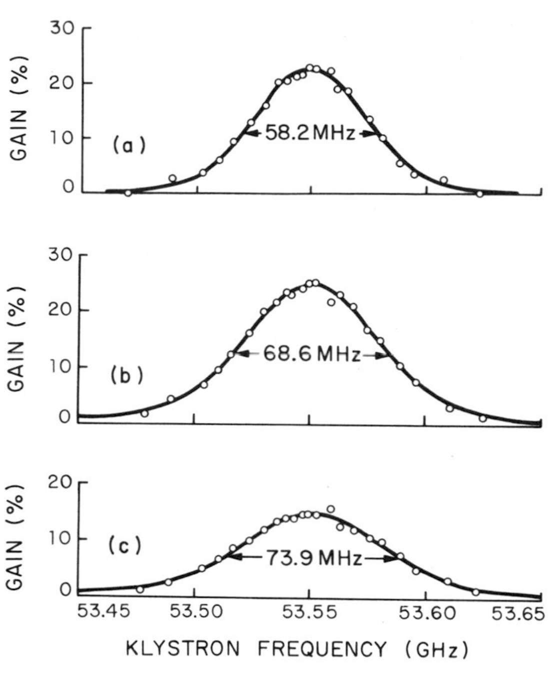

Fig. 2.17a–c. Gain lineshapes of CO_2 laser amplifier using discharge pressures and temperatures of (a) 1.64 Torr, 317 K; (b) 4.87 Torr, 341 K; and (c) 7.65 Torr, 354 K. (After CORCORAN et al. [2.10])

also made of the CO loss lineshapes as well as the transformation from strong gain to strong loss produced by varying the buffer gas pressure or discharge current.

Recently, CORCORAN et al. [2.10] reported measurements of the gain lineshape of a CO_2 laser in the 10.6-μm region using a TPM as the tunable source (see *Two-Photon Mixer* above). By mixing in a crystal of GaAs, 1-W emission from the 949.191 cm^{-1} $P(20)$ line of a CO_2 laser with a 0.3-W signal from a microwave klystron having 6-GHz tunability [2.45], they were able to scan the $P(22)$ CO_2 transition under varying conditions of temperature and pressure. Because of the low tunable laser-power from the TPM ($\lesssim 1\,\mu W$), the spectra had to be recorded point by point using an integration time of a few seconds to achieve a reasonable signal-to-noise ratio. Representative spectra are shown in Fig. 2.17 for three different conditions of discharge pressure and temperature. The solid curves are theoretical lineshapes fit to the data, involving convolutions of Doppler- and collision-broadening formulas [2.129, 145]. At the relatively low discharge pressure of 1.64 Torr, the profile shown in Fig. 2.17a is mostly Gaussian since Doppler-broadening is dominant. For Fig. 2.17c, corresponding to 7.65 Torr pressure, the lineshape is tending toward Lorentzian as collision-broadening becomes significant.

Detailed investigations of molecular lasers with tunable lasers should yield valuable information concerning inversion mechanisms and the physics of gas-laser operation. This technique may prove useful for the identification of chemically-formed species or the monitoring of chemical reactions in discharge and chemical lasers. It would also be interesting to study the single-pass gain of an *operating* laser in order to observe nonlinear "hole burning" (saturation) at the output wavelength. Because of the high-resolution and frequency-calibration capabilities of tunable lasers based upon microwave spectroscopy techniques, these lasers should be particularly useful for obtaining spectral information in the vicinity ($\lesssim 2\,\mathrm{cm}^{-1}$) of established laser lines, and there appears to be no fundamental reason why TPM techniques cannot be extended to all types of gas lasers.

Line Intensities

Having measured the shape of a spectral line, a determination of its intensity would appear to be a simple matter. However, since the intensity of a line is the area under its absorption-coefficient curve, i.e., $I=\int_{-\infty}^{\infty}\alpha(\nu)d\nu$, serious errors may arise because of difficulties in distinguishing between the background signal and absorption in the "wings" of the line. The error can be particularly serious for a Lorentzian line, which has a rather significant portion of its area well away from the line center. Measurements with tunable lasers of line intensities have been reported for SO_2 [2.76], NH_3 [2.115], CO [2.21], NO [2.104, 105], and H_2O [2.58, 119, 120]. The measured values are generally in reasonable agreement with those determined using conventional techniques or calculated from known molecular data.

2.3.2. Line Identification and Band Analysis

The utility of tunable lasers for line identification and band analysis depends largely on the spectral line density of the molecule of interest and the continuous tuning range of the laser. Due to mode hopping (see Section 2.2 and Figs. 2.3 and 2.5), narrow-linewidth lasers at present have restricted increments of continuous tuning, typically less than $1\text{–}2\,\mathrm{cm}^{-1}$. As a result, it is difficult to make a multiple-line study of molecules which have average line spacings of $0.5\,\mathrm{cm}^{-1}$ or more. The

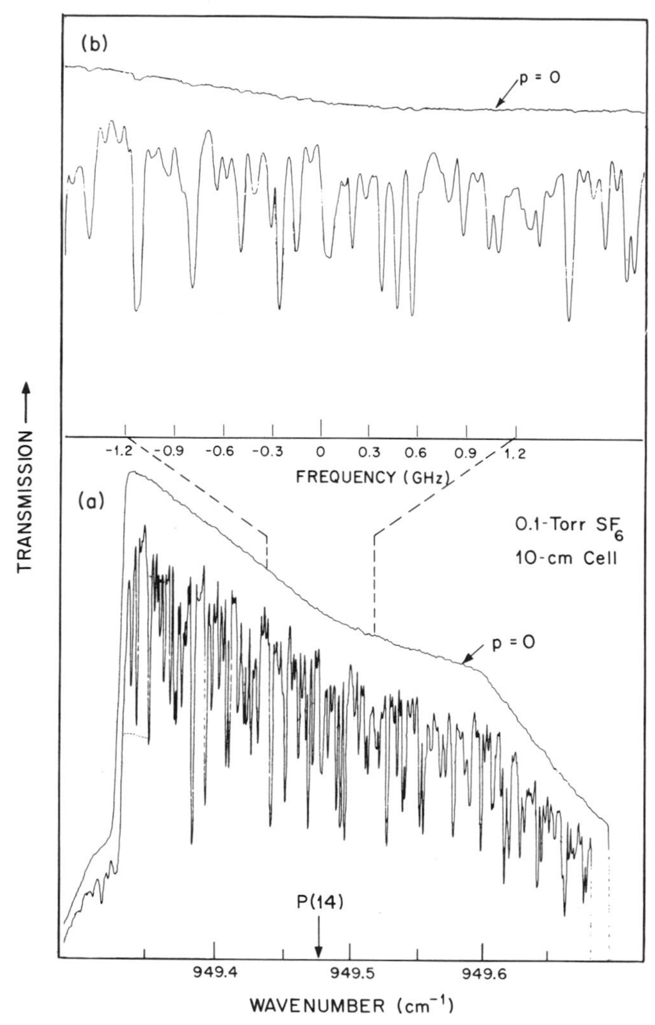

Fig. 2.18a and b. Absorption spectrum in the ν_3 band of SF_6 near 10.5 μm using (a) spectrometer calibration of laser wavelength, and (b) heterodyne calibration using the $P(14)$ CO_2 laser line as reference. The SF_6 pressure was 0.1 Torr, and cell length 10 cm [2.144]

gaps in spectral coverage lead to omitted lines, and it is difficult to calibrate the absolute laser frequency as it moves from one mode to another.

Typical diatomic molecules have rotational constants of 1–3 cm^{-1}, which give vibration-rotation line spacings of 2–6 cm^{-1}. Such simple molecules have been adequately studied from the band-analysis point of view using conventional techniques. *Lineshapes* and *fine structure* are the most suitable targets for tunable lasers studies of these molecules.

More complex molecules generally have rich infrared spectra with lines spaced 0.01–0.1 cm^{-1} apart, so that many of them fall within the continuous-tuning range of a narrow-linewidth tunable laser. Figure 2.18 shows tunable laser spectra of a portion of the ν_3 band of SF_6 in the vicinity of the 944.477 cm^{-1} $P(14)$ CO_2 laser line. These scans were taken with a current-tuned $Pb_{0.88}Sn_{0.12}Te$ diode laser. The lower trace a) is 0.35 cm^{-1} wide and reveals approximately 70 Doppler-broadened lines—an average separation of 0.005 cm^{-1}. Wavelength calibration was made with the aid of a grating spectrometer. The upper trace b) is the second laser scan within 1.2 GHz of the $P(14)$ CO_2 laser line. In this case the heterodyne/absorption calibration technique discussed earlier was used. (The 1.2-GHz frequency limit corresponds to the cutoff frequency of the Ge:Cu infrared detector used for the heterodyne measurement.) Since the Doppler width of SF_6 at room temperature is 29 MHz in this region, partial overlap of adjacent lines occurs. Similar measurements have been made around all the CO_2 laser lines between $P(12)$ and $P(24)$. Progress in obtaining line assignments has, however, been extremely slow because of the complex nature of this molecule, and a full analysis will be a formidable task even with the tunable laser results.

Figure 2.18b is an example of the technique of using an established infrared frequency standard, in this case the $P(14)$ CO_2 laser line [1.146], for absolute calibration of the tunable laser frequency. However, once the SF_6 spectrum has been calibrated, it can serve as a universal secondary frequency standard since its unique absorption structure can be readily reproduced and identified. The entire infrared region could eventually be calibrated in this manner using a variety of gases.

Another molecule which has a high line density, but which *is* amenable to analysis, is SO_2. Using a series of $Pb_{1-x}Sn_xTe$ diode lasers (with $0.058 \leqq x \leqq 0.074$), HINKLEY et al. [2.76, 112] performed an extensive analysis of the ν_1 band of SO_2 centered around 8.7 μm. Approximately 10% of some 3000 lines in the 1100–1200 cm^{-1} region were analyzed, and positive identification was made of 95% of the observed lines. A new value for the band center of 1151.71 ± 0.01 cm^{-1} was established on the basis of these measurements, and the band strength extrapolated from the individual line intensity measurements was consistent with those obtained from more conventional measurements.

Figure 2.19 represents a detailed comparison in the 1147 cm^{-1} region between results of a $Pb_{0.93}Sn_{0.07}Te$ SDL scan and theoretical intensities and positions of lines in this region predicted [2.76] on the basis of accurate microwave measurements [2.147]. The curves are

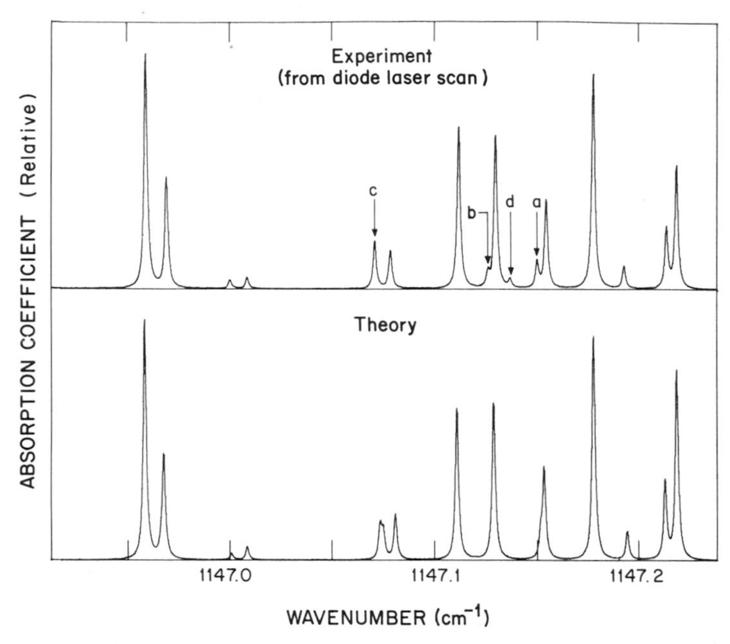

Fig. 2.19. Comparison between measured (upper trace) and calculated (lower trace) absorption coefficients and positions for several lines in the ν_1 vibration-rotation band of SO_2 [2.76]

Lorentzian fits to the intensities, incorporating a linewidth which best reproduces the shape of the experimental curve at a pressure of 1 Torr. (At this pressure, the SO_2 line profile is actually in the Voigt regime.) Minor differences between the two curves are indicated at *a*, *b*, and *c*, where transitions involving higher values of J and K_A (relative to adjacent lines) are noticeably shifted from their calculated positions. At *d* is a line which is not predicted—it may be due to a "hot" band transition, other isotope, or weaker transition not included in the calculation of the theoretical curve. Using the fractional contributions of several rather isolated lines in the 1100–1200 cm^{-1} region to the total band intensity, a projected total ν_1 band intensity of $358 \pm 20 \times 10^{-20}\,cm^{-1}$ molecule^{-1} cm^2 was obtained [2.69]. This compares favorably with a value of $371 \pm 20 \times 10^{-20}\,cm^{-1}$ molecule^{-1} cm^{-1} obtained by BURCH et al. [2.148] employing the conventional integrated-band technique.

As indicated above, tunable lasers with narrow linewidths and high-frequency stability invariably have narrow continuous tuning ranges. If one tunes such a laser over a broad range in an attempt to perform wideband molecular spectroscopy, the resulting spectra are incomplete

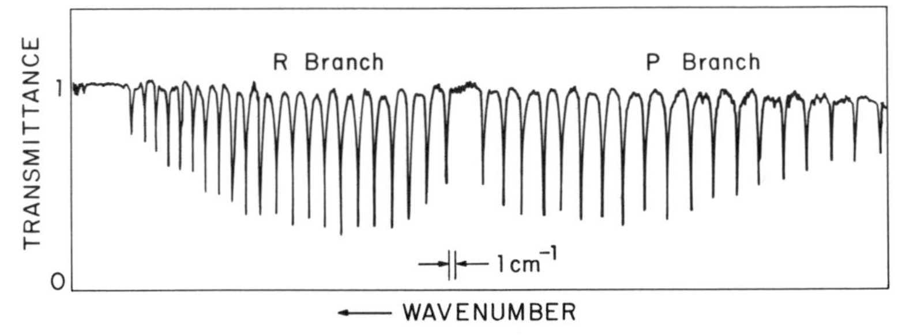

Fig. 2.20. Absorption spectrum of the first overtone ($v=0 \rightarrow v=2$) of CO taken with a $LiNbO_3$ parametric oscillator. CO pressure was 600 Torr. (After BYER [2.96])

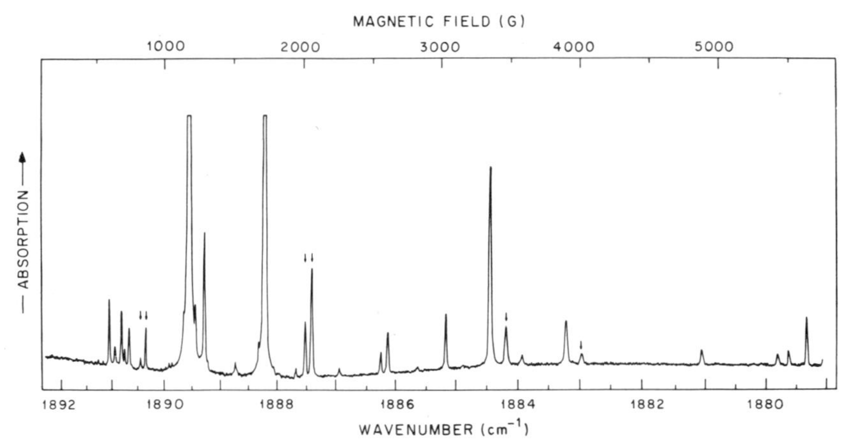

Fig. 2.21. Absorption spectrum of a portion of the H_2O band around 1886 cm^{-1}, obtained with an SFR laser. 20 ppm of NO were contained in the cell at 76 Torr total pressure, and the NO lines are indicated by the arrows. The SFR pump laser was at 1892.25 cm^{-1}. (After PATEL [2.108])

due to the gaps left by laser mode hopping. If one relaxes the resolution requirements through either frequency instabilities or cavity detuning, it is possible to obtain complete spectra with 100% frequency coverage and a resolution as good as that of many grating spectrometers. Both the OPO and the SFR lasers have been used in this manner to obtain broadband spectra of molecules in a manner more consistent with conventional band analysis spectroscopy. BYER [2.96] and SCHLOSSBERG [2.92] have recorded spectra of the first overtone of CO at 2.3 μm. In Fig. 2.20 are shown BYER'S results, taken with a $LiNbO_3$ parametric oscillator pumped by a doubled Nd:YAG laser. The resolution was 0.2 cm^{-1}, and the entire scan over ≈ 160 cm^{-1} required 10 min to take. Note that about 20 lines are covered in each of the *P* and *R* branches.

Wideband spectra have been obtained by PATEL [2.108] and ALLWOOD et al. [2.109, 110] for H_2O and NO around 5.3 μm using SFR lasers. Figure 2.21 shows a portion of the H_2O spectrum near 1886 cm^{-1} obtained by PATEL [2.108]. The gas sample in this experiment contained 20 ppm NO, whose absorption lines are identified in the figure by arrows. Approximately 20 H_2O lines were observed over the 13 cm^{-1} scan, but the estimated resolution is only 0.1 cm^{-1}. This problem of competition between resolution and wavelength coverage is a recurrent one. It is a matter which deserves more attention in the development of tunable infrared lasers.

2.3.3. Rotational Fine Structure

To a first approximation, the energy levels of a simple molecule are given by the expression

$$E(v, J) = \sum_i [h\nu_{v_i}(v + 1/2) + 4\pi\hbar c B_i J(J+1)] , \qquad (2.19)$$

where for the ith vibrational band v is the vibrational quantum number, $h\nu_{v_i}$ the vibrational energy, B_i the rotational constant, and J the quantum number for the total angular momentum. Each vibration-rotation band has a rotational series of lines. There are a number of physical effects which add terms to the energy expression of (2.19) and split each of these rotational lines into several members. Such rotational fine structure has been studied for many years in the microwave and millimeter wave regimes [2.149]. These measurements were usually made on vibrational ground states, and typically involved a rotational transition. Because the line splittings produced by rotational fine structure lie in the radio and microwave range, observations in the infrared had been precluded because of the limited resolution of conventional instrumentation. Tunable lasers are ideally suited to measurements of fine structure and, indeed, a number of interesting measurements have already been made. The effects observed include: Λ-type doubling, nuclear hyperfine splitting, and Zeeman and Stark splitting.

Λ-Type doubling in NO

Nitric oxide is a simple molecule which is rich in rotational fine structure. It is the only stable diatomic molecule with both spin (***S***) and orbital (***L***) electronic momenta in the electronic ground state. This ground state is $^2\Pi$, with multiplet states $^2\Pi_{1/2}$ and $^2\Pi_{3/2}$, each having an infrared vibration-rotation band near 1900 cm^{-1}. The coupling of the various

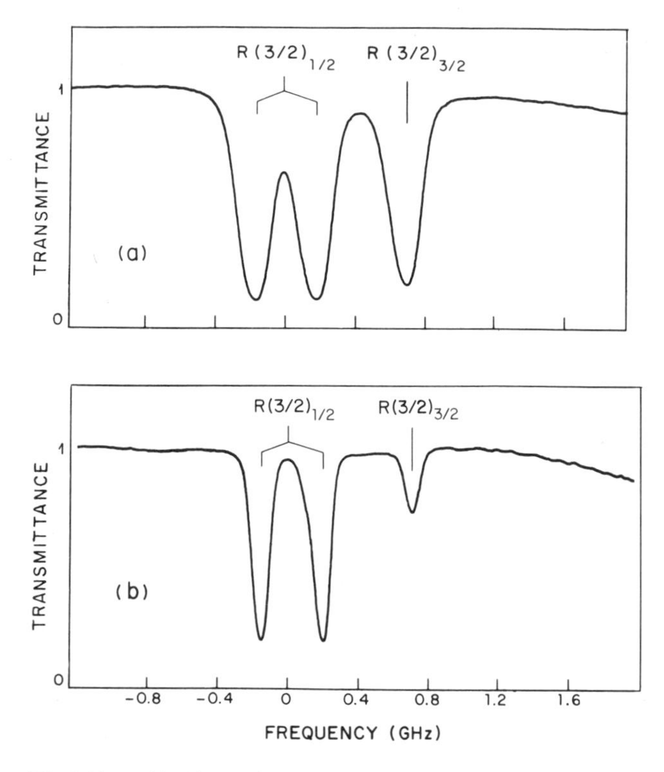

Fig. 2.22a and b. Absorption spectrum of the $R(3/2)_{3/2}$ and $R(3/2)_{1/2}$ lines of NO, illustrating Λ-type doubling of the $R(3/2)_{1/2}$ line at (a) 295 K, 4 Torr; and (b) 77 K, 1 Torr. Cell length was 5 cm. (After NILL et al. [2.104, 122])

angular momenta in NO is almost that of Hund's case (a), with a strong coupling of $\boldsymbol{S}$ and $\boldsymbol{L}$ to the internuclear axis. The magnitude of the components of $\boldsymbol{S}$ and $\boldsymbol{L}$ along this axis are $\Sigma = 1/2$ and $\Lambda = 1$, respectively. For a pure Hund's case (a), the $^2\Pi_{3/2}$ states would exhibit an electronic Zeeman effect, while the $^2\Pi_{1/2}$ states would not, and states with $M_L = \pm\Lambda$ would be doubly degenerate. Because of molecular rotation, there is a slight uncoupling of $\boldsymbol{L}$ and $\boldsymbol{S}$ from the internuclear axis. The $\boldsymbol{S}$ uncoupling yields a situation intermediate between Hund's cases (a) and (b), and gives the $^2\Pi_{1/2}$ states a sizable Zeeman effect. The $\boldsymbol{L}$ uncoupling splits the degeneracy of the configurations $+\Lambda$ and $-\Lambda$, yielding Λ-type doubling of the energy levels. In addition, the electronic rotational levels are significantly perturbed (split) by magnetic dipole coupling to the spin ($I = 1$) of the N^{14} nucleus. All of these effects in NO have been observed in microwave experiments [2.149–151]. However, until recently none had been fully resolved or completely studied in the infrared.

Observations of these effects using SDL's [2.104, 107, 122, 123] and SFR lasers [2.69, 108] have now been made in the infrared with a new degree of clarity.

Using tunable $PbS_{0.6}Se_{0.4}$ diode lasers, spectra showing Λ-type doubling were obtained on several R and Q branch lines by NILL et al. [2.104, 122]. Figure 2.22 shows absorption spectra of the $R(3/2)_{1/2}$ and $R(3/2)_{3/2}$ lines, which are almost coincident in the fundamental band of NO at 5.3 μm. Spectra are shown for both room temperature (a) and 77 K (b). The line splitting due to Λ-type doubling for the $2\Pi_{3/2}$ transitions is much smaller than the Doppler width, and cannot be resolved unless nonlinear (saturation) experiments similar to those discussed earlier are performed. The measured splitting of the $R(3/2)_{1/2}$ line is 350 ± 18 MHz, which agrees with a value of 349.9 MHz calculated using molecular parameters measured in microwave experiments on the vibrational ground state.

Nuclear Hyperfine Structure

The first observation of nuclear hyperfine splitting in the infrared vibration-rotation spectrum of NO was made with a $PbS_{0.6}Se_{0.4}$ diode laser by BLUM et al. [2.107, 122]. A fully-resolved absorption spectrum of the first transition in the fundamental $Q_{1/2}$ branch is shown in Fig. 2.23a. The absorption is due to the $Q(1/2)_{1/2}$ line near 1876.1 cm^{-1}, which occurs in the $^2\Pi_{1/2}$ electronic ground state between energy levels with the minimum rotational angular momentum of $J = 1/2$ (exclusive of nuclear spin). The $Q(1/2)_{1/2}$ transition exhibits two types of fine structure, which, in turn, yield four major absorption lines. The gross splitting into symmetric pairs of lines results from Λ-type doubling discussed above. The smaller splittings into symmetric pairs results from a coupling of the electrons to the spin ($I = 1$) of the N^{14} nucleus of $N^{14}O^{16}$.

It is known from microwave measurements on the vibrational ground state ($v = 0$) that the electron-nuclear coupling in NO is predominantly magnetic-dipole in origin, the electric quadrupole contribution being smaller. In fact, for $J = 1/2$, the electric quadrupole terms vanish identically. The electronic moment couples the molecular momentum J to the N^{14} spin I, yielding the resultant momentum F. The appropriate quantum numbers are Λ, J, and F. For $J = 1/2$, the allowed values of F are 1/2 and 3/2, and each of the Λ-doublet energy levels splits into two levels, as shown schematically in Fig. 2.23. The selection rules for Q-branch electric dipole transitions are $\Delta J = 0$, $\Delta F = 0, \pm 1$. In addition, the initial and final Λ-doublet states must have opposite symmetry, i.e., $\Lambda_+ \rightarrow \Lambda_-$ or $\Lambda_- \rightarrow \Lambda_+$. The four allowed hyperfine transitions are shown schematically in the energy level diagram of Fig. 2.23 for one of the two

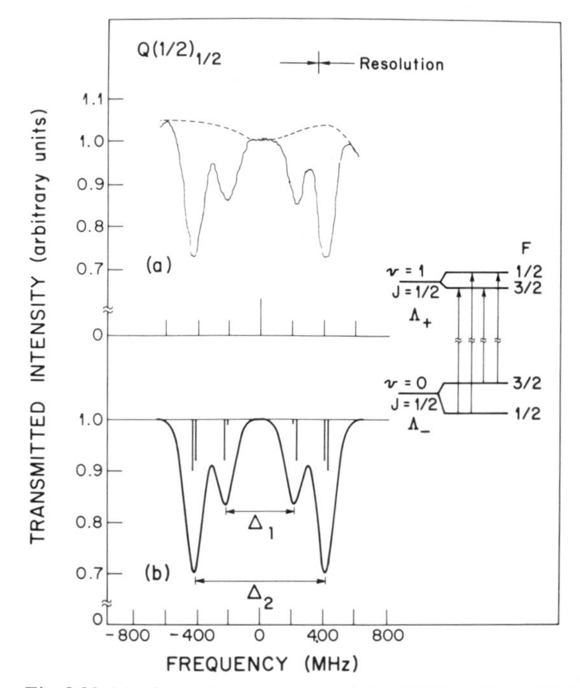

Fig. 2.23. (a) Absorption spectrum of the $Q(1/2)_{1/2}$ line of NO, illustrating nuclear hyperfine splitting. (b) Theoretical spectrum assuming Gaussian lineshapes. The inset shows the energy levels and associated transitions for a Λ-doublet pair. (After BLUM et al. [2.107]

allowed Λ-doublet transitions. Neglecting the vibrational dependence of the fine structure constants, the theoretical absorption spectrum of Fig. 2.23b was calculated [2.107]. There are eight hyperfine lines whose relative strengths and positions are given by the vertical lines. The agreement between the measured and calculated spectra is excellent.

More recently, HALL and BORDÉ [2.121] resolved the hyperfine components of the $F_2^{(2)}$ component of the $P(7)$ line of CH_4 at 3.39 μm using a piezoelectrically-tuned He–Ne laser. They achieved an unprecedented resolution of 12 kHz by expanding the laser beam to 5-cm diameter, and cooling the CH_4 gas to 77 K in a saturation spectroscopy experiment. Three main lines with $\Delta F = \Delta J = -1$, and two weak lines with $\Delta F = 0$ were observed, in addition to some level-crossing peaks. With this enormous resolving power of nearly 1×10^{10}, they were able to establish completely the hyperfine structure of both the ground ($v = 0, J = 7$) and the first vibrationally excited states ($v = 1, J = 6$).

Other observations [2.86, 152] of hyperfine structure have been restricted to studies of electronic spectra in the near infrared and visible with either conventional or fixed-frequency laser techniques.

Zeeman Splitting

For a weak applied magnetic field B, the rotational energy levels of a molecule are perturbed by an additive amount

$$\Delta E_Z(J) = g_J \mu B M_J , \tag{2.20}$$

where g_J is the effective splitting factor for states with angular momentum $\boldsymbol{J}$, μ is the appropriate characteristic magnetic moment (either the nuclear or Bohr magneton), and M_J is the component of $\boldsymbol{J}$ along $\boldsymbol{B}$. For molecules in the $^1\Sigma$ states, μ is the nuclear magneton, and ΔE is very small indeed (radio frequencies). For more complex molecules, μ can be the Bohr magneton, g can vary with rotational direction, and ΔE can be in the microwave region.

Two studies of the molecular Zeeman effect with tunable infrared lasers have been reported. UZGIRIS et al. [2.98] used saturation spectroscopic techniques to study the Zeeman splitting of the $P(7)$ line of CH_4 using a 3.39-μm He–Ne laser which was piezoelectrically tuned. They observed a novel level-crossing, nonlinear peak first predicted by SCHLOSSBERG and JAVAN [2.153], and the extremely high resolution and frequency precision of their technique permitted measurement of frequency splittings as small as 200 kHz for this $^1\Sigma$ state molecule. Also, the Zeeman effect in NO has been studied by NILL et al. [2.104, 122] using SDL techniques. The unpaired electron in NO gives it a magnetic moment which is on the order of the Bohr magneton. Uncoupling of the electronic spin from the inter-nuclear axis makes NO strongly paramagnetic in both the $2\Pi_{1/2}$ and $^2\Pi_{3/2}$ states. The g-factor is strongly J-dependent particularly for the $^2\Pi_{3/2}$ states. A derivative transmission spectrum of the Zeeman-split $R(15/2)_{3/2}$ line of NO at 31 kG is shown in Fig. 2.24b. There are 14 major Zeeman components due to the high J value and the rapid variation of g_J with J. The measured line positions are shown (open circles) as a function of magnetic field in Fig. 2.24a. The data agree reasonably well with the calculated (solid) lines, but show a small, systematic deviation at high fields which may be due to second-order (B^2) Zeeman effects.

Associated with the Zeeman effect in a molecular gas are infrared circular birefringence and dichroism. The magnetic rotation signal [2.154], which is the intensity of radiation transmitted through a gas cell placed between crossed linear polarizer plates, is affected by both these effects. Doppler-limited magnetic rotation spectra of the $R(13/2)_{1/2}$ line of NO were obtained by BLUM et al. with a tunable $PbS_{0.6}Se_{0.4}$ diode laser, and qualitative agreement between theory and experiment was found.

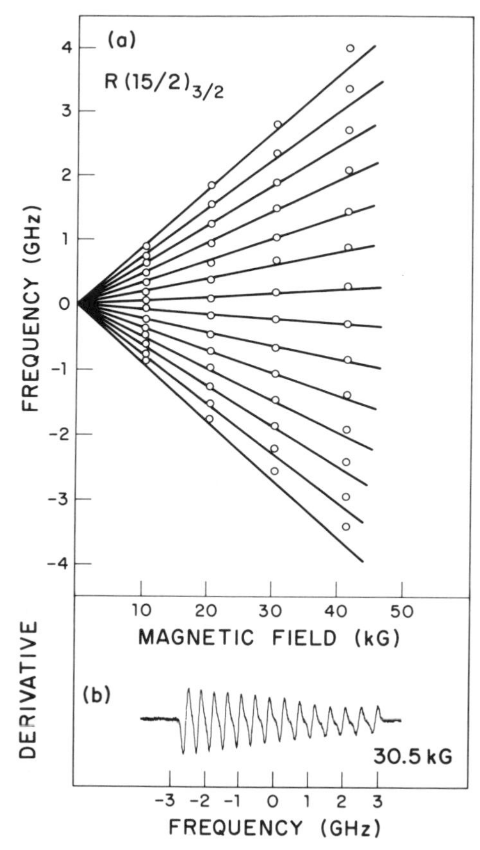

Fig. 2.24a and b. For the $R(15/2)_{3/2}$ line of NO: (a) Zeeman line frequencies vs. magnetic field given by experiment (open circles) and linear Zeeman theory (solid lines); and (b) derivative absorption spectrum at 30.5 kG. (After NILL et al. [2.104, 122])

Stark Splitting

For a weak applied electric field, the molecular rotational energy levels are perturbed by

$$\Delta E_S(J) = \bar{\mu}_S(J)\varepsilon\,, \tag{2.21}$$

where ε is the electric field strength, and $\bar{\mu}_S$ the average electric moment along ε. For molecules with a permanent electric moment and non-zero angular momentum about their symmetry axis, $\bar{\mu}_S$ is independent of ε to a first approximation and the Stark effect is linear in ε. Otherwise, $\bar{\mu}_S$ is linear in ε to first order, and ΔE_S is quadratic in ε.

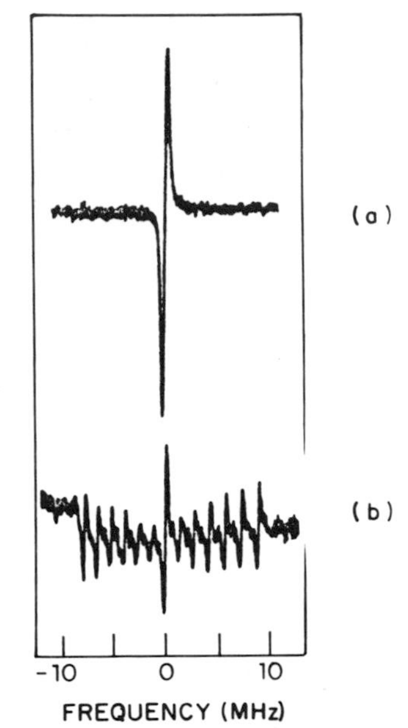

Fig. 2.25a and b. Saturation spectra of the E component of the $P(7)$ line in the ν_3 band of CH_4: (a) no Stark field, and (b) with a Stark field of 1660 V/cm. (After LUNTZ and BREWER [2.99])

Tunable laser have been used to obtain high-resolution Stark spectra of several gases: 1) a Zeeman-tuned 3.39-μm He–Ne laser was used by LUNTZ and BREWER [2.99] to study the Stark effect in CH_4; 2) Zeeman-tuned He–Xe and He–Ne lasers were used by UEHARA et al. [2.102] to study H_2CO and other organic molecules in the 3- and 5-μm regions; 3) CORCORAN et al. [2.117] used the TPM technique to study methyl bromide and methyl chloride; and 4) FRANCKE et al. [2.124] measured Stark splitting in NH_2D in the 10-μm region using a tunable $Pb_{0.88}Sn_{0.12}Te$ diode laser. In each case, modulation of the Stark field was used to obtain derivative spectra.

Saturation (Lamb-dip) spectra of the E-component of the $P(7)$ line of the ν_3 band of CH_4, obtained by LUNTZ and BREWER [2.99], are shown in Fig. 2.25. The spectrum of Fig. 2.25a is that without a Stark field, and shows a single, isolated line. Figure 2.25b was obtained for a Stark field of 1660 V/cm, in which thirteen lines are resolved; this number is equal to $2J+1$, where J is the upper-state rotational quantum number. Note that the total frequency scale covers only 20 MHz, and that the linewidths are about 100 kHz. These data are an excellent example of the power of the combination of tunable infrared laser and the saturation spectroscopy technique.

2.3.4. Spectroscopic Detection of Trace Gases—Application to Air Pollution Monitoring

Most of the atmospheric pollutant gases have characteristic spectral lines in the 2–20 μm region of the infrared [2.155], and a great deal of attention has recently been given to the use of laser spectroscopic techniques to monitor their concentrations in the ambient air or at effluent sources [2.156–160]. We shall briefly consider three basic monitoring techniques involving tunable infrared lasers: reduced-pressure point sampling for high specificity; *in situ* ambient air and source monitoring by resonance absorption; and several single-ended remote sensing techniques. Each of these has its particular usefulness and has specific demands on the tunable laser.

Point Sampling at Reduced Pressure

Conventional instruments for the detection of trace gases usually require the taking of a sample (from the ambient air or the effluent of a pollutant source) for either on-site or later laboratory analysis. Although sensitivities of such instruments are usually adequate, interfering reactions may be caused by other gases invariably present in the sample, resulting in erroneous readings. High-resolution infrared spectroscopy, when used as a monitoring technique, greatly reduces the chance of interference. That is, by transmitting tunable laser radiation through a gas cell containing the specimen to be analyzed at a pressure of a few Torr or less, Doppler-broadened spectra similar to those shown earlier in this chapter are obtained, and these infrared "fingerprints" provide unequivocal identification of a particular gas specie and allow its concentration to be determined.

The laser power P transmitted through a gas cell of length L is related to the power P_0 through an empty cell by the expression

$$P = P_0 \exp(-\alpha p L), \tag{2.22}$$

where α is the absorption coefficient per unit pressure of the gas to be detected, and p its partial pressure. The derivative technique is very useful for low-pressure trace gas analysis, as illustrated in Fig. 2.10, and its signal is highest when the total gas pressure is around 5 Torr. Since the number of molecules for analysis is smaller at the lower pressure, it is illustrative to perform a calculation to ensure a useful detection sensitivity. If we assume that a 0.1% change in laser power can be detected as the laser frequency tunes through an absorption line, and that the absorption coefficient for this line is 1.3 $\mathrm{cm^{-1}\,Torr^{-1}}$ [as it is for the $P(7)$

line of CO and strong lines of several other gases], then the minimum detectable ($S/N = 1$) concentration over a pathlength of 100 cm is 7.7×10^{-6} Torr (out of 5 Torr), or 1.5 ppm (parts per million). This sensitivity is usually adequate for analyzing source effluents, which have pollutant concentrations typically in the hundreds of ppm, and which can present severe interference difficulties with conventional instrumentation.

Tunable lasers have been used to perform several trace gas detection experiments using the reduced-pressure sampling technique. Semiconductor diode lasers tailored to strongly-absorbing regions of C_2H_4 (10.5 μm), CO (4.7 μm), and NO (5.3 μm) were used to detect these gases in samples of automobile exhaust [2.79, 161, 162], and SO_2 has been detected in samples of smokestack gas from an operating power plant [2.112]. Using direct *absorption* measurements by an opto-acoustic detector of amplitude-modulated radiation from a 1-W tunable SFR laser [2.163], KREUZER and PATEL [2.164a] measured concentrations of NO in air samples taken from several locations. This technique has also been used to measure stratospheric NO from a balloon-borne platform [2.164b]. The major advantages of opto-acoustic detection are that a cooled infrared detector is unnecessary, that the signal is proportional to *absorption* (rather than *transmission*), and that with a tunable laser having moderate power it is capable of detecting concentrations well below 1 ppb for many gases [2.165].

In situ *Monitoring by Resonance Absorption*

Optical techniques have a capability that point-sampling instruments do not have—i.e., monitoring can be carried out without disturbing the region under study. At atmospheric pressure the very high specificity of Doppler-limited spectroscopy is no longer possible due to pressure broadening, and interferences from other gases may occur with this *in situ* technique, necessitating careful selection of the spectral line for monitoring a particular gas specie. Such measurements can be performed on the exhaust emissions of automobiles, in stack gases of power or heating plants, or over long paths in the ambient air. In general, the equation for transmitted laser power can be written as follows:

$$P = P_0 \exp\left\{-\int_0^L [\sigma(\nu)N(x) + \beta]\right\} dx\,, \tag{2.23}$$

where $\sigma(\nu)$ is the molecular absorption cross section for the gas to be monitored, $N(x)$ its density distribution, and β is an effective scattering or extinction factor caused by aerosol scattering or turbulence, and is assumed to vary slowly with wavelength. If the laser is tuned alternately between a frequency ν_1 at line center and ν_2 where $\sigma = 0$, then the ratio $P(\nu_2)/P(\nu_1)$ is independent of the extinction parameter. That is, if we assume that $\bar{N}$ is the integrated pollutant density over the distance L, then

$$\ln[P(\nu_2)/P(\nu_1)] = \sigma_{max} \bar{N} L, \tag{2.24}$$

and thus depends only on the peak absorption cross section, pathlength, and average pollutant concentration over the total path.

Figure 2.26 shows a set of scans of the $P(7)$ line of CO taken by HINKLEY et al. [2.161] with a current-tuned $PbS_{0.82}Se_{0.18}$ SDL, using a 10-cm cell with the gas at atmospheric pressure. Samples were taken from the exhaust of a 1972 automobile (a) and a 1973 station wagon (c), and compared with calibration traces corresponding to 1970 ppm CO in air (b) and 6050 ppm CO in air (d). From the calibration traces, $\sigma_{max} = 1.8 \times 10^{-18}$ cm^2, indicating CO concentrations of 1350 ppm (0.14%) and 4980 ppm (0.5%) for (a) and (c), respectively. Perhaps the most striking observation from Fig. 2.26 is that there are no apparent interferences from hydrocarbons, water vapor, and other gases which are present in the raw exhaust specimens, as these would change the shapes of the spectral lines. This is not necessarily the case for gases such as SO_2 which have both weak and overlapping absorption lines at atmospheric pressure [2.166].

Although there are 150 times more molecules at atmospheric pressure than at the 5 Torr pressure used for the derivative technique, the improvement in sensitivity is not nearly as great due to pressure broadening of the spectral lines. To show this, we analyze (2.24) on the same basis as (2.22); i.e., a cell length of 100 cm, minimum detectable power change of 0.1%, and the same line whose cross section is 1.8×10^{-18} cm^2 in the atmosphere. The minimum detectable concentration is found to be 5.6×10^{12} cm^{-3}, or 2 ppm, which is only 7.5 times better than the value at reduced pressure.

Long atmospheric paths are required to detect low concentrations of pollutant gases present in the ambient air, and for this type of measurement a power discrimination of 1% would be more reasonable, since atmospheric scintillation effects are usually present. For a 1-km pathlength, therefore (to a detector 1 km away or to a retro-reflector 0.5 km away), the minimum detectable CO concentration using the $P(7)$ line would be 2 ppb (parts per billion); and barring interferences, this is the same value that would obtain for any gas having an absorption line with the same cross section. As a demonstration of this technique, a 100-μW tunable SDL was used to measure CO in the ambient air over a 610-m path. The resulting data showed an integrated-path value of around

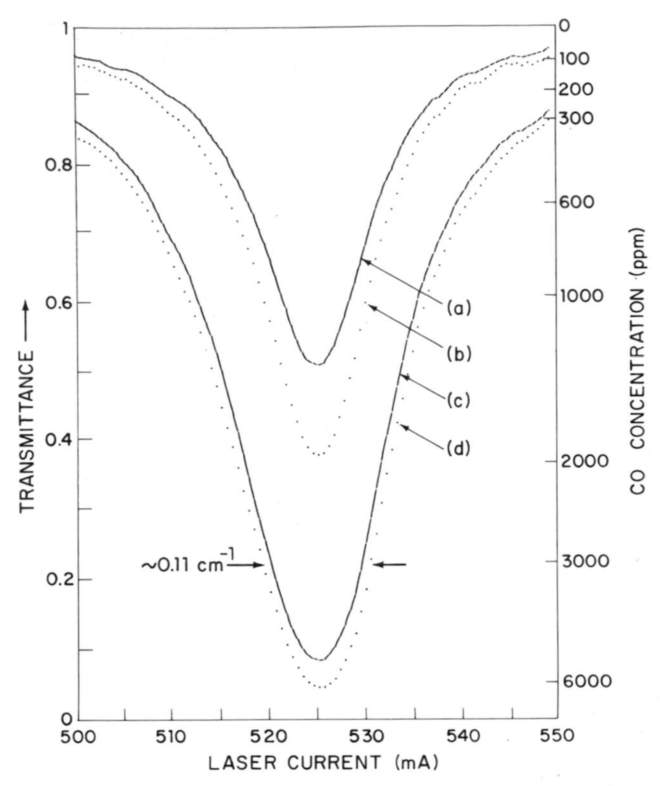

Fig. 2.26. SDL scans of the $P(7)$ CO transition at $2115.63\,\mathrm{cm}^{-1}$ at atmospheric pressure using a 10 cm cell, illustrating the usefulness of this technique for *in situ* monitoring of pollutant gases. (a) and (c) are scans of the raw exhaust emission from a 1972 sedan and 1973 station wagon, respectively; (b) and (d) are calibration traces of pure CO (1.5 Torr and 4.6 Torr, respectively) in 1 atm of air [2.158]

400 ppb in a semirural area away from the highway [2.161]. Using a single-pulsed SDL technique, C_2H_4 was monitored across a parking lot, and it exhibited a concentration rise to 1 ppm averaged over the 250-m path as the traffic volume peaked [2.162].

Single-Ended Remote Sensing

There are several optical monitoring techniques which do not require a remotely situated reflector or detector. In each case to be described the pollutant detection sensitivity is less than that for double-ended resonance absorption, and the required laser power is considerably larger.

a) Backscatter absorption using a topological reflector was proposed by BYER and GARBUNY [2.167a] as a way to avoid the need for a remote mirror or retro-reflector. Although in principle the sensitivity of this approach depends upon the same cross section as that for (2.24), the substantially smaller reflected power from "uncooperative" targets such as buildings must be compensated for by a higher laser power. In the first reported application of this technique, HENNINGSEN et al. [2.167b] used the reflected signal from foliage 107 m away to detect atmospheric CO with an OPO tuned to the 2.3 μm first overtone region of this gas.

b) Backscatter absorption using Mie scattering from the atmosphere has been discussed by several authors [2.167–169] as a technique to combine the sensitivity of resonance absorption with the versatility afforded by distributed atmospheric backscattering. Range information can be obtained by time-gating the infrared detector as in LIDAR (laser radar). A mathematical formulation of this technique (also known as DAS for "differential absorption and scattering") is beyond the scope of this chapter, but thorough treatments may be found in the references cited above. BYER and GARBUNY [2.167a] showed that it is theoretically possible to detect a cloud of 20 ppm CO with a depth resolution of 15 m at a range of 1 km using 10^6-W peak power (100-mJ, 100-ns) laser pulses alternately tuning "on" and "off" a strong line in the infrared. However, the first detection of a pollutant by this technique was reported by ROTHE et al. [2.170] who used a tunable dye laser to measure atmospheric NO_2 at distances up to 4 km. (See also Fig. 1.2.)

c) Resonance fluorescence occurs when a molecule returns radiatively to its original state after being excited by an incident laser beam of the appropriate frequency. The magnitude of the fluorescence indicates the concentration of the molecule. Most resonance fluorescence measurements have been carried out with tunable dye lasers because the cross sections for absorption of the laser light are orders of magnitude larger in the visible and near ultra-violet than in the infrared. In particular, GELLWACHS and BIRNBAUM [2.171] demonstrated the detection of NO_2 in the ambient air to 1 ppb sensitivity using laser radiation at 0.4416 and 0.4880 μm, with the fluorescence monitored at 0.7–0.8 μm. With regard to the possibility of using resonance fluorescence in the infrared, KILDAL and BYER [2.157] predicted that with 1000 pulses of laser radiation in the fundamental CO band, each having 1-mJ energy, it should be possible to detect 10–10^4 ppm of the gas (depending upon its physical depth along the laser beam) at a range of 1 km.

d) Heterodyne detection offers a remote, single-ended technique for the detection of atmospheric pollutant gases, which is entirely passive, i.e., no laser radiation is transmitted into the atmosphere [2.79]. The apparatus can also be far removed (≈ 1 km) from the pollutant source. As described in detail in *Heterodyne Detection of Infrared Emission Lines* above, a tunable laser can be used as the local oscillator for sensitive heterodyne detection of characteristic gas emission lines in the infrared. By scanning the laser frequency through that of the emission line, a beat frequency is produced whenever the difference frequency is within the bandwidth of the detector-amplifier combination. The signal-to-noise ratio is given by [2.79]

$$\frac{S}{N}=(1-\mathrm{e}^{-N\sigma L})\left(\frac{1}{\mathrm{e}^{h\nu/kT_{\mathrm{g}}}-1}-\frac{\varepsilon_{\mathrm{b}}}{\mathrm{e}^{h\nu/kT_{\mathrm{b}}}-1}\right)(B\tau)^{1/2}\,, \tag{2.25}$$

where L is the thickness of the plume, T_{g} its temperature, ε_{b} the background emissivity, and T_{b} the background temperature. The other parameters have been defined previously. This equation is based on the following assumptions: 1) that the system bandwidth B is less than the emission linewidth; 2) that the emission and absorption due to the pollutant in the ambient atmosphere is negligible; 3) that background attenuation from other molecular absorption lines is negligible; and 4) that the local oscillator has sufficient power to overcome the other sources of noise. Under these conditions, it should be possible to detect concentrations of a few ppm of NO and CO in stack gas using emission lines in the 4–5 μm region, and to detect even smaller quantities of C_2H_4 and NH_3 using their emission lines in the 10-μm region [2.159]. Although the use of *tunable* infrared lasers for remote heterodyne detection of pollutant gases has not yet gone beyond the laboratory stage, several experiments using fixed-wavelength lasers have been reported [2.77, 81, 82]. Particularly noteworthy are the recent results of MENZIES and SHUMATE [2.82], which show sensitive detection of several pollutant gases (NO, SO_2, O_3, C_2H_4, and NH_3) using partial coincidences between their characteristic spectral lines and those of the CO or CO_2 laser local oscillators. Further advances in this technique of pollution monitoring will require stable, tunable lasers having ≈ 1 mW of mode power in conjunction with fast (≈ 1 GHz), efficient infrared detectors.

2.4. Conclusion

It is evident from the rapid pace with which developments are being made in the field of tunable infrared lasers that we are on the threshold of many new discoveries in infrared spectroscopy, both fundamental and applied. Each type of laser has its own advantages and limitations for such applications. Tunable semiconductor diode lasers are the simplest to operate, but require a high level of solid-state technology for fabrication. Although the present diode lasers require cryogenic temperatures, it appears that closed-cycle coolers can be successfully utilized, and will be particularly useful for field application of these devices. Due to its complexity, the spin-flip Raman laser will probably remain a laboratory instrument despite advances in closed-cycle cooling and reductions in the magnetic field requirements. With its high power and good mode quality, the SFR laser will be an important source for nonlinear spectroscopy experiments. All of the other tunable lasers described in this chapter operate at room temperature, although a superconducting magnet is sometimes used for the Zeeman-tuned gas laser. The recent developments of optical parametric oscillators using proustite and CdSe as the nonlinear material have opened up the infrared region beyond 4 μm for this tunable source. But the intra-cavity etalon required for narrow-line operation of the OPO complicates the procedure for continuously tuning this source. Difference frequency generation will be useful over restricted regions of the infrared, and tunable sources based on two-photon-mixing of a fixed-frequency laser and a klystron will be able to perform precision spectroscopy around the laser line employed. The four-photon mixer, involving an alkali metal vapor as the nonlinear material and two tunable dye lasers as pumps, has already shown an impressive performance in the laboratory, but it is probably too complicated an instrument for field use. The Zeeman-tuned gas lasers are limited to spectral regions around fixed-frequency laser lines which can be shifted with a magnetic field; their coverage is, therefore, quite limited. Finally, high-pressure gas lasers may eventually extend the Doppler-limited tuning ranges of conventional gas lasers to the entire vibration-rotation band of the particular gas being used. The development of simple pump schemes may lead to widespread utilization of the HPG laser, particularly if cw operation can be achieved.

Tunable lasers have brought to infrared spectroscopists new ultra-high-resolution techniques, including derivative and nonlinear spectroscopy, sensitive opto-acoustic detection, and heterodyne detection of spectral emission lines. The heterodyne technique, in particular, promises to have an exciting future in the field of astrophysics, permitting the detection of weak instellar spectral lines, as well as a correlation of signals in the wide-baseline interferometer configuration. The infrared experiments already performed have yielded much new information with regard to spectral-line parameters, hyperfine structure, precise line location, and band analysis. Work in nonlinear spectroscopy, like that of heterodyne detection, is just in its infancy.

Some types of tunable lasers will be particularly useful in field applications such as the monitoring of atmospheric pollutant gases. Several active and passive systems using tunable lasers have been proposed, and some are now being implemented. For regional air monitoring, pollutant gas concentrations averaged over distances of the order of 1 km represent valuable information for modeling programs aimed at forecasting pollutant levels, and this type of information can best be obtained with tunable lasers.

Finally, a textbook on molecular spectroscopy [2.172] published in 1966 contained the following observation concerning dispersive instruments: "There comes a point when decreasing the slit width results in such weak signals that they become indistinguishable from the background noise.... Thus, spectroscopy is a continual battle to find the minimum slit width consistent with acceptable signal-to-noise values." This restriction for high-resolution infrared spectroscopy has now been removed by tunable lasers, as shown by illustrations contained in this chapter. Furthermore, we are confident that within a few years tunable infrared laser spectroscopy will be commonly used in both laboratory and field applications throughout the world.

Acknowledgment

This work was supported by the National Science Foundation (RANN), the Environmental Protection Agency, and the Department of the Air Force. We would like to thank our colleagues who graciously contributed information and figures. A special note of appreciation is expressed to Mrs. C. A. MAWDSLEY and Miss KATHY MOLLOY for typing the manuscript.

References

2.1. W. DEMTRÖDER: *Laser Spectroscopy* (Springer, Berlin, Heidelberg, New York, 1971).
2.2. K. SHIMODA, T. SHIMIZU: *Nonlinear Spectroscopy of Molecules*, ed. by J. H. SANDERS and S. STENHOLM (Pergamon Press, Oxford, 1972).
2.3. For a recent discussion of tunable laser applications to chemistry, including an extensive list of references, see C. B. MOORE, P. F. ZITTEL: Science **182**, 541 (1973).
2.4. H. J. GERRITSEN: *Physics of Quantum Electronics*, ed. by P. L. KELLEY, B. LAX, and P. E. TANNENWALD (McGraw-Hill, New York, 1966), p. 581.
2.5. E. D. HINKLEY: Appl. Phys. Letters **16**, 351 (1970).
2.6. E. D. HINKLEY, C. FREED: Phys. Rev. Letters **23**, 277 (1969).
2.7. C. K. N. PATEL, E. D. SHAW, R. J. KERL: Rev. Letters **25**, 8 (1970).
2.8. C. F. DEWEY, Jr., L. O. HOCKER: Appl. Phys. Letters **18**, 58 (1971).
2.9. R. W. WALACE: See H. R. SCHLOSSBERG, P. L. KELLEY: Physics Today **25**, 36 (1972), Fig. 5.
2.10. V. J. CORCORAN, J. M. MARTIN, W. T. SMITH: Appl. Phys. Letters **22**, 517 (1973).
2.11. I. M. BETEROV, V. P. CHEBOTAYEV, A. S. PROVOROV: Opt. Commun. **7**, 410 (1973).
2.12. See, for example, K. N. RAO: *Molecular Spectroscopy: Modern Research* (Academic Press, New York, 1972), p. 346.
2.13. L. SCHAWLOW, C. H. TOWNES: Phys. Rev. **112**, 1940 (1958).

2.14. I. MELNGAILIS: J. Phys. Colloq. C-4 (Suppl.), No. 11, 12 (1968), pp. C4-84 to C4-94; T.C. HARMAN: J. Phys. Chem. Sol. Suppl. **32**, 363 (1971). See also [2.116].
2.15. R.B. DENNIS, C.R. PIDGEON, S.C. SMITH, B.S. WHERRETT, R.A. WOOD: Proc. Roy. Soc. London A **331**, 203 (1972).
2.16. S.E. HARRIS: Proc. IEEE **57**, 2096 (1969).
2.17. D.C. HANNA, B. LUTHER-DAVIES, R.C. SMITH: Appl. Phys. Letters **22**, 440 (1973).
2.18. P.P. SOROKIN, J.J. WYNNE, J.R. LANKARD: Appl. Phys. Letters **22**, 342 (1973).
2.19. T. KASUYA: Japanese J. Appl. Phys. **11**, 1575 (1972). See also [2.2].
2.20. N.G. BASOV, E.M. BELENOV, V.A. DANILYCHEV, A.F. SUCHKOV: JETP Letters **14**, 375 (1971).
2.21. K.W. NILL, F.A. BLUM, A.R. CALAWA, T.C. HARMAN: Appl. Phys. Letters **19**, 79 (1971).
2.22. K.W. NILL, A.J. STRAUSS, F.A. BLUM: Appl. Phys. Letters **22**, 677 (1973).
2.23. A.R. CALAWA: J. Luminescence **7**, 477 (1973).
2.24. R.W. RALSTON, J.N. WALPOLE, A.R. CALAWA, T.C. HARMAN, J.P. MCVITTIE: Solid Device Research Conference, Boulder, Colorado, June, 1973 (unpublished). See also J.N. WALPOLE, A.R. CALAWA, R.W. RALSTON, T.C. HARMAN: J. Appl. Phys. **44**, 2905 (1973);
K.W. NILL and A.J. STRAUSS: Private communication.
2.25. E.D. HINKLEY: MIT Lincoln Laboratory Solid-State Research Report **3**, 19 (1968); T.C. HARMAN, A.R. CALAWA, I. MELNGAILIS, J.O. DIMMOCK: Appl. Phys. Letters **14**, 333 (1969).
2.26. J.F. BUTLER, A.R. CALAWA: *Physics of Quantum Electronics*, ed. by P.L. KELLEY, B. LAX, and P.E. TANNENWALD (McGraw-Hill, New York, 1966), p. 458;
A.R. CALAWA, J.O. DIMMOCK, T.C. HARMAN, I. MELNGAILIS: Phys. Rev. Letters **23**, 7 (1969).
2.27. K.W. NILL, F.A. BLUM, A.R. CALAWA, T.C. HARMAN: Appl. Phys. Letters **21**, 132 (1972).
2.28. J.M. BESSON, J.F. BUTLER, A.R. CALAWA, W. PAUL, R.H. REDIKER: Appl. Phys. Letters **7**, 206 (1965);
J.M. BESSON, W. PAUL, A.R. CALAWA: Phys. Rev. **173**, 699 (1968).
2.29a. F.A. BLUM, K.W. NILL: In *Proceedings of the Laser Spectroscopy Conference* (Plenum Press, New York, 1974).
2.29b. S.H. GROVES, K.W. NILL, A.J. STRAUSS: IEEE Device Research Conference, Santa Barbara, California, June 1974.
2.29c. K.W. NILL, S.H. GROVES, A.J. STRAUSS: Postdeadline paper R.5 at VIII Internat'l Quant. Electr. Conf. San Francisco, Calif. (1974);
S.H. GROVES, K.W. NILL, A.J. STRAUSS: Appl. Phys. Letters **25**, 331 (1974).
2.30. C.K.N. PATEL, E.D. SHAW: Phys. Rev. Letters **24**, 451 (1970); Phys. Rev. B**3**, 1279 (1971).
2.31a. A. MOORADIAN, S.R.J. BRUECK, F.A. BLUM: Appl. Phys. Letters **17**, 481 (1970).
2.31b. R.S. ENG, A. MOORADIAN, H.R. FETTERMAN: Postdeadline paper R.2 at VIII Internat'l Quant. Electr. Conf., San Francisco, Calif. (1974); also Appl. Phys. Letters (to be published).
2.32a. C.K.N. PATEL: Phys. Rev. Letters **28**, 1458 (1972).
2.32b. S.R.J. BRUECK, A. MOORADIAN: J. Quant. Electr. (to be published).
2.33. S.R.J. BRUECK, A. MOORADIAN: Appl. Phys. Letters **18**, 229 (1971).
2.34. C.S. DESILETS, C.K.N. PATEL: Appl. Phys. Letters **22**, 543 (1973).
2.35. G. NATH, G. PAULI: Appl. Phys. Letters **22**, 75 (1973).
2.36. L.S. GOLDBERG: Appl. Phys. Letters **17**, 489 (1970).
2.37. A.I. IZRAILENKO, A.I. KOVRIGIN, P.V. NIKLES: JETP Letters **12**, 331 (1970).
2.38. R.C. MILLER, W.A. NORDLAND: Appl. Phys. Letters **10**, 53 (1967).

2.39. R. W. Wallace: Appl. Phys. Letters **17**, 497 (1970).
2.40a. A. A. Davydov, L. A. Kulevskii, A. M. Prokhorov, A. D. Savel'ev, V. V. Smirnov: JETP Letters **15**, 513 (1972).
2.40b. J. A. Weiss, L. S. Goldberg: Appl. Phys. Letters **24**, 389 (1974).
2.41. R. L. Herbst, R. L. Byer: Appl. Phys. Letters **21**, 189 (1972).
2.42. C. D. Decker, F. K. Tittel: Appl. Phys. Letters **22**, 411 (1973).
2.43. D. W. Meltzer, L. S. Goldberg: Opt. Comm. **5**, 209 (1972).
2.44. D. C. Hanna, R. C. Smith, C. R. Stanley: Opt. Comm. **4**, 300 (1971).
2.45. V. J. Corcoran, R. E. Cupp, J. J. Gallagher, W. T. Smith: Appl. Phys. Letters **16**, 316 (1970).
2.46. J. J. Wynne, P. P. Sorokin: In *Proceedings of the Laser Spectroscopy Conference* (Plenum Press, New York, 1974).
2.47. J. A. Giordmaine, R. C. Miller: Phys. Rev. Letters **14**, 973 (1965).
2.48. J. Pinard, J. F. Young: Opt. Comm. **4**, 425 (1972).
2.49. A. S. Pine: Private communication.
2.50. T. Kasuya: Appl. Phys. **3**, 223 (1974).
2.51. C. Freed, A. H. M. Ross, R. G. O'Donnell: J. Mol. Spectrosc. (to be published).
2.52. A. A. Beaulieu: Appl. Phys. Letters **16**, 504 (1970).
2.53. N. G. Basov, V. A. Danilychev, O. M. Kerimov, A. S. Podsosonnyi: ZhETF Pis. Red. **17**, 147 (1973).
2.54. V. N. Bagratashvili, I. N. Knyazev, Yu. A. Kudryavtsev, V. S. Letokhov: To be published.
2.55. A. J. Alcock, K. Leopold, M. C. Richardson: Appl. Phys. Letters **23**, 562 (1973).
2.56. J. S. Levine, A. Javan: Appl. Phys. Letters **22**, 55 (1973).
2.57a. R. L. Abrams, W. B. Bridges: IEEE J. Quant. Electron. QE-**9**, 940 (1973).
2.57b. R. L. Abrams: VIII Internat'l Quantum Electron. Conf., San Francisco, Calif. (1974).
2.58. F. A. Blum, K. W. Nill, P. L. Kelley, A. R. Calawa, T. C. Harman: Science **177**, 694 (1972).
2.59. J. Connes, P. Connes: J. Opt. Soc. Am. **56**, 896 (1966).
2.60. T. S. Jaseja, A. Javan, C. H. Townes: Phys. Rev. Letters **10**, 165 (1963); A. E. Siegman, B. Daino, K. R. Manes: IEEE J. Quant. Electron. QE-**3**, 180 (1967).
2.61. R. S. Eng: Private communication.
2.62. C. Freed: IEEE J. Quantum Electron. JQE-**4**, 404 (1968).
2.63. A. Pine, K. W. Nill: Private communication.
2.64. W. R. Bennett, Jr.: Appl. Opt. Suppl. **1**, 24 (1962); See also G. Birnbaum: *Optical Masers* (Academic Press, New York 1964).
2.65. A. E. Siegman: *An Introduction to Lasers and Masers* (McGraw-Hill, New York, 1971), p. 370.
2.66. E. D. Hinkley, T. C. Harman, C. Freed: Appl. Phys. Letters **13**, 49 (1968).
2.67. C. K. N. Patel: Appl. Phys. Letters **19**, 400 (1971).
2.68. C. K. N. Patel: Phys. Rev. **141**, 71 (1966); See also A. W. Mantz, E. R. Nichols, B. D. Alpert, K. N. Rao: J. Mol. Spectrosc. **35**, 325 (1970).
2.69. M. A. Guerra, S. R. J. Brueck, A. Mooradian: IEEE J. Quant. Electron. (to be published).
2.70. S. R. J. Brueck, A. Mooradian: Private communication.
2.71. S. R. J. Brueck: Private communication.
2.72. K. W. Nill, F. A. Blum: Unpublished.
2.73. See, for example R. J. Keyes, T. M. Quist: In *Semiconductors and Semimetals*, Vol. 5, ed. by R. K. Willardson and A. C. Beer (Academic Press, New York, 1970).

2.74. C. K. N. PATEL: VII Internat. Quant. Electron. Conf., Montreal, May 8-11, 1972, Paper S-5.
2.75. D. L. SPEARS, C. FREED: Appl. Phys. Letters **23**, 445 (1973).
2.76. E. D. HINKLEY, A. R. CALAWA, P. L. KELLEY, S. A. CLOUGH: J. Appl. Phys. **43**, 3222 (1972).
2.77. E. D. HINKLEY, R. H. KINGSTON: 1971 IEEE/OSA CLEA, Washington, D.C., (1971) and Paper 71—1079 at the Joint Conf. on Sensing of Environmental Pollutants, Palo Alto, Calif. (1971).
2.78. R. T. MENZIES: Appl. Opt. **10**, 1532 (1971).
2.79. E. D. HINKLEY, P. L. KELLEY: Science **171**, 635 (1971).
2.80. R. T. MENZIES: Opto-Electronics **4**, 179 (1972).
2.81. R. T. MENZIES: Appl. Phys. Letters **24**, 389 (1974).
2.82. R. T. MENZIES, M. S. SHUMATE: Science **184**, 570 (1974).
2.83. R. G. BREWER: Science **178**, 247 (1972).
2.84. G. HERZBERG: *Molecular Spectra and Molecular Structure*, Vols. 1 and 2 (D. Van Nostrand Co., New York, 1950).
2.85. S. SIAHATGAR, V. E. HOCHULI: IEEE J. Quant. Electron. **5**, 295 (1969).
G. SINGH, F. DILAVORE, C. O. ALLEY: IEEE J. Quant. Electron. **7**, 196 (1971).
2.86. T. W. HANSCH, M. D. LEVENSON, A. L. SCHAWLOW: Phys. Rev. Letters **26**, 946 (1971); T. W. HANSCH, I. S. SHAHIN, A. L. SCHAWLOW: Phys. Rev. Letters **27**, 707 (1971).
2.87. Descriptions of both linear and nonlinear spectroscopic experiments of this type can be found in the review articles cited in [2.1, 2, and 83].
2.88. K. SHIMODA, T. SHIMIZU: Progr. Quant. Electron. **2**, 45 (1972).
2.89. See, for example, S. M. FREUND, T. OKA: Appl. Phys. Letters **21**, 60 (1972).
2.90. A. SZOKE, A. JAVAN: Phys. Rev. Letters **10**, 521 (1963)
R. A. MCFARLANE, W. R. BENNETT, W. E. LAMB: Appl. Phys. Letters **2**, 189 (1963).
See also [2.83] and [2.88].
2.91. M. A. POLLACK, T. J. BRIDGES, A. R. STRAND: Appl. Phys. Letters **10**, 182 (1967);
C. BORDE, L. HENRY: IEEE J. Quant. Electron. QE-**4**, 874 (1968);
T. KAN, H. T. POWELL, G. J. WALGA: IEEE J. Quant. Electron. QE-**5**, 299 (1969);
C. FREED, H. A. HAUS: IEEE J. Quant. Electron. QE-**6**, 617 (1972).
2.92. H. SCHLOSSBERG: Private communication
2.93. K. W. NILL: Unpublished.
2.94. S. R. LEONE, C. B. MOORE: Opt. Soc. of America Meeting, San Francisco, 1972 (unpublished).
2.95. P. V. AMBARTSUMYAN, V. M. APATIN, V. S. LETOKHOV: JETP Letters **15**, 237 (1972).
2.96. R. L. BYER: Private communication.
2.97. T. KASUYA: Japan J. Appl. Phys. **11**, 1575 (1972).
2.98. E. E. UZGIRIS, J. L. HALL, R. L. BARGER: Phys. Rev. Letters **26**, 289 (1971).
2.99. A. C. LUNTZ, A. G. BREWER: J. Chem. Phys. **54**, 3641 (1971).
2.100. B. A. ANTIPOV, V. E. ZUEV, P. D. PYRSIKOVA, V. A. SAPOZHNIKOVA: Opt. and Spectrosc. **31**, 488 (1971).
2.101. R. L. BARGER, J. L. HALL: Phys. Rev. Letters **22**, 4 (1969).
2.102. M. TAKAMI, K. SHIMODA: Japan. J. Appl. Phys. **11**, 1648 (1972);
K. UEHARA, T. SHIMIZU, K. SHIMODA: IEEE J. Quant. Electron. QE-**4**, 728 (1968).
2.103. F. A. BLUM: Unpublished.
2.104. K. W. NILL, F. A. BLUM, A. R. CALAWA, T. C. HARMAN: Chem. Phys. Letters **14**, 234 (1972).
2.105. G. A. ANTCLIFFE, S. G. PARKER, R. T. BATE: Appl. Phys. Letters **21**, 505 (1972).
2.106. R. A. WOOD, R. B. DENNIS, J. W. SMITH: Opt. Comm. **4**, 383 (1972).

2.107. F. A. BLUM, K. W. NILL, A. R. CALAWA, T. C. HARMAN: Chem. Phys. Letters **15**, 144 (1972).
2.108. C. K. N. PATEL: Private communication.
2.109. R. L. ALLWOOD, R. B. DENNIS, R. G. MELLISH, S. D. SMITH, B. S. WHERRETT, R. A. WOOD: J. Phys. C**4**, L126 (1971).
2.110. R. G. MELLISH, R. B. DENNIS, R. L. ALLWOOD: Opt. Comm. **4**, 249 (1971).
R. A. WOOD, R. B. DENNIS, J. W. SMITH: Opt. Comm. **4**, 383 (1972).
2.111. F. A. BLUM, K. W. NILL: Unpublished.
2.112. E. D. HINKLEY: "Development and Application of Tunable Diode Lasers to the Detection and Quantitative Evaluation of Pollutant Cases." Final Technical Report by MIT Lincoln Laboratory for the Environmental Protection Agency (September, 1971).
2.113. G. A. ANTCLIFFE, J. S. WROBEL: Appl. Opt. **11**, 1548 (1972).
2.114. P. NORTON, P. CHIA, T. BRAGGINS, H. LEVINSTEIN: Appl. Phys. Letters **18**, 158 (1971).
2.115. E. D. HINKLEY: Phys. Rev. A**3**, 833 (1971).
2.116. G. A. ANTCLIFFE, S. G. PARKER: J. Appl. Phys. **44**, 4145 (1973).
2.117. V. J. CORCORAN: 1973 IEEE CLEA, Washington, D.C. (June, 1973), Paper 7.7.
2.118. K. W. NILL, F. A. BLUM, E. D. HINKLEY: Unpublished.
2.119. R. S. ENG, P. L. KELLEY, A. MOORADIAN, A. R. CALAWA, T. C. HARMAN: Chem. Phys. Letters, (to be published).
2.120. R. S. ENG, A. R. CALAWA, T. C. HARMAN, P. L. KELLEY, A. JAVAN: Appl. Phys. Letters **21**, 303 (1972).
2.121. J. L. HALL, C. BORDE: Phys. Rev. Letters **30**, 1101 (1073).
2.122. K. W. NILL, F. A. BLUM, A. R. CALAWA, T. C. HARMAN, A. J. STRAUSS: To be published.
2.123. F. A. BLUM, K. W. NILL, A. J. STRAUSS: J. Chem. Phys. **58**, 4968 (1973).
2.124. R. E. FRANCKE, M. S. FELD, E. D. HINKLEY: Unpublished;
See, P. L. KELLEY, E. D. HINKLEY: In *Fundamental and Applied Laser Physics—Proceedings of the Esfahan Symposium*, ed. by M. S. FELD, A. JAVAN, N. A. KURNIT (Academic Press, New York, 1973).
2.125. E. D. HINKLEY, D. G. SUTTON, J. I. STEINFELD: 25th Symp. on Molec. Struct. and Spectrosc., Columbus, Ohio, September, 1970.
2.126. F. A. BLUM, K. W. NILL, A. R. CALAWA, T. C. HARMAN: Appl. Phys. Letters **20**, 377 (1972).
2.127. See, for example, K. N. RAO, C. J. HUMPHREYS, D. H. RANK: *Wavelength Standards in the Infrared* (Academic Press, New York, 1966).
2.128. See, for example, D. R. SOKOLOFF, A. SANCHEZ, R. M. OSGOOD, A. JAVAN: Appl. Phys. Letters **17**, 257 (1970), and references therein.
2.129. S. S. PENNER: *Quantitative Molecular Spectroscopy and Gas Emissivities* (Addison-Wesley, Reading, Mass., 1959).
2.130. W. S. BENEDICT, R. F. CALFEE: *Line Parameters for the 1.9 and 6.3 Micron Water Bands* (Government Printing Office, Washington, D.C., 1967).
2.131. See L. TRAFTON: J. Quant. Spectrosc. Radiat. Transfer **13**, 821 (1973), and references contained therein.
2.132. R. H. DICKE: Phys. Rev. **89**, 472 (1953).
2.133. L. GALATRY: Phys. Rev. **122**, 1218 (1961);
N. NELKIN, A. GHATAK: Phys. Rev. **135**, A4 (1964);
J. I. GERSTEN, H. M. FOLEY: J. Opt. Soc. Am. **58**, 933 (1968).
2.134. G. R. BIRD: J. Chem. Phys. **38**, 2678 (1963).
2.135. V. G. COOPER, A. D. MAY, E. H. HARA, H. F. P. KNAPP: Can. J. Phys. **46**, 2019 (1968).
2.136. See J. R. MURRAY, A. JAVAN: J. Mol. Spectrosc. **42**, 1 (1972), and references therein.

2.137. C. P. SLICHTER: *Principles of Magnetic Resonance* (Harper and Row, New York, 1964), Chapter 5.
2.138. W. E. LAMB, Jr.: Phys. Rev. **134**, A1429 (1964).
2.139. C. FREED, A. JAVAN: Appl. Phys. Letters **17**, 53 (1970).
2.140. K. M. EVANSON, J. S. WELLS, F. R. PETERSEN, B. L. DANIELSON, G. W. DAY, R. L. BARGER, J. L. HALL: Phys. Rev. Letters **29**, 1346 (1972).
2.141. K. M. EVENSON, J. S. WELLS, F. R. PETERSEN, B. L. DANIELSON, G. W. DAY: Appl. Phys. Letters **22**, 192 (1973).
2.142. R. L. BARGER, J. L. HALL: Appl. Phys. Letters **22**, 196 (1973).
2.143. M. A. POLLACK: *Handbook of Lasers*, ed. by R. J. PRESSLEY (The Chemical Rubber Co., Cleveland, Ohio, 1971), p. 298.
2.144. E. D. HINKLEY: Unpublished.
2.145. A. C. MITCHELL, M. W. ZEMANSKY: *Resonance Radiation and Excited Atoms* (Cambridge University Press, Cambridge, England, 1961), pp. 154—170.
2.146. T. K. MCCUBBIN, Jr.: Report 67-0437 of the Air Force Cambridge Research Laboratories (October, 1966).
2.147. G. STEENBECKELIERS: Ann. Soc. Sci. (Bruxelles) **82**, 331 (1968).
2.148. D. E. BURCH, J. D. PEMBROOK, D. A. GRYVNAK: Final Technical Report of the Philco-Ford Corporation to the U. S. Environmental Protection Agency (July, 1971).
2.149. W. GORDY, R. L. COOK: *Microwave Molecular Spectra* (Interscience, New York, 1970).
2.150. J. J. GALLAGHER, C. M. JOHNSON: Phys. Rev. **103**, 1727 (1956);
J. J. GALLAGHER, F. D. BEDARD, C. M. JOHNSON: Phys. Rev. **93**, 729 (1954).
2.151. M. MIZUSHIMA, J. T. COX, W. GIORDY: Phys. Rev. **98**, 1034 (1955);
P. FAVERO, A. M. MIRRI, W. GORDY: Phys. Rev. **114**, 1534 (1954).
2.152. G. R. HANES, C. E. DALHSTROM: Appl. Phys. Letters **14**, 362 (1969);
R. S. ENG, J. T. LATOURETTE: Bull. Am. Phys. Soc. **16**, 43 (1971);
J. D. KNOX, Y. H. PAO: Appl. Phys. Letters **18**, 360 (1971).
2.153. H. R. SCHLOSSBERG, A. JAVAN: Phys. Rev. **150**, 267 (1966).
2.154. For a review, see A. D. BUCKINHAM, P. J. STEPHENS: Ann. Rev. Phys. Chem. **17**, 399 (1966).
2.155. See, for example, P. L. HANST: In *Advances in Environmental Science and Technology*, Vol. 2, ed. by J. N. PITTS and R. L. METCALF (John Wiley, New York, 1971), Chapter 4.
2.156. P. L. HANST: Appl. Spectrosc. **24**, 161 (1970).
2.157. H. KILDAL, R. L. BYER: Proc. IEEE **59**, 1644 (1971).
2.158. I. MELNGAILIS: IEEE Trans. Geosci. Electron. GE-**10**, 7 (1972).
2.159. E. D. HINKLEY: Opto-Electronics **4**, 69 (1972).
2.160. T. H. MAUGH, II: Science **177**, 1090 (1972).
2.161. E. D. HINKLEY, R. T. KU, J. O. SAMPLE: Unpublished.
2.162. E. D. HINKLEY, H. A. PIKE: 1973 Annual Meeting of the Air Pollution Control Association, Chicago, JU. (1973).
2.163. L. B. KREUZER: J. Appl. Phys. **42**, 2934 (1971).
2.164a. L. B. KREUZER, C. K. N. PATEL: Science **173**, 45 (1971).
2.164b. C. K. N. PATEL: VIII Internat'l Quantum Electron. Conf., San Francisco, Calif. (1974).
2.165. L. B. KREUZER, N. D. KENYON, C. K. N. PATEL: Science **177**, 347 (1972).
2.166. E. D. HINKLEY: "Development of In Situ Prototype Diode Laser System to Monitor SO_2 Across the Stack", Final Technical Report by MIT Lincoln Laboratory to the U.S. Environmental Protection Agency (May, 1973).

2.167a. R. L. BYER, M. GARBUNY: Appl. Opt. **12**, 1496 (1973).
2.167b. T. HENNINGSEN, M. GARBUNY, R. L. BYER: Appl. Phys. Letters **24**, 242 (1974).
2.168. R. M. MEASURES, G. PILON: Opto-Electronics **4**, 141 (1972).
2.169. S. A. AHMED: Appl. Opt. **12**, 901 (1973).
2.170. K. W. ROTHE, U. BRINKMANN, H. WALTHER: Appl. Phys. **3**, 115 (1974).
2.171. J. A. GELBWACHS, M. BIRNBAUM, A. W. TUCKER, C. L. FINCHER: Opto-Electronics **4**, 155 (1972).
2.172. C. N. BANWELL: *Fundamentals of Molecular Spectroscopy* (McGraw-Hill Book Company, New York, 1966), p. 23.
2.173. J. KUHL, W. SCHMIDT: Appl. Phys. **3**, 251 (1974).
2.174. H. G. HÄFELE: Appl. Phys. **5**, 97 (1974).

3. Double-Resonance Spectroscopy of Molecules by Means of Lasers[1]

K. SHIMODA

With 35 Figures

Double-resonance effects are manifestations of the interaction between a two-frequency electromagnetic field and three levels of atoms or molecules. Double-resonance experiments have been performed on free atoms since the optical-radiofrequency double-resonance experiment on Hg atoms by BROSSEL, KASTLER, and BITTER [3.1] in 1949—1952. (see also Sect. 1.4). Double resonance of magnetic dipole transitions of electrons and nuclei in condensed matter has been extensively studied in electron-nuclear double-resonance (ENDOR) experiments [3.2] since 1953.

Double-resonance experiments in molecules were first reported in 1955. The resonance modulation effect in the *l*-type doubling transitions of OCS was observed and theoretically explained by AUTLER and TOWNES [3.3]. Sensitive detection of weaker transitions in an ammonia-beam maser was studied by means of a double-resonance technique by SHIMODA and WANG [3.4]. The theory of microwave double resonance in three levels under strong pumping on the one hand weak signal radiation on the other was given by JAVAN [3.5] in 1957. Experimental studies showing coherence effects in microwave double resonance were performed by BATTAGLIA et al. [3.6] and by YAJIMA [3.7].

These double-resonance experiments have some common features, i.e. high resolution of narrow natural linewidth in the presence of broad Doppler broadening, and very high sensitivity that enables spectroscopy of excited states in particular. The double-resonance technique is also closely related to level-crossing and optical pumping techniques. Such experiments employing non-laser light sources have been carried out by a number of researchers during the last twenty years [3.8, 9a].

The laser is a source of strong coherent light; it introduces a variety of nonlinear effects into double-resonance phenomena in the infrared and visible regions. The nonlinear double-resonance effects had previously been investigated only in the radiofrequency and microwave regions.

Lasers are not tunable in general except within their rather narrow linewidths, although tunable lasers and other types of tunable coherent sources are currently being developed. Until such tunable sources

[1] When this article has been completed for publication, the first report of double-resonance spectroscopy with a tunable cw dye laser has already appeared [3.9b].

become readily accessible to spectroscopy, matching of the laser frequency with the spectral frequency of a molecule must rely upon lucky coincidence. Moreover, a double-resonance experiment with two fixed-frequency lasers is very difficult to set to work. The microwave oscillator, on the other hand, can easily be tuned and its frequency can be readily measured with sufficient accuracy.

The infrared-microwave double-resonance experiment where a laser and a microwave oscillator are used has therefore been investigated to some extent, as described in this article. The same technique will be immediately applied to double-resonance experiments in a wider variety of molecular levels, when tunable sources are available in the visible, infrared, and far infrared regions.

The method of infrared-microwave or infrared-infrared double resonance has been considered to be useful for the following applications [3.10, 11]:

1) assignment of vibration-rotation transitions,
2) observation of very weak transitions,
3) studies of relaxation processes and collision-induced transitions,
4) microwave spectroscopy in the vibrationally excited state,
5) coherent generation, amplification, modulation, and detection of infrared and far-infrared,
6) measurement of the infrared and microwave field intensities.

The feasibility of these applications has recently been demonstrated, as described in this article.

The sensitivity in laser spectroscopy is extremely high and has been investigated in detail by the author [3.12]. Hence the infrared double-resonance signal is more easily observed than the microwave-microwave double-resonance signal. Furthermore, saturation and other higher-order effects are observable with moderate laser power (of the order of milliwatts). A large variety of molecular double-resonance experiments, therefore, can be performed by means of the infrared laser.

3.1. Theory of Double Resonance

Consider a three-level system of a molecule with allowed electric dipole transitions between levels $1 \leftrightarrow 3$ and levels $2 \leftrightarrow 3$ but not between levels $1 \leftrightarrow 2$. In the presence of near-resonant radiations at two frequencies ω_p and ω_s, interaction between the molecule and the electromag-

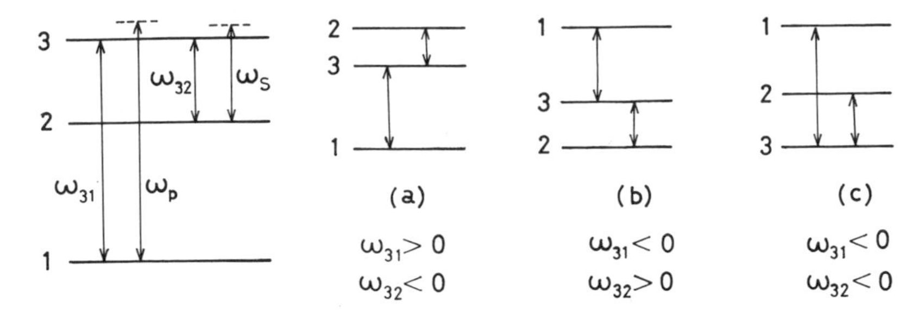

Fig. 3.1. A typical three-level system for double resonance

Fig. 3.2a–c. Three other kinds of three-level systems

netic field represents double-resonance effects[2]. In double-resonance experiments using coherent radiations spontaneous emission can be neglected and need only be considered in a phenomenological decay constant. This allows one to treat the problem by a semiclassical theory in which the radiation is described by a classical electromagnetic wave. In the following, however, a quantum-theoretical treatment is also described briefly in order to make the processes fully understandable.

Three levels 1, 2, and 3 are typically in this order as shown in Fig. 3.1. Other cases, shown by a, b, and c in Fig. 3.2, can, however, readily be treated by considering that the negative frequency term of the perturbing field is resonant, as shown for each case.

In any case, because the two frequencies are nearly resonant to two transition frequencies of the molecule, small differences are expressed by

$$\Delta\omega_p = \omega_p - \omega_{31} \tag{3.1}$$

and

$$\Delta\omega_s = \omega_s - \omega_{32}\,. \tag{3.2}$$

It is noted here that the observation of double resonance is actually performed with an ensemble of molecules. Molecules in a gas are moving at random and the transition frequencies ω_{31} and ω_{32} are distributed

[2] In some experiments a strong radiation at frequency ω_p is employed to pump the molecules from level 1 to 3 and the signal is observed at the frequency ω_s. The notations p and s stand for pumping and signal, respectively, but they are used in this paper as labels for the two radiations in general, even when their roles may not be pumping and signal.

according to their Doppler shifts by the molecular velocity. The effects of Doppler broadening on double resonance depend on experimental conditions; these are discussed in Subsections 3.1.3 and 3.2.1.

3.1.1. Semiclassical Theory

A semiclassical analysis of double-resonance effects in a three-level system was given by JAVAN [3.5] in 1957 and MACKE et al. [3.13] in 1969. The electromagnetic field at the position of the molecule is considered classically and it is written as

$$E(t) = E_p \exp(i\omega_p t) + E_s \exp(i\omega_s t) + c \cdot c\,, \tag{3.3}$$

where E_p and E_s are complex amplitudes of the electric fields.

The wavefunction of a molecule under the time-dependent perturbation as given by

$$H = H_0 - \mu \cdot E(t) \tag{3.4}$$

is expressed by a superposition of eigenfunctions ψ_i in the form

$$\psi = a_1(t)\psi_1 + a_2(t)\psi_2 + a_3(t)\psi_3\,, \tag{3.5}$$

where H_0 is the unperturbed Hamiltonian. The equations of motion for $a_i(t)$ are obtained by substituting (3.3), (3.4), and (3.5) into the Schrödinger equation to become

$$\begin{aligned} da_1/dt &= ix_p a_3 \exp(i\Delta\omega_p t) \\ da_2/dt &= ix_s a_3 \exp(i\Delta\omega_s t) \\ da_3/dt &= i[x_p^* a_1 \exp(-i\Delta\omega_p t) + x_s^* a_2 \exp(-i\Delta\omega_s t)]\,, \end{aligned} \tag{3.6}$$

where nonresonant terms have been ignored in the so-called rotating wave approximation and

$$x_p = \mu_{13} E_p/\hbar\,, \quad x_s = \mu_{23} E_s/\hbar\,. \tag{3.7}$$

μ_{13} and μ_{23} are the matrix elements of the electric dipole moment operator μ in (3.4). Any relaxation effects are neglected so far.

When x_p and x_s are constant or change slowly with time, the particular solution of the set of Eq. (3.6) is obtained as

$$a_1 = A_1 \exp[i(\Delta\omega_p - \lambda)t], \quad a_2 = A_2 \exp[i(\Delta\omega_s - \lambda)t],$$
$$a_3 = A_3 \exp(-i\lambda t),$$

where λ is the root of

$$\begin{vmatrix} \Delta\omega_p - \lambda & 0 & x_p \\ 0 & \Delta\omega_s - \lambda & x_s \\ x_p^* & x_s^* & -\lambda \end{vmatrix} = 0$$

or

$$\begin{aligned} &\lambda^3 - (\Delta\omega_p + \Delta\omega_s)\lambda^2 + (\Delta\omega_p \Delta\omega_s - |x_p|^2 - |x_s|^2)\lambda \\ &\quad + |x_p|^2 \Delta\omega_s + |x_s|^2 \Delta\omega_p = 0 . \end{aligned} \tag{3.8}$$

For arbitrary values of x_p and x_s, the roots of the above cubic equation cannot be expressed in a closed form except for $\Delta\omega_p = \Delta\omega_s$, which is discussed below under case b). When x_s is small and $x_s \ll x_p$, a solution to the first order in x_s can be found easily as shown below.

a) When $x_s \ll x_p$.

In the case when the signal field is weak, the probability amplitudes of ψ_3 and ψ_2 for the initial conditions of $a_1 = 1$ and $a_2 = a_3 = 0$ at $t = 0$ are given by neglecting higher-order terms of x_s:

$$a_{1\to 3} = i(x_p^*/\Omega_p) \exp(-i\Delta\omega_p t/2) \sin \Omega_p t \tag{3.9a}$$

and

$$a_{1\to 2} = \frac{x_p^* x_s}{2\Omega_p} \left\{ \frac{\exp[i(\Omega_p + \delta)t] - 1}{\Omega_p + \delta} + \frac{\exp[-i(\Omega_p - \delta)t] - 1}{\Omega_p - \delta} \right\}, \tag{3.9b}$$

where

$$\Omega_p^2 = (\Delta\omega_p/2)^2 + |x_p|^2 \tag{3.10}$$

and

$$\delta = \Delta\omega_s - \Delta\omega_p/2 .$$

Then the probability for the double quantum transition corresponding to simultaneous absorption of $\hbar\omega_p$ and emission of $\hbar\omega_s$ is given by

$$|a_{1\to 2}|^2 = \frac{|x_p|^2 |x_s|^2}{\Omega_p^2 (\Omega_p^2 - \delta^2)^2} |\Omega_p[\exp(-i\delta t) - \cos \Omega_p t] + i\delta \sin \Omega_p t|^2 . \tag{3.11}$$

The probability amplitude of ψ_2 for the initial conditions of $a_3=1$ and $a_1=a_2=0$ at $t=0$ is similarly calculated to be

$$a_{3\to 2}=\frac{x_s}{4\Omega_p}\left\{\frac{2\Omega_p+\Delta\omega_p}{\Omega_p+\delta}\{\exp[i(\Omega_p+\delta)t]-1\}\right.$$
$$\left.-\frac{2\Omega_p-\Delta\omega_p}{\Omega_p-\delta}\{\exp[-i(\Omega_p-\delta)t]-1\}\right\}. \tag{3.12}$$

The probability for the single quantum transition at ω_s in the presence of a strong pump field at ω_p is therefore

$$|a_{3\to 2}|^2=|x_s|^2\Omega_p^{-2}(\Omega_p^2-\delta^2)^{-2}\,|\Omega_p(\Delta\omega_s-\Delta\omega_p)$$
$$\cdot[\exp(-i\delta t)-\cos\Omega_p t]$$
$$+i(\Omega_p^2-\Delta\omega_p\delta/2)\sin\Omega_p t|^2\,. \tag{3.13}$$

A simple case of relaxation is assumed such that the relaxation times of the three states and the transverse relaxation times are all equal to τ. Then the statistical average of transition probability for the three-level molecules in gas is calculated to become

$$\langle|a_{1\to 2}|^2\rangle=\frac{|x_p|^2|x_s|^2\tau^2}{2\Omega_p^2}\left[\frac{1}{1+(\Omega_p+\delta)^2\tau^2}+\frac{1}{1+(\Omega_p-\delta)^2\tau^2}\right.$$
$$\left.+\frac{2(1+2\Omega_p^2\tau^2)(\Omega_p^2-\delta^2)\tau^2-2}{(1+4\Omega_p^2\tau^2)[1+(\Omega_p+\delta)^2\tau^2][1+(\Omega_p-\delta)^2\tau^2]}\right], \tag{3.14}$$

$$\langle|a_{3\to 2}|^2\rangle=\frac{|x_s|^2\tau^2}{8\Omega_p^2}\left[\frac{(2\Omega_p+\Delta\omega_p)^2}{1+(\Omega_p+\delta)^2\tau^2}+\frac{(2\Omega_p-\Delta\omega_p)^2}{1+(\Omega_p-\delta)^2\tau^2}\right.$$
$$\left.+\frac{8|x_p|^2[1-(1+2\Omega_p^2\tau^2)(\Omega_p^2-\delta^2)\tau^2]}{(1+4\Omega_p^2\tau^2)[1+(\Omega_p+\delta)^2\tau^2][1+(\Omega_p-\delta)^2\tau^2]}\right]. \tag{3.15}$$

The radiation at the signal frequency is emitted either in the double quantum transition from level 1 to 2 or in the single quantum transition $3\to 2$, while it is absorbed in the reverse transitions. Since $|a_{2\to 1}|^2=|a_{1\to 2}|^2$ and $|a_{2\to 3}|^2=|a_{3\to 2}|^2$, the net induced emission of signal power is given by

$$P_{\text{emi}}(\omega_s)=\hbar\omega_s\tau^{-1}[(N_1-N_2)\langle|a_{1\to 2}|^2\rangle-(N_2-N_3)\langle|a_{3\to 2}|^2\rangle]\,, \tag{3.16}$$

where N_1, N_2, and N_3 are the unperturbed populations of the three levels. When $N_1 > N_2 > N_3$, as in the case of thermal equilibrium, the first term in (3.16) always shows emission and the second term absorption.

When $N_1 \gg N_2$ and $N_1 \gg N_3$, the power emitted at frequency ω_s is simply

$$P_{emi}(\omega_s) = N_1 \hbar \omega_s \tau^{-1} \langle |a_{1\to 2}|^2 \rangle . \tag{3.17}$$

The power absorbed at frequency ω_p by the double quantum transition is given by

$$P_{abs}(\omega_p) = (\omega_p/\omega_s) P_{emi}(\omega_s) . \tag{3.18}$$

Spectral lineshapes of the double quantum transition as functions of the signal frequency are shown in Fig. 3.3 for $|x_p|\tau = 1$ and $|x_p|\tau = 3$ in the case when $N_2 = N_3 = 0$. It is seen that the line is in general split into two components.

When the pumping is on resonance at $\omega_p = \omega_{31}$, the emission signal becomes a doublet centered at ω_{32}. The splitting of this doublet for strong pumping is given by $2|x_p|$. This is called the modulation doubling or coherence splitting of the resonant modulation effect. Under weak pumping of $|x_p|\tau \ll 1$ the signal line shows a single peak.

When the pumping frequency is off-resonant by a large amount so that $\Delta\omega_p \gg |x_p|$, the double-resonance signal shows two peaks at frequencies $\omega_s = \omega_{32}$ and $\Delta\omega_s = \Delta\omega_p$. The latter is the Raman-type transition, in which the Stokes radiation at $\omega_s = \omega_p - \omega_{21}$ is emitted while the strong incident radiation at $\omega_p \neq \omega_{31}$ is absorbed.

b) When $\Delta\omega_p = \Delta\omega_s$.

In the Raman-frequency condition, when $\omega_p - \omega_s = \omega_{31} - \omega_{32} = \omega_{21}$, the three-level problem can be solved for arbitrary values of x_p and x_s. The solutions of (3.8) in this case are

$$\lambda = -2\delta , \quad -\delta \pm \Omega , \tag{3.19}$$

where

$$\delta = \Delta\omega_p/2 = \Delta\omega_s/2 , \quad \Omega^2 = \delta^2 + |x_p|^2 + |x_s|^2 .$$

Then the probability amplitudes of the states at time t are expressed in the form [3.14]

$$\begin{pmatrix} a_1(t) \\ a_2(t) \\ a_3(t) \end{pmatrix} = \frac{1}{\theta^2} \begin{pmatrix} |x_s|^2 + |x_p|^2 F & x_p x_s^* (F-1) & x_p \theta G \\ x_p^* x_s (F-1) & |x_p|^2 + |x_s|^2 F & x_s \theta G \\ -x_p^* \theta G^* & -x_s^* \theta G^* & \theta^2 F^* \end{pmatrix} \begin{pmatrix} a_1(0) \\ a_2(0) \\ a_3(0) \end{pmatrix}, \tag{3.20}$$

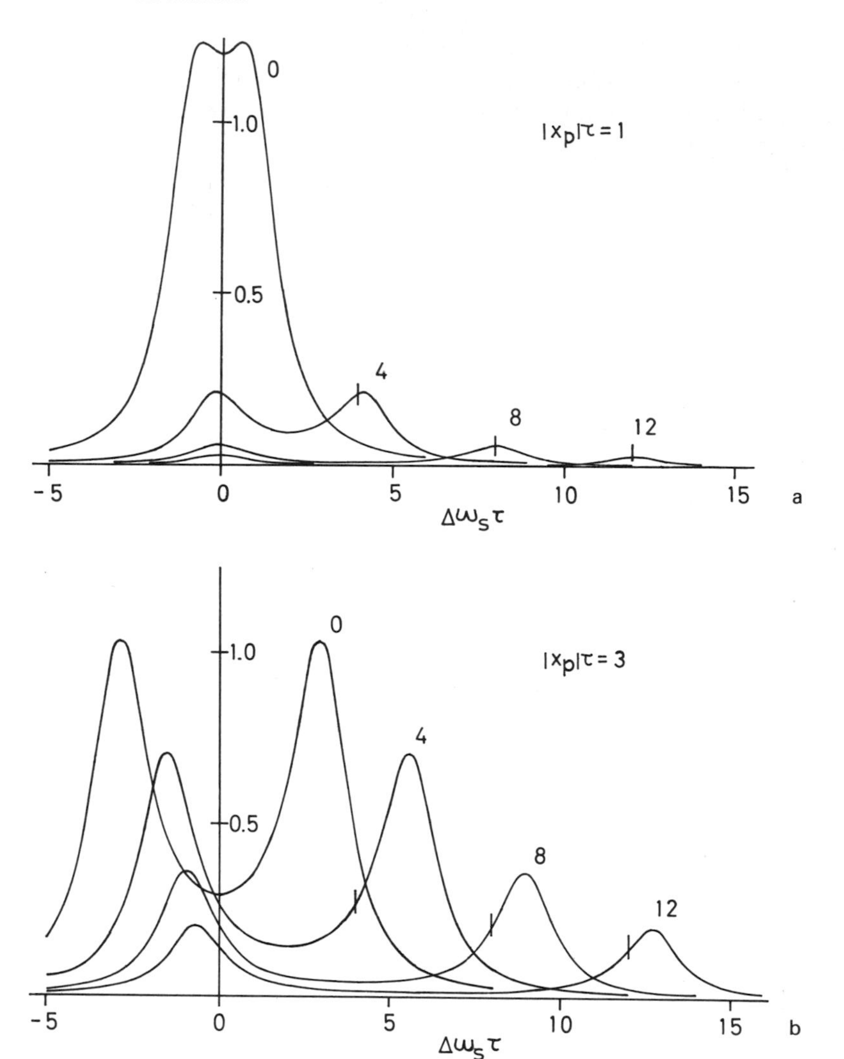

Fig. 3.3a and b. Calculated lineshapes of the double-resonance signal $\langle|a_{1\to2}|^2\rangle$ for different values of detuning. The parameter is $\Delta\omega_p\tau = (\omega_p - \omega_{31})\tau$. a) $|x_p|\tau = 1$. b) $|x_p|\tau = 3$. Short vertical lines indicate the frequencies for $\Delta\omega_p = \Delta\omega_s$

where[3]

$$\theta^2 = |x_p|^2 + |x_s|^2 ,$$

$$F = \exp(i\delta t)\{\cos\Omega t - i\delta\Omega^{-1}\sin\Omega t) , \tag{3.21}$$

$$G = i\theta\Omega^{-1}\exp(i\delta t)\sin\Omega t . \tag{3.22}$$

[3] The notations used in [3.14] are $\delta \to -\delta$ and $G \to -G^*$.

For the initial condition that $a_1 = 1$ and $a_2 = a_3 = 0$ at $t = 0$, one obtains

$$a_1(t) = \theta^{-2}[|x_s|^2 + |x_p|^2 \exp(i\delta t)(\cos\Omega t - i\delta\Omega^{-1}\sin\Omega t)], \tag{3.23a}$$

$$a_2(t) = x_p^* x_s \theta^{-2}[\exp(i\delta t)(\cos\Omega t - i\delta\Omega^{-1}\sin\Omega t) - 1], \tag{3.23b}$$

$$a_3(t) = i x_p^* \Omega^{-1} \exp(-i\delta t)\sin\Omega t. \tag{3.23c}$$

In the case when off-resonance is large so that $\delta^2 \gg \theta^2$, Ω may be expanded in the form

$$\Omega = \delta + \theta^2/2\delta.$$

Thus the population of level 2 varies sinusoidally as [3.15]

$$|a_2(t)|^2 = 4|x_p|^2 |x_s|^2 \theta^{-4} \sin^2(ft/2), \tag{3.24}$$

where

$$f = \theta^2/2\delta. \tag{3.25}$$

Hence the double quantum transition, often called the two-photon transition, exhibits a nutation-like behavior when both radiations are strong. The amplitude of sinusoidal modulation is maximal when $|x_s| = |x_p|$.

In the presence of random collisions, as described by a single relaxation time τ, the probability of the double quantum transition from level 1 to 2 is given by

$$\langle |a_2(t)|^2 \rangle = \frac{2|x_p|^2 |x_s|^2}{\theta^4} \cdot \frac{f^2\tau^2}{1 + f^2\tau^2} = \frac{2|x_p|^2 |x_s|^2 \tau^2}{\Delta\omega^2 + \theta^4\tau^2}. \tag{3.26}$$

It is easy to see that (3.26) agrees with (3.14) for $\delta^2 \gg |x_p|^2$ so that $\Omega_p \simeq \delta = \Delta\omega/2$.

3.1.2. Quantum-Mechanical Analysis

A perturbative quantum treatment of double resonance in a three-level system was given by Di Giacomo et al. [3.16] in 1969. The exact solution of a fully quantum-mechanical treatment given by Walls [3.17] is described below.

The three-level system of Fig. 3.1 interacting with two modes of the electromagnetic field operator at frequencies ω_p and ω_s is considered.

Then the Hamiltonian of the process takes the form

$$
\begin{aligned}
H &= H_0 + H_1 \\
H_0 &= \hbar\omega_1 a_1^\dagger a_1 + \hbar\omega_2 a_2^\dagger a_2 + \hbar\omega_3 a_3^\dagger a_3 + \hbar\omega_p b_p^\dagger b_p + \hbar\omega_s b_s^\dagger b_s \\
H_1 &= \hbar\kappa_p (a_1 b_p a_3^\dagger + a_1^\dagger b_p^\dagger a_3) + \hbar\kappa_s (a_2 b_s a_3^\dagger + a_2^\dagger b_s^\dagger a_3)\,,
\end{aligned}
\tag{3.27}
$$

where $\hbar\omega_1$, $\hbar\omega_2$, and $\hbar\omega_3$ are the energies, and a_i is the annihilation operator for the molecule in state i, while b_i is the photon annihilation operator obeying boson commutation relations. Commutation relations for a_i's are anti-commuting and

$$
\begin{aligned}
a_i^\dagger a_j + a_j a_i^\dagger &= \delta_{ij} \\
a_i a_j + a_j a_i &= 0\,.
\end{aligned}
$$

The coupling constants κ_p and κ_s are proportional to the squares of matrix elements of the molecular dipole moment μ_{13} and μ_{23}, respectively.

We construct the following two operators

$$
\begin{aligned}
I_0 &= \hbar(\omega_1 - \Delta\omega_p) a_1^\dagger a_1 + \hbar(\omega_2 - \Delta\omega_s) a_2^\dagger a_2 \\
&\quad + \hbar\omega_3 a_3^\dagger a_3 + \hbar\omega_p b_p^\dagger b_p + \hbar\omega_s b_s^\dagger b_s \\
I_1 &= H_1 + \hbar\Delta\omega_p a_1^\dagger a_1 + \hbar\Delta\omega_s a_2^\dagger a_2
\end{aligned}
\tag{3.28}
$$

such that $H = I_0 + I_1$. The commutation relations are written as

$$
[I_0, I_1] = [I_0, H] = [I_1, H] = 0\,. \tag{3.29}
$$

Thus I_0 and I_1 are constants of motion and a representation may be found in which the total Hamiltonian H is diagonal.

We choose as a complete set of basis states

$$
\psi^\dagger = (|1, n_p + 1, n_s\rangle, |2, n_p, n_s + 1\rangle, |3, n_p, n_s\rangle)\,,
$$

where 1, 2, and 3 are the molecular levels, and n_p and n_s are the photon numbers. The Schrödinger equation in the interaction picture then becomes

$$
I_1 \psi = \hbar A \psi\,, \tag{3.30}
$$

where

$$A=\begin{pmatrix} \Delta\omega_p & 0 & g_p \\ 0 & \Delta\omega_s & g_s \\ g_p & g_s & 0 \end{pmatrix}, \tag{3.31}$$

$$g_p=\kappa_p\sqrt{n_p+1}\,, \quad g_s=\kappa_s\sqrt{n_s+1}\,.$$

The eigenstates and eigenvalues of the system are obtained by diagonalizing the matrix A. The eigenvalues are given by the roots of

$$\lambda^3-(\Delta\omega_p+\Delta\omega_s)\lambda^2+(\Delta\omega_p\Delta\omega_s-g_p^2-g_s^2)\lambda+g_p^2\Delta\omega_s+g_s^2\Delta\omega_p=0. \tag{3.32}$$

This cubic equation is exactly the same as (3.8) under the following correspondence

$$|x_p|^2\sim g_p^2=\kappa_p^2(n_p+1) \tag{3.33}$$

and

$$|x_s|^2\sim g_s^2=\kappa_s^2(n_s+1)\,. \tag{3.34}$$

The quantum-theoretical results for cases a) and b) of Subsection 3.1.2 have been given [3.17, 18]. These results are therefore almost identical to the results of a semiclassical calculation. The important difference is the $+1$ factors in (3.33) and (3.34), which may be considered to correspond to spontaneous emission from level 3 by the transitions $3\rightarrow1$ and $3\rightarrow2$. Thus the quantum theory of the Raman effect [3.18] naturally gives a unified interpretation of the spontaneous and stimulated Raman effects.

3.1.3. Rate-Equation Approximation

A semiclassical theory is adequate to describe double-resonance effects. An alternative expression to the analysis in Subsection 3.1.1 is the equation of motion of the ensemble-averaged density matrix, which is more convenient in introducing phenomenological rate constants for longitudinal and transverse relaxations.

The element of ensemble-averaged density matrix is defined by

$$\varrho_{mn}=\langle a_m(t)a_n^*(t)\exp(i\omega_{nm}t)\rangle\,, \tag{3.35}$$

where $\hbar\omega_{nm} = \hbar\omega_n - \hbar\omega_m$ is the difference between energies of state n and m. Then the equations of motion for a three-level system of Fig. 3.1 under a two-frequency perturbation $\hbar V = -\mu \cdot E(t)$ are written as

$$d\varrho_{11}/dt + (\varrho_{11} - \varrho_{11}^0)/\tau_1 = -iV_{13}\varrho_{31} + i\varrho_{13}V_{31}, \tag{3.36a}$$

$$d\varrho_{22}/dt + (\varrho_{22} - \varrho_{22}^0)/\tau_2 = -iV_{23}\varrho_{32} + i\varrho_{23}V_{32}, \tag{3.36b}$$

$$d\varrho_{33}/dt + (\varrho_{33} - \varrho_{33}^0)/\tau_3 = -iV_{31}\varrho_{13} - iV_{32}\varrho_{23} + \text{c.c.}, \tag{3.36c}$$

$$d\varrho_{12}/dt - i\omega_{21}\varrho_{12} + \gamma_{21}\varrho_{12} = -iV_{13}\varrho_{32} + i\varrho_{13}V_{32}, \tag{3.36d}$$

$$d\varrho_{13}/dt - i\omega_{31}\varrho_{13} + \gamma_{31}\varrho_{13} = iV_{13}(\varrho_{11} - \varrho_{33}) + i\varrho_{12}V_{23}, \tag{3.36e}$$

$$d\varrho_{23}/dt - i\omega_{32}\varrho_{23} + \gamma_{32}\varrho_{23} = iV_{23}(\varrho_{22} - \varrho_{33}) + i\varrho_{21}V_{13}, \tag{3.36f}$$

where $V_{12} = V_{21} = 0$ is assumed. Here τ_1, τ_2, and τ_3 are the lifetimes of states 1, 2, and 3, respectively; γ_{21}, γ_{31}, and γ_{32} are the relaxation rates of the off-diagonal elements of the density matrix; and ϱ_{ii}^0 is the unperturbed value of ϱ_{ii}. The incident radiation at the position of the molecule is expressed by (3.3).

Exact solutions of (3.36) for arbitrary values of E_p, E_s, ω_p, and ω_s cannot be obtained in closed forms, as discussed in Subsection 3.1.1. In order to discuss higher-order effects of E_s and E_p in double resonance, therefore, a rate-equation approximation can be employed in the following condition.

For convenience, the off-diagonal elements are written as

$$\varrho_{13} = p_p \exp(i\omega_p t) \quad \text{and} \quad \varrho_{23} = p_s \exp(i\omega_s t). \tag{3.37}$$

If ω_p is close to ω_{31} and ω_s is close to ω_{32}, while their amplitudes E_p and E_s can vary only slowly, p_p and p_s do not vary rapidly with time and appear in the first-order perturbation. Then ϱ_{12} is seen from (3.36d) to appear in the second order. On condition that the third-order terms $\varrho_{12}V_{23}$ in (3.36e) and $\varrho_{21}V_{13}$ in (3.36f) may be ignored and that

$$|dp_p/dt| \ll |\gamma_p p_p|, \quad \text{and} \quad |dp_s/dt| \ll |\gamma_s p_s|, \tag{3.38}$$

one obtains from (3.36e)

$$\varrho_{13} = p_p \exp(i\omega_p t) = \frac{ix_p(\varrho_{33} - \varrho_{11})}{i\Delta\omega_p + \gamma_p} \exp(i\omega_{31} t), \tag{3.39}$$

where $\Delta\omega_p$ and x_p were defined by (3.1) and (3.7), respectively, and the transverse relaxation rates γ_{31} and γ_{32} are rewritten as γ_p and γ_s, respectively. Substitution of (3.39) into (3.36a) gives a rate equation in the form

$$\frac{d\varrho_{11}}{dt} + \frac{\varrho_{11} - \varrho_{11}^0}{\tau_1} = \frac{2|x_p|^2 \gamma_p}{\Delta\omega_p^2 + \gamma_p^2} (\varrho_{33} - \varrho_{11}) . \tag{3.40}$$

The rate equation for ϱ_{22} is similarly obtained from (3.36f) and (3.36b):

$$\frac{d\varrho_{22}}{dt} + \frac{\varrho_{22} - \varrho_{22}^0}{\tau_2} = \frac{2|x_s|^2 \gamma_s}{\Delta\omega_s^2 + \gamma_s^2} (\varrho_{33} - \varrho_{22}) . \tag{3.41}$$

The rate equation for ϱ_{33} then becomes

$$\begin{aligned} \frac{d\varrho_{33}}{dt} + \frac{\varrho_{33} - \varrho_{33}^0}{\tau_3} &= \frac{2|x_p|^2 \gamma_p}{\Delta\omega_p^2 + \gamma_p^2} (\varrho_{11} - \varrho_{33}) \\ &+ \frac{2|x_s|^2 \gamma_s}{\Delta\omega_s^2 + \gamma_s^2} (\varrho_{22} - \varrho_{33}) . \end{aligned} \tag{3.42}$$

The molecules in the gas are moving with their thermal velocities. Their Doppler effect must be considered in the theory of double-resonance effects of molecules in gas. The resonant frequency ω_{31} of a molecule moving at velocity v along the direction of the pumping radiation is shifted to become $\omega_{31}(1 - v/c)$. When the signal radiation is in the same direction, the frequency ω_{32} is shifted to become $\omega_{32}(1 - v/c)$.

Then the rate equations for the moving molecule with the axial velocity component v are expressed by using the transition probabilities

$$S_p = \frac{2|x_p|^2 \gamma_p}{(\omega_p - \omega_{31} + k_p v)^2 + \gamma_p^2} , \tag{3.43}$$

$$S_s = \frac{2|x_s|^2 \gamma_s}{(\omega_s - \omega_{32} + k_s v)^2 + \gamma_s^2} , \tag{3.44}$$

in the form

$$\begin{aligned} dn_1/dt &= -(n_1 - n_3)S_p - (n_1 - n_1^0)/\tau_1 \\ dn_2/dt &= -(n_2 - n_3)S_s - (n_2 - n_2^0)/\tau_2 \\ dn_3/dt &= (n_1 - n_3)S_p + (n_2 - n_3)S_s - (n_3 - n_3^0)/\tau_3 , \end{aligned} \tag{3.45}$$

where $n_1 = \varrho_{11} n$, $n_2 = \varrho_{22} n$, and $n_3 = \varrho_{33} n$ are populations of the moving molecule, and $k_p = \omega_p/c$, $k_s = \omega_s/c$.

The number of molecules moving with the velocity between v and $v+\mathrm{d}v$ is $n\mathrm{d}v=w(v)N\mathrm{d}v$, where N is the total number of the three-level molecules in a unit volume, and $w(v)$ is the velocity distribution function. The Maxwellian distribution is given by

$$w(v)=(1/\sqrt{\pi}u)\exp(-v^2/u^2), \tag{3.46}$$

where u is the most probable velocity. Then the double-resonance behavior of the gas can be calculated by integrating the solution of (3.45) over the whole velocity distribution.

In the case when the signal radiation is in the opposite direction to the pumping radiation, the problem can be treated by changing the sign of k_s in (3.44).

It should be mentioned that the rate-equation approximation is valid on the assumption of (3.38), which is rewritten using (3.39), and the corresponding equation for ϱ_{23} in the form $|x_p| \ll \gamma_p$ and $|x_s| \ll \gamma_s$. However, it is also correct generally when the integrated line intensity is observed. If the double-resonance effect is observed with a low resolution γ_{eff} that is larger than $|x_p|$ and $|x_s|$, the rate-equation approximation is equivalent to the semiclassical theory. In the presence of the Doppler effect, therefore, the rate equation is valid provided that $|x_p|$ and $|x_s|$ are smaller than the effective Doppler broadenings.

It is noted finally that integration over the molecular velocity distribution should also apply to the semiclassical and quantum-mechanical analysis. The actual computation, however, is usually very tedious in these cases even when the relaxation constants are independent of the velocity.

3.2. Infrared-Infrared Double Resonance

Because of the limited tunability of infrared gas lasers, the infrared-infrared double-resonance experiment has so far been performed only in a few exceptional cases: one is the molecular three-level system involving the laser transition, and the other is the two coupled transitions within the Doppler width. The recent development of tunable coherent sources will make it possible to perform double-resonance experiments on a wide variety of molecules. Investigations of the kind that have been developed in infrared microwave double resonance, as described in Section 3.3, are expected to appear with fruitful results in near future. Thus preliminary experiments with infrared-infrared double resonance using the fixed-frequency laser are briefly described below.

3.2.1. Doppler Effect in Infrared-Infrared Double Resonance

Doppler broadening in the microwave transition is normally smaller than other kinds of broadening so that it may be neglected. At higher frequencies, however, the spectral line of molecules in gas at low pressure is dominantly broadened by the Doppler effect.

In the infrared double-resonance experiment, therefore, the Doppler shifts for two coupled transitions must be considered. The Doppler effect in double resonance is strongly dependent on the relative directions of the two radiations.

First, let us assume that the two radiations are traveling in the same direction. Then the double-resonance signal in a three-level system, as shown in Fig. 3.4a or b, is sharp. This is because the Doppler shift of a moving molecule for one transition is compensated for the most part by the Doppler shift of the other transition when $\omega_{31}-\omega_{32} \ll \omega_{31}$. In other words, the magnitude of $\Delta\omega_p - \Delta\omega_s$ is almost independent of the molecular velocity. Then the Raman-type double quantum transition, as discussed under case b) in Subsection 3.1.2, occurs consecutively for all molecular velocities. In the cascade transitions in Fig. 3.4c, on the other hand, the double-resonance line is broader, because Doppler shifts of the two transitions are superposed for the Raman condition.

Secondly, when the two radiations are traveling in the opposite direction,the double resonance linc is broad in the case of Fig. 3.4a or b, but sharp in the case of Fig. 3.4c. This laser-induced line-narrowing effect has been discussed in a number of papers [3.19–22].

Figure 3.5 illustrates the spectral line profile for the energy levels of Fig. 3.4a in the presence of a strong laser field at frequency ω_p, which is higher than ω_{31} within the Doppler width. When a hole is burnt on the high-frequency side of the absorption line around ω_{31} by the pump ω_p, a sharp hole appears on the high-frequency side of the signal line in Fig. 3.5a, whereas a broad hole appears on the low-frequency side in Fig. 3.5b. Then it shows a sharp gain for the signal wave in the same

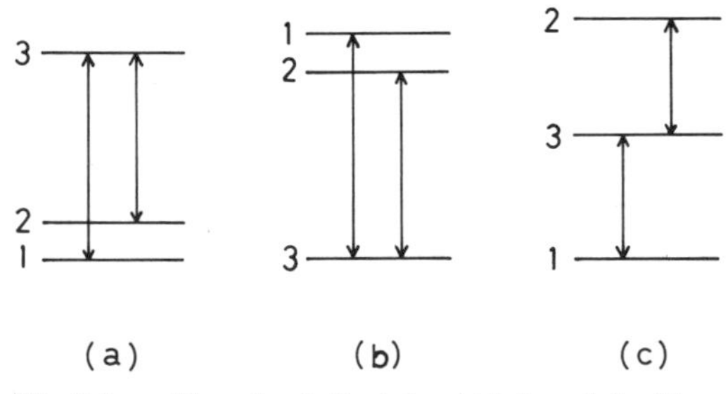

Fig. 3.4a–c. Three levels for infrared-infrared double resonance

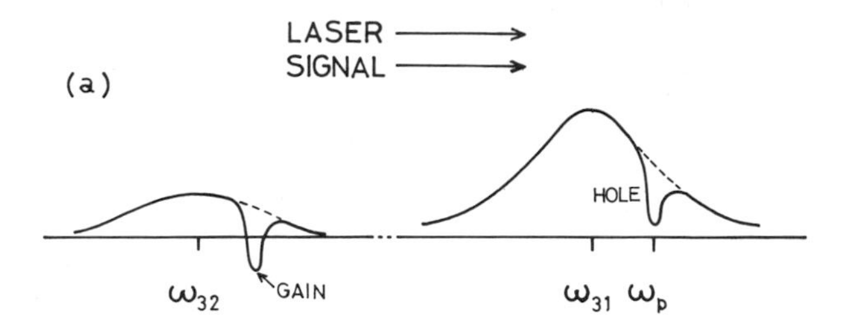

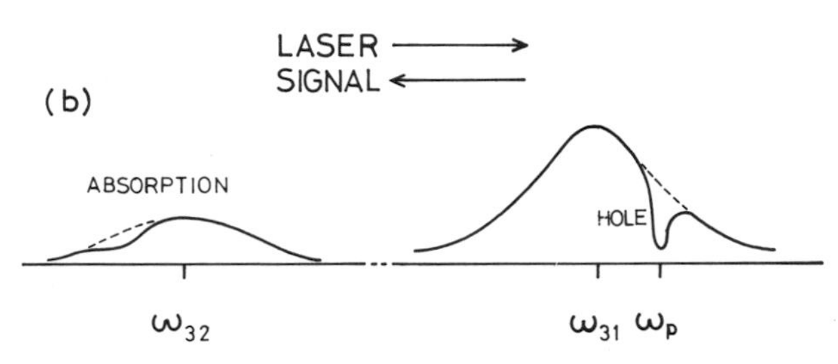

Fig. 3.5a and b. Absorption line profiles influenced by a strong pumping radiation at ω_p in the three-level system of Fig. 3.4a. a) Pumping and signal radiations are in the same direction. b) Pumping and signal radiations are in the opposite direction

direction (a) and a broad dip in absorption for the signal wave in the opposite direction (b).

The narrow holewidth for the small-signal double resonance in Fig. 3.5a and b is expressed by

$$\gamma_N = \gamma_2 + [(\omega_s/\omega_p)(\gamma_1 + \gamma_3) - \gamma_3](1 + 4|x_p|^2/\gamma_1\gamma_3)^{1/2} \tag{3.47}$$

and the broad holewidth is

$$\gamma_B = \gamma_2 + [(\omega_s/\omega_p)(\gamma_1 + \gamma_3) + \gamma_3](1 + 4|x_p|^2/\gamma_1\gamma_3)^{1/2} . \tag{3.48}$$

If the saturation broadening shown by the square root in (3.47) and (3.48) can be ignored, the two linewidths become [3.21] $\gamma_N = \gamma_1 + \gamma_2$ and $\gamma_B = \gamma_1 + \gamma_2 + 2\gamma_3$ for $\omega_s \approx \omega_p$.

3.2.2. Unidirectional Amplification in Optically Pumped HF Gas

The directional properties of infrared double resonance have been fully utilized to build a unidirectional amplifier with laser-pumped HF gas by

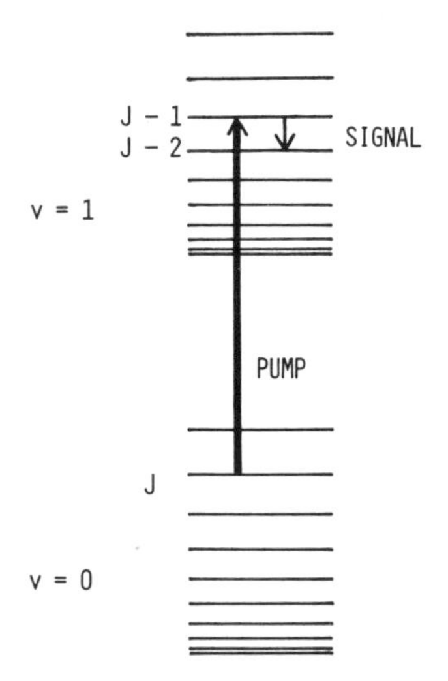

Fig. 3.6. Vibration rotation levels of a diatomic molecule such as HF

Table 3.1. Wavelengths of far-infrared rotational oscillations for P-branch transitions in HF

Pump transition		Coupled transition ($v=1$ excited state)	
Designation $P(J)$	Wavelength [μm]	$J-1 \rightarrow J-2$	Wavelength [μm]
$P(3)$	2.608	$2 \rightarrow 1$	126.5
$P(4)$	2.639	$3 \rightarrow 2$	84.4
$P(5)$	2.672	$4 \rightarrow 3$	63.4
$P(6)$	2.707	$5 \rightarrow 4$	50.8
$P(7)$	2.744	$6 \rightarrow 5$	42.4
$P(8)$	2.782	$7 \rightarrow 6$	36.5

JAVAN and his collaborators [3.23, 24]. The HF gas in an absorption cell 12 cm long at a pressure of 0.05–6 Torr was pumped by one of the 2.7 μm $P(J)$ lines of a pulsed HF laser with a peak power of 3–4 kW for 1 μs duration. The laser on the $P(J)$ line excites molecules that are in the J rotational state in the ground state to the $J-1$ rotational state in the vibrationally excited state, as shown in Fig. 3.6. Laser gain was observed at the coupled rotational $(J-1) \rightarrow (J-2)$ transition in the $v=1$ vibrational state. SKRIBANOWITZ et al. [3.23] observed a large difference in gain of these far-infrared transitions between forward and backward directions.

The observed rotational transitions and the pumping transitions are shown in Table 3.1. The gain in these far-infrared lines is very large and they oscillate easily without mirrors in the manner of "superradiance". Although the pumping and signal frequencies differ considerably in their magnitudes, the theoretical results of (3.47) and (3.48) predict a large anisotropy in gain. The observed intensity in the forward direction is found to be 5–10 times larger than that in the backward direction.

The pulse shape and the delay of this "superradiance" emission have recently been observed and explained by SKRIBANOWITZ et al. using a semiclassical approach of coupled Maxwell-Schrödinger equations [3.25]. Since the emitted pulse has been found to obey an area theorem similar to the one in self-induced transparency [3.26], it should be called "self-induced emission" rather than "superradiance".

SKRIBANOWITZ et al. have also observed unidirectional amplification in HF at 2.6–2.7 μm of the $P(J)$ transition by pumping with the $R(J-2)$ laser line at 2.4–2.5 μm [3.24]. The forward gain and the backward gain were measured with an HF cell placed in a ring resonator. Normally, in the absence of feedback of forward radiation into the backward direction, the backward gain was found to be much less below the laser threshold whereas the forward gain was considerably higher.

3.2.3. Double Resonance within the Doppler Width

Another type of infrared-infrared double-resonance effect which has been observed so far is the experiment with two laser frequencies which are only slightly separated within the Doppler width. The laser may generate two frequencies ω_1 and ω_2 separated by approximately $\omega_1 - \omega_2 = \pi c/L$, where L is the length of the laser resonator.

The difference in frequency is tens of megahertz for a cavity a few meters in length and is less than the Doppler width $\Delta\omega_D$. Double resonance occurs when

$$\omega_1 - \omega_{31} = \omega_2 - \omega_{32} \tag{3.49}$$

in the three levels of Fig. 3.4a. Two frequencies ω_1 and ω_2 may be produced either by modulating the output of a single-frequency laser or by using two similar lasers.

In any case, even though the two frequencies are not stable, the difference between the two can be kept quite constant. Then the above Raman-type resonance condition is satisfied for any variation in laser frequency within the Doppler width and for any velocity of molecules, because the Doppler effect does not change the value of $\omega_{31} - \omega_{32}$ by any appreciable amount.

This type of double-resonance effect was first observed in hyperfine components of Xe by SCHLOSSBERG and JAVAN [3.27]. The technique was called mode-crossing and was extended for the study of Stark components of molecules by BREWER [3.28]. The ${}^qQ(12, 2)$ line of the ν_3 band of CH_3F nearly coincides with the $P(20)$ line of the 9.6 μm CO_2 laser. Since CH_3F is a symmetric top molecule, the first-order Stark effect allows tuning of molecular frequencies to satisfy (3.49) when

$$\frac{2\mu K}{\hbar J(J+1)} E_0 = \omega_1 - \omega_2 , \tag{3.50}$$

where E_0 is the static field, μ the molecular dipole moment, and $J=12$, $K=2$ in this case.

Two stable CO_2 lasers of identical construction were used at a frequency difference of 39.629 MHz to observe the double-resonance signals of Fig. 3.7 in CH_3F at a pressure of 4 mTorr, while the Stark field was swept.

Although the frequency difference was controlled by piezoelectric tuning, a 20–30 kHz acoustic jitter limited the accuracy. The observed line on the left of Fig. 3.7 is the Stark-modulated zero-field level-crossing signal. The next two lines appear at the condition (3.50) and correspond to the two different Stark effects in the ground and vibrational states. Thus the dipole moment of CH_3F in the ground state ($J=12$, $K=2$) is found to be 1.8596 ± 0.0010 Debye; that of the excited state ($J=12$, $K=2$) is slightly different, being 1.9077 ± 0.0010 Debye [3.28].

The linewidth observed in this type of double resonance is determined by the relaxation times of the three levels, including the effects of the duration of coherent interaction between the molecule and the radiation. An observed half-width of 100 kHz demonstrates the capability of this double-resonance technique in high-resolution spectroscopy. A result of particular importance is the observation of a double-resonance signal corresponding to the Stark effect in the vibrationally excited state as well as in the ground state.

The effect described above is quite similar to that observed in level-crossing of "dressed" atoms with a radiofrequency field or in radio-frequency-optical double resonance. Molecular analogues of this effect were performed in CH_4 with the 3.39 μm He–Ne laser by LUNTZ [3.29] and in PH_3 with the $R(33)$ line of the 10.4 μm N_2O laser by SHIMIZU [3.30]. Stark splittings of the E component in $J=6$, $L=7$ in the ν_3 vibrational state of CH_4 and the ν_2 vibrational state of PH_3 respectively were observed.[4]

[4] For levels of CH_4, see Subsection 3.3.3.

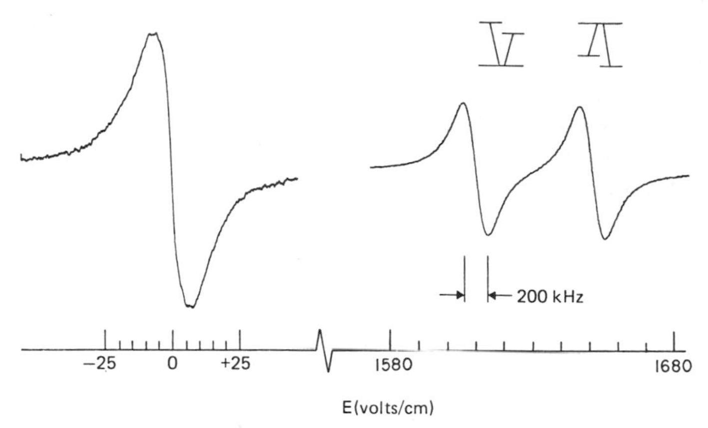

Fig. 3.7. Stark-tuned infrared-infrared double resonance in CH_3F. (Reproduced from [3.84] by permission of the author)

3.3. Infrared-Microwave Double Resonance

The infrared-microwave double-resonance experiment is performed using one laser and one microwave oscillator, both resonant to the vibration-rotation transitions of a molecule. It is characterized by the fact that the energy of an infrared photon is higher than the thermal energy, while the microwave photon energy is much lower than that. Firstly, this is the reason for the large difference in thermal populations as between vibrational and rotational excited states. Secondly, it produces large differences in relaxation rates between vibrational and rotational states. Thirdly, the Doppler broadening of the microwave transition is smaller than other broadenings in gas at a pressure of 10^{-2} to 10 Torr, while the Doppler broadening of the infrared transition is a dominant factor of its linewidth.

The first observation of an infrared-microwave double resonance was reported by RONN and LIDE [3.31] in 1967. A slight change in microwave absorption of the $J = 2 \leftarrow 1$ line of CH_3Br was detected when the gas was irradiated by a powerful CO_2 laser that was not resonant within the Doppler width of the infrared transition. Similar observation of the $J = 3 \leftarrow 2$ line of CH_3Br pumped with a CO_2 laser was reported by LEMAIRE et al. [3.32] in 1969.

Some double-resonance effect in NH_3 was observed in 1970 by FOURRIER et al. [3.33] who monitored the microwave ($J = 4$, $K = 4$) and ($J = 5$, $K = 3$) inversion lines of NH_3 under CO_2 laser pumping. Large, unambiguous signals of double resonance with pumping at exact

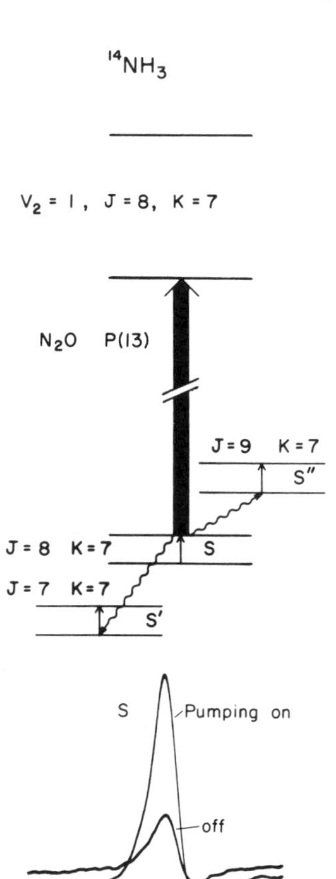

Fig. 3.8. Energy levels of $^{14}NH_3$ and observed traces of the microwave ($J=8, K=7$) line with and without pumping (Reproduced from [3.34] by permission of the author)

resonance were first observed by SHIMIZU and OKA [3.34] in 1970. Figure 3.8 shows energy levels and observed signals in $^{14}NH_3$. The $\nu_2{}^qQ_-(8, 7)$ line of $^{14}NH_3$ is in close coincidence with the $P(13)$ line of the 10.78 µm N_2O laser within 10 MHz. This is one of the rare cases when the discrepancy is smaller than the Doppler width.

A few millitorr of $^{14}NH_3$ in a 30 cm long absorption cell of a microwave spectrometer was irradiated with the N_2O laser output of 80 mW. The microwave absorption of the ($J = 8$, $K = 7$) inversion line shown as S in Fig. 3.8 was observed to increase with infrared pumping, as indicated in the lower part of Fig. 3.8. Weak change signals were observed in the S' and S'' lines; this result will be discussed below in *double resonance in four levels*.

Double-resonance effects on the ($J=4, K=4$) line of $^{15}NH_3$ at 23.0 GHz with the $\nu_2{}^qQ_-(4, 4)$ transition pumped by the $P(15)$ line of the 10.80 µm N_2O laser were also observed [3.34].

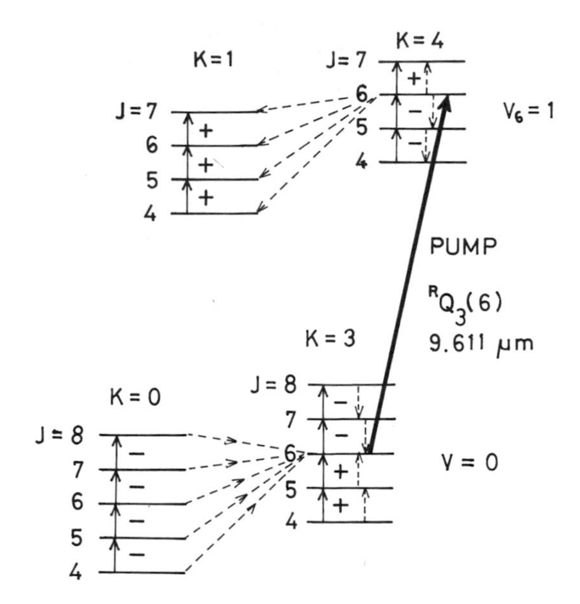

Fig. 3.9. Infrared-microwave double resonance in CH_3Cl

Effects of collisional relaxation in $CH_3\,^{35}Cl$ were recorded with a double-resonance technique by FRENKEL et al. [3.35] in 1971. The $^{R}Q_3(6)$ transition was pumped with the $P(26)$ line of the 9.61 μm CO_2 laser, and the resultant changes in rotational transitions, as shown by arrows in Fig. 3.9, were observed. The infrared pumping depletes the population of the $J=6$, $K=3$ level in the ground state and increases the population of the $J=6$, $K=4$ level in the $v_6=1$ state. Microwave absorptions of the transitions marked + in Fig. 3.9 were found to increase while those marked − were found to decrease with infrared pumping. This behavior was explained by the reasonable appearance of preferred transitions between rotational states induced by collisions.

The following sections describe recent progress in double-resonance experiments which has revealed higher-order effects and their application to the detection of extremely weak transitions in molecules. Preliminary investigation of the coherent modulation effect and of the two-photon transition in double resonance are shown in addition.

3.3.1. Infrared-Microwave Double Resonance in H_2CO

A strong absorption line of formaldehyde H_2CO, observed at a frequency about 180 MHz higher than the center frequency of the 3.51 μm He–Xe laser, was assigned by SAKURAI et al. [3.36] as the $6_{06}(v_5=1)\leftarrow 5_{15}(v=0)$

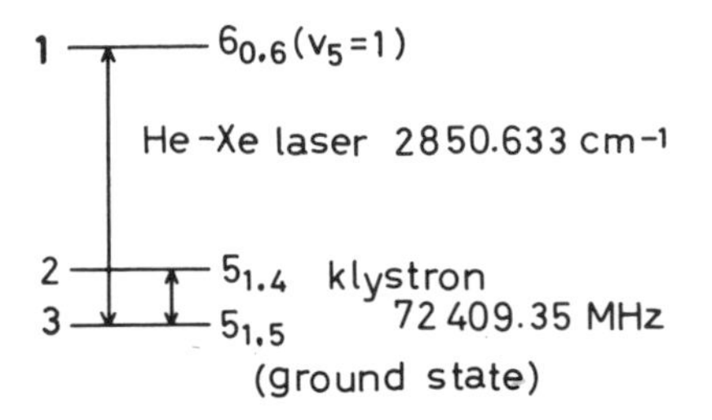

Fig. 3.10. A three-level system in H_2CO

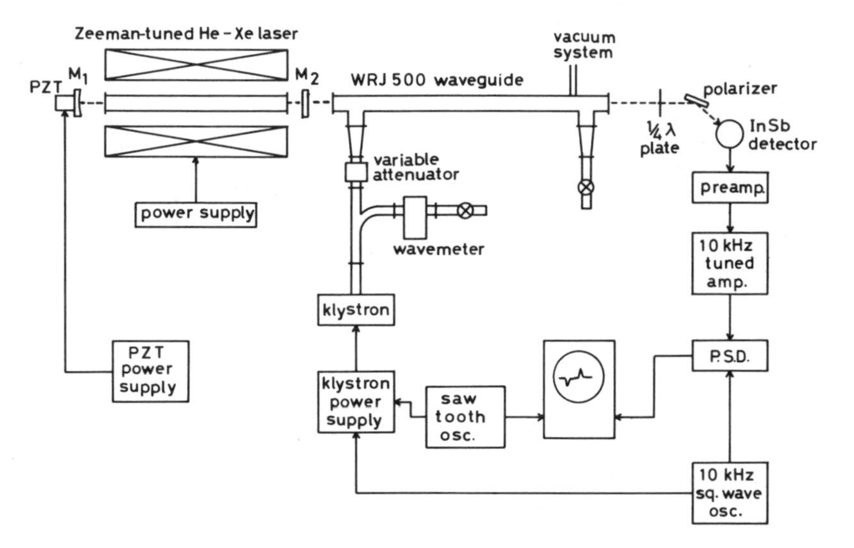

Fig. 3.11. Experimental arrangement for infrared-microwave double resonance in H_2CO

transition. The lower state is coupled to the K-type doubling transition $5_{14} \leftarrow 5_{15}$ at 72.4 GHz, as shown in Fig. 3.10.

This assignment has now been confirmed by the double-resonance experiment of TAKAMI and SHIMODA [3.37]. In their preliminary experiment some change in the infrared absorption of H_2CO under microwave pumping was observed. Further study of its pressure dependence revealed a similarly interesting behavior, which was explained by the higher-order effects of both infrared and microwave fields [3.38, 39]. Theoretical analysis with the rate-equation approximation is in satisfactory agreement with experimental observations.

The double-resonance effect on the inverted Lamb dip of H_2CO in the internal cell is described in Subsection 3.3.4, since it cannot be discussed with the rate equations.

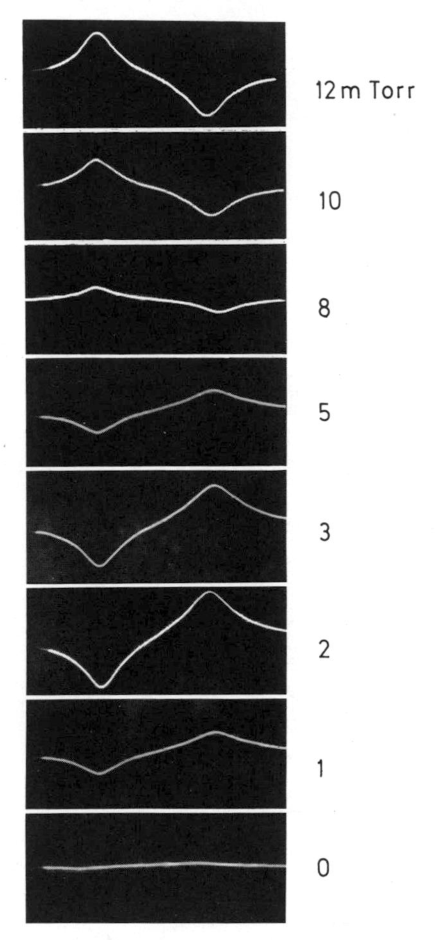

Fig. 3.12. Oscilloscope pictures of double resonance in H_2CO at different pressures

Experimental Results

A block diagram of the experimental apparatus is shown in Fig. 3.11. The absorption cell is 1 m in length with quartz windows at both ends. The laser tube is in a solenoid 60 cm long. An axial magnetic field of about 160 Oe is required to tune the He–Xe laser into the center of the absorption line of H_2CO. The laser output passes through the cell, a quarter-wave plate, and a polarizer. The higher-frequency Zeeman component is detected by a cooled InSb photoconductive detector.

A microwave power of a few milliwatt is fed through a side arm of the absorption cell by an Oki-70 V11A klystron which is frequency-modulat-

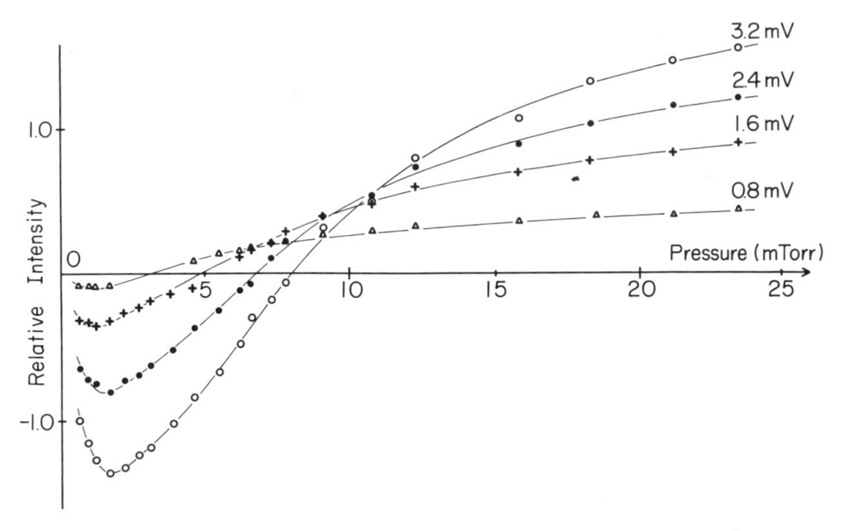

Fig. 3.13. Observed pressure dependence of double resonance in H_2CO at low pressure

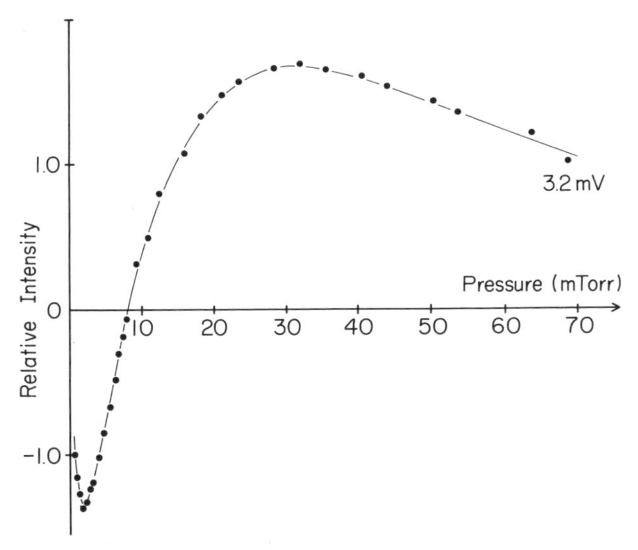

Fig. 3.14. Observed pressure dependence up to 70 mTorr pressure of H_2CO

ed by square waves at 10 kHz. The signal from the infrared detector is amplified, phase-sensitively detected and displayed on a cathode-ray oscilloscope or a recorder while the microwave frequency is swept.

Preliminary observations were made with an absorption cell inside the laser cavity in order to make use of the stronger infrared field. Although the result was theoretically explained by taking into account the decrease in laser power with the pressure of H_2CO [3.39], the experiment where an external cell was used allowed a simpler comparison with theory.

A series of oscilloscope traces of double resonance at different pressures is shown in Fig. 3.12. The peak intensity is plotted in Fig. 3.13 against the pressure of H_2CO, and the parameter in mV shows the power level of the laser measured by the InSb detector. The double-resonance signal is found to change its sign at a pressure of several millitorr. The pressure dependence up to about 70 mTorr is shown in Fig. 3.14.

Theoretical Analysis

The experimental results exhibited in Figs. 3.13 and 3.14 cannot be explained by the weak-signal theory. So a rate-equation approximation as described in Subsection 3.1.3 for three levels is used. Even though the saturation parameters for infrared and microwave fields are large, so that $|x_\mathrm{p}| \gtrsim \gamma_\mathrm{p}$ and $|x_\mathrm{s}| \gtrsim \gamma_\mathrm{s}$, the use of rate equations is justified for the low-resolution observation, as discussed in Subsection 3.1.3.

The probability of an induced transition between level 3 and level 1 in the three-level system of Fig. 3.10 is given by (3.43) if ω_{31} is replaced by ω_{13}. If we neglect the Doppler effect for the microwave transition between levels 2 and 3, the microwave transition probability is now expressed by

$$S_\mathrm{s} = \frac{2\gamma_\mathrm{s}|x_\mathrm{s}|^2}{(\omega_\mathrm{s} - \omega_{23})^2 + \gamma_\mathrm{s}^2}\,. \tag{3.51}$$

Then the rate equations for the molecule moving with an axial velocity between v and $v + \mathrm{d}v$ are given by (3.45).

If the laser and the microwave power do not vary faster than the relaxation rates of the three levels, the steady-state solution of (3.45) may be employed. The induced emission and absorption are proportional to $n_3 - n_1$, which is calculated to become

$$n_3 - n_1 = \frac{[1 + (\tau_2 + \tau_3)S_\mathrm{s}]\,(n_3^0 - n_1^0) - \tau_3 S_\mathrm{s}(n_3^0 - n_2^0)}{[1 + (\tau_2 + \tau_3)S_\mathrm{s}]\,[1 + (\tau_1 + \tau_3)S_\mathrm{p}] - \tau_3^2 S_\mathrm{p} S_\mathrm{s}}\,. \tag{3.52}$$

On the assumption of Maxwellian velocity distribution in the absence of radiation, the unperturbed value of n_i^0 is given by

$$n_i^0 = N_i^0 w(v) = (N_i^0/\sqrt{\pi}\,u)\exp(-v^2/u^2)\,, \tag{3.53}$$

where N_i^0 is the unperturbed population of level i, and u is the most probable velocity.

The increase in transmitted power or the decrease in infrared absorption, when the microwave is modulated, can be calculated by using

$$\Delta(n_3-n_1)=[n_3(S_s)-n_1(S_s)]-[n_3(0)-n_1(0)]$$

and (3.43) in the form

$$\Delta P=-2\hbar\omega_p|x_p|^2\gamma_p\int_{-\infty}^{\infty}\frac{\Delta(n_3-n_1)}{(\omega_p-\omega_{13}-k_pv)^2+\gamma_p^2}dv\,. \tag{3.54}$$

In the "Doppler-limit" approximation for $k_pu\gg\gamma_p$, (3.54) can be easily integrated. The result for $\omega_p\simeq\omega_{13}$ and $\omega_s=\omega_{23}$ becomes

$$\Delta P=2\sqrt{\pi}\hbar c|x_p|^2u^{-1}[(N_3^0-N_2^0)B^{-1}C-(N_3^0-N_1^0)(B^{-1}-A^{-1})], \tag{3.55}$$

where

$$A^2=1+2|x_p|^2(\tau_1+\tau_3)/\gamma_p$$
$$B^2=1+2|x_p|^2(\tau_1+\tau_3-\tau_3C)/\gamma_p$$
$$C=2|x_s|^2\tau_3/[\gamma_s+2|x_s|^2(\tau_2+\tau_3)]\,.$$

Since $N_1^0\ll N_3^0$, (3.55) may be rewritten as

$$\Delta P=(2\sqrt{\pi}N_3^0\hbar c|x_p|^2C/uB)[\delta-2|x_p|^2\tau_3/\gamma_pA(A+B)]\,, \tag{3.56}$$

where $\delta=(N_3^0-N_2^0)/N_3^0=\hbar\omega_s/kT$ is the relative population difference between levels 3 and 2.

When the laser field is weak, the second term in the square brackets in (3.56) can be neglected and the double-resonance signal that is observed as the increase in infrared power ΔP is proportional to $|x_s|^2|x_p|^2\delta/[\gamma_s+2(\tau_2+\tau_3)|x_s|^2]$. This is the ordinary change signal due to microwave pumping.

When the infrared transition is saturated, on the other hand, so that $|x_p|^2/\gamma_p^2$ cannot be small, the second term in the square brackets predominates over the first term. Thus the double-resonance signal changes its sign as the laser power is increased.

The pressure dependence of the double-resonance signal can be calculated from (3.55) or (3.56) by taking into account the relaxation constants as functions of the gas pressure p. Collisional relaxation is

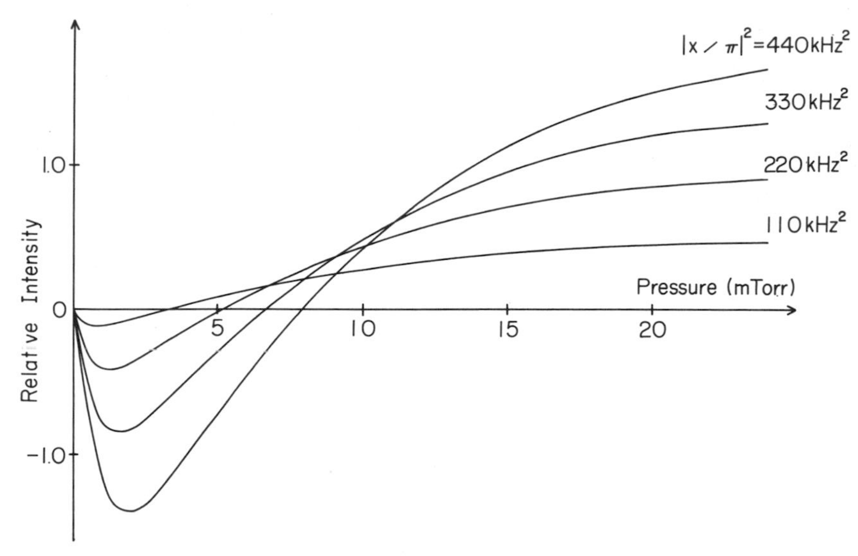

Fig. 3.15. Calculated pressure dependence at low pressure

assumed to predominate in τ_1, τ_2, τ_3, and γ_s: for the simplest case it is written as

$$\tau_1^{-1} = \tau_2^{-1} = \tau_3^{-1} = \gamma_s = C_1 p, \tag{3.57}$$

whereas the transverse relaxation rate of the infrared transition is

$$\gamma_p = C_0 + C_1 p. \tag{3.58}$$

Here C_1 is the pressure-broadening constant and C_0 is the residual linewidth due to the lifetime and the finite transit time of the molecule across the laser beam.

The theoretical pressure dependence calculated by using values of $C_1/2\pi = 10$ MHz/Torr and $C_0/2\pi = 36$ kHz is shown in Figs. 3.15 and 3.16, where the microwave field is chosen to be $|x_s|/2\pi = 145$ kHz to give the best fit to the experimental results. It corresponds to a microwave power of 1 mW in the waveguide cell of 4.78×2.39 mm^2 cross-section. The agreement between theoretical and experimental results is very good [3.38].

Evaluation of Infrared Field Intensity

If the dipole matrix element μ_{31} averaged over the degenerate levels in H_2CO is known, the effective value of $|x_p| = |\mu_{31} E_p|/\hbar$ that is found by

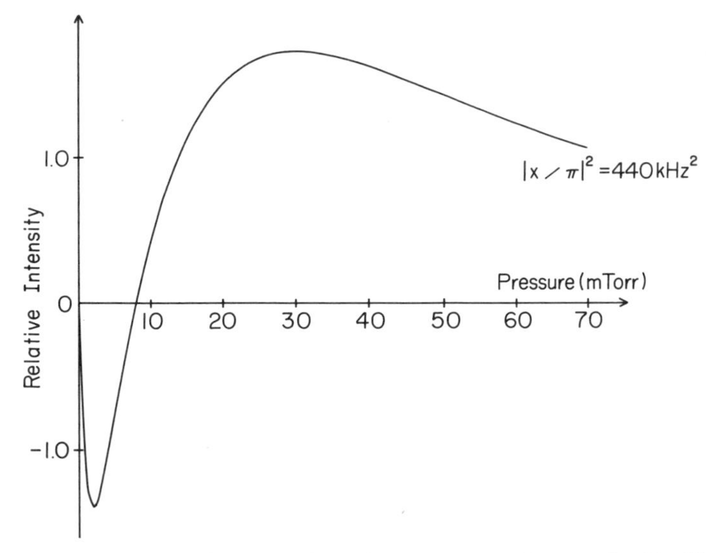

Fig. 3.16. Calculated pressure dependence up to $p = 70$ mTorr for $(2|x_p|/2\pi)^2 = 440$ kHz^2

comparing Figs. 3.13 and 3.15 will give a value of the amplitude of the infrared field $2|E_p|$. An average value of $2|E_p|$ corresponding to $2|x_p|/2\pi = \sqrt{440}$ kHz is 1.4 V/cm, which gives a power of 90 μW through a circular cross-section of 1 mm radius [3.39]. The estimated value is subject to change if the difference in saturation for degenerate levels is taken into consideration.

Evaluation of the infrared field intensity has previously been made with the saturation broadening of the inverted Lamb dip [3.40]. This is, however, less reliable and more difficult experimentally than the method mentioned above.

3.3.2. Rotational Transitions of Vibrationally Excited Molecules

When the vibrational frequency of a molecule exceeds 2000 cm^{-1}, the population of the vibrationally excited state is very small at room temperature. Pumping by a laser to enhance the population of the excited state to the extent that it permits microwave spectroscopy of the excited molecules is not very promising, because only the molecules within its narrow homogeneous width among the broad Doppler width are pumped. The double-resonance method, on the other hand, has been shown to be a powerful tool for microwave spectroscopy of vibrationally excited molecules.

The principle of sensitive detection of rotational transition in a vibrationally excited state is explained in the three levels of Fig. 3.1. Since $N_3^0 \simeq 0$ and $N_2^0 \simeq 0$ in this case, the double-resonance signal for $\omega_p = \omega_{31} = -\omega_{13}$ is obtained from (3.55) by changing the sign to become

$$\Delta P = -4\sqrt{\pi}\hbar c N_1^0 |x_p|^4 \tau_3 C / u\gamma_p AB(A+B) . \qquad (3.59)$$

This is exactly the same as the second term in (3.56), since N_1^0 in (3.59) is the ground-state population as is N_3^0 in (3.56). Thus only the higher-order effects appear in this case. The lowest-order term in (3.59) is $|x_p|^4|x_s|^2$, because the lowest-order term in C is $|x_s|^2$.

The double-resonance signal in this case appears only when ω_s is resonant to the rotational transition $3 \leftrightarrow 2$ in the vibrationally excited state, and it is almost as large as that in the three-level of Fig. 3.10 at low pressure, even though the thermal population of the vibrationally excited state is completely negligible.

Double-Resonance Experiments in HDCO

The above method of microwave spectroscopy of vibrationally excited molecules has been demonstrated in four levels of formaldehyde d_1, HDCO, as shown in Fig. 3.17 [3.41]. The infrared transition $3_{12}(v_1 = 1) \leftarrow 2_{11}(v = 0)$ is coupled to the microwave transitions in the ground state at 16 GHz and in the vibrationally excited state at 33 GHz. The increase in saturated absorption of the infrared transition is observed when the microwave transition is resonantly excited either in the ground or the excited state.

The experimental apparatus for HDCO was similar to that described in *Experimental results*, except that a K-band cell 1 m in length was placed inside the laser resonator. The laser was tuned by a magnetic field of 1.07 kOe into the HDCO line at a frequency 1.33 GHz lower. The frequency of the $2_{11} \leftarrow 2_{12}$ rotational transition in the ground state was known to be 16.03806 GHz. Then the pressure dependence of the double-resonance signal obtained with the 3.51 μm laser and the 16 GHz Varian X-12 klystron was observed as shown in Fig. 3.18.

Because the common level for the two transitions is the middle one of three levels in this case, the two terms in the square brackets of (3.55) or (3.56) have the same (negative) sign. The contribution of the higher-order effects as given by the second term is seen in Fig. 3.18 to dominate that of the first term at a pressure below about 30 mTorr. Thus double resonance with the microwave transition in the excited state is expected to be observable with comparable intensity.

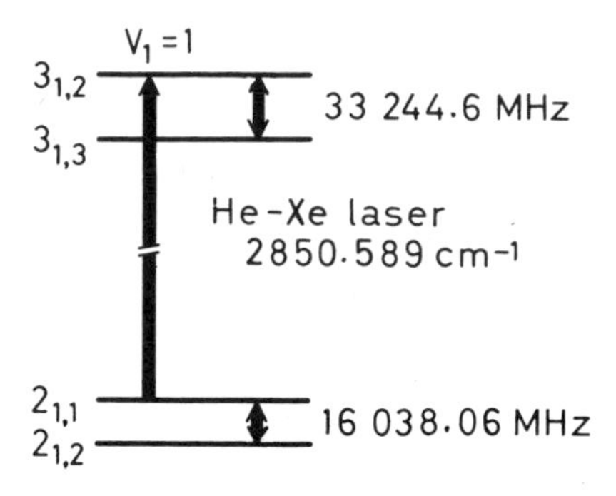

Fig. 3.17. Four levels of HDCO for two kinds of infrared-microwave double resonance

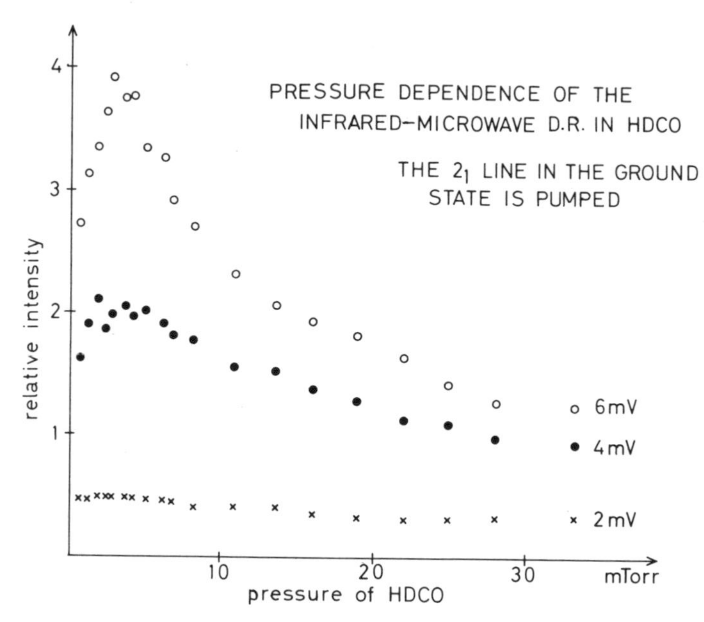

Fig. 3.18. Observed pressure dependence of double resonance in HDCO with the 16 GHz rotational transition in the ground state

A search for the unknown $3_{12} \leftrightarrow 3_{13}$ line in the $v_1 = 1$ excited state in the estimated frequency range between 31.5 and 33.5 GHz resulted in a large double-resonance signal at 33.2446 GHz. The change in saturated absorption of the infrared transition had the same sign as that of Fig. 3.18. The pressure dependence of the double-resonance signal with the $3_{12} \leftrightarrow 3_{13}$ rotational transition in the $v_1 = 1$ state is shown in Fig. 3.19. The signal in Fig. 3.19 is seen to disappear at high gas pressure, but it is almost as strong as that in Fig. 3.18 at low pressure.

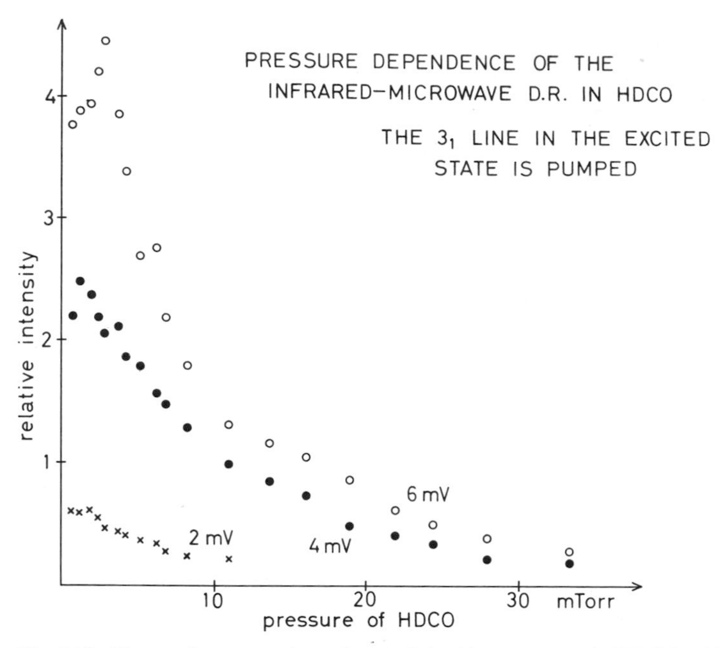

Fig. 3.19. Observed pressure dependence of double resonance in HDCO with the 33 GHz rotational transition in the vibrationally excited state

Microwave Spectroscopy of HCOOH in the Excited State

The microwave transition in the vibrationally excited state of negligible population is now observable with the double-resonance technique. Most of the polyatomic molecules exhibit complicated unresolved rotation-vibration spectra in the infrared. The double-resonance method picks up only a few lines that provide a definite assignment for precision spectroscopy.

For example, formic acid HCOOH is known to have many overlapping absorption lines in the 3.39 μm range. The rotational constant of HCOOH is precisely known in the ground state and in a few low-energy vibrational states, but no precise value is known for the excited state of the CH stretching mode of vibration by non-laser spectroscopy.

The double-resonance technique was applied by TAKAMI [3.42] in order to observe microwave transitions in the excited state of HCOOH. Eleven lines of the CH stretching-vibration band of HCOOH had been observed within the tuning range of the 3.39 μm He–Ne laser and analysed by UEDA [3.43]. Most of the observed lines constitute three-level systems for double resonance. An example of the $13_{2,11}(v=1) \leftarrow 13_{1,12}(v=0)$ transition near the line center of the 3.39 μm He–Ne

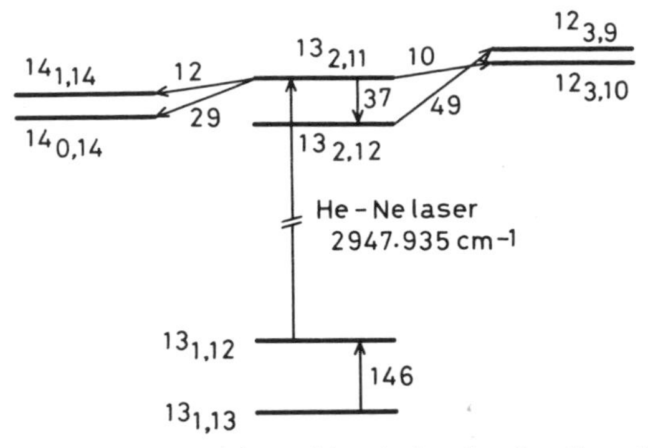

Fig. 3.20. Microwave transitions in the vibrationally excited state of HCOOH to be observed by the double- or triple-resonance method with the 3.39 μm He–Ne laser. Microwave frequencies are given approximately in GHz

Table 3.2. Observed frequencies of microwave rotational transitions of HCOOH in the excited state of CH-vibration

Infrared transition	Microwave transition in the $v=1$ state	Observed frequency [MHz]
$13_{2,11}(v=1) \leftarrow 13_{1,12}(v=0)$	$13_{2,11} \rightarrow 13_{2,12}$	37337.677 ± 0.010
	$13_{2,11} \rightarrow 14_{1,14}$	12740.852 ± 0.010
	$13_{2,11} \rightarrow 14_{0,14}$	29101.520 ± 0.010
	$12_{3,10} \leftarrow 13_{2,11}$	10567.333 ± 0.010
$12_{2,10}(v=1) \leftarrow 12_{1,11}(v=0)$	$12_{2,10} \rightarrow 12_{2,11}$	28146.616 ± 0.010
	$12_{2,10} \rightarrow 13_{1,13}$	13478.732 ± 0.010
	$12_{2,10} \rightarrow 13_{0,13}$	33150.797 ± 0.010
	$11_{3,9} \leftarrow 12_{2,10}$	39578.180 ± 0.010
$14_{2,12}(v=1) \leftarrow 14_{1,13}(v=0)$	$14_{2,12} \rightarrow 14_{2,13}$	48198.147 ± 0.010
	$14_{2,12} \rightarrow 15_{1,15}$	14573.848 ± 0.010
	$14_{2,12} \rightarrow 15_{0,15}$	28039.904 ± 0.010
	$14_{2,12} \rightarrow 13_{3,11}$	19516.661 ± 0.010
$11_{2,9}(v=1) \leftarrow 11_{1,10}(v=0)$	$11_{2,9} \rightarrow 11_{2,10}$	20596.882 ± 0.010
	$11_{2,9} \rightarrow 12_{1,12}$	16777.857 ± 0.010
	$11_{2,9} \rightarrow 12_{0,12}$	40181.770 ± 0.010
	$10_{3,8} \leftarrow 11_{2,9}$	67479.058 ± 0.020

laser is shown in Fig. 3.20. With a technique very much like the one described in *Double-Resonance Experiments in* HDCO above, the rotational $13_{2,11} \leftrightarrow 13_{2,12}$ transition in the $v=1$ excited state was found at about 37.34 GHz, whereas the value of the frequency is calculated to be 37.16 GHz by using the rotational constants of the molecule in the excited state that have been determined by UEDA [3.43].

Fifteen other rotational transitions in the vibrationally excited state have recently been observed with the double resonance method. The observed microwave frequencies are shown in Table 3.2. The $13_{3,9} \leftarrow 13_{2,12}$ transition at about 49 GHz shown in Fig. 3.20 is not directly connected to the infrared transition induced by the laser. The transition has, however, been observed successfully with a technique of triple resonance by employing the 3.5 μm, 37 GHz and 49 GHz radiations. Rotational constants of HCOOH in the CH-vibrational state can thus be calculated from these frequencies. The results can not, however, be very accurate because of the Coriolis and other kinds of interaction, which will be the subject of further investigation.

3.3.3. Observation of Very Weak Microwave Transitions in CH_4

In the previous section it was shown that weak microwave transitions can be observed between levels of very small populations in which the dipole matrix element is of considerable magnitude. The double resonance method is likewise useful in observing very weak transitions having a very small dipole moment. As an example of this type, microwave spectroscopy of methane CH_4 by using the 3.39 μm He–Ne laser is described below.

The 3.39 μm line of CH_4 corresponding to the $F_1^{(2)}(v_3=1) \leftarrow F_2^{(2)}$ $(v_3=0)$ component of the $P(7)$ branch of the ν_3 mode of vibration is particularly important as an absolute standard of both wavelength and frequency, since the proposal by the author [3.44] in 1968 and the excellent experiments by BARGER and HALL [3.45] in 1969.

The CH_4 molecule is tetrahedral and has no electric dipole moment in the ground state. Thus no rotational transition of CH_4 is allowed. It has a small dipole moment in the triply degenerate vibrational state ν_3, however, and the J state is split into three states[5] of $L=J-1, J$, and $J+1$, because of the vibration-rotation coupling [3.46, 48]. The $J=6$, $L=7$, $v_3=1$ state is further split into six sublevels by Coriolis interaction [3.47], as shown in Fig. 3.21. The dipole moment of the $v_3=1$ excited state was measured by the Stark effect of the E component as 0.02 Debye [3.48, 49]. Weak transitions, therefore, may occur between the F_1 and F_2 sublevels [3.46, 48], as shown by arrows in Fig. 3.21.

The $F_1^{(2)}(v_3=1) \leftarrow F_2^{(2)}(v=0)$ component of the $P(7)$ line has been denoted by the $F_1^{(2)}$ component in most previous references. It is now recommended to denote it the $F_2^{(2)}$ component.

---

[5] These are denoted by R instead of L according to HECHT [3.47] and others.

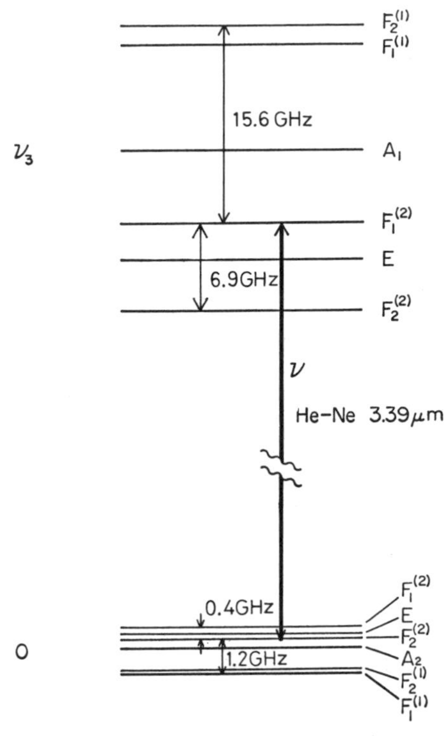

Fig. 3.21. Energy levels of CH_4 involved in the 3.39 μm $P(7)$ branch. The $F_2^{(2)}$ component of the $P(7)$ line is shown by a heavy arrow, and weak microwave transitions between sublevels are shown by light arrows

Among four microwave transitions between the F_1 and F_2 sublevels in the $J=6$, $L=7$, $v_3=1$ state, two transitions $F_2^{(1)} \leftrightarrow F_1^{(2)}$ and $F_1^{(2)} \leftrightarrow F_2^{(2)}$ which share the $F_1^{(2)}$ sublevel in common with the infrared transition at the frequency of the He–Ne laser can be observed by the double-resonance technique. The He–Ne laser cannot be tuned into the other F components. The frequencies of the above two transitions calculated from the infrared data [3.50, 51] are 15.62 GHz and 6.89 GHz. Their intensities of absorption are many orders of magnitude smaller than those of ordinary absorption, because both the population and the dipole moment are small in this case.

The $F_1^{(2)} \leftrightarrow F_2^{(2)}$ transition in the $v_3=1$ state of CH_4 was initially observed by CURL and OKA [3.52] using a microwave cavity resonator placed inside the laser resonator in order to make the perturbation as strong as possible. The frequency was measured as 6895.3 ± 0.3 MHz. A refined measurement with a waveguide 1 m long was carried out by

TAKAMI et al. [3.53] using an apparatus similar to that described in *Experimental Results* above. In order to obtain higher resolution with sufficient sensitivity, a lower power of some ten milliwatts and a lower pressure of a few millitorr were used. The $F_1^{(2)} \leftrightarrow F_2^{(2)}$ transition in the $v_3 = 1$ state was observed to be 6895.204 ± 0.010 MHz.

The $F_2^{(1)} \leftrightarrow F_1^{(2)}$ transition in the $v_3 = 1$ state was similarly observed at 15601.846 ± 0.010 MHz, and the half width at half intensity was about 100 kHz at a pressure of 4 mTorr.

The same technique was applied by CURL et al. [3.54] to observe extremely weak microwave transitions in the ground state of CH_4. The $J = 7$ ground state was known to be split into six components as shown in Fig. 3.21. Centrifugal distortion effects in such a tetrahedral molecule produce a small dipole moment in the ground state. The estimated value of the rotation-induced electric dipole moment in CH_4 [3.55, 56] is very small to be 0,0018 Debye for $J = 7$. Because of the low transition frequency and the small dipole moment, the $F_1^{(2)} \leftrightarrow F_2^{(2)}$ transition in the ground state could only be observed with the double resonance method. The absorption cell of a coaxial cavity resonator was about 8 cm long, and placed inside the 3.39 μm He–Ne laser resonator of 50 cm length.

The $F_1^{(2)} \leftrightarrow F_2^{(2)}$ transition in the ground state of CH_4 was observed at 423.02 ± 0.02 MHz [3.54, 57a]. The $F_2^{(2)} \leftrightarrow F_1^{(1)}$ transition in the ground state was lately observed to be 1246.55 ± 0.02 MHz with the same method [3.57a]. In either case, the pressure of CH_4 was about 10 mTorr, and the microwave field was of the order of 500 V/cm.

3.3.4. Modulation Doubling of the Inverted Lamb Dip

The double resonance signal in a non-degenerate three-level system becomes a doublet, when the pumping is strong, as shown in Fig. 3.3. This is called modulation doubling or the resonant modulation effect and has been previously observed in radiofrequency and microwave transitions. At optical frequencies, however, the line is broadened by a Doppler effect of the order of 100 MHz, so that a small modulation doubling in double resonance cannot be observed.

When the frequency of a stable laser with an internal absorption cell is swept slowly, a sharp inverted Lamb dip caused by nonlinear absorption is observed at the center of the Doppler broadened absorption line. The inverted Lamb dip in molecules was first observed in CH_4 by BARGER and HALL [3.45] with a linewidth as narrow as 1 MHz or less. Splitting of the inverted Lamb dip under resonant modulation can thus be observed even when relatively low power is employed.

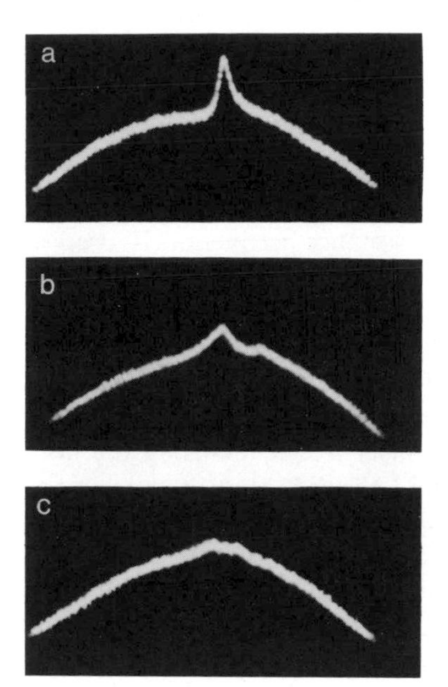

Fig. 3.22a–c. Inverted Lamb dip of H_2CO at 3.5 μm showing the resonant modulation effect. a) Microwave is more than 20 MHz off-resonance. b) It is about 1 MHz off-resonance. c) It is at resonance

The double resonance effect on the inverted Lamb dip in H_2CO was observed by TAKAMI and SHIMODA [3.39]. The oscilloscope pictures of the inverted Lamb dip of H_2CO inside the 3.51 μm He–Xe laser are shown in Fig. 3.22. When the strong microwave at 72.4 GHz is applied to the three-level system of Fig. 3.10, the infrared line splits into eleven components corresponding to the $2J+1$ different M states. Since the magnitude of separation depends on the M value, inverted Lamb dip becomes weaker. At a pressure of 2 m Torr the full width in Fig. 3.22a is about 0.4 MHz and the coherence splitting of resonant modulation is evident in Fig. 3.22c, although M components are not clearly resolved.

It is also possible to resolve M components of coherence splitting in infrared-infrared double resonance, in principle, as proposed by SKRIBANOWITZ et al. [3.57b]. Because the modulation doubling is also smeared by the effect of inhomogeneous distribution of the infrared field, in addition to the Doppler effect, it is difficult, in general, to resolve the above-mentioned high-frequency Stark components.

The double resonance effect on the inverted Lamb dip in CH_4 has lately been observed by CURL and OKA [3.52] with the 3.39 μm He–Ne laser and a powerful klystron of 5 W at 6895 MHz. Almost complete destruction of the Lamb dip is found in the presence of this resonant microwave perturbation as described in Section 3.3.3.

3.3.5. Raman-Type Double Quantum Transition

When the strong pumping at an off-resonance frequency is used, a Raman-type double resonance of two-photon transition occurs as discussed in Section 3.1. This type of two-photon transition with one infrared photon from a fixed-frequency laser and the second photon from a tunable microwave oscillator can be effectively applied to off-resonance spectroscopy of molecules. In effect, the microwave frequency is either "added" or "subtracted" from the fixed laser frequency using the non-linear process of the molecular system. Needless to say, the parity selection rule is different from the single-photon transitions.

OKA and SHIMIZU [3.58] observed such a two-photon transition in a three-level system of $^{15}NH_3$, as shown in Fig. 3.23. The frequency of the $P(15)$ line of the 10.80 μm N_2O laser is lower than that of the $\nu_2{}^qQ_-(4,4)$ line of $^{15}NH_3$ by about 300 MHz. With a microwave power of 20 W at a frequency 300 MHz higher than the $J=K=4$ inversion doubling at 23046 MHz, absorption of the laser radiation by the two-photon transition was observed.

By means of this technique FREUND and OKA [3.59] have recently recorded a number of two-photon transitions in $^{14}NH_3$ and $^{15}NH_3$. Two-photon transitions with a magnitude of off-resonance as large as 12 GHz were observed.

The probability of two-photon transition is given by (3.26), while that of the single-photon transition at resonance $\omega_p=\omega_{31}$ is known to be $2|x_p|^2\tau^2/(1+4|x_p|^2\tau^2)$. Thus the two-photon transition is weaker by a ratio of

$$R=[|x_s|^2/(\Delta\omega^2+\theta^4\tau^2)](1+4|x_p|^2\tau^2)\,. \tag{3.60}$$

When the two radiations are not very strong so that $|x_p|^2\tau^2\ll 1$ and $\theta^4\tau^2\ll\Delta\omega^2$, the ratio becomes simply

$$R=|x_s|^2/\Delta\omega^2\,. \tag{3.61}$$

At a microwave power of 10 W in a K-band waveguide, for example, a microwave field of $|E_s|=80$ V/cm is applied. With a value of $|\mu_{13}|=1$ Debye, it gives $|x_s|/2\pi=40$ MHz. Thus the two-photon transition is weaker by a factor of $(40/300)^2=1.8\times10^{-2}$ for $\Delta\omega/2\pi=300$ MHz, and $(40/12000)^2=1\times10^{-5}$ for $\Delta\omega/2\pi=12$ GHz; both are well above the minimum detectable absorption of the laser spectrometer [3.12].

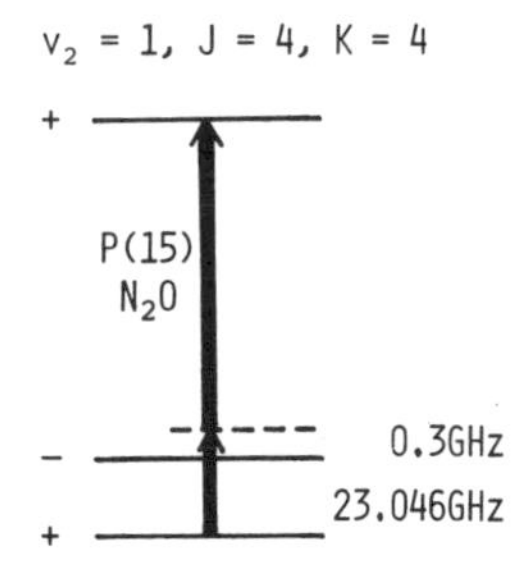

Fig. 3.23. Energy levels for the two-photon transition in $^{15}NH_3$

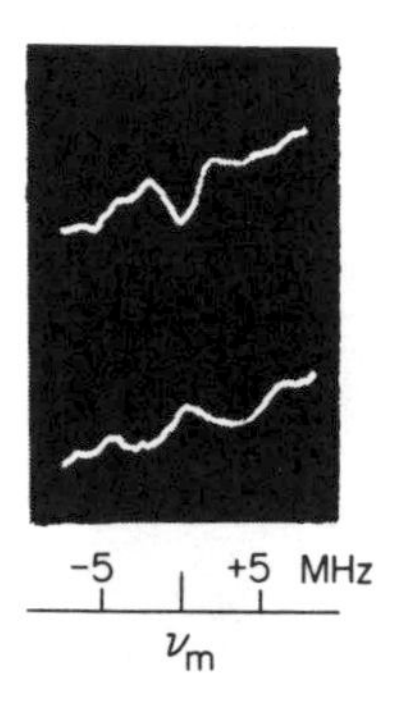

Fig. 3.24. Observed two-photon Lamb dip involving the $\nu_2{}^qQ_-(4, 4)$ transition in $^{15}NH_3$. The lower trace shows coherence splitting caused by the additional microwave at resonance. (Reproduced from [3.60] by permission of the author)

The inverted Lamb dip can also be observed in the infrared-microwave two-photon transition when saturation is achieved. From (3.26) it is concluded that saturation appears at $2|x_p|^2|x_s|^2\tau^2 \simeq \Delta\omega^2$. Take $|x_p|/2\pi \simeq |x_s|/2\pi \simeq 40$ MHz, $\Delta\omega/2\pi = 1$ GHz, and $(2\pi\tau p)^{-1} = 30$ MHz/Torr, for example, then the Lamb dip will be observable at a pressure of about 10 mTorr or below.

The two-photon Lamb dip was observed in $^{14}NH_3$ and $^{15}NH_3$ by FREUND and OKA [3.60] in 1972. The absorption cell is inside the CO_2 or the N_2O laser cavity. In contrast to the ordinary experiment of the Lamb dip, the laser frequency is fixed and the microwave frequency is swept in order to observe the infrared-microwave two-photon Lamb dip. In the three-level system of Fig. 3.23, the two-photon Lamb dip is found, as shown in the upper trace in Fig. 3.24 at a microwave frequency of 23.360 GHz. Thus the frequency difference between the $\nu_2{}^qQ_-(4, 4)$ line of $^{15}NH_3$ and the $P(15)$ line of the 10.80 μm N_2O laser is determined to be 314 MHz. Similar observation with the $\nu_2{}^qQ_+(5, 4)$ line of $^{14}NH_3$ and the $R(6)$ line of the 10.35 μm CO_2 laser gives a difference of 558 MHz. The observed half-width at half-maximum is approximately 0.8 MHz in either case.

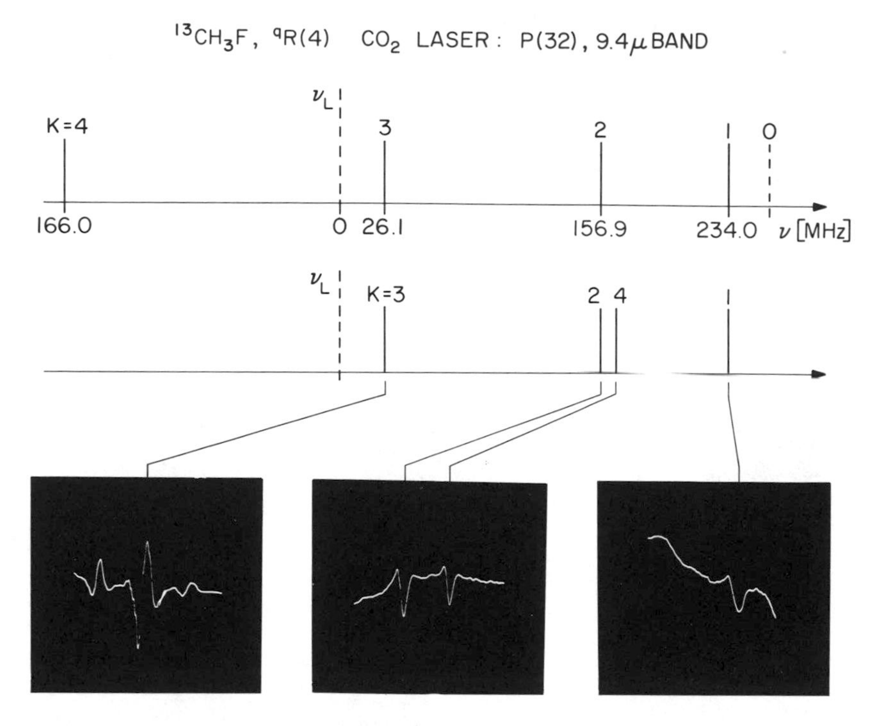

Fig. 3.25. K components of the $\nu_3{}^qR(4)$ line of $^{13}CH_3F$ resolved by the technique of two-photon Lamb dip with the $P(32)$ 9.66 μm CO_2 laser line. (By courtesy of T. OKA) Satellite lines in the left picture are not identified

When the second microwave radiation at the resonant frequency of 23.046 GHz shown in Fig. 3.23 is applied, the two-photon Lamb dip reveals a coherence splitting of the resonant modulation effect. This is shown in the lower trace in Fig. 3.24. An interesting feature of this splitting for the two-photon transition is that the relative intensities of the M components have a higher power dependence on M than does that for a single-photon transition. Therefore, the modulation doubling of the strongest M component is primarily visible without much overlap of other M components in this case, which is different from the case of Subsection 3.3.4.

The method of high-resolution spectroscopy by two-photon transition has recently been applied to resolve K components of the $\nu_3{}^qR(4)$ line of $^{13}CH_3F$ by using the $P(32)$ line of the 9.66 μm CO_2 laser. The observed spectrum and oscilloscope traces of the two-photon Lamb dip are shown in Fig. 3.25. The high resolution and ease of microwave-frequency measurement will make this method one of the powerful tools of spectroscopy. Since the absolute frequencies of the CO_2 and N_2O laser lines are now known, absolute energy differences between molecular levels can be determined very accurately by this two-photon technique.

3.4. Relaxation and Transient Effects

The double-resonance technique is a useful tool for the study of collisional and other relaxation processes in molecules. In Subsection 3.3.1 a relaxation time determined by pressure broadening in microwave lines was assumed to explain the pressure dependence of the steady-state double-resonance signal. If the pressure dependence and power dependence of double resonance are measured in some detail, the theoretical analysis in *Theoretical Analysis* above will be utilized to determine the relaxation rates involved in the transitions. Thus τ_1, τ_2, τ_3 and γ will be measured separately. The lineshape and the width will also provide some information on relaxation.

The study of relaxation with this steady-state double-resonance method, however, is indirect and often shows misleading results. Hence very accurate and careful measurements are required. In order to study a relaxation time of the order of microseconds, for example, by lineshape observation, an experimental resolution of much better than 1 MHz is necessary; this is not easy to attain at infrared frequencies.

The method of time-resolved observation of infrared-microwave double resonance, on the other hand, gives more direct and reliable information on relaxation. A pulsed infrared pumping radiation suddenly depletes the population of the lower level and increases that of the upper level. Any subsequent change in population is monitored by weak microwave radiation at the frequency corresponding to the transition from the level to be studied.

Since the Doppler broadening is much larger than the homogeneous broadening in the optical region, the change in population caused by a monochromatic pumping radiation occurs within a narrow range of molecular velocities. Thus the pumping burns a hole in the lower state and piles it up in the upper state in the molecular velocity distribution. Hence the velocity dependence of collisional relaxation can be observed by the double-resonance or triple-resonance technique.

The laser and the microwave oscillator cause coherent perturbations upon the molecule and establish molecular coherence. The use of a pulsed radiation, therefore, enables transient coherent effects in the three-level system, similar to those observed in the two-level system, to be observed by the double-resonance technique.

3.4.1. Lifetime Measurement of Laser-Excited Levels

A Q-switched gas laser produces output pulses of about 100 nsec duration. The pulsewidth can be narrowed to a few nanoseconds by the mode-locking method. A few watts of laser power is sufficient in many cases to pump an appreciable fraction of the lower-state molecules to the upper level at a low gas pressure.

An exponential decay of the upper-level population following the excitation pulse can most easily be observed by fluorescence from the upper level. The laser-excited fluorescence method has been used to study vibrational, rotational and electronic energy transfer, and photochemical reactions in molecules [3.61–64]. This method furnishes ample information on rate constants for individual processes connecting rotational, vibrational and electronic states.

State selection by both the laser and the fluorescence wavelength is a very useful technique for the study of the transfer of molecular energy between rotation, vibration and translation. But the fluorescence light is incoherent so that it is not suitable for the study of line-broadening nor of molecular coherence effects, although this is possible in principle.

We describe below some double-resonance techniques that are experimentally more difficult than the laser-excited fluorescence technique but promise to yield more detailed knowledge on the physics of relaxation processes.

Relaxation Study of Molecular Laser Levels

The study of laser levels is essential for understanding and improving laser performance. Since the molecular levels are intrinsically coincident with the laser line of the same molecule, no tuning of the laser is required, it is only necessary to select the laser line.

An infrared-infrared double-resonance experiment on CO_2 with two lasers was performed by RHODES et al. [3.65] in 1968. In this experiment, relaxation of CO_2 in an absorption cell at an arbitrary pressure was observed using pulsed pumping of the 9.6 μm CO_2 laser while a cw power of the 10.6 μm CO_2 laser was used for the signal transition.

The temporal change in the molecular population of the laser level was seen after the pumping pulse by means of an apparatus shown schematically in Fig. 3.26. The laser tube No. 1 is Q-switched and excites the sample gas in tube No. 3. Population changes in the (00^01) and (10^00)

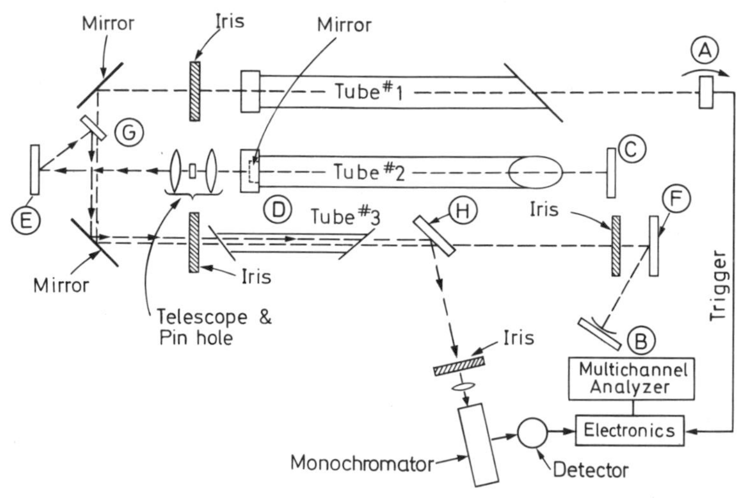

Fig. 3.26. Apparatus for time-resolved observation of infrared-infrared double resonance in CO_2

states are observed by irradiating the tube No. 3 with the cw power at 10.6 μm from the laser tube No. 2. A is a rotating mirror, F is a grating for selecting a line at 9.6 μm, and another grating E for 10.6 μm serves to isolate laser tubes Nos. 1 from 2. Beam splitters G and H are used to couple in and couple out the signal radiation.

The observed initial fast decay of the order of 10^{-6} sec is ascribed to the collisional relaxation between the $(10^0 0)$ and the $(02^0 0)$ states, which is depleted by pumping. The slower relaxation of the order of 10^{-5} sec corresponds to the lifetime of the $(10^0 0)$ state, while the slowest relaxation of the order of 10^{-3} sec corresponds to that of the $(00^0 1)$ upper state.

A similar study on N_2O laser levels was carried out with N_2O lasers by Bates et al. [3.66].

Investigations of this kind on dye-laser molecules by means of pulsed solid-state and/or dye lasers will be actively performed in the near future. Since the dye laser has a high gain, one part of the dye in the cell may generate weak signal power to measure the gain in the other part of the cell, as in the experiment of Shank et al. [3.67] for example.

Time-Resolved Infrared Double Resonance

Relaxation rates in three molecular levels can readily be studied by the method of time-resolved infrared double resonance, once a good coincidence has been found between the molecular frequency and the laser frequency. The infrared-infrared double-resonance experiment is more difficult than the infrared-microwave double-resonance experiment unless a coherent infrared source having a wide tuning range is utilized. Since SF_6 has many absorption lines in the 10.6 μm range of the CO_2 laser, the time-resolved infrared-infrared double-resonance experiment on SF_6 was first carried out by BURAK et al. [3.68] in 1969 and by STEINFELD et al. [3.69] in 1970. The pulsed output of several kilowatts of the *P*(20) line of the 10.6 μm CO_2 laser with some of the *P*(18) and *P*(22) lines was used for pumping, while the cw power of 0.1 W of any line from *P*(12) to *P*(28) was used to monitor the exponential relaxation following the pulse. Relaxation rates in pure SF_6 and in mixtures with other molecules were measured at different pressures.

An experiment of the same sort on BCl_3 was carried out in 1972 by HOUSTON et al. [3.70a] for the study of vibration-vibration and vibration-translation relaxation in laser-excited molecules. The vibration-vibration relaxation rate was found to be faster than the vibration-translation rate, in agreement with the results of other investigators. More recently, a time-resolved infrared-microwave double-resonance in $^{13}CH_3F$ with the pulsed 9.66 μm CO_2 laser was observed by JETTER et al. [3.70b]. Time-resolved signals at the 3.39 μm wavelength of the He–Ne laser in ethylene, propylene, and vinyl chloride following pulsed pumping with a CO_2 laser were recorded by GRABINER et al. [3.71].

In these preliminary studies, however, state selection and the assignment of the observed transition were imperfect. In addition, a heating effect of the laser pulse is involved in the observed results of these experiments.

The time-resolved study of infrared-microwave double resonance in a well-defined three-level system has been pursued quite recently by LEVY et al. [3.72]. It revealed the transient nutation effect described in Subsection 3.4.3.

3.4.2. Collision-Induced Transitions

A molecule in a quantum state i collides with another molecule and changes its state to j. The difference in energy between rotational states of the molecule is normally smaller than the thermal energy kT at room temperature. Thus a rather weak intermolecular interaction is sufficient to cause a rotational transition by collision. The change in vibrational state by collisions, on the other hand, requires much higher energy than kT and occurs at a much lower rate.

The dominant transitions in molecules induced by collisions, therefore, are the rotational transitions. Because of lack of knowledge about the processes that occur in molecular collisions, it was previously assumed that changes of the rotational state occur in a more or less random manner. By using the method of microwave-microwave double-resonance, OKA and his collaborators [3.73a] have established that there exists a large difference in transition probabilities between the rotational states induced by collisions. A sort of "selection rules" governing collision-induced transitions has been found. The selection rules for preferred transitions induced by molecular collisions are primarily of dipolar type, in agreement with the expectation that the long-range Coulomb interaction must predominate in the rotational energy transfer in polar molecules.

Since the frequency of rotational transition falls in the microwave region, the microwave spectrum is an appropriate probe to study the rotational states. The steady-state microwave-microwave double-resonance experiment, and later its time-resolved version, have so far been performed for the study of collision-induced transitions. The rate constants for collisional transitions between each state have been measured for particular collision partners (colliding molecules or atoms).

The aforementioned development of infrared-microwave double resonance is providing new and powerful methods for the study of collisional energy transfer. The advantages of the infrared-microwave double-resonance method are: easier isolation of the signal from the pump, availability of short laser pulses for pumping, high sensitivity and high resolution. On the other hand, the infrared method has a drawback in that its application is limited to the coincidence of frequencies between the transition and the infrared laser, because the tuning range of the conventional laser is narrow.

Double Resonance in Four Levels

Consider four levels of a molecule as shown in Fig. 3.27, where levels *1*, *2*, and *3* are in the ground state whereas level *4* is in the vibrationally excited state. When the microwave transition $2\leftrightarrow 1$ is saturated, no change in the infrared absorption $4\leftarrow 3$ would arise in the absence of collision-induced transitions. Nor could a signal be observed in such four-level double resonance if the collisional transition probabilities were equal among the rotational states having energy differences less than kT.

Thus any observed signal of double resonance in four levels is a strong indication of some difference between the probabilities of rotational energy transfer between $3\leftrightarrow 1$ and $3\leftrightarrow 2$. One may also pump the infrared transition $4\leftarrow 3$ and monitor the microwave absorption $2\leftarrow 1$ in the four-level system of Fig. 3.27.

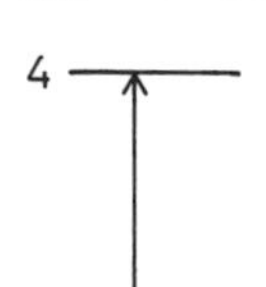

Fig. 3.27. Double resonance in four levels and collision-induced transitions

The observation of steady-state double-resonance signal in four levels of NH_3 was reported by SHIMIZU and OKA [3.34] in their initial infrared-microwave double-resonance experiment, as described in Subsection 3.3.1. Microwave absorptions of the $(J=7, K=7)$ line S' and the $(J=9, K=7)$ line S'' in Fig. 3.8 were observed to decrease when the upper-level population of the $(J=8, K=7)$ inversion doublet was depleted by the N_2O laser.

The experimental result is in agreement with the selection rules for preferred transitions by collision as established by microwave-microwave double resonance experiments [3.73a]. The preferred transitions with $\Delta J = \pm 1$, $\Delta K = 0$, and $+ \leftrightarrow -$ (parity) are shown by wavy arrows in Fig. 3.8.[6] Changes in absorption of other microwave lines corresponding to the inversion doublets of different values of J and K are being studied by KANO et al. [3.73b].

The infrared-microwave double-resonance effect in four levels of H_2CO was recently observed by TAKAMI and SHIMODA [3.74a]. In this experiment the $4_{13} \leftarrow 4_{14}$ transition in the ground state at 48.285 GHz was pumped while the change in infrared absorption was monitored with the Zeeman-tuned 3.51 μm He–Xe laser. Some decrease in population of the 4_{14} level by microwave absorption is selectively transferred to the 5_{15} level, as shown by broken arrows in Fig. 3.28. The population of the 5_{14} level likewise increases.

Instead of simply observing the change in infrared absorption, the double-resonance signal with the $5_{14} \leftrightarrow 5_{15}$ transition at 72.4 GHz (Subsection 3.3.1) was employed to detect the population change of the 5_{14} and 5_{15} levels. The 3.5 μm-72 GHz double-resonance signal I was observed to diminish. The amount of decrease of the double resonance signal ΔI with saturation of the $4_{13} \leftarrow 4_{14}$ transition gives the rate of collision-induced transitions between the 4_1 and 5_1 levels relative to the relaxation rate of the 5_1 levels.

One of the advantages of using the infrared laser is that it can probe a group of molecules within a narrow range of their whole velocity

[6] The $\Delta J = 0$, $\Delta K = 0$, $+ \leftrightarrow -$ collision-induced transitions are more frequent.

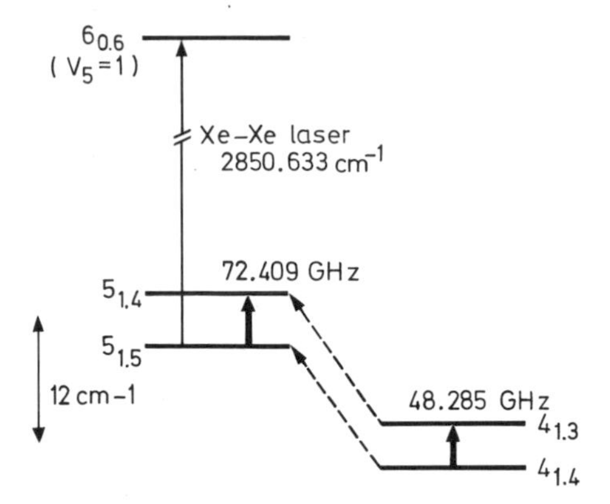

Fig. 3.28. Energy levels in H_2CO for the study of collision-induced transitions by an infrared-microwave double resonance method

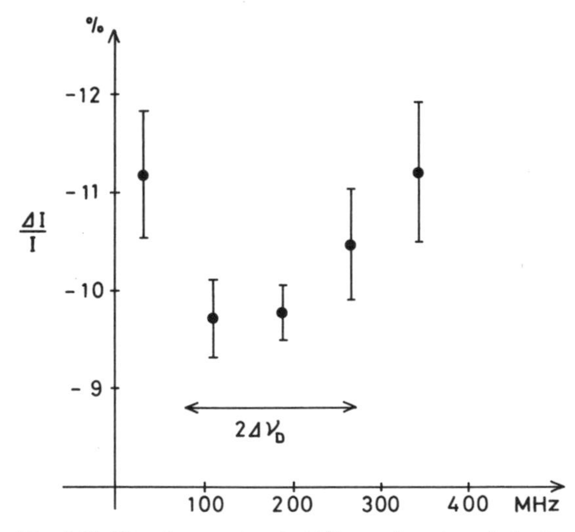

Fig. 3.29. The change signal $\Delta I/I$ as a function of the frequency of the He-Xe laser. The Doppler linewidth of H_2CO is shown by $2\Delta\nu_D$

distribution because their Doppler shift is larger than the homogeneous width.

Thus information on the velocity dependence of collision-induced transitions may be obtained by observing the relative change in signal $\Delta I/I$ as a function of the frequency of the laser. Figure 3.29 shows a preliminary result. Although the experimental scatter is rather large, as seen by error bars which are three times the standard deviations, the observed result favors the interpretation that the faster molecules have the greater probability of collision-induced transitions. The frequency corresponding to the minimum value of $\Delta I/I$ in Fig. 3.29 corresponds to the frequency difference of 180 MHz between the center frequencies of the He–Xe laser and the absorption line of H_2CO.

Time-Resolved Study of Rotational Energy Transfer

The technique of time-resolved infrared-microwave double resonance is applied to the direct observation of the rates of collision-induced transitions of molecules in gas. TAKAMI and AMANO [3.74b] are observing the transient infrared-microwave double-resonance signals in H_2CO.

A time-resolved study of infrared-microwave double resonance in $^{14}NH_3$ involving the levels of Fig. 3.8, as used in the initial experiment by SHIMIZU and OKA [3.34], has been carried out by LEVY et al. [3.75].

In this experiment the upper ($J=8, K=7$) inversion level of the ground state was depleted by a Q-switched $P(13)$ line of the N_2O laser. Temporal variations of microwave absorption were observed on the ($J=9, K=7$), ($J=8, K=6$), and ($J=7, K=4$) inversion lines in the 20 GHz region. The transients were also found in admixtures with helium in order to study NH_3–He collisions.

The time-resolved signal observed in the ($J=9, K=7$) line is shown in Fig. 3.30, which is the X-Y recorder display of the time-averaged output of a boxcar integrator of 5 nsec resolution. The signal in Fig. 3.30 corresponds to a transient decrease in microwave absorption. Addition of helium was observed to increase the effect, but the signal decreased at higher helium pressure.

A transient increase was found in the ($J=7, K=4$) line, but no signal was found in the ($J=8, K=6$) line [3.75]. All these results are in agreement with what is expected from the known selection rules for NH_3–NH_3 and NH_3–He collisions. Rate constants of transfer between the involved rotational states can be obtained.

Double-Resonance Experiments Using a Two-Photon Technique

By the use of the Raman-type two-photon transition with a fixed-frequency laser and a tunable microwave oscillator (Subsection 3.3.5) an off-resonant infrared transition can either be saturated or monitored. As the microwave frequency is tuned, a hole is burnt at different positions of the molecular velocity distribution, as described in *Double Resonance in Four Levels* above but this time without tuning the laser. This makes it possible to conveniently monitor the molecular velocity as well as the rotational states.

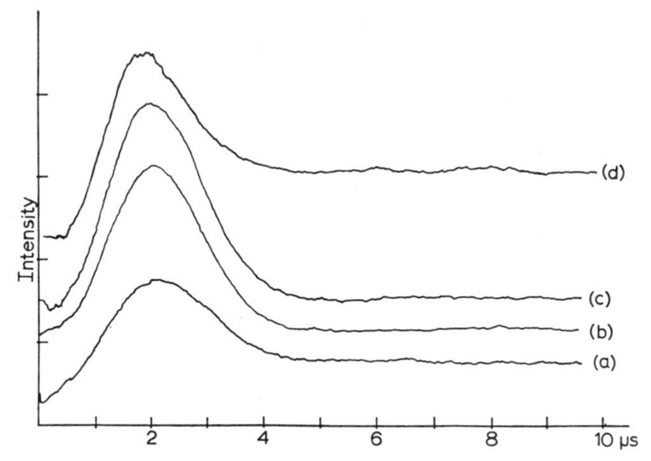

Fig. 3.30a–d. Time-resolved absorption in the $(J=9, K=7)$ line, obtained with pulsed pumping of the $\nu_2{}^qQ_-(8, 7)$ transition. Pressure of ammonia in all traces is 5.68 mTorr. Helium pressures are (a) zero, (b) 60 mTorr, (c) 209 mTorr, and (d) 543 mTorr. (Reproduced from [3.75] by permission of the author)

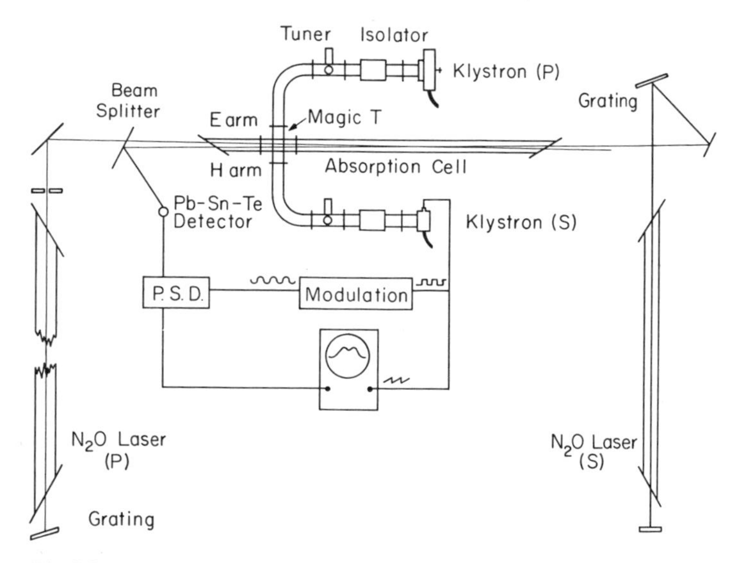

Fig. 3.31. Apparatus for two-photon double-resonance experiments

Interesting applications of double resonance, i.e. applying a technique of two-photon transition to the study of collision-induced transitions, have recently been demonstrated by FREUND et al. in NH_3 [3.76]. Figure 3.31 exhibits their experimental apparatus. A cw N_2O laser (P)

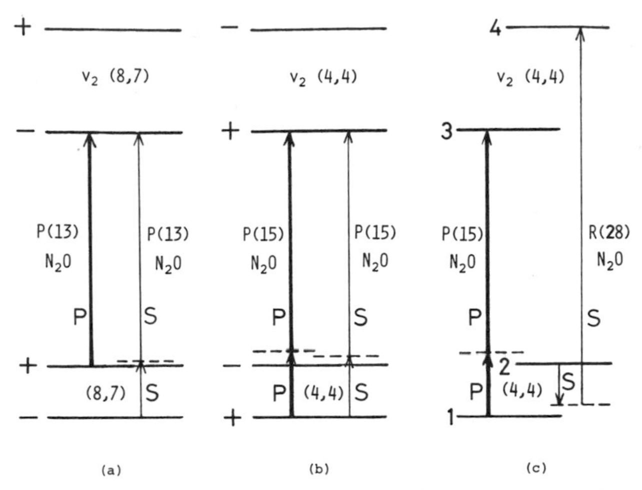

Fig. 3.32a–c. Energy levels of $^{14}NH_3$ and $^{15}NH_3$ in the $v_2 = 1$ and 0 states, showing two-photon transitions. a) $J = 8$, $K = 7$ in $^{14}NH_3$. $\Delta\omega/2\pi = 7.4$ MHz with the $P(13)$ line of N_2O. The single-photon transition is used for pumping. b) $J = 4$, $K = 4$ in $^{15}NH_3$. $\Delta\omega/2\pi = 312.1$ MHz with the $P(15)$ line of N_2O used for pumping and signal. c) $J = 4$, $K = 4$ in $^{15}NH_3$. $\Delta\omega/2\pi = 1253.1$ MHz with the $R(28)$ line of N_2O for signal. The $P(15)$ line is used for pumping

of 5 W and a klystron (P) of 10 W are used for two-photon pumping, while a smaller N_2O laser (S) of 0.5 W and a klystron (S) of 0.5 W are used to probe the signal of a two-photon transition. The absorption cell is a K-band waveguide 1.5 m in length.

Relevant energy levels of $^{14}NH_3$ and $^{15}NH_3$ in the experiment of FREUND et al. [3.76] are shown in Fig. 3.32, where the infrared transition are all in the Q-branch v_2 band, and inversion splittings are exaggerated.

The hole-burning affect of a single-photon transition P in the case of Fig. 3.32a, which is the same as Fig. 3.8, was first observed when using the two-photon transition for the probe. The hole was burnt by a two-photon transition in the case of Fig. 3.32b, which is the same as Fig. 3.23, and was monitored by another two-photon transition. Two N_2O lasers P and S were oscillating on the $P(15)$ line with a small difference in their frequencies. Directional properties were observed similar to those discussed in Subsections 3.2.1 and 3.2.2. Rotational energy transfer was involved in the observations of these experiments, but it was more directly studied in their third experiment described below.

In the level scheme of Fig. 3.32c, the $P(15)$ line of the N_2O laser P and the klystron P were used for the two-photon pumping $3 \twoheadleftarrow 1$, while the $R(28)$ line of the N_2O laser S and the klystron S were monitoring the $4 \twoheadleftarrow 2$ two-photon transition. Evidently, in this case, the presence of

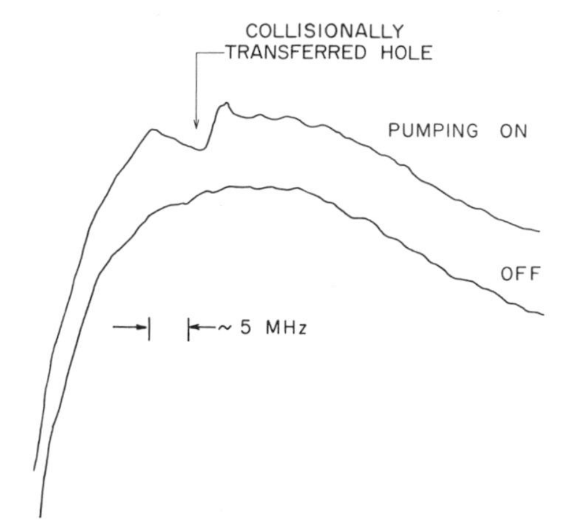

Fig. 3.33. Two-photon signals in the level scheme of Fig. 3.32c, showing the dip due to the collisionally transferred hole. (Reproduced from [3.76] by permission of the author)

the $4 \twoheadleftarrow 2$ signal transition is the result of collision-induced transitions $2 \leftrightarrow 1$ and $4 \leftrightarrow 3$ in this four-level system. Figure 3.33 shows recorder traces of the two-photon signal with and without the two-photon pumping. The observed dip moved with the frequency of either the pump laser or the pump klystron. The dip was largest at a pressure of about 25 mTorr.

One may conclude, therefore, that the hole (and spike) burnt by the two-photon pumping in the velocity distribution of the + parity state is transferred to the − parity state by collisional processes. The appearance of a sharp dip in Fig. 3.33 indicates the interesting fact that the molecule can transfer its rotational energy through weak collisions without much affecting its translational velocity. This confirms the theoretical expectation of ANDERSON [3.77].

Finally, the $(J=8, K=7)$ levels were pumped with the $P(13)$ line of the N_2O laser P, while the $(J=7, K=7)$ levels were observed with the $P(32)$ line of the N_2O laser S and the klystron S. A small signal was observed over the entire Doppler profile. Hence, the $\Delta J = \pm 1$ collision-induced transitions were found to have lower probabilities than those of $\Delta J = 0$. When the pumping was done on the $(J=8, K=7)$ levels by the $P(13)$ N_2O laser line, the signal was observed on almost all transitions. Hence these results were ascribed to the heating effect of the pumping radiation. Similar effects had been observed previously with the single-photon pumping on CH_3Cl [3.35], SF_6 [3.68, 69], and BCl_3 [3.70].

Time-resolved observations obtained by using a technique of pulsed two-photon pumping will be performed elsewhere. It may be mentioned here that the experiment of Fig. 3.32b is related to the four-wave parametric mixing experiment [3.78].

3.4.3. Coherent Transients in Double Resonance

Coherent transient effects were initially found at radiofrequencies in nuclear magnetic resonance. Many of the optical coherent transient effects have recently been observed in two-level systems with the resonant laser radiation (see also Sect. 1.4).

Self-induced transparency in gaseous molecules was observed by PATEL and SLUSHER [3.79] in 1967 in the resonant absorption of the 10.6 μm CO_2 laser pulse by SF_6. They also observed photo-echoes in SF_6 [3.80]. Optical nutation in SF_6 under a step-function excitation with the 10.6 μm CO_2 laser was observed by HOCKER and TANG [3.81] in 1968. BREWER and SHOEMAKER [3.82, 83] in 1971 developed a Stark-

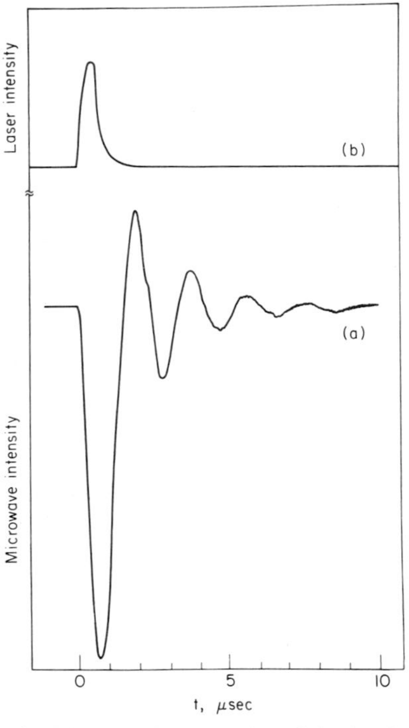

Fig. 3.34. Transient nutation of the ($J = 8$, $K = 7$) line following the Q-switched N_2O laser pulse in ammonia. The transverse relaxation time is measured as 1.8 μsec. (Reproduced from [3.72] by permission of the author)

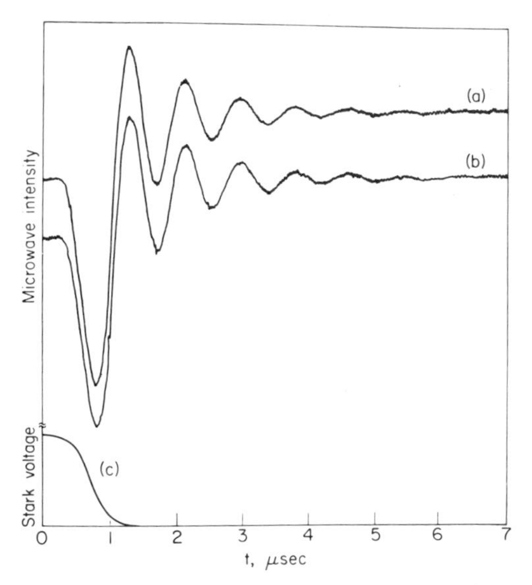

Fig. 3.35a–c. Transient nutation in ammonia following switching off of the Stark field shown by (c). The pressure of ammonia is (a) 8.8 mTorr, and (b) 8.5 mTorr. (Reproduced from [3.72] by permission of the author)

pulse technique for observing transient effects of molecular coherence at optical frequencies with much ease. Optical nutation in $^{13}CH_3F$ was observed when using a square Stark pulse while the gas was excited by a cw 9.6 μm CO_2 laser [3.82]. For a nondegenerate transition of NH_2D at the frequency of the $P(20)$ line of the CO_2 laser, optical free-induction decay was clearly observed following a step-function Stark field [3.84]. Photo-echoes were observed in $^{13}CH_3F$ when using two Stark pulses in succession [3.82].

Such coherent transients may be observed in a three-level system in which the Stark pulse is replaced by a pump laser pulse. Transient nutation in double resonance was observed in the three levels of $^{14}NH_3$ (Fig. 3.8) by LEVY et al. [3.72].

The 10.78 μm N_2O laser was Q-switched by a rotating mirror to produce pulses of less than 1 μsec duration. About 0.5 mTorr of $^{14}NH_3$ in a 1.2 m length of waveguide was irradiated by the laser pulse and the cw microwave at a frequency differing by a few MHz from the ($J=8$, $K=7$) line at 23.2322 GHz.

An example of the observed microwave nutation is shown in Fig. 3.34, where the microwave frequency offset is 1.0 MHz. For comparison, transient nutation on the same transition observed with the Stark-pulse technique of SHOEMAKER and BREWER is shown in Fig. 3.35, in which the microwave-frequency offset is 1.3 MHz. The Stark pulse cannot be switched in a time much shorter than 1 μsec, whereas the laser pulse can be as short as 1 nsec.

In the above experiment the observed coherent transient effect is essentially the phenomenon in a two-level system. The optical nutation of a real double-resonance effect will appear as expressed by (3.24). It was conjectured that this double-resonance nutation might occur in the stimulated Raman effect in the self-focused region in liquids such as CS_2 [3.14, 18]. A recent study of two-photon self-induced transparency in potassium vapor by TAN-NO et al. [3.85] is closely related to this. But no cxperimental study involving direct observation of double-resonance nutation in molecules has yet been done. Since the nutation frequency is given by (3.25), in contrast to that of the single resonance given by (3.10), it would be difficult to observe it in degenerate levels of a molecule.

Acknowledgements

The author is grateful to Dr. T. OKA, Dr. R. G. BREWER, and Professor J. I. STEINFELD for information about their work. The assistance of his colleagues at the University of Tokyo and the Institute of Physical and Chemical Research, in particular, Professor T. SHIMIZU, Dr. K. UEHARA, and Dr. M. TAKAMI, is acknowledged.

References

3.1. J. BROSSEL, A. KASTLER: Compt. Rend. **29**, 1213 (1949).
J. BROSSEL, F. BITTER: Phys. Rev. **86**, 308 (1952).
3.2. T. R. CARVER, C. P. SLICHTER: Phys. Rev. **92**, 212 (1953);
A. W. OVERHAUSER: Phys. Rev. **92**, 411 (1953);
G. FEHER: Phys. Rev. **103**, 500 and 834 (1956); **105**, 1122 (1956), Physica **24**, 580 (1958); Phys. Rev. **114**, 1219 (1959);
G. FEHER, G. A. GARE: Phys. Rev. **103**, 501 (1956); **114**, 1245 (1959);
H. SEIDEL: Z. Phys. **165**, 218 and 239 (1961).
3.3. S. H. AUTLER, C. H. TOWNES: Phys. Rev. **100**, 703 (1955).
3.4. K. SHIMODA, T. C. WANG: Rev. Sci. Instr. **26**, 1148 (1955).
3.5. A. JAVAN: Phys. Rev. **107**, 1579 (1957).
3.6. A. BATTAGLIA, A. GOZZINI, E. POLACCO: Nuovo Cimento **14**, 1076 (1959).
3.7. T. YAJIMA: J. Phys. Soc. Japan **16**, 1594 and 1709 (1961).
3.8. See, for example, A. KASTLER: In *Progress in Optics*, Vol. 5, ed. by E. WOLF (North Holland Publishing Co., Amsterdam, 1966);
R. BERNHEIM: *"Optical Pumping"* with a list of papers and reprints of selected papers (W. A. Benjamin, Inc., New York and Amsterdam, 1965).
3.9a. G. W. SERIES: In *Quantum Optics*, ed. by S. M. KAY and A. MAITLAND (Academic Press, London and New York, 1970).
3.9b. R. W. FIELD, A. D. ENGLISH, T. TANAKA, D. U. HARRIS, D. A. JENNINGS: J. Chem. Phys. **59**, 2191 (1973).

3.10. K. SHIMODA, T. SHIMIZU: "Nonlinear Spectroscopy of Molecules" in *Progress in Quantum Electronics*, Vol. 2, No. 2, ed. by J. H. SANDERS and S. STENHOLM (Pergamon, Oxford 1972).
3.11. K. SHIMODA: IEEE J. Quant. Electron. QE-**8**, 603 (1972).
3.12. K. SHIMODA: Appl. Phys. **1**, 77 (1973).
3.13. B. MACKE, J. MESSELYN, R. WERTHEIMER: J. Phys. (Paris) **30**, 665 (1969).
3.14. K. SHIMODA: Z. Physik **234**, 293 (1970).
3.15. K. SHIMODA: Japan. J. Appl. Phys. **6**, 620 (1967).
3.16. A. DI GIACOMO, S. SANTUCCI: Nuovo Cimento **63**B, 407 (1969).
3.17. D. F. WALLS: J. Phys. A (Gen. Phys.) **4**, 638 (1971).
3.18. D. F. WALLS: Z. Physik **244**, 117 (1971).
3.19. H. R. SCHLOSSBERG, A. JAVAN: Phys. Rev. **150**, 267 (1966).
3.20. G. E. NOTKIN, S. G. RAUTIAN, A. A. FEOKTISTOV: Zh. Eksper. I. Teor. Fiz. **52**, 1673 (1967) (English transl.: Soviet Phys.-JETP **25**, 112 (1967)].
3.21. M. S. FELD, A. JAVAN: Phys. Rev. **177**, 540 (1969).
3.22. T. YA. POPOVA, A. K. POPOV, S. G. RAUTIAN, R. I. SOKOLOVSKII: Zh. Eksper. I. Teor. Fiz. **57**, 850 (1969) [English transl.: Soviet Phys.-JETP **30**, 466 (1970)].
3.23. N. SKRIBANOWITZ, I. P. HERMAN, R. M. OSGOOD, Jr., M. S. FELD, A. JAVAN: Appl. Phys. Letters **20**, 428 (1972).
3.24. N. SKRIBANOWITZ, I. P. HERMAN, M. S. FELD: Appl. Phys. Letters **21**, 466 (1972).
3.25. N. SKRIBANOWITZ, I. P. HERMAN, J. C. MACGILLIVRAY, M. S. FELD: Phys. Rev. Letters **30**, 309 (1973).
3.26. S. L. MCCALL, E. L. HAHN: Phys. Rev. **183**, 457 (1969).
3.27. H. R. SCHLOSSBERG, A. JAVAN: Phys. Rev. Letters **17**, 1242 (1966).
3.28. R. G. BREWER: Phys. Letters **25**, 1639 (1970);
A. C. LUNTZ, J. D. SWALEN, R. G. BREWER: Chem. Phys. Letters **14**, 512 (1972).
3.29. A. C. LUNTZ: Chem. Phys. Letters **11**, 186 (1971).
3.30. F. SHIMIZU: Chem. Phys. Letters **17**, 620 (1972).
3.31. A. M. RONN, D. R. LIDE, Jr.: J. Chem. Phys. **47**, 3669 (1967).
3.32. J. LEMAIRE, J. HOURIEZ, J. BELLET, J. THIBAULT: Compt. Rend. **268**, 922 (1969).
3.33. M. FOURRIER, M. REDON, A. VAN LERBERGHE, C. BORDÉ: Compt. Rend. **270**, 537 (1970).
3.34. T. SHIMIZU, T. OKA: Phys. Rev. A**2**, 1177 (1970).
3.35. L. FRENKEL, H. MARANTZ, T. SULLIVAN: Phys. Rev. A**3**, 1640 (1971).
3.36. K. SAKURAI, K. UEHARA, M. TAKAMI, K. SHIMODA: J. Phys. Soc. Japan **23**, 103 (1967). Note that the notation of the normal vibration has been changed into the one now generally accepted.
3.37. M. TAKAMI, K. SHIMODA: Japan. J. Appl. Phys. **10**, 658 (1971).
3.38. K. SHIMODA, M. TAKAMI: Opt. Commun. **4**, 388 (1972).
3.39. M. TAKAMI, K. SHIMODA: Japan. J. Appl. Phys. **11**, 1648 (1972).
3.40. J. L. HALL: In *Proc. 3rd Intern. Conf. Atomic Physics*, Boulder, August 1971 (Plenum Press, New York, to be published).
3.41. M. TAKAMI, K. SHIMODA: Japan. J. Appl. Phys. **12**, 603 (1973).
3.42. M. TAKAMI: Personal communication.
3.43. Y. UEDA: Personal communication.
3.44. K. SHIMODA: IEEE Trans. Instr. Meas. IM-**17**, 343 (1968).
3.45. R. L. BARGER, J. L. HALL: Phys. Rev. Letters **22**, 4 (1969).
3.46. M. MIZUSHIMA, P. VENKATESWARLU: J. Chem. Phys. **21**, 705 (1953).
3.47. K. T. HECHT: J. Mol. Spectrosc. **5**, 355 and 390 (1960).
3.48. K. UEHARA, K. SAKURAI, K. SHIMODA: J. Phys. Soc. Japan **26**, 1018 (1969).
3.49. A. C. LUNTZ, R. G. BREWER: J. Chem. Phys. **56**, 3641 (1971).
3.50. W. L. BARNES, J. SUSSKIND, R. H. HUNT, E. K. PLYLER: J. Chem. Phys. **56**, 5160 (1972).

3.51. A.J.DORNEY, J.K.G.WATSON: J. Mol. Spectrosc. **42**, 135 (1972).
3.52. R.F.CURL, Jr., T.OKA: J. Chem. Phys. **58**, 4908 (1973).
3.53. M.TAKAMI, K.UEHARA, K.SHIMODA: Japan J. Appl. Phys. **12**, 924 (1973).
3.54. R.F.CURL, Jr., T.OKA, D.S.SMITH: J. Mol. Spectrosc. **46**, 518 (1973).
3.55. J.K.G.WATSON: J. Mol. Spectrosc. **40**, 536 (1971).
3.56. K.FOX: Phys. Rev. Letters **27**, 233 (1971).
3.57a. R.F.CURL: J. Mol. Spectrosc. **48**, 165 (1973).
3.57b. N.SKRIBANOWITZ, M.J.KELLY, M.S.FELD: Phys. Rev. A**6**, 2302 (1972).
3.58. T.OKA, T.SHIMIZU: Appl. Phys. Letters **19**, 88 (1971).
3.59. S.M.FREUND, T.OKA: IEEE J. Quant. Electron. QE-**8**, 604 (1972).
3.60. S.M.FREUND, T.OKA: Appl. Phys. Letters **21**, 60 (1972).
3.61. M.MARGOTTIN-MACLON, L.DOYENNETTE, L.HENRY: Appl. Opt. **10**, 1768 (1971).
3.62. J.I.STEINFELD: MTP International Review of Science, Physical Chemistry, Series I, vol. 9: *Gas Kinetics*, ed. by J.C.POLANYI (Butterworth, London, 1972), p. 247.
3.63. C.B.MOORE: Adv. Chem. Phys. **23**, 41 (1973).
3.64. E.S.YEUNG, C.B.MOORE: J. Chem. Phys. **58**, 3988 (1968).
3.65. C.K.RHODES, M.J.KELLY, A.JAVAN: J. Chem. Phys. **48**, 5730 (1968).
3.66. R.D.BATES, Jr., G.W.FLYNN, A.M.RONN: J. Chem. Phys. **49**, 1432 (1968).
3.67. C.V.SHANK, A.DIENES, W.T.SILFVAST: Appl. Phys. Letters **17**, 307 (1970).
3.68. I.BURAK, A.V.NOWAK, J.I.STEINFELD, D.G.SUTTON: J. Chem. Phys. **51**, 2275 (1969).
3.69. J.I.STEINFELD, I.BURAK, D.G.SUTTON, A.V.NOVAK: J. Chem. Phys. **52**, 5421 (1970).
3.70a. P.L.HOUSTON, A.V.NOWAK, J.I.STEINFELD: J. Chem. Phys. **58**, 3373 (1973).
3.70b. H.JETTER, E.F.PEARSON, C.L.MCGURK, W.H.FLYGARE: J. Chem. Phys. **59**, 1796 (1973).
3.71. F.R.GRABINER, D.R.SIEBERT, G.W.FLYNN: Bull. Am. Phys. Soc. **17**, 573 (1972).
3.72. J.M.LEVY, J.H.S.WANG, S.G.KUKOLICH, J.I.STEINFELD: Phys. Rev. Letters **29**, 395 (1972).
3.73a. T.OKA: J. Chem. Phys. **48**, 4919 (1968);
T.OKA: J. Chem. Phys. **49**, 3135 (1968);
P.W.DALY, T.OKA: J. Chem. Phys. **53**, 3572 (1970);
T.OKA: In *Advances in Atomic and Molecular Physics*, Vol. 9 (1973), p. 127.
3.73b. S.KANO, T.AMANO, T.SHIMIZU: Chem. Phys. Letters **25**, 119 (1974).
3.74a. M.TAKAMI, K.SHIMODA: Japan. J. Appl. Phys. **12**, 934 (1973).
3.74b. M.TAKAMI, T.AMANO: Private communication.
3.75. J.M.LEVY, J.H.S.WANG, S.G.KUKOLICH, J.I.STEINFELD: Chem. Phys. Letters **21**, 598 (1973).
3.76. S.M.FREUND, J.M.C.JOHNS, A.R.W.MCKELLER, T.OKA: J. Chem. Phys. **59**, 3445 (1973).
3.77. P.W.ANDERSON: Phys. Rev. **76**, 647 (1949).
3.78. P.P.SOROKIN, J.J.WYNNE, J.R.LANKARD: Appl. Phys. Letters **22**, 342 (1973).
3.79. C.K.N.PATEL, R.E.SLUSHER: Phys. Rev. Letters **19**, 1019 (1967).
3.80. C.K.N.PATEL, R.E.SLUSHER: Phys. Rev. Letters **20**, 1087 (1968).
3.81. G.B.HOCKER, C.L.TANG: Phys. Rev. Letters **21**, 1151 (1968).
3.82. R.G.BREWER, R.L.SHOEMAKER: Phys. Rev. Letters **27**, 631 (1971).
3.83. R.G.BREWER, R.L.SHOEMAKER: Phys. Rev. **6**A, 2001 (1972).
3.84. R.G.BREWER: Science **178**, 247 (1972). This is a review article on nonlinear spectroscopy.
3.85. N.TAN-NO, K.YOKOTO, H.INABA: Phys. Rev. Letters **29**, 1211 (1972).

4. Laser Raman Spectroscopy of Gases

J. M. CHERLOW and S. P. S. PORTO

With 12 Figures

In 1928, RAMAN [4.1] in India, and LANDSBERG and MANDELSTAN [4.2], in Russia, discovered what is now known as the Raman effect. That is, they observed the existence of light scattered from a material which had a wavelength different from that of the exciting light. In fact, such an effect had already been predicted by SMEKAL [4.3] in 1923. The earliest Raman work concerned liquids and solids, but soon after the initial discovery RAMDAS [4.4] observed vibrational scattering in ethyl ether vapor and RASETTI [4.5] reported observations of the rotational Raman spectra of the simple gases, N_2, O_2, and H_2. Almost at the same time WOOD [4.6] reported studies of the dipolar gas HCl. Many investigations of other simple systems soon followed. The theoretical study of these phenomena culminated in the polarizability theory of PLACZEK [4.7] and PLACZEK and TELLER [4.8]. This theory satisfactorily explained the observations up till that time and has provided an invaluable framework for further work. In addition to the original sources this theory is described by the texts of BHAGAVANTAM [4.9], HERZBERG [4.10], and WILSON et al. [4.11].

After the period of early interest, little was published in the field of Raman spectroscopy of gases during the years 1935 through 1951. This lack of activity resulted from the fundamental experimental problem of this field: Raman scattering from gases is a weak effect. Primarily because of the reduced density, but also because of the absence of local field effects, Raman scattering from gases is much weaker than that from liquids and solids. The resurgence in interest in the early 1950's resulted from the development of the high intensity mercury lamp called the "Toronto Lamp" by WELSH [4.12] and his collaborators. With this lamp used at the sharp Hg 4358 Å line, rotation and vibration spectra of low pressure gases could be photographed with high-resolution spectrographs and exposure times of a few hours; observation of the much weaker vibration-rotation spectra required both higher pressures and longer exposures. STOICHEFF [4.13, 14] has reviewed the work of this period, which resulted in the accurate measurement of molecular rotational and vibrational constants for a number of molecules. Also during this period BERNSTEIN and collaborators [4.15] were active in measuring absolute cross-sections of gases.

Another significant innovation during this decade was the first use of reasonable quality photomultipliers for the detection of Raman scattering. Although photographic detection remains far superior for very high resolution Raman spectroscopy, photoelectric detection with a scanning spectrometer possesses great advantages in convenience as well as in the quantitative measurement of intensities. Further development has brought about the current high-quality photomultiplier tubes which have greatly increased sensitivities.

The invention of the laser revolutionized the whole field of Raman spectroscopy. (PORTO and WOOD [4.16] in 1962 reported the first observation of Raman scattering using a laser source.) Its high power, narrow linewidth, directionality, and polarization gave the laser a tremendous advantage over the mercury lamp as an exciting source for Raman spectroscopy. The narrow linewidths are particularly helpful in the study of gases, since the transitions under study are much sharper in gases than in liquids or solids. The Hg 4358 Å line has a width of $0.2\,cm^{-1}$ while the single mode gas lasers have linewidths of $0.001\,cm^{-1}$. Hence, the limiting factor of the resolution is no longer the width of the exciting line. Also the complete polarization of the laser has permitted the accurate measurement of the low depolarization ratios of almost completely polarized Raman vibrational lines. With mercury lamp sources such accurate measurements were impossible. Further, the high power available with laser sources permitted the possibility of measurements using small volumes of gas.

In this chapter, we will discuss the various types of Raman scattering which have been performed with laser sources in gases. We divide these into two classes. The first includes those experiments which utilize the high power and narrow linewidth of the laser for high resolution measurements. The second includes those which take particular advantage of the directionality and polarization properties for the accurate measurement of total cross-sections and depolarization ratios. For both of these classes we emphasize the various experimental methods. Specific results are discussed only for simple gases. The analysis of complex molecules by the combined application of Raman and infra-red spectroscopy is beyond the scope of this review even though such analysis was the ultimate goal during the development of some of the techniques described here.

4.1. High Resolution Spectroscopy

4.1.1. Techniques

The first reports by WEBER and collaborator of Raman experiments in gases using laser sources were devoted to the demonstration of the feasibility of the technique [4.17, 18]. In 1965 their first paper [4.17] reported the successful use of the He–Ne laser to produce pure rotational Raman spectra of gases and exhibited the spectrum of methylacetylene. They employed a relatively simple experimental system. A Raman tube with Brewster angle windows was placed inside the laser cavity and the laser beam traversed the gas held in the tube. A Dove prism and condensing lenses were used to produce a vertical image of the horizontal luminous gas column on the spectrograph slit. It was necessary for them to place a cylindrical lens in front of the photographic plate in order to photograph pure rotational Raman spectra with reasonable exposure times. With this simple system they obtained spectra with about 1/10 of the intensity available when using mercury lamp techniques. They pointed out that the major difference between their method and the then-conventional system of mercury lamps and a multipass cell was in the effective sample volume used. Their method used an effective volume of 0.59 cm^3 while the mercury lamp method required an effective volume of the order of 3000 cm^3.

In their second paper WEBER et al. [4.18] described more sophisticated techniques which enabled them to record improved spectra both photographically and photoelectrically. For photographic measurements they replaced their earlier single-pass Raman cell with a multiple pass tube that was used inside the cavity of a He–Ne laser. This multiple pass tube illustrated in Fig. 4.1 used two flat mirrors to fold the laser beam about 9 times through the tube and also a set of four concave spherical mirrors to collect the scattered radiation. These spherical mirrors were arranged in a configuration similar to that in multiple pass Raman gas cells used with mercury lamp sources. These have been described by WELSH et al. [4.19, 20], STOICHEFF [4.14, 21], and will be discussed below. WEBER et al. [4.18] pointed out that the intensity of the strong Rayleigh line can be greatly reduced with respect to the weak rotational Raman lines by observing the scattered radiation in a direction parallel to the polarization (electric vector) direction of the laser radiation. In this direction, the Rayleigh line will vanish to a degree described by its low depolarization ratio [4.7, 9]. Photographic spectra of pure rotational spectra taken with these techniques were superior to those obtained with mercury lamp sources.

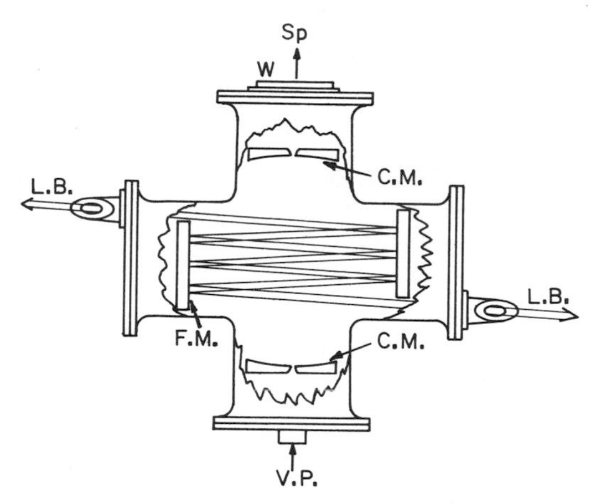

Fig. 4.1 Multiple-pass Raman tube. Laser beam enters and leaves through Brewster-angle windows. CM are concave spherical mirrors which collect the scattered light [4.18]

Weber et al. [4.18] also reported photoelectric measurements of the pure rotational Raman spectra of simple gases (N_2, O_2, CO_2) using the more powerful (4880 nm and 514.5 nm) argon ion laser as the exciting source. This laser was just then coming into use. Those measurements were made with a simple unfolded Raman tube and either a single or double monochrometer; phase sensitive detection was employed. They obtained an increase in signal-to-noise ratio of about a factor of seven by placing the cell outside the cavity and having it resonate externally with the laser. In both cases the laser beam was focussed into the Raman cell. Two of their experimental configurations are shown in Fig. 4.2. The results of this paper demonstrated the obsolescence of the mercury arc as the source of exciting radiation for Raman experiments.

In 1968, Barrett and Adams [4.22] reported further improvements in optical and electronics techniques for laser Raman scattering in gases. They circumvented the major difficulty in the use of a multi-pass cell-collecting *a* the Raman scattered light over a large volume—by strongly focussing laser beam into a small volume gas sample. They showed, both theoretically and experimentally, that, for this case, it is possible to obtain a magnified image of the Raman sample so that all of the Raman light emitted into a large solid angle can be transmitted through the slit and aperture stop of a spectrometer. Their calculations demonstrated that the amount of Raman scattered radiation which was collected increased with increasing magnification of the collection optics. The calculation further showed that there is an optimum value for the

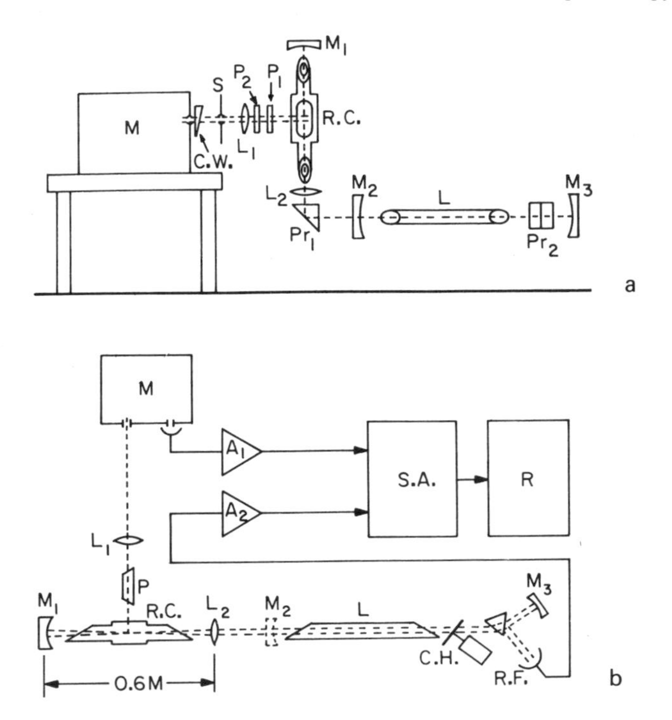

Fig. 4.2a and b. Photoelectric Raman apparatus. In (a) the Raman cell RC is positioned vertically and in (b) horizontally. In both L is an argon laser, M a monochromator, L_1 and L_2 lenses, P_1 and P_2 polarizers. In each case the Raman cell could be inside or outside the laser cavity, (inside-cavity formed by M_1 and M_3, outside-cavity formed by M_2 and M_3). For external operation the mirror M_1 may be adjusted to externally resonate the beam through the Raman cell. The calcite wedge cw depolarizes the radiation before it enters the monochromator. In (b) D is a Dove prism. In (a) the polarization of the laser is perpendicular to the plane of the figure and in (b) parallel [4.18]

focusing angle; this value corresponds to fairly strong focusing. We should point out here that this method of strong focusing and large magnification is inappropriate for the measurement of depolarization ratios of nearly completely polarized bands because it introduces a large systematic error.

In addition to their optical system, these authors describe a photon-counting detection system (based on a suggestion by PORTO [4.23]) which gave superior results to the previous phase-sensitive detection scheme. This system enabled them to make photoelectric observation of resolved vibration-rotation spectra of gases. Figure 4.3 shows their spectrum for nitrogen and clearly indicates the alternation of intensities for the rotational lines. (These vibration-rotation bands had been

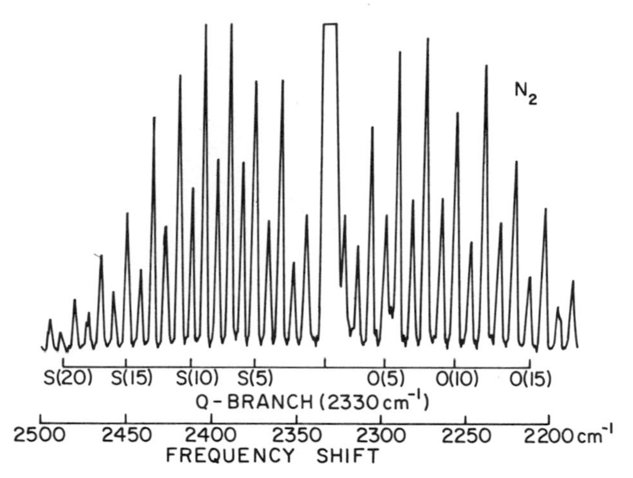

Fig. 4.3. Rotation-vibration Raman band in N_2. The spectrum clearly shows the alternation of intensities for the rotational lines [4.22]

studied earlier by STOICHEFF [4.24] using photographic detection and a mercury lamp source.)

For those measurements it was not necessary to contain the common atmospheric gases N_2, O_2, and CO_2 in a Raman cell; a nozzle shot a stream of the gas through that point in the laser cavity at which the beam was brought to a focus. The authors estimated that the effective scattering volume was only about 10^{-8} cm^3 and contained about 10^{11} molecules.

BARRETT and WEBER [4.25] further refined this experimental technique for a study of the pure rotational Raman scattering in a CO_2 electric discharge. The results of this study will be discussed in the next section. For this experiment a focusing angle of 0.1 rad was optimal; however, in order to have a large enough working space in the Raman sample area, a negative lens was used to expand the 3 mm diameter laser beam to a diameter of 15 mm at the focusing lens. The total length of the cavity was 2.8 m and the laser power in the Raman sample was estimated to be 20 W.

The authors [4.25] describe a sophisticated method to eliminate errors arising from the fluctuation of the laser power due to mechanical instabilities of the long optical resonator. A thin beam splitter was positioned just behind the entrance slit of the monochromator so that a small fraction of the incoming light was sampled by a reference photomultiplier. The signal from this photomultiplier was then used to control the motion of the steppable grating. Each step of the grating corresponded to an equal number of photons entering the entrance slit of the

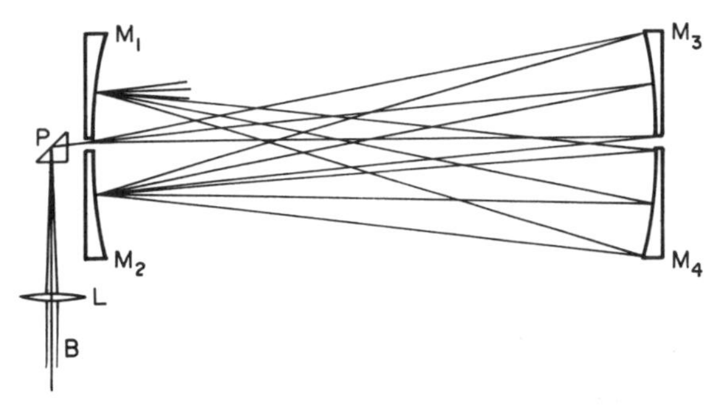

Fig. 4.4. Schematic diagram of multiple reflection Raman tube before completion and alignment. (Plan view, not to scale) B is parallel beam from laser, L: lens, P: total reflection prism, M_1, M_2, M_3, M_4: semicircular concave mirrors [4.27]

spectrometer. These authors also describe a method for computer smoothing of the experimental data.

In 1971, RICH and WELSH [4.26, 27] reported an improved multiple reflection Raman cell. The mirror system illustrated in Fig. 4.4 is used to multiply pass the exciting laser beam lengthwise through the cell, and at the same time, to bring a large amount of scattered light into the cone of the spectrometer. The four semicircular concave mirrors are placed such that the "front" pair (M_1, M_2) have their centers of curvature coincident at the midpoint of the narrow slot between the back pair (M_3, M_4) whose centers of curvature lie at the diagonally opposite edges of the narrow slot formed by the separation of the straight edges of the front pair. The figure illustrates the cell with the mirror not in their positions of final adjustment, but at an intermediate stage. The incoming laser beam is focussed at the front slit after reflection by the prism P. The diverging beam fills the mirror at the back (successive traversal are shown for only one-half of the cone). Focussed successive reflections at the front end approach the outer edges of the mirrors until the beam finally "walks off" the edge. Their cell allowed $\simeq 50$ traversals; the limiting factor (besides mirror reflectivity) is the ratio of the front slot width to the mirror diameter. Scattered light originating at any point in the cell which strikes the mirrors at either end emerges through the front slot. A series of lenses images this slot onto the entrance slit of the spectrometer. (The collimator of the spectrometer is simultaneously imaged on M_3 and M_4.) The prism P is thin with respect to the slot and blocks very little of the scattered radiation. This complex, large (1 m long) cell is necessarily used external to the laser cavity; RICH and WELSH point out that the advantages of a small sample volume stressed by BARRETT and ADAMS [4.22] are not important for common easily available gases. They reported excellent results in resolving the electronic Raman structure of NO as well as in resolving the vibration-rotation structure of methane, cyclopropane, and acetylene. The vibration-rotation spectra which they exhibit were obtained photoelectrically. As a result the frequency calibration was not sufficiently accurate over a wide range in order for them to obtain truly useful new information about molecular structure.

WEBER and SCHLUPP [4.28] have recently reported the development of an extremely high resolution Raman spectrometer useful for studying low pressure gases. They employ a multiple pass Raman cell placed within the cavity; the cell is similar to that used earlier by WEBER et al. [4.18] (Fig. 4.1). One major innovation in their method is using an argon-ion laser in a single longitudinal mode configuration. Placing a Fabry-Perot etalon within the cavity reduces the linewidth of the laser from 0.15 cm^{-1} to less than 0.001 cm^{-1}. This reduction in linewidth yields the possibility of greatly increased resolution; the width of the exciting line is no longer the limiting factor. Careful temperature and humidity control are required to maintain the frequency stability of the etalon. Barometric pressure control, however, over the long (10 hrs) experimental periods is impossible; these authors solve this problem by varying the etalon temperature so as to compensate the effects of a change in barometric pressure. This procedure allows compensation over a préssure range of 10 Torr. One disadvantage in the use of the single mode argon laser is the lack of *a priori* knowledge of the exact value of the exciting frequency. Averaging of Stokes and anti-Stokes displacements eliminate this as a concern in pure rotational scattering. However, for an exact determination of a vibrational frequency it is necessary to measure the wave number of the exciting line for each exposure independently.

The measured resolving power of their spectrograph at 488.0 nm is 700000. With this spectrograph and the frequency stabilized argon laser WEBER and SCHLUPP [4.28] were able to observe the resolved rotation spectra of cyanuric fluoride, 1,3,5-trifluorobenzene, hexafluorobenzene, tetramethylallene, and other substances. The resolution of the J-odd lines of the R brand of hexafluorobenzene with a spacing of 0.07 cm^{-1} represents the best resolution obtained thus far in Raman spectroscopy with a 90° scattering configuration. They exhibit the rotational spectrum of cyclohexane and demonstrate its superiority to a similar spectrum taken with mercury lamp excitation. For this and the other rotational spectra the Rayleigh scattering is discriminated against by the method described previously.

Several other authors have also reported systems which are designed to enhance the generally weak Raman scattering from gases. NEELY et al.

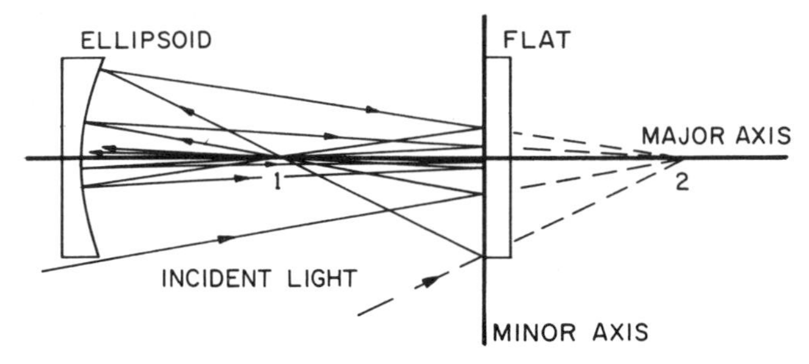

Fig. 4.5. Schematic of light trapping cell [4.30]

[4.29] have described a simple method for modifying a commercial argon-ion laser so that it can be used for intra-cavity excitation of gas samples. Their system is similar to the unfolded intra-cavity cell described earlier by WEBER et al. [4.18], and BARRETT and ADAMS [4.22]; their report includes, however, a useful procedure for aligning the optical components used to extend the laser cavity. The essential parts of the system are: 1) an antireflection coated quartz lens placed at a distance equal to its focal length from the sample cell; 2) gas cell with Brewster's angle windows; and 3) a dielectrically coated spherical mirror placed at a distance equal to its radius of curvature away from the sample center. Although not critical, the mirror radius is usually made approximately equal to the focal length of the lens. A calculation has shown that 10–15 cm is an optimum focal length of the lens, compromising between tight focussing and laser stability. One or more dielectrically coated plane mirrors may be placed in the system so as to orient the gas cell in the proper configuration.

HARTLEY and HILL [4.30] have reported the design of a light trapping cell which increases the amount of Raman scattered light by producing a very high flux of laser light at the point of observation. Their system makes use of a unique property of ellipsoidal mirrors; that is, light brought to one focus will be reflected alternately between the two foci and eventually collapse to the major axis. A schematic of the light trapping cell is shown in Fig. 4.5. It consists of an on-axis ellipsoidal mirror and a flat mirror positioned such that its face is coincident with the minor axis of the ellipsoid and its normal is parallel to the major axis. Because the flat mirror reflects the rays that approach focus 2 back through focus 1, a large number of images are focussed at a single focus. The light trapping cell can then be set up so that the bright focal region is imaged onto the spectrometer slit (major axis imaged parallel to slit). The authors give a mathematical description of the behavior of the cell. With the use of the cell they reported a gain of a factor of 23 in the intensity of the Q branch vibrational Raman line of atmospheric nitrogen, when compared with a measurement not using the cell.

In a Raman experiment, to a first approximation the scattering volume can be taken as a cylinder whose axis lies along the laser beam. On this basis MORET-BAILLY and BERGER [4.31] have pointed out that the most efficient scattering direction is only a few degrees from the direction of the laser beam; that is for isotropic scattering the scattering volume observed by a spectrograph is inversely proportional to the sine of the angle between the incident and scattered directions. Using this principle they constructed a long 1 m cell which they placed inside the cavity of an argon-ion laser and constructed a WHITE [4.32] type multiple reflection collection system of three mirrors oriented only a few degrees from the laser axis [4.32]. An advantage of this system is that the collection volume can be increased simply by increasing the length of the cell. With this system they obtained good rotational spectra.

CHAPPUT et al. [4.33] modified this idea somewhat and constructed a "semi-longitudinal" Raman scattering cell to be used external to the laser cavity. An external cell is preferable for the study of corrosive gases since in an intra cavity configuration any slight degradation of the cell window would cause the laser power to decrease greatly. With the external system forward and backward scattering can be separated while for the intra-cavity system they are observed simultaneously.

Recently, BERGER et al. [4.34] have reported an improvement of their original experimental system. In order to increase the flux of the scattered light they used a set-up in which the laser is folded inside the scattering cell. That is, the laser cavity itself forms a White's system which lies within the collection system. This configuration is very large; the scattering cell is 7.5 m long and the effective laser cavity using five passes through the White system is 30 m. With this optical configuration and with two Fabry-Perot etalons inside the laser cavity they have been able to study the weak vibration-rotation spectrum of methane. They estimated their working resolution as 0.15 cm^{-1}.

KIEFER et al. [4.35] have very recently reported an external multi-pass cell, as illustrated in Fig. 4.6. The system uses an off-axis external resonator formed by the two concentric spherical mirrors M_1 and M_2. The incident laser beam is focused very close to the center of M_1 and M_2 and is then reflected and refocussed by M_2. The beam reflects alternately from the two mirrors and passes two focal points, one on each side of the center of the two spheres. The gas cell is a cylinder with Brewster-angle windows. The scattered light is collected at 90° with the spherical mirror M_3 and the lens L_2. The gain factor of the whole system was about 20 relative to a single focussed beam. With this

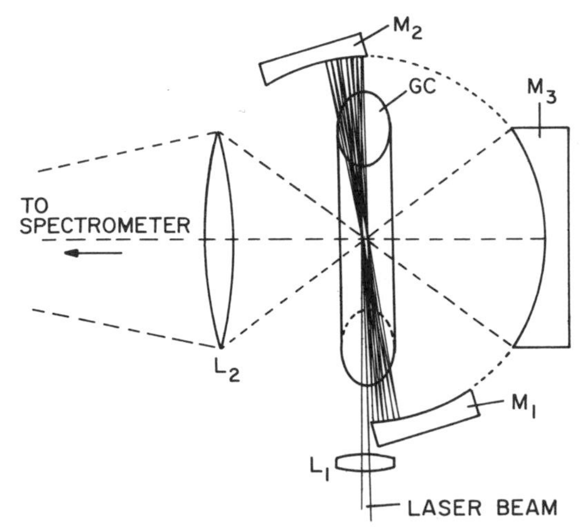

Fig. 4.6. Schematic diagram of Raman multipass system. Spherical mirrors M_1, M_2 with radii: 4 cm and diameter: 2.5 cm, M_3 with radius: 4 cm and diameter: 5 cm. Lens L_1, L_2 with focal length of 5 cm. GC: gas cell [4.35]

system they studied the vapor phase Raman spectrum of trimethylene oxide.

A commercial multipass gas cell and mirror assembly is available from the Cary division of Varian. This system is also based on the concept of a resonating pair of off-axis spherical mirrors.

The prime problem in high resolution Raman spectroscopy is that of achieving very high resolution without sacrificing the sensitivity necessary to observe the weak Raman signals. As an alternative to the conventional double spectrometer or spectrograph a possible solution to this problem lies in the use of a Fabry-Perot interferometer as the dispersing element. Several different methods for using a Fabry-Perot interferometer have been developed. These have been reviewed by JONES [4.36], but we shall describe them briefly here. One method, described by BUTCHER et al. [4.37] uses a fixed path-length Fabry-Perot etalon in conjunction with a spectrograph. This method increases the precision in the measurement of individual Raman shifts while retaining the full aperture of the spectrograph of relatively short focal length. It allows the coverage of a large spectral range although it does not make full use of the resolution capability of the Fabry-Perot etalon.

The apparatus of BUTCHER et al. [4.37] is shown schematically in Fig. 4.7. An unfolded Raman tube was used inside the cavity. The lens L_1 collimates the radiation scattered from the Raman cell; the two lens arrangement gives a point to point correspondence between the source

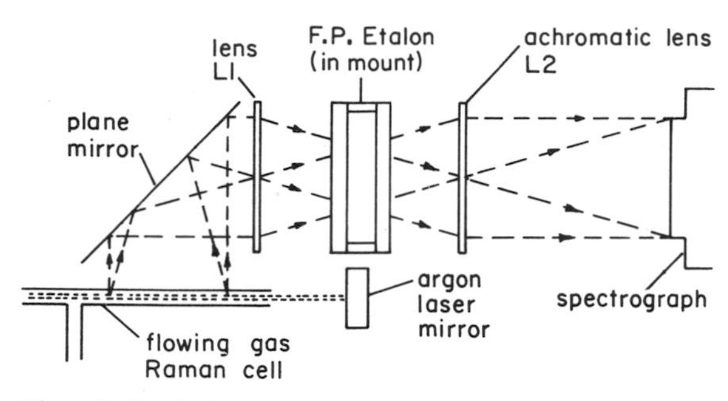

Fig. 4.7. Optical arrangement of Fabry-Perot etalon and spectrograph used for photographing rotational Raman spectra of oxygen and nitrogen. Lenses L_1 and L_2 have focal length of $\cong 2.5$ cm [4.36]

and the slit, and ensures that the Fabry-Perot etalon is illuminated by parallel light from the laser. The 476.5 nm line of the argon laser was used in these measurements because the inherent width of this line is less than that of any of the other argon laser lines. Since in the experiments of Butcher et al. the wavelengths of the various Raman rotational lines studied were quite well known to start with, the only purpose of the spectrograph is to separate the individual lines so that the set of fringes from one line does not overlay and mask the set of fringes of an adjacent line. The calculation of the wavelength of the scattered radiation depends only on the separation of the fringes corresponding to a single line. The calculation is based on the simple equation which gives the condition for the production of a maximum intensity in the fringe pattern

$$n\lambda = 2d\cos\theta\,. \tag{4.1}$$

Here n is the order of interference, d the etalon spacing and λ the air wavelength of the emission line.

The etalon used had a free spectral range of $0.4\,\mathrm{cm}^{-1}$. This range was larger than the width of the Raman lines under study. The etalon had a relatively low (70–80%) reflectivity in order not to reduce the transmission of the weak signal. The observed spectral resolution of this system was $0.07\,\mathrm{cm}^{-1}$.

With this system Butcher et al. [4.37] measured the rotational constants of oxygen and nitrogen. Their values for the B_0 and D_0 rotational constants for nitrogen were an order of magnitude more accurate than previous Raman measurements, and their results for oxygen are in good agreement with results obtained by other techniques.

The study of the profile of spectral lines is better carried out using photoelectric rather than photographic recording techniques. For these types of measurements a pressure-scanned Fabry-Perot interferometer, as used by CLEMENTS and STOICHEFF [4.38] is well suited. It is particularly useful when the transition studied can be isolated from neighboring transitions. In this instrument measurements are usually made in the forward direction with $\theta = 0$. By varying the gas pressure between the plates of the interferometer, the index of refraction can be varied. Hence, by (4.1) the air wavelength transmitted through the interferometer can be continuously changed. Different frequencies can be transmitted through the aperture free of overlay from these same frequencies until the index of refraction has changed sufficiently so that they will be transmitted in adjacent orders. Measurements with a scanning interferometer yield good spectral profiles since the instrumental profile can be determined very precisely by studying a narrow spectral line.

CLEMENTS and STOICHEFF [4.38] used a specially designed He–Ne laser for their measurements; it was a low-gain oscillator with a wide (15 mm) tube and a 2% transmission output mirror. The laser produced 400 mW with a line width of 0.025 cm^{-1}. They employed either transmission filters and/or a monochromator between the sample cell and the interferometer in order to reduce the background. With this system CLEMENTS and STOICHEFF studied the line profile of the breathing mode in methane as well as pressure broadening of the rotational lines in hydrogen. Some of their results will be discussed further below. Current developments of piezoelectrically driven scanning interferometers may well lead to large improvements in this technique.

BARRETT and MYERS [4.39] have developed a means of studying the rotational Raman spectra of gases which takes advantage of a particular property of a Fabry-Perot interferometer. Regularly spaced spectral lines can be simultaneously transmitted by a Fabry-Perot interferometer when the free spectral range is equal to the frequency difference between adjacent spectral lines. If the small effects of centrifugal distortion are ignored the rotational Raman lines satisfy this criterion. The individual rotational lines occur at wavenumbers

$$\nu = \nu_0 + 4B(J + 3/2) \qquad J = 0, 1, 2, \ldots . \tag{4.2}$$

Adjacent lines in the Stokes and anti-Stokes branches have constant separation $4B$ and the separation between the first Stokes and first anti-Stokes Raman line is $12B$, an integral multiple of $4B$. However, the unshifted Rayleigh line is separated from the adjacent rotational Raman lines by $6B$, a half integral multiple. Hence, with the proper setting of the free spectral range, that is of the plate separation, the Fabry-Perot interferometer can be made to pass all of the rotational Raman lines but to reject completely the Rayleigh line. From a measurement of the plate separation corresponding to the maximum transmission of the rotational Raman lines the rotational constant B can be calculated. One advantage of this technique is that the signal being measured, the total intensity of all the rotational Raman lines, is fairly large. To demonstrate the accuracy of the method BARRETT and MYERS measured the rotational constant for nitrous oxide and obtained a value in good agreement with microwave resonance results.

4.1.2. New Results

The study of DICKE [4.40], narrowing of the Raman lines of hydrogen, provides an excellent example of the use of high resolution Raman spectroscopy of gases to obtain new physical information. Dicke, or motional, narrowing causes a decrease in the normal Doppler width of a spectral line when the mean free path in the gas becomes comparable to the wavelength of the emitted radiation. The narrowing is a manifestation of the effect of collisions on the transitional states of the molecules and within a certain density region increases with increasing density. Dicke narrowing of the rotational Raman lines of hydrogen has been studied in detail by MAY and collaborators [4.41–44], and by CLEMENTS and STOICHEFF [4.38], while narrowing of the vibrational line has been investigated by MURRAY and JAVAN [4.45]. In some of these experiments the complementary phenomenon of collisional broadening was also studied. With higher pressures the effect of collisions is to broaden the line by perturbing the "internal motion" of the molecules. This effect arises primarily from anisotropic intermolecular forces which can, in collisions, modify the phase or frequency of the molecular motion. Rotationally inelastic collisions or collisions which change the orientation of the angular momentum or shift the phase of rotation broaden the rotational lines. A similar phenomenon exists for vibrational lines and also because of the rotation-vibration interaction, rotationally inelastic collisions can indirectly lead to a broadening of the Q branch. The linewidth measurements in all these experiments were performed with pressure-scanning Fabry-Perot interferometers similar to those described above.

The combined effects of Dicke narrowing and collisional broadening for Raman scattering from gases at density ϱ such that the product of the mean free path Λ and the momentum transfer q is much less than one yield a Lorentzian lineshape with a half width $\Delta\nu_{\frac{1}{2}}$ given by

$$\Delta\nu_{\frac{1}{2}} = \Delta\nu_{di} + \Delta\nu_{\text{coll}} = q^2 D_0/2\pi\varrho + B\varrho\,, \tag{4.3}$$

where D_0 and B are the diffusion constant at 1 amagat and the broadening coefficient, respectively. From (4.3) we see that the two effects can be separated by their different density dependence. The work of MAY and collaborators showed that the simple expression in (4.3) qualitatively explained the data observed in a 90° scattering experiment. A minimum in the line width occurred at a density where both effects were small. A general conclusion reached from this work is that there is no correlation in the effects of collisions on the translational and rotational motions of the molecules. This group also measured the anisotropy of the Dicke narrowing and showed, within a simple collision model, how this anisotropy is related to the nonsphericity of the molecular cross section. They demonstrated that, in general, the diffusion constant depends on the polarizations of the incident and scattered light and on the scattering angle.

This group also performed detailed measurements of the change in frequency of the rotational lines as a function of pressure—both hydrogen pressure and inert gas pressure. In addition, they accurately measured the displacements of the $S_0(0)$ and $S_0(1)$ lines of H_2 as 354.390 and $587.060 \pm 0.002\ \mathrm{cm}^{-1}$, respectively. This is an order of magnitude more accurate than any previous measurement.

From (4.3) it is seen that the effects of Dicke narrowing are more pronounced for small q, that is, for forward scattering. A reduction of the observed linewidth of the $S_0(1)$ line in H_2 at 2 atm from $\Delta\nu_{\frac{1}{2}} = 0.15\ \mathrm{cm}^{-1}$ for 90° scattering to $\Delta\nu_{\frac{1}{2}} = 0.04\ \mathrm{cm}^{-1}$ for forward scattering was reported by CLEMENTS and STOICHEFF [4.38], as shown in Fig. 4.8. This low value, though, was instrumentally broadened and they estimated the true width arising only from pressure broadening as $0.006\ \mathrm{cm}^{-1}$.

MURRAY and JAVAN [4.45] used a specially constructed argon-ion laser with a linewidth of about $0.0085\ \mathrm{cm}^{-1}$ to study the motional narrowing of the $Q(1)$ line in hydrogen as a function of pressure. They observed the narrowing in both the forward and backward directions and saw an effect similar to that discussed in the previous paragraph.

DION and MAY [4.44] have shown that the Q branch of HD exhibits a different kind of motional narrowing at high pressures. This phenomenon was first predicted by ALEKSEEV and SOBELMAN [4.46] and occurs when the rotationally inelastic collision frequency is large compared with the vibration-rotation interaction frequency. The experimental results reveal that the individual components of the Q branch broaden at low densities, but at high densities they eventually overlap and the band then collapses. The results are in good agreement with the theoretical prediction.

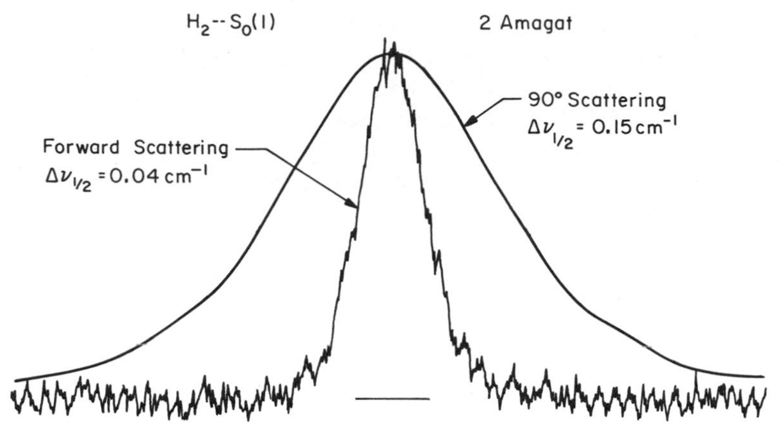

Fig. 4.8. Profiles of the S(1) rotational line of H_2 gas at 2 atm pressure observed in forward and 90° scattering [4.38]

GRAY [4.47] has theorized that the collisional broadening of the rotational Raman lines of dipolar gases has rather different properties from those observed in nonpolar gases such as hydrogen. For dipolar molecules his theory predicts that dipole interactions and resonant collisions make the largest contributions to the line broadening. The probability of a resonant collision is highest if both molecules have the same rotational quantum number J. Hence, the most intense rotational lines (those arising from the most populous states) should experience the greatest amount of self-broadening. Further, this self-broadening should be proportional to the pressure. His theory predicts that the self-broadening coefficient should decrease rapidly with increasing J; however, for large J the self-broadening coefficient should oscillate with increasing J. (The self-broadening coefficient can be written as $\Delta\sigma/p$, where $\Delta\sigma$ is the half width [cm^{-1}] at half the peak intensity and p is the pressure.)

Several groups have experimentally studied the pressure broadening of the highly dipolar molecule HCl. RICH and WELSH [4.48], FABRE et al. [4.49], and PERCHARD et al. [4.50], (who also studied HBr, DCl, and DBr), all confirmed that the rotational linewidths varied with the pressure. Further, they confirmed the general functional dependence of the self-broadening coefficient on J which Gray predicted. However, the predicted alternation of the self-broadening coefficient could not be checked because the weak intensity of the rotational Raman scattering from HCl did not allow the accurate measurement of linewidth for J greater than 8. Also, the observed values for the self-broadening were significantly less than the calculated ones. The high chemical re-

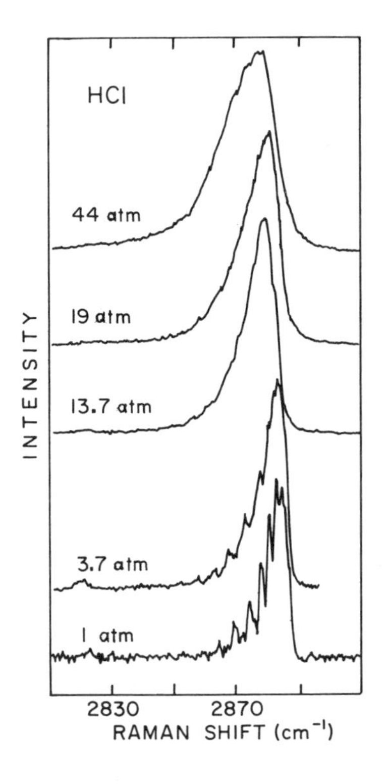

Fig. 4.9. Total scattering intensity for Raman band of HCl gas between 1 and 44 atm. Multiple pass cell used for 1 atm and 3.7 atm [4.50]

activity of HCl produced certain experimental problems. The large multiple pass cells developed by WELSH et al. and described earlier could not be used.

The measurements of FABRE et al. [4.49] and PERCHARD et al. [4.50] were extended to higher pressure ($\simeq 40$ atm) and revealed a feature not seen by RICH and WELSH [4.48] who only went to 14 atmospheres. For pressures greater than 30 atm a significant broadening of the Rayleigh scattering occurs, enough to obscure the first rotational line. In addition, at high pressures, there is significant overlap of the rotational lines.

PERCHARD et al. [4.50] also studied the self-broadening of the vibrational Q branch of the molecules HCl, HBr, DCl, and DBr, as well as the foreign broadening induced by the chemically inert gases SF_6 and C_2F_6 (Fig. 4.9). Their work showed a clear difference between the two types of broadening. They concluded that, for inert gas broadening each component of the Q branch retains a Lorentzian increase with pressure, and also increases linearly with J in agreement

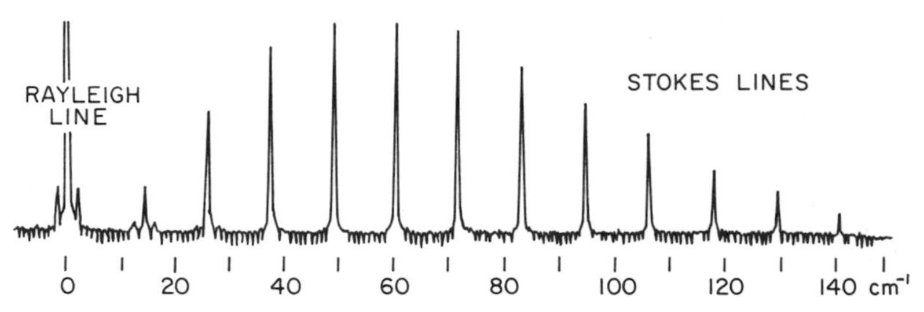

Fig. 4.10. Rotational Raman spectrum of oxygen showing triplet structure (Photograph of recorder trace) [4.59]

with theoretical prediction [4.51, 52]. For self-broadening, however, the data did not fit this model. The authors suggested that the existence of a polymeric state could explain this discrepancy as well as the broadening of the Rayleigh line.

High resolution Raman spectroscopy of gases, using laser sources, has permitted more detailed studies of the fine structure of the Raman spectrum of simple gases. In the oxygen molecule the spin-spin and spin-rotation interactions split the $^3\Sigma_g^-$ molecular ground state. JAMMU et al. [4.53] using Hg lamp techniques first noted the effects of this splitting on the Raman spectrum of O_2, and it has been studied in more detail by RENSCHLER et al. [4.54], and RICH and LEPARD [4.55] using laser sources. The splitting affects the Raman spectra in two important ways. First, there is the appearance of satellites on both sides of the Rayleigh line and the Q branch. Second is the presence of satellites on both sides of the lowest rotational transition (Fig. 4.10). In the first experiment RENSCHLER et al. [4.54] using an intracavity cell with a He–Ne laser resolved the Rayleigh and rotational satellites. Using the much stronger argon laser and the multiple pass cell discussed earlier RICH and LEPARD [4.55] were able to observe the Q branch satellites. All of these satellites were shown to come from expected transitions between well-known energy levels. Recent theoretical calculations [4.56] predict an intensity ratio for the rotational satellites which is a function of the rotational quantum number. These calculations differ somewhat from the Hund's case approximation and fit the data within the experimental error.

The Raman spectrum of nitric oxide is unique among those of gases because of the presence of a Raman transition between molecular electronic levels. RASETTI [4.57] in 1930, first observed this feature at $1200\,\mathrm{cm}^{-1}$ and interpreted it as a transition between the two spin components of the $^2\Pi$ ground state. Several recent experiments have studied this phenomenon in more detail and with greater resolution.

SHOTTEN and JONES [4.58] using a small multi-reflection intracavity cell with an argon laser clearly resolved the allowed $S(J=1)$ and $R(J=1)$ bands of the rotational spectrum of this symmetric top molecule. Each of the S band transitions is split by the spin-orbit interaction into two components with their separation increasing with increasing J. The R branch lines should also be split; however, theory predicts that one set of components should be much weaker than the other, and only one set was seen. The rapid decrease in intensity of the R branch with increasing J also followed theoretical predictions. However, they could not satisfactorily resolve the $^2\Pi_{\frac{3}{2}} \rightarrow {}^2\Pi_{\frac{1}{2}}$ pure electronic transition. (The early work by RASETTI did not come close to resolving this line and barely resolved the main S branch lines.) RENSCHLER et al. [4.59] also reported measurements of the Raman spectrum of NO, and gave a somewhat detailed discussion of the molecular levels of NO. They, too, could not satisfactorily resolve the purely electronic transition. This feat was achieved by RICH and WELSH [4.27] who reported the result as a demonstration of the capabilities of their multiple pass cell.

The use of electrical discharge to excite gases to higher vibrational states has allowed the observation of Raman transitions arising from those higher states. WEBER and collaborators [4.25, 60] have measured the molecular rotational constants for excited states of CO_2 and CS_2 while NELSON et al. [4.61] have reported observing vibrational transitions originating from higher vibrational levels in nitrogen.

Both photoelectric and photographic studies of CO_2 were performed. The photoelectric measurements were carried out with an argon-ion laser and the detection scheme described in the previous section. They clearly showed pure-rotational Raman scattering originating in the $01'0$ vibrational state of CO_2 superimposed on the ground-state pure rotational spectrum. Since the spins of both the ^{32}S and ^{16}O nuclei are zero, all rotational levels of the $^1\Sigma_g^+$ ground states of $C^{16}O_2$ and $C^{32}S_2$ which are antisymmetric in the simultaneous exchange of nuclei are absent. Hence for the vibrational ground states the pure rotational spectrum consists of $J=$ even transitions only [4.9]. However, for the $01'0$ vibrational state with angular momentum $l=1$ the vibration-rotation interaction splits each J-level into a doublet consisting of one symmetric and one antisymmetric member. Thus, rotational Raman transitions in the $01'0$ state are possible for all values of J and will appear approximately superimposed upon, and midway between, transitions in the $00^\circ 0$ state.

The same effect was seen to a much smaller extent in unexcited CO_2. Excited and unexcited spectra are illustrated in Fig. 4.11. The effective rotational temperature of the excited gas was estimated at 750 K. The

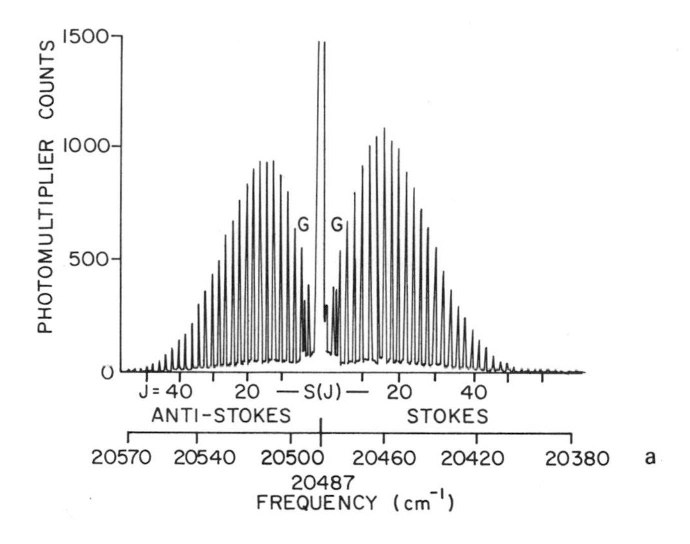

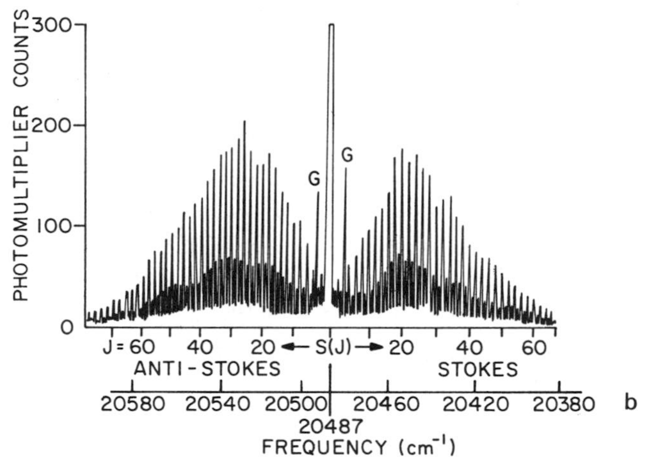

Fig. 4.11a and b. Computer-plotted smoothed rotational Raman spectrum of CO_2 at a pressure of 40 Torr: (a) No discharge, (b) electrical discharge. In (b) the odd J-value rotational lines from the 01′0 vibrational levels are enhanced by the discharge, while in (a) they show up only as base line noise [4.25]

photographic measurements performed with a He–Ne laser yielded very accurate values for the rotational constants for both the 00°0 ground state and the 01′0 excited state. A weak suggestion of the allowed R branch ($\Delta J = +1$) of the excited state was seen in the photographic measurements. Only photographic measurements were performed on CS_2.

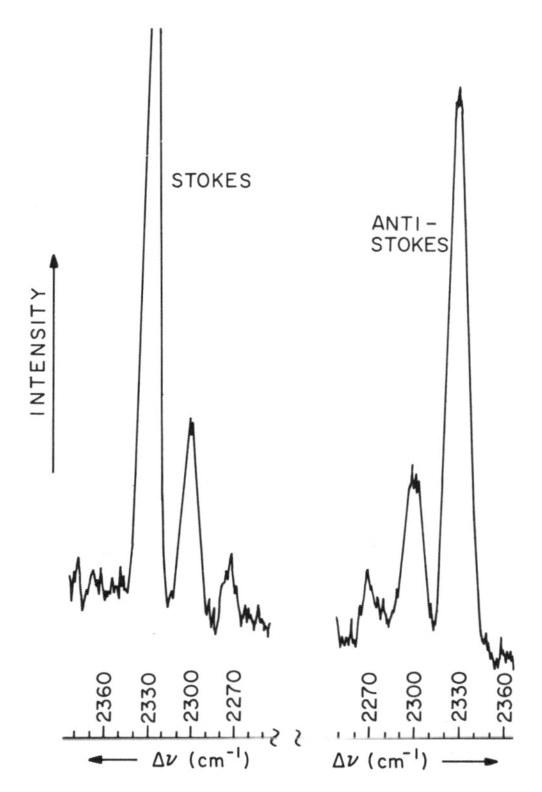

Fig. 4.12. Stokes and anti-Stokes Raman spectra of vibrationally excited nitrogen. (Conditions: 200 Torr N_2, 1000 counts/s, $\tau = 2$ sec, slits 9 cm^{-1} in Stokes region; and 500 counts/sec, $\tau = 2$ sec, slits 11 cm^{-1} in anti-Stokes region) [4.61]

NELSON et al. [4.61] recorded Raman spectra from electrically excited nitrogen gas flowing through an intracavity Raman cell. Both Stokes and anti-Stokes transitions originating from several excited states were observed, as illustrated in Fig. 4.12. The observed excited vibrational transition frequencies agreed well with those obtained from infra-red data. A Boltzmann distribution corresponding to a vibrational temperature of 1950 K fit the observed amplitudes well. This observation implies a rapid $V-V$ energy exchange process due to the near-perfect resonance between transitions and the relatively high pressures used. An electric discharge might not be expected to produce an initial Boltzmann distribution.

HOCHENBLEICHER and SCHROTTER [4.62] have investigated the "hot" (vibrationally-excited) bands of chlorine and carbon tetrachloride without the benefit of electrical excitation. Their observation during a room temperature experiment of hot bands in chlorine with a vibrational temperature of 450 K was attributed to absorption of the laser beam by the gas. The hot bands in CCl_4 vapor were studied by comparing spectra taken at various temperatures.

New information about Fermi resonance in carbon dioxide has been brought out by two recent investigations with laser sources. (The nature of this resonance phenomenon is described by HERZBERG [4.10] and can be understood in terms of perturbation theory.) Briefly, this phenomenon is a result of the interaction through the anharmonic terms in the potential of two molecular vibrational states with the same symmetry and similar energies. As a result of the coupling there is a mixture of the two vibrational states, and they cannot be unambiguously designated. Also, as a result of the coupling, the two energy levels are repelled from each other and their spacing is greater than if no resonance were present. Because of the mixture the relative intensities of Raman transitions to the two states are of the same order. This occurs even though one transition may be a fundamental and the other an overtone.

In carbon dioxide, there is a strong Fermi resonance between the $10^{\circ}0$ and $02^{2}0$ states, both of Σ_g^+ symmetry. In the Raman spectrum of $^{12}CO_2$ the higher frequency component has a higher intensity. Hence, traditionally the unperturbed $10^{\circ}0$ level has been taken to have higher energy than the unperturbed $02^{2}0$ level, because the fundamental should be more intense than the overtone. However, HOWARD-LOCK and STOICHEFF [4.63], motivated by infrared measurements, have demonstrated that the reverse is the case of $^{12}CO_2$ while the above conclusion does hold for $^{13}CO_2$. Their experimental measurements were simply a careful measurement of the intensity ratios for the Fermi diads in $^{12}CO_2$ and $^{13}CO_2$. Their analysis was based on the assumption that the ratio α_u/α_l is not appreciably changed by isotopic substitution. Here α_u/α_l is the ratio of the matrix elements of the polarizability for the transition from the gound level to the upper and lower levels of the Fermi diad. Since this ratio is related to the observed intensity ratio of the diad through parameters which depend on the ordering of the vibrational states, this method of analysis can be carried out.

WRIGHT and WANG [4.64] have studied the pressure dependence of Raman scattering from the Fermi diad in CO_2 over the density range 15–534 amagat. Both components shifted linearly to lower frequencies with increasing density and the splitting between the components increased with increasing density.

4.2. Cross Sections and Depolarization Ratios

The coherence, very small divergence, and practically complete polarization of laser radiation have greatly simplified the measurement of absolute Raman cross-sections and depolarization ratios. The complicated geometrical correction factors required when doing such experiments with mercury lamp excitation [4.15] are no longer necessary. Especially for the determination of low depolarization ratios laser measurements are far more accurate, in almost all cases yielding lower (hence, more accurate) values than prelaser measurements.

In principle the differential Raman cross-section for a particular band in a gas can be measured directly using

$$I = (\mathrm{d}\sigma/\mathrm{d}\Omega)\, i_0 \,. \tag{4.4}$$

Here I is the integrated scattered radiant intensity (power per unit solid angle per molecule), $\mathrm{d}\sigma/\mathrm{d}\Omega$ the differential Raman cross-section for the band, and i_0 the incident irradiance (power per unit area). However, since Raman scattering in gases is such a weak effect, direct determination represents a difficult measurement. Hence, two other types of absolute measurement have also been used. One is a comparison of the weak Raman scattering with the scattering from a stronger scatterer whose scattering cross-section can be measured directly. The intermediate substance usually used is liquid benzene. The other is a comparison of the band under study with one whose scattering cross-section can be calculated from other known parameters. All these quantitative measurements of intensity, of course, utilize photoelectric detection.

The cross-sections which have been most frequently measured in laser Raman studies are those of Q branch (pure vibrational) transitions in relatively simple gases. Usually these measurements employ a right angle geometry with the direction of observation perpendicular to both the incident laser beam and the direction of its electric vector. For these experimental conditions PLACZEK [4.7] has shown that within the polarizability theories the differential cross-section and the depolarization ratios of the Q branch are related to molecular parameters by

$$\frac{\mathrm{d}\sigma}{\mathrm{d}\Omega} = \frac{(2\pi)^4}{45} \frac{b_j^2(\tilde{\nu}_0 - \tilde{\nu})^4}{1 - \exp(-h\nu c/kT)} g(45\,\alpha'^2 + 4\chi\gamma'^2)\,, \tag{4.5}$$

$$\varrho = \frac{3\chi\gamma'^2}{45\,\alpha'^2 + 4\chi\gamma'^2}\,. \tag{4.6}$$

Here $\tilde{\nu}_0$ and $\tilde{\nu}$ are the wavenumbers of the incident radiation and the vibrational mode of the molecule, $b=(h/8\pi^2 \nu_c c)^{1/2}$ is the zero point vibrational amplitude of the mode, g is the degree of degeneracy, $3\alpha'$ and γ'^2 are the trace and anisotropy of the derived polarizability tensor associated with the normal coordinate, and χ is the fraction of the anisotropic scattering which occurs in the Q branch. This factor has been calculated for various types of symmetric top molecules and for linear molecules with sufficiently small rotational constants $\chi=0.25$. (The relative advantages of various techniques for depolarization ratio measurements using this geometry have been discussed by a number of workers, among them CLAASSEN et al. [4.65] and PROFFITT and PORTO [4.66].)

In all of the determinations of cross-sections by laser methods, the cross sections are quoted relative to that of the Q branch of nitrogen. The choice of this standard does not arise from any fundamental physical grounds, but simply because nitrogen is the predominant gas in the atmosphere. Hence, it is a useful standard for the use of Raman scattering in monitoring atmospheric constituents. Since the absolute cross section of nitrogen thus becomes a quantity of fundamental interest, its measurement by various methods will be discussed below.

PENNEY et al. [4.67] measured the nitrogen Q branch cross section by two methods. The first was a comparison of Raman and Rayleigh scattering data for the same experimental conditions. The polarized Rayleigh cross section is proportional to the square of the mean molecular polarizability; this quantity in turn is related to the refractive index of the gas by the well-known formula

$$n-1=2\pi N\alpha \,. \tag{4.7}$$

The second method utilized an absolute calibration of the Raman spectrometer based on standard lamps. We note that the first method required only a knowledge of the relative spectral response of the system. Both of the measurements were performed with a laser wavelength of 514.5 nm. The results of the two were in good agreement and are given in Table 4.1 along with the results of other workers.

FOUCHE and CHANG [4.68] measured the cross section for the nitrogen Q branch also for 514.5 nm incident light by comparing the Raman intensities from benzene and from N_2. In their measurements the spectrometer slits were set wide enough so that the line shape appeared trapezoidal. The intensities were then proportional to the height of the trapezoid. For the total cross section of benzene they used a value obtained from a measurement of the peak differential cross section and linewidth of benzene, carried out by SKINNER and NILSEN

Table 4.1. Differential Raman cross-section for Q branch vibrational transition in N_2. (Cross-sections in units of 10^{-31} cm^2/mol · sterad)

λ_L	PGL [a]	FC [b]	FC [c]	FHKP [d]	FHKP [e]	HC [f]	MHB [g]	SCW [h]
488.0 nm	5.5 ± 0.3	5.6 ± 2.2	4.9 ± 1.9	4.3 [i]	3.3	5.4 ± 0.3 [i]	5.5	4.8
514.5 nm	4.4 ± 0.2 [i]	4.4 ± 1.7 [i]	3.9 ± 1.5	3.4	2.6	4.3 ± 0.2 [i]	4.4	3.8

[a] [4.67].
[b] [4.68].
[c] [4.68].
[d] [4.72].
[e] [4.72].
[f] [4.88].
[g] [4.78].
[h] [4.77].
[i] Measured value. All others adjusted using a ν_R^4 frequency dependence.

[4.69]. This value, however, differs by about 20% from a more recent determination of the benzene cross section at 514.5 nm by KATO and TAKUMA [4.70, 71]. Hence, in Table 4.1 we have listed two values for the results of FOUCHE and CHANG, the value they give and one based on the result of KATO and TAKUMA for benzene.

FENNER et al. [4.72], in our laboratory, have measured the nitrogen cross section at cm incident wavelength of 488.0 nm by both the calibrated spectrometer and comparison with benzene methods. Their results for the two methods were not in good agreement, they are listed in Table 4.1, too.

More recently here, HYATT et al. [4.73] measured the Q branch cross section in N_2 by comparison with the cross-section for the $J=1 \rightarrow 3$ rotational transition in hydrogen. GOLDEN and CRAWFORD [4.74] proposed the use of the hydrogen transition as a primary intensity standard since its magnitude is proportional to the square of the molecular anisotropic polarizability and for hydrogen this quantity can be calculated. Correction must, of course, be made for the rotational population factor. FORD and BROWNE [4.75, 76] have calculated accurate tables for the rotational Raman cross-section and these values were used by HYATT et al. Their measurements were performed for incident wavelengths of both 488.0 nm and 514.5 nm and are also tabulated in Table 4.1. Also included in Table 4.1 are two prelaser measurements, one by STANSBURY et al. [4.77] and the other taken from the compilation of prelaser data by MURPHY et al. [4.78] as derived from a measurement by YOSHINO and BERNSTEIN [4.15] which was ultimately based on a comparison with hydrogen. These results have been restated in terms of the incident wavelengths 488.0 nm and 514.5 nm by applying a cor-

Table 4.2. Differential Raman cross-sections for Stokes Q-branch vibrational lines relative to that for N_2. (All values adjusted for incident 488.0 nm radiation using assumed ν_R^4 frequency dependence)

Gas	ω_R [cm^{-1}]	ϱ	$d\sigma/d\Omega$
N_2	2331	0.022[a], 0.021[f]	1.0
O_2	1556	0.047[a], 0.051[f]	1.20 ± 0.06[a], 1.2[b], 1.3[c], 1.2[d], 1.2[e]
O_3	1103		4.0[b]
H_2 (sum)	4161	0.0125[f]	2.4[b], 2.4[c], 2.7[d], 2.8[e]
$H_2(Q(1))$	4161	0.019[f]	1.6[c]
CO	2145	0.038[a]	0.99 ± 0.005[a], 0.91[b], 1.0[c], 0.94[d]
CO_2 (ν_1)	1388	0.027[a]	1.5 ± 0.1[a], 1.5[b], 1.4[c], 1.3[d], 4.0[e]
CO_2 ($z\nu_2$)	1286		1.0[b], 0.89[c], 0.92[d]
NO	1877	0.046[a], 0.070[f]	0.46 ± 0.03, 0.55[b], 0.27[c], 0.5[g], 0.54[h]
N_2O (ν_1)	1285		2.7[b], 2.2[c]
N_2O (ν_3)	2224		0.53[b], 0.51[c]
N_2O_4	1360		0.6[i]
SO_2 (ν_1)	1151	0.018[a]	5.3 ± 0.5[a], 5.4[b], 5.2[c], 2.4[g]
H_2O	3652	0.025[a]	2.6 ± 0.3[a]
HF	3962	0.03[f]	0.7[j]
HCl	2886	0.056[k]	2.0[d], 2.4[j]
HBr	2558	0.067[k]	3.5[d], 4.1[j]
HI	2230		6.0[j]
NH_3 (ν_1)	3334		5.0[c], 3.6[d]
ND_3 (ν_1)	2420		3.0[c]
CH_4 (ν_1)	2914	0[a]	7.9 ± 0.4[a], 8.2[b], 6.0[c], 7.0[d], 8.0[e]
C_2H_6 (ν_3)	993		1.6[c], 3.5[d]
C_6H_6 (ν_2)	942		9.1[c], 16.6[d]
C_6H_6 (ν_1)	3062		7.0[c], 10.0[d]
F_2	892	0.005[h], 0.058[h]	0.44[h]
ONF	1844	0.35[h]	0.73[h]

[a] [4.67]—(514.5 nm).
[b] [4.68]—(514.5 nm).
[c] [4.72]—(488.0 nm).
[d] [4.78]—(Hg lamp).
[e] [4.77]—(Hg lamp).
[f] [4.90].
[g] [4.79]—(337.1 nm).
[h] [4.83]—(488.0 nm).
[i] [4.91]—(694.3 nm).
[j] [4.88]—(488.0 nm).
[k] [4.50]—(488.0 nm).

rection term of the form ν_R^4, as indicated by (4.5). The validity of this wavelength dependence was checked by HYATT et al. [4.73] in the region 457.9 nm to 514.5 nm and was found accurate to within an experimental error of 6%.

As seen from Table 4.1, the HYATT et al. [4.73], PENNEY et al. [4.67], and MURPHY et al. [4.78] measurements are all in relatively good agreement and this value can be taken as an accepted standard. With regard to the experimental techniques all are subject to various errors. However, the hydrogen comparison method avoids several experimental difficulties inherent in the others. The comparison is between two signals with similar magnitude and no refractive index corrections are necessary. The hydrogen comparison method, though, depends on the accuracy of the theoretical calculation.

In Table 4.2 we list the values of Q branch cross sections of numerous other gases relative to that of nitrogen. These values are taken from the work of all the groups above. Some of these workers measured the cross section only for polarized scattering while others reported the total cross section for polarized plus depolarized scattering. These measurements are all grouped together, since, in general, the Q branch depolarization ratios are small compared to the experimental uncertainties in the measurements. Also included in the table are the depolarization ratios which have been measured. Results from the pre-laser compilation of MURPHY et al. are included as well as results by LEONARD [4.79] obtained with a pulsed nitrogen laser for NO and SO_2. The signal/noise for these measurements was lower than that for measurements made with continuous lasers. (We should note that Raman studies of the atmosphere have been conducted by several authors using pulsed lasers [4.80–82]. A discussion of these techniques is beyond the scope of chapter.) All of the measurements were performed using a simple gas cell and a right angle geometry except for the measurements of F_2, NO, and ONF by HOELL et al. [4.83] who performed these measurements in the flowing chamber of a chemical laser. These measurements were used for diagnostic purposes in the study of the chemical laser.

The accurate depolarization ratio measurements possible with laser Raman spectroscopy has allowed HOLZER et al. [4.84] to verify a prediction of PLACZEK and TELLER [4.7] regarding a variation with J of the depolarization ratio of the components of the Q branch. They made this measurement in H_2 and D_2 since because of the small moment of inertia of these molecules the Q branch components are completely resolvable. Their results fully verified the theoretical predictions, in particular with regard to the completely isotropic characteristic of the $0 \rightarrow 0$ transition.

The only gases for which resonance Raman effects have been observed are the heavy non-transparent halogen gases Cl_2, Br_2, I_2, BrCl, ICl, and IBr. HOLZER et al. [4.85] studied these gases using the various output wavelengths of the argon-ion laser from 514.5 nm to 457.9 nm and investigated the distinction between resonance Raman scattering and resonance fluorescence. The case of I_2 has been investigated by several groups. Among them KIEFER and BERNSTEIN [4.86] have studied the fine structure of the overtone progression while WILLIAMS and ROUSSEAU [4.87] have given a clear understanding of the resonance phenomenon. In these gases the resonance phenomenon manifests itself in the appearance of many overtones, all of which preserve the polarized character of the fundamental band. No resonance phenomena have been observed in non-transparent gases. CHERLOW et al. [4.88] measured the total Raman cross section for hydrogen iodide as a function of wavelength from 514.5 nm to 457.9 nm and found no deviation from a ν_R^4 dependence of the cross section. This molecule has a strong continuous absorption in the near uv starting at 27500 cm^{-1} and peaking at 48000 cm^{-1}, and one might expect some pre-resonance behavior [4.89]. The absence of such behavior indicates that the important electronic intermediate states for the Raman process are located at energies near the ionization potential. This conclusion holds for both one intermediate state and two intermediate states processes [4.89].

Acknowledgement. This work was supported by the National Science Foundation.

References

4.1. C.V.RAMAN: Indian J. Phys. **2**, 387 (1928).
4.2. G.LANDSBERG, L.MANDELSTAN: Naturwissenschaften **16**, 557, 772 (1928).
4.3. A.SMEKAL: Naturwissenschaften **11**, 873 (1923).
4.4. L.A.RAMDAS: Indian J. Phys. **3**, 131 (1928).
4.5. F.RASETTI: Proc. Nat. Acad. Sci. Am. **15**, 515 (1929); Phys. Rev. **34**, 367 (1929).
4.6. R.W.WOOD: Phys. Rev. **33**, 1097 (1929); **35**, 1355 (1930); Phil. Mag. **7**, 744 (1929).
4.7. G.PLACZEK: In *Marx Handbuch der Radiologie*, Vol. VI, Part 2, p. 209 (Akademische Verlagsgesellschaft, Leipzig, 1934); UCRL-Trans-526L.
4.8. G.PLACZEK, E.TELLER: Z. Physik **81**, 209 (1933).
4.9. S.BHAGAVANTAM: *Scattering of Light and the Raman Effect* (Andrha University, Waltair, India, 1940).
4.10. G.HERZBERG: *Molecular Spectra and Molecular Structure, I. Spectra of Diatomic Molecules*, 2nd ed. (Van Nostrand, Princeton, N. J., 1950) and *II. Infrared and Raman Spectra of Polyatomic Molecules* (Van Nostrand, Princeton, N. J., 1945).
4.11. E.B.WILSON, J.C.DECIUS, P.C.Cross: *Molecular Vibrations: The Theory of Infrared and Raman Vibrational Spectra* (McGraw Hill, New York, 1955).
4.12. H.L.WELSH, M.F.CRAWFORD, R.T.THOMAS, G.R.LOVE: Canad. J. Phys. **30**, 577 (1952).
4.13. B.P.STOICHEFF: In *Advances in Spectroscopy I*, ed. by H. W. THOMPSON (Interscience, New York, 1959), pp. 91–174.
4.14. B.P.STOICHEFF: In *Experimental Physics: Molecular Physics* (ed. by D. WILLIAMS), Vol. 3 (Academic Press, New York, 1902), pp. 111–155.
4.15. E.g., T.YOSHINO, H.J.BERNSTEIN: J. Mol. Spectr. **2**, 213 (1958).
4.16. S.P.S.PORTO, D.L.WOOD: J. Opt. Soc. Am. **53**, 1446 (1962).
4.17. A.WEBER, S.P.S.PORTO: J. Opt. Soc. Am. **55**, 1033 (1965).
4.18. A.WEBER, S.P.S.PORTO, L.E.CHEESMAN, J.J.BARRETT: J. Opt. Soc. Am. **57**, 19 (1967).

4.19. H. L. Welsh, C. Cumming, E. J. Stansbury: J. Opt. Soc. Am. **41**, 712 (1951).
4.20. H. L. Welsh, E. J. Stansbury, J. Romanko, T. Feldman: J. Opt. Soc. Am. **45**, 338 (1955).
4.21. B. P. Stoicheff: Canad. J. Phys. **32**, 330 (1954).
4.22. J. J. Barrett, N. I. Adams: J. Opt. Soc. Am. **58**, 311 (1968).
4.23. S. P. S. Porto: Bull. Am. Phys. Soc. **11**, 79 (1966).
4.24. B. P. Stoicheff: Canad. J. Phys. **32**, 630 (1954); **36**, 218 (1958).
4.25. J. J. Barrett, A. Weber: J. Opt. Soc. Am. **60**, 70 (1970).
4.26. N. H. Rich, H. L. Welsh: J. Opt. Soc. Am. **61**, 977 (1971).
4.27. N. H. Rich, H. L. Welsh: Indian J. Pure Appl. Phys. **9**, 944 (1971).
4.28. A. Weber, J. Schlupp: J. Opt. Soc. Am. **62**, 428 (1972).
4.29. G. O. Neely, L. Y. Nelson, A. B. Harvey: Appl. Spectrosc. **25**, 360 (1971).
4.30. D. L. Hartley, R. A. Hill: J. Appl. Phys. **43**, 4134 (1972).
4.31. J. R. Moret-Bailly, H. Berger: Compt. Rend. Acad. Sci. **269**, 416 (1969).
4.32. J. U. White, N. L. Albert, A. G. de Bell: J. Opt. Soc. Am. **45**. 154 (1954).
4.33. A. Chapput, M. Delhaye, J. Wrobel: Compt. Rend. Acad. Sci. **272**, 461 (1971).
4.34. H. Berger, M. Faivre, J. P. Champion, J. Moret-Bailly: J. Mol. Spectrosc. **45**, 298 (1973).
4.35. W. Kiefer, H. J. Bernstein, H. Wieser, M. Danylak: J. Mol. Spectrosc. **43**, 393 (1973).
4.36. W. J. Jones: Contemp. Phys. **13**, 419 (1972).
4.37. R. J. Butcher, D. V. Willets, W. J. Jones: Proc. Roy. Soc. (London) A **324**, 231 (1971).
4.38. W. R. L. Clements, B. P. Stoicheff: J. Mol. Spectrosc. **33**, 183 (1970).
4.39. J. J. Barrett, S. A. Myers: J. Opt. Soc. Am. **61**, 1246 (1971).
4.40. R. H. Dicke: Phys. Rev. **89**, 472 (1953).
4.41. V. G. Cooper, A. D. May, E. H. Hara, H. F. P. Knapp: Canad. J. Phys. **46**, 2019 (1968).
4.42. V. C. Cooper, A. D. May, B. K. Gupta: Canad. J. Phys. **48**, 725 (1970).
4.43. B. K. Gupta, S. Hess, A. D. May: Canad. J. Phys. **50**, 778 (1972).
4.44. P. Dion, A. D. May: Canad. J. Phys. **51**, 36 (1973).
4.45. J. R. Murray, A. Javan: J. Mol. Spectrosc. **29**, 502 (1969).
4.46. V. A. Alekseev, J. I. Sobelman: Report of the Lebedev Inst. Inte. Moscow, No. 58, Acta. Phys. Polon **34**, 579 (1968); IEEE J. Quant. Electron. **4**, 654 (1968).
4.47. C. G. Gray: Chem. Phys. Letters **8**, 528 (1971).
4.48. N. H. Rich, H. L. Welsh: Chem. Phys. Letters **11**, 292 (1971).
4.49. D. Fabre, G. Widenlocher, H. Vu: Opt. Commun. **4**, 421 (1972).
4.50. J. P. Perchard, W. F. Murphy, H. J. Bernstein: Molec. Phys. **23**, 535 (1971).
4.51. J. Finvak, J. van Kranendonk: Canad. J. Phys. **41**, 21 (1963).
4.52. J. van Kranendonk: Canad. J. Phys. **41**, 433 (1963).
4.53. K. S. Jammu, G. E. St. John, H. L. Welsh: Canad. J. Phys. **44**, 797 (1966).
4.54. D. L. Renschler, J. L. Hunt, T. K. McCubbin, S. R. Polo: J. Mol. Spectrosc. **31**, 173 (1969).
4.55. N. H. Rich, D. W. Lepard: J. Mol. Spectrosc. **38**, 549 (1971).
4.56. D. W. Lepard: Canad. J. Phys. **48**, 1664 (1970).
4.57. F. Rasetti: Z. Physik **66**, 646 (1930).
4.58. K. C. Shotten, W. J. Jones: Canad. J. Phys. **48**, 632 (1970).
4.59. D. L. Renschler, J. L. Hunt, T. K. McCubbin, S. R. Polo: J. Mol. Spectrosc. **32**, 347 (1969).
4.60. W. J. Walker, A. Weber: J. Mol. Spectrosc. **39**, 57 (1971).
4.61. L. Y. Nelson, A. W. Saunder, A. B. Harvey, G. O. Neely: J. Chem. Phys. **55**, 5127 (1971).

4.62. G. HOCHENBLEICHER, H. W. SCHROTTER: Appl. Spectrosc. **25**, 360 (1971).
4.63. H. HOWARD-LOCK, B. STOICHEFF: J. Mol. Spectrosc. **37**, 321 (1971).
4.64. R. B. WRIGHT, C. H. WANG: J. Chem. Phys. **58**, 2893 (1973).
4.65. H. H. CLAASSEN, H. SELIG, J. SHAMIR: Appl. Spectrosc. **23**, 8 (1969).
4.66. W. PROFFITT, S. P. S. PORTO: J. Opt. Soc. Am. **63**, 77 (1973).
4.67. C. M. PENNEY, L. M. GOLDMAN, M. LAPP: Nature Phys. Sci. **235**, 110 (1972).
4.68. D. G. FOUCHE, R. K. CHANG: Appl. Phys. Lett. **18**, 579 (1971); **20**, 256 (1972).
4.69. J. G. SKINNER, W. G. NILSEN: J. Opt. Soc. Am. **58**, 113 (1968).
4.70. Y. KATO, H. TAKUMA: J. Opt. Soc. Am. **61**, 347 (1971).
4.71. Y. KATO, H. TAKUMA: J. Chem. Phys. **54**, 5398 (1971).
4.72. W. R. FENNER, H. A. HYATT, J. M. KELLAM, S. P. S. PORTO: J. Opt. Soc. Am. **63**, 73 (1973).
4.73. H. A. HYATT, J. M. CHERLOW, W. R. FENNER, S. P. S. PORTO: J. Opt. Soc. Am. **63**, 1604 (1973).
4.74. D. M. GOLDEN, B. CRAWFORD: J. Chem. Phys. **36**, 1654 (1962).
4.75. A. L. FORD, J. C. BROWNE: Phys. Rev. A **7**, 418 (1973).
4.76. A. L. FORD, J. C. BROWNE: Atomic Data **5**, 305 (1973).
4.77. E. J. STANSBURY, M. F. CRAWFORD, H. L. WELSH: Canad. J. Phys. **31**, 954 (1953).
4.78. W. F. MURPHY, W. HOLZER, H. J. BERNSTEIN: Appl. Spectrosc. **23**, 211 (1969).
4.79. D. A. LEONARD: J. Appl. Phys. **41**, 4238 (1970); see also N. M. REISS: J. Appl. Phys. **43**, 739 (1972).
4.80. J. A. COONEY: Appl. Phys. Letters **12**, 40 (1968).
4.81. J. A. COONEY: Nature **224**, 1098 (1969).
4.82. S. H. MELFI, J. D. LAWRENCE, M. P. MCCORMACK: Appl. Phys. Letters **15**, 245 (1969).
4.83. J. M. HOELL, F. ALLARIO, O. JARRET, R. K. SEALS: J. Chem. Phys. **58**, 2896 (1973).
4.84. W. HOLZER, Y. LEDUFF, K. ALTMANN: J. Chem. Phys. **58**, 642 (1973).
4.85. W. HOLZER, W. F. MURPHY, H. J. BERNSTEIN: J. Chem. Phys. **52**, 399 (1970).
4.86. W. KIEFER, H. J. BERNSTEIN: J. Mol. Spectrosc. **43**, 366 (1973).
4.87. P. F. WILLIAMS, D. L. ROUSSEAU: Phys. Rev. Letters **30**, 951 (1973).
4.88. J. M. CHERLOW, H. A. HYATT, S. P. S. PORTO: Unpublished.
4.89. A. C. ALBRECHT, M. C. HUTLEY: J. Chem. Phys. **55**, 4438 (1971).
4.90. W. HOLZER, Y. LEDUFF: In *Advances in Raman Spectroscopy*, Vol. 1, ed. by J. P. MATHIEU (Heydon, London, 1973), p. 109.
4.91. C. J. CHEN, F. WU: Appl. Phys. Letters **19**, 452 (1971).

5. Linear and Nonlinear Phenomena in Laser Optical Pumping

B. DECOMPS, M. DUMONT, and M. DUCLOY

With 27 Figures

The use of a laser source in an optical pumping experiment was proposed by JAVAN [5.1] in 1964. The high power of a laser beam concentrated within a narrow spectral width and a small volume creates a very intense optical electric field, which is able to induce resonant transitions between two excited atomic levels. JAVAN's proposal allowed an extension of the methods and the results of optical pumping with conventional sources to atoms first excited by a discharge to very excited levels, which cannot be optically reached from a long-lived level, such as the fundamental state or a metastable state.

The interactions between the laser light and the excited atoms must be resonant. In atomic physics, the difficulty of finding perfect coincidences meant that, during the 1960's, the same atom had to be used as the laser source and in the pumped medium. This is why so many papers deal with a helium-neon laser pumping excited neon atoms [5.2–6].

By contrast, in the very rich spectrum of molecules, coincidences were found with the argon-ion or the krypton-ion laser. In this field we quote the work of ZARE and coworkers [5.7] and LEHMANN and coworkers [5.8] on the resonance line of Na_2 and I_2 (see also Subsect. 1.1.1).

More recently, the considerable progress in dye laser technology [5.9] has opened up a very wide field of optical pumping experiments in any atomic or molecular level. The first work in this field was reported by SVANBERG et al. [5.10] (see Subsect. 1.4.1).

Up to now, the present authors have done the major part of their work with the helium-neon laser pumping excited neon atoms. These experiments appear to be a good example of optical pumping with a laser beam, on account of the diversity of the results that can be obtained and of the relative simplicity of the corresponding atomic spectrum. The results are sufficiently numerous to allow a complete theoretical analysis of all the phenomena that occur in this kind of optical pumping.

For these reasons, this paper is devoted essentially, in all its experimental references, to neon. Here, linear effects and nonlinear phenomena can be readily separated, yet the theoretical analysis is clearly applicable to many other cases.

5.1. Experimental Features of Laser Optical Pumping of a Resonant Atomic Medium

5.1.1. General Survey of the Observed Phenomena

The study of the fluorescence light emitted by excited atoms optically pumped by a laser beam provides a wide range of experiments. Many of them, dealing with atomic physics and spectroscopy (Hanle effect [5.11], magnetic resonances [5.12], level crossings [5.13]) do not use the specific properties of the laser beam. This laser beam only represents a convenient light source, and similar experiments have already been performed with conventional light sources. As will be shown later in the theoretical analysis of optical pumping with a laser beam, these effects appear in the *linear response* of the medium, i.e. they are linear with respect to the laser intensity. Most of the linear effects could be predicted equally well from a theoretical analysis devoted to optical pumping with a conventional light source. Nevertheless, the spectral features of the laser radiation can produce characteristic phenomena. For instance, with a multimode laser the fluorescence light exhibits modulations [5.2, 14] at the beat frequencies between modes, which are resonant when their frequency is equal to the Zeeman splitting. This effect is also *linear* with respect to laser intensity.

In general, however, the characteristic phenomena of laser optical pumping appear in *nonlinear effects* with respect to the laser intensity. Except at a very low pumping rate, nonlinear effects interfere with preceding linear effects (broadening of the Hanle effect [5.5], light shifts in double-resonance experiments [5.12]). This interference between linear and nonlinear effects complicates the interpretation of experimental results, and an extrapolation procedure has to be carried out at vanishing laser intensity in order to get reliable results. Besides these rather obvious effects, nonlinearities in laser optical pumping provide many original effects, such as the very well-known Lamb dip [5.15] in monomode laser operation, which can also be observed on the fluorescence light [5.14]. In multimode laser operation, saturation resonances have been obtained in zero magnetic field [5.16–18] and in nonzero magnetic field, and each time the beat frequency between modes is equal to the Zeeman splitting [5.19]. The nonzero field resonances provide means to measure Landé g factors.

Furthermore, the strong pumping rates involved in laser optical pumping experiments are responsible for the appearance of very-high-order nonlinear effects. In contrast to optical pumping using conventional light sources, which is only able to induce orientation and alignment in atomic levels—i.e. tensorial quantities of order not higher than two—laser optical pumping can produce higher-order tensorial quantities and render them observable on the fluorescence light [5.20], e.g. the hexadecapole moment of the $2p_4$ level of neon [5.21].

5.1.2. Experimental Apparatus

All the experiments described below have been performed on neon levels using a He–Ne laser. This laser was chosen because of its technological simplicity and its ability to oscillate on many transitions, which allows optical pumping in several levels of different J values. Furthermore, the neon spectrum remains relatively simple so that the fluorescence light can be studied with rudimentary monochromators.

The experimental arrangement consists of three main elements: the laser source, the cell containing the atoms to be optically pumped, and the detection system of fluorescence light. In most cases, the cell is situated *inside* the laser cavity. Nevertheless, in order to induce orientation in an atomic level, it is more convenient to put the cell *outside* the cavity.

Let us describe first a typical arrangement including an internal cell (Fig. 5.1). The other arrangements will be depicted later by reference to this one.

Internal Cell Configuration

Between the two mirrors of the laser cavity [5.11, 14, 22] a tube L filled with a helium-neon mixture insures laser oscillation on a single line λ_L of the neon spectrum. Numerous precautions must be taken to avoid simultaneous oscillations on several lines. In particular, methane eliminates the 3.39 μm line. Due to Brewster-angle windows, the laser light is purely linearly polarized. Its intensity inside the cavity is deduced from a knowledge of the transmission coefficient of the mirrors and the detection of the laser light behind it. Whenever it is important, the spectral features of the laser beam are analysed either with a Fabry-Perot interferometer or by the use of beat frequencies between modes detected with a fast detector D and measured with a spectrum analyser calibrated with a reference oscillator.

A cell C, independent of L, is also placed in the cavity. C contains the atoms that are to be studied (in general, a mixture of neon and other noble gases). The gas contained in C is excited by a dc discharge of intensity i_c and irradiated by the linearly polarized line λ_L, resonant for

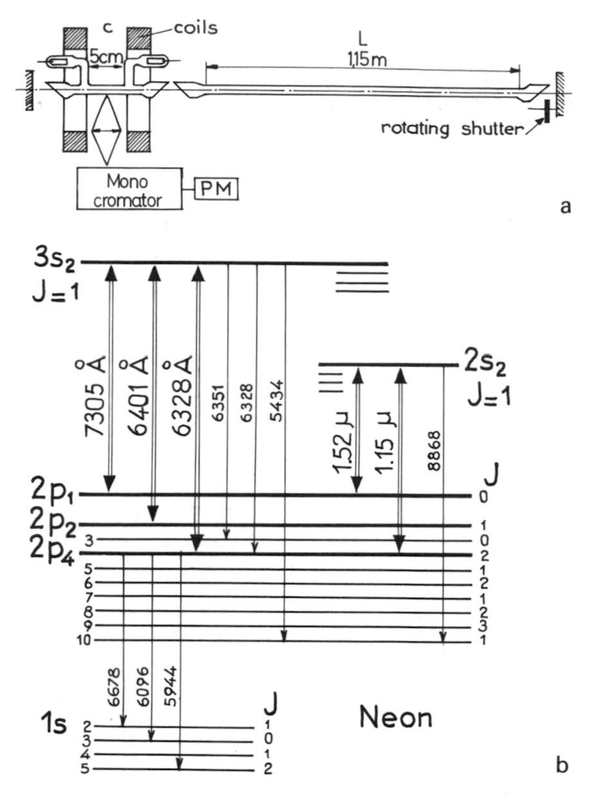

Fig. 5.1. (a) Typical experimental set-up with the cell inside the laser cavity. (b) Neon energy levels with the laser lines (double arrows) and fluorescence lines used

a transition between an upper level b and a lower level a. The atoms of the cell are also subjected to an axial dc magnetic field[1].

The fluorescence light detected with a monochromator and a photomultiplier is analyzed perpendicular to the axis of the cell, both parallel to and perpendicular to the polarization of the laser light. In each direction, a polarizer allows the detection of the π or σ component. In most cases, we are only interested in the laser-induced variations $\Delta L_{\parallel}^{\sigma \text{ or } \pi}$ and $\Delta L_{\perp}^{\sigma \text{ or } \pi}$ of the fluorescence light in different directions and polarization components. Therefore, to improve the signal-to-noise ratio, it is useful to introduce a periodic shutter inside the cavity. The synchronous modulations of the photocurrents of the photomultipliers are then detected by lock-in amplifiers.

[1] To perform a double-resonance experiment, one must add a radio-frequency magnetic field of appropriate frequency perpendicular to the dc magnetic field, i.e. perpendicular to the axis of the cell.

The main advantage of this experimental configuration is the strong pumping rate that can be obtained inside the cavity, resulting in significant laser induced variations of the fluorescence light and easily detectable nonlinear effects. Nevertheless, it has two drawbacks:

1) The necessarily linear polarization of the laser limits the variety of optical pumping experiments. In particular, orientation cannot be induced in an excited atomic level in this kind of experiment.

2) It is impossible to change the total intensity i_λ of the laser beam without simultaneously changing the geometry of the pumping beam and, in general, the number and frequency of oscillating modes. This greatly complicates the analysis of the nonlinear effects, the study of reliable observables, and the extrapolation of results to vanishing laser intensities.

External Cell Configuration

One would suppose that bringing the cell out of the cavity would avoid these complications. This arrangement has been used [5.14, 23] to study the orientation in $3s_2$ and $2p_4$ levels of neon. But even with a rather intense cw gas laser source (85 mW) the laser-induced variations in the fluorescence light remain small. To improve the signal-to-noise ratio, polarization modulation of the pumping light is used: the laser beam crosses a quarter-wavelength plate rotating in its plane at a frequency Ω. Whenever the axes of the quarter-wavelength plate are parallel to or perpendicular to the laser polarization, i.e. 4 times per period, the polarization remains linear. Therefore, it is easy to see that the signal observed at 4Ω with a lock-in amplifier will provide the same signal as that obtained in the internal cell configuration. Exactly as in optical pumping experiments with a conventional linearly polarized light source, transverse alignment is then detected on the fluorescence light. By contrast, whenever the angle between the axis of the quarter-wavelength plate and the laser polarization is $\pi/4$, the laser polarization becomes σ^+ or σ^- (2 times σ^+ and 2 times σ^- per period). Hence, a new kind of optical pumping is induced at frequency 2Ω. Fluorescent light studied at that frequency through a circular analyzer will provide the orientation signal.

In these kinds of experiments the two drawbacks of the internal cell arrangement disappear, but as the signal-to-noise ratio remains relatively low, the method is limited to strong laser lines (the 6328 Å line of neon, for example). For this reason, the internal cell arrangement has been used whenever possible.

5.1.3. Observation of Linear Effects

Hanle Effect

α) Alignment and Population with Internal Cell

By analogy with optical pumping with conventional light sources, the σ pumping in the internal cell arrangement should induce variations in the populations of the upper and lower levels and create alignment in both levels. To measure these quantities, it is convenient to observe the σ components $\Delta L_{\parallel}^{\sigma}$ and $\Delta L_{\perp}^{\sigma}$ of the fluorescence light.

Typical curves of $\Delta L_{\parallel}^{\sigma}$ and $\Delta L_{\perp}^{\sigma}$ versus dc magnetic field H are shown in Fig. 5.2. All the curves appear to be of Lorentzian shape. From this kind of result two quantities are deduced: the width ΔH of the Lorentzian curve, and the anisotropy R in zero magnetic field defined as

$$R = \frac{\Delta L_{\parallel}^{\sigma}(0) - \Delta L_{\perp}^{\sigma}(0)}{\Delta L_{\parallel}^{\sigma}(0) + \Delta L_{\perp}^{\sigma}(0)} . \tag{5.1}$$

To improve the precision in the determination of the width one should take the difference between $\Delta L_{\parallel}^{\sigma}$ and $\Delta L_{\perp}^{\sigma}$

$$\Delta L^{\sigma}(H) = \Delta L_{\parallel}^{\sigma}(H) - \Delta L_{\perp}^{\sigma}(H) . \tag{5.2}$$

This experimentally determined curve is known as the Hanle curve.

For the sake of simplicity let us focus on the study of the upper level b of the laser transition $b \to a$. Measurements are done on several fluorescent lines $b \to g$. We denote the related kinetic momenta by J_b, J_a, and J_g. As will be shown in Section 5.3, at the limit of a very low pumping rate the width ΔH, independent of the fluorescence line involved, is simply connected to the alignment relaxation rate $\Gamma_b(2)$ of the b level

$$\Delta H = \frac{\Gamma_b(2)}{g_b \beta} , \tag{5.3}$$

where β is the Bohr magneton and g_b the Landé factor.

On the other hand, the anisotropy R is simply connected to the ratio between the population relaxation rate $\Gamma_b(0)$ and the alignment rate

$$R = \frac{(-)^{2J_b+J_a+J_g} \dfrac{1}{2} \begin{Bmatrix} 2 & 1 & 1 \\ J_a & J_b & J_b \end{Bmatrix} \begin{Bmatrix} 2 & 1 & 1 \\ J_g & J_b & J_b \end{Bmatrix}}{\dfrac{1}{g(2J_b+1)} \dfrac{\Gamma_b(2)}{\Gamma_b(0)} + \dfrac{1}{6} (-)^{2J_b+J_a+J_g} \begin{Bmatrix} 2 & 1 & 1 \\ J_a & J_b & J_b \end{Bmatrix} \begin{Bmatrix} 2 & 1 & 1 \\ J_g & J_b & J_b \end{Bmatrix}} \tag{5.4}$$

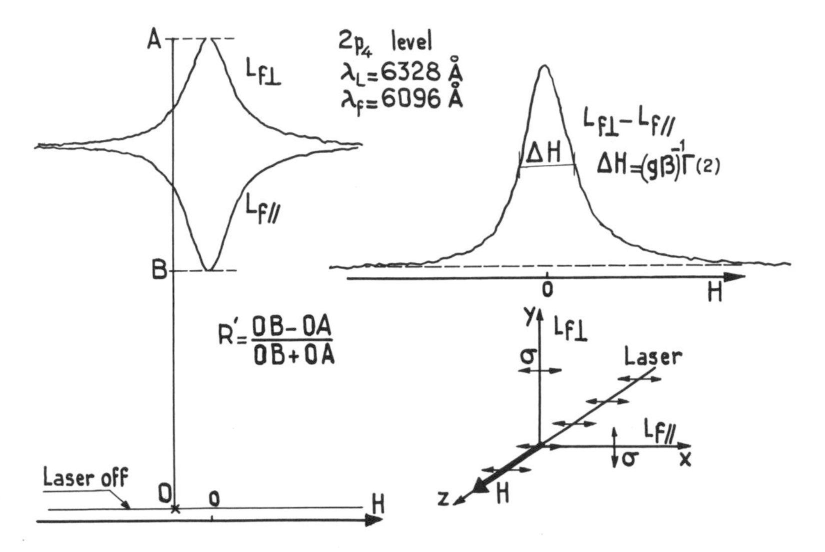

Fig. 5.2. Traces of the fluorescent light in the two detection directions as functions of the magnetic field; they illustrate the anisotropy ratio in zero magnetic field. The difference between the two signals (upper right) gives directly the Hanle effect

so that the experimental determination of the width ΔH and the anisotropy R of the σ components of the fluorescence light allows the measurement of the two relaxation rates of population and alignment[2].

A wide range of relaxation studies of excited levels of neon relies upon these two results [5.11, 22]. For instance, by varying the partial pressures of the gases in the cell, one can deduce collision cross-sections for the alignment and for the population (quenching collision). Important experimental and theoretical developments have resulted from these investigations [5.22]. The method also gives rise to a new study of coherent trapping of the fluorescent line. At the present time, research of this kind is being developed for the study of Penning collisions, which may occur in discharges.

It is to be noted that the results discussed above have been obtained with a multimode laser. As will be shown in the theoretical analysis, if the results were interpreted in terms of relaxation rates with a monomode pumping laser, large errors could arise.

[2] The ratio $\Gamma_b(2)/\Gamma_b(0)$ giving rise to the knowledge of the population relaxation rate can also be deduced from the polarization rate $(\Delta L^\sigma - \Delta L^\pi)/(\Delta L^\sigma + \Delta L^\pi)$ of the fluorescence light in zero magnetic field. This method has been recently used by TOSCHEK and coworkers, for the study of Ne*–Ar collision processes [5.24].

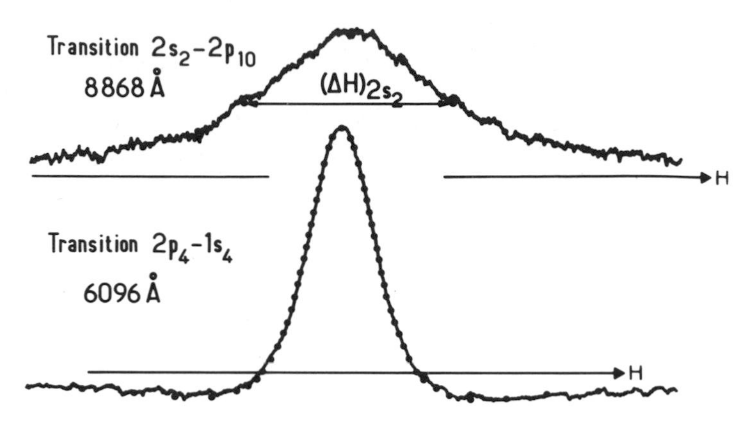

Fig. 5.3. Comparison of the direct Hanle effect (upper curve) and of the indirect Hanle effect (lower curve) obtained by spontaneous emission from the $2s_2$ to the $2p_4$ levels. The $2s_2$ level is directly pumped by the 1.52 μm laser line ($2s_2-2p_1$). Points show the best fit of the theoretical shape with the experimental results and give a measure of $\alpha=\Gamma^{(2)}_{2p4}/\Gamma^{(2)}_{2s_2}$

β) The Case of an External Cell [5.14, 23]

Besides results identical to those obtained with an internal cell, the σ^+ or σ^- pumping that can be performed with an external cell gives rise to an orientation signal, as indicated in *External cell configuration* above.

The 2Ω modulated component of the fluorescence line detected through a circular polarizer also gives a Lorentzian shaped curve as a function of the magnetic field. The width of this Hanle curve determines the relaxation rate of orientation $\Gamma_b(1)$. This method allowed the study of orientation relaxation in $3s_2$ and $2p_4$ levels of neon and provided an interesting comparison with alignment relaxation. Unfortunately, the signal-to noise ratio is too poor to encourage an extension of the method to the various laser transitions that can be used in the internal cell arrangement.

γ) Cascade Effects

The changes in population, alignment, or orientation induced by the laser in the upper and lower levels of the corresponding laser transition are to some extent transferred by spontaneous emission to lower levels g. The fluorescence lines emitted by those levels ($g\rightarrow h$ transitions) also exhibit, around the zero-magnetic-field, cascade Hanle-effect curves (Fig. 5.3) and anisotropy of the σ component [5.25, 26].

A detailed analysis of the shape of the cascade curve [5.26, 27] gives a measure of the alignment or orientation relaxation rates of the level g. Furthermore, the anisotropy in zero magnetic field gives the ratio between the relaxation rate of alignment and the relaxation rate of popula-

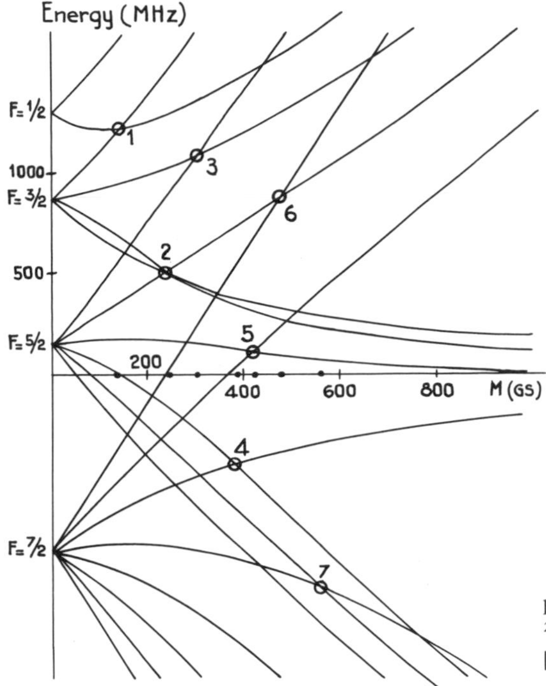

Fig. 5.4. Energy diagram of hyperfine sublevels of the ^{21}Ne $2p_4$ level (from GIACOBINO [5.60]). The circles give the $\Delta m = 2$ crossings

tion. The method is in principle very rich in results since, with a convenient laser pumping between two highly excited levels, it allows one to obtain the relaxation rates of all the lower levels susceptible to coupling by spontaneous emission to the two previous ones. The validity of the method has been demonstrated quantitatively from a comparison with the results obtained in direct laser optical pumping experiments.

δ) Cascade Effect on the Lower Levels of the Laser Transition

Among the various lower levels that can be coupled by spontaneous emission to the upper level b of the laser transition is the lower level a. Consequently, for a lower level a, direct interaction with the laser beam and cascade effect from the upper level interfere. It can be shown in most cases that the measurement of $\Gamma_a(2)$ or $\Gamma_a(1)$ from the width of the Hanle curves is not appreciably modified by this complication. On the other hand, the interpretation of the anisotropy in zero magnetic field is greatly affected. To neglect the cascade contribution would give rise to enormous errors in the measurement of $\Gamma_a(0)$. It can be shown [5.22] that the transition probability of the laser line appears in the detailed correc-

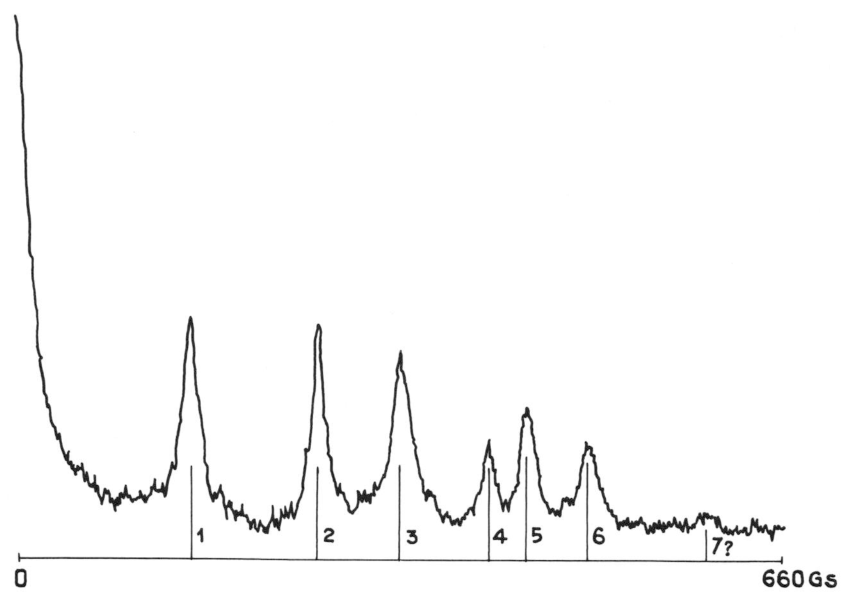

Fig. 5.5. Recording of the $2p_4$ level-crossings (from GIACOBINO [5.60])

tions and that its measurement can be deduced from the results of R in a lower level.

Magnetic Resonance and Level-Crossing Experiments

α) Measurement of g Landé Factor [5.12]

Adding a radio-frequency field perpendicular to the dc magnetic field allows double resonance experiments. Due to the short lifetime of the excited levels (10^{-8} sec or less) a rf field of several gauss must be applied to the cell. The resonance curves observed on the fluorescence light give the g Landé factor. Experiments have been performed with a pure even isotope of neon on the $3s_2$ level with the 7305 Å laser line, and on the $2p_4$ level with the 6328 Å laser line; this has improved the determination of the corresponding g values.

β) Level-Crossing Experiments

If fine structure level-crossing lies at too high a magnetic field to be observed in these kinds of experiments, several hyperfine level-crossings of ^{21}Ne excited levels can be observed in a magnetic field of some hundred gauss. Recently, $\Delta m = 2$ level-crossings have been observed on $3s_2$ and $2p_4$ levels [5.13] (see Figs. 5.4 and 5.5). Measurements of the hyperfine structure in zero field can be deduced from these experiments.

Modulation Phenomena at the Beat Frequency between Modes

Each of the linear effects reported up to now is completely similar to effects observed in optical pumping experiments with a conventional light source. However, the appearance of modulations at the beat frequencies between modes in the fluorescence light is characteristic of the multimode laser optical pumping. This modulation is observable when the laser is mode-locked (either imposed mode-locked operation [5.2] or spontaneous mode locking [5.14, 19]) but not in free-running operation. The shape of the amplitude of modulation of the σ component versus magnetic field shows a resonance when the frequency of modulation is equal to the Zeeman splitting.

These effects will be discussed and interpreted in detail in Section 5.3. Let us remark that mode-locked operation is equivalent to an amplitude modulation of the laser intensity. This is very similar to experiments made with modulated conventional light sources [5.28].

5.1.4. Nonlinear Effects

Pure Nonlinear Effects

All the results identified as pure nonlinear effects have been observed up to now in the internal-cell experimental configuration. Due to the linearly polarized pumping light, an alignment is here induced. It is well known that the transverse alignment thus induced can be observed only on the σ component of the fluorescence light. When a conventional light source is used, there is no signal on the π component of the fluorescence light.

However, when a laser pumping source is employed, a dip generally appears in the π component of the fluorescence light. This dip, centered in zero magnetic field, is approximately Lorentzian shaped; it is typical of a nonlinear effect. It can be interpreted as an increase of saturation processes that affect populations and coherences when the σ^+ component and the σ^- component of the σ laser light interact with the same atoms. This intuitive interpretation justifies the name given to this effect: *zero magnetic field saturation resonance*. In fact, as discussed in Sections 5.3–7 it is rather complicated to understand the shape, width, and amplitude of this saturation resonance. For example, in the special case of $J=1 \leftrightarrow J=2$ laser transition, the saturation resonance observed on a $J=2 \rightarrow J=2$ fluorescence line appears to be negative when the pumping rate is sufficient. This special effect, which cannot be understood by a perturbative method, is discussed in Section 5.7.

Similarly, in multimode operation, whenever the σ^+ component of one laser mode interacts the same atoms as the σ^- component of another mode, a nonzero magnetic field saturation resonance must appear. In fact, the dips are only visible in mode-locking operation, like the resonances of the modulated fluorescence light (see *Modulation phenomena at the beat frequencies between modes*). The exact position of these dips that appear in the π fluorescence light is discussed and interpreted in Section 5.3. In particular, in some special cases, the g_J Landé factor can be measured from this effect. Saturation resonances appear also on the σ components of the fluorescence light but they are isotropic (identical on $\Delta L_{\parallel}^{\sigma}$ and $\Delta L_{\perp}^{\sigma}$) and can be cancelled out by the difference method defined by (5.2).

Furthermore, the Hanle curve itself is distorted by saturation phenomena; in many cases these appear experimentally as a simple broadening of the Lorentzian curve. Nevertheless, large distortions are observed in one case: when the laser transition is a $J=1 \leftrightarrow J=2$ transition, the Hanle effect that can be observed on lines coming from the $J_b=1$ level decreases very rapidly with increasing laser intensities; the shape is no longer Lorentzian, but with sufficient laser intensities shows three separate bumps. As discussed in Section 5.7, this special Hanle shape denotes the influence of the laser-induced hexadecapole electric moment in the $J_a=2$ level [5.21].

Method for Obtaining Reliable Experimental Results with Linear Effects

Nonlinear effects appear quite frequently as superimposed on the linear effects. All the linear results described in Subsection 5.1.3 are valid only at vanishing laser intensities and must be corrected at nonzero i_λ values.

For instance, in general[3], the width of Hanle curves increases with increasing pumping intensities. Experimentally, in almost all cases (except, as described below, for a $J=1$ level pumped with a $J=1-J=2$ laser line of high intensity) it is not possible to detect any departure from the Lorentzian line shape, while the width directly measured from the Hanle curve increases considerably with i_λ. Whereas, in many cases, the anisotropy R is laser-intensity dependent, it remains unchanged at varying laser intensities for a $J=1-J=0$ laser transition or for a $J=1-J=1$ laser transition. This particular effect is discussed in Section 5.7.

In double-resonance experiments or in level-crossing experiments care must be taken to ensure that possible light shifts induced by the laser do not introduce systematic errors in the measurements [5.12].

[3] In the external cell arrangement, due to its particular modulation of polarization, the width of the Hanle curve for the alignment decreases with increasing laser intensities [5.14].

Table 5.1

	1968 Internal cell Linear extrapolation [5.11, 22]	1969 External cell Linear extrapolation [5.14]	1971–72 Internal cell Extrapolation deduced from BLA theory Section 5.4 [5.16, 17, 29]
$2p_4$ level: γ	9.35 ± 0.30	8.64 ± 0.40	
$3s_2$ level			
$\gamma(0)$	2.45 ± 0.12		2.21 ± 0.15
$\gamma(1)$	5.41 ± 0.29[a]	4.70 ± 0.45	4.86 ± 0.23[a]
$\gamma(2)$	4.22 ± 0.10		3.80 ± 0.14
γ	8.35 ± 0.45[a]		7.50 ± 0.44[a]

γ is the natural width. $\gamma(0)$, $\gamma(1)$, $\gamma(2)$ are the extrapolated widths of population, orientation, and alignment at very low pressure but when the multiple scattering of the resonance line remains complete [5.22]. All the results are given in MHz.

Values[a] are deduced from the other results by calculation and not from direct measurements [5.14, 29].

Clearly, in all cases, the dependence of the observed signal upon the nonlinearities induced by the laser optical pumping imposes the following alternatives: Obtaining a correct experimental extrapolation at vanishing laser intensities, or having a good theory of nonlinearities in order to be able to interpret experimental results found at low, but finite intensity.

In fact, both terms of the alternative are necessary for reliable and precise measurements. To demonstrate this necessity, the results obtained by the present authors at different times with two experimental techniques and two extrapolation methods are shown in Table 5.1.

Although there is no significant discrepancy between the last two columns, due in particular to the low laser intensity and to the regular change of laser intensity in the external cell arrangement, the large differences between the first and the last columns show the need for a correct theory of nonlinearities, even to obtain extrapolation of results at vanishing laser intensities.

5.2. Semiclassical Equations: Hypotheses and Notations

To calculate the intensity and polarization of the fluorescence lines emitted by the optically pumped gas we have only to determine its average density matrix

$$\bar{\varrho}(\boldsymbol{r}, t) = \int \varrho(\boldsymbol{v}, \boldsymbol{r}, t) d^3 \boldsymbol{v} \tag{5.5}$$

and to apply the classical formulae for the spontaneous emission (Appendix 1). As the detection set-up does not resolve the spectral shape of a fluorescence line, and as the detection is perpendicular to the laser beam, it is not necessary to take into account the frequency correlation between the laser and the fluorescence light[4].

With conventional light sources, the linewidth of the pumping light is generally very broad and all atoms, whatever their velocities have an equal probability of being pumped. Furthermore, due to the incoherence of the pumping light, off-diagonal elements of the density matrix between two different levels (the so-called "optical coherences") vanish. In particular, there is no macroscopic electric dipole at optical frequencies. Hence, the absorption of a photon being instantaneous compared to all atomic lifetimes, it is possible to find a rate equation for the averaged density matrix of atomic levels, $\bar{\varrho}$. This theory of optical pumping with ordinary light sources has been developed by COHEN-TANNOUDJI [5.33] with a quantum-mechanical description of light.

With laser light sources, the pumping beam is made up of one or several monochromatic modes, each with a very long coherence time, so that a macroscopic electric dipole appears in the medium. Therefore it is easier to represent the laser light as a classical electric field in a semiclassical method like that used by LAMB [5.15] in his theory of gas laser. Because of the Doppler effect, atoms of different velocities do not see the same optical frequency. Therefore it is not possible as in COHEN-TANNOUDJI's theory to write a single equation for $\bar{\varrho}$ (except in some special cases, cf. Section 5.4), and a separate equation must be written for each value of the velocity.

To obtain the equation for $\varrho(\boldsymbol{v}, \boldsymbol{r}, t)$ we can choose a particular atom to write the Schrödinger equation in a frame moving with it, and to average the result for all atoms. This is the method used by LAMB [5.15] for a collisionless motion of atoms, and by GYORFFY et al. [5.34] for atoms randomly deviated by collisions. An alternative method, completely equivalent to the previous one with regard to classical trajectories employed for atomic motion[5], is to write the Schrödinger equation for the density matrix directly in the laboratory frame, utilizing a formulation like the Boltzmann equation [5.35]. We adopt this latter method because it is easier to write.

We assume the gas excitation to be spatially homogeneous and the laser beam to be approximated by a plane wave. Then the density matrix

[4] If the fluorescence light is detected in the direction of the laser beam with a Fabry-Perot interferometer, it is possible to resolve the velocity distribution of pumped atoms [5.30]. In that case (which is beyond the scope of this paper) the frequency correlation between the laser and the fluorescent light is of utmost importance [5.31, 32].

[5] The difficulties arising from the use of classical trajectories for collisions of atoms in a coherent superposition of two states have been studied by BERMAN and LAMB [5.36]. This problem, though very interesting for laser theory, is of little importance in this context.

depends only on the projection, on the laser axis, of the position (r) and of the velocity (v). This assumption implies that the distance covered by any excited atom during its lifetime is negligible compared to the beam diameter. (This hypothesis fails for long-living vibrational levels of molecules, as used in saturated absorption experiments.)

In what follows we shall call ${}^{(0)}\varrho$ the density matrix without optical pumping and ϱ the laser-induced modifications of this density matrix, so that the total density matrix is

$$^{(t)}\varrho = {}^{(0)}\varrho + \varrho\,.$$

It is given by

$$\begin{aligned} {}^{(t)}\dot{\varrho} &= \left(\frac{\partial}{\partial t} + v\frac{\partial}{\partial r}\right)^{(t)} \varrho(v,r,t) \\ &= -i[\mathscr{H}(r,t), {}^{(t)}\varrho(v,r,t)] + \Lambda(v) + \left[\frac{d}{dt}\,{}^{(t)}\varrho(v,r,t)\right]_{\text{relax}} \end{aligned} \tag{5.6}$$

where Λ expresses the excitation by the discharge. $\mathscr{H}$ is the Hamiltonian and $(d\varrho/dt)_{\text{relax}}$ contains all relaxation phenomena. We will now study all terms of this equation and indicate our approximations. For ϱ, $\mathscr{H}$, and Λ, the formalism has been defined by DUMONT and DECOMPS [5.37].

5.2.1. Density Matrix

The density matrix is restricted to the two levels connected to the laser transition (a is the lower level, b the upper one). It is composed of four submatrices visualized by

$$\varrho = \begin{pmatrix} {}_a\varrho & {}_{ab}\varrho \\ {}_{ba}\varrho & {}_b\varrho \end{pmatrix}. \tag{5.7}$$

The nondiagonal submatrices ${}_{ab}\varrho$ and ${}_{ba}\varrho$ represent optical macroscopic quantities. Following COHEN-TANNOUDJI [5.33], we call them "optical coherences". The submatrices ${}_a\varrho$ and ${}_b\varrho$ represent the state of atoms in the a and b levels; with the standard basis $|JM\rangle$, their diagonal elements are the "populations" of the Zeeman sublevels and their off-diagonal elements are the "Zeeman coherences".

Due to the symmetry of the relaxation processes, we shall represent the density operator on the basis of normalized irreducible tensors[6] [5.37–39].

$$_{\alpha\beta}\varrho = \sum_{kQ} {}_{\alpha\beta}\varrho^k_Q \, {}_{\alpha\beta}T^k_Q ; \qquad {}_{\alpha\beta}\varrho^k_Q = \mathrm{Tr}(\varrho \cdot {}_{\alpha\beta}T^{k\dagger}_Q) , \tag{5.8}$$

where α and β stand for a or b.

As is well known $_{\alpha}\varrho^0_0$ defines the total population of the level $\alpha(n_\alpha = \sqrt{2J_\alpha+1}\,{}_{\alpha}\varrho^0_0)$, the $_{\alpha}\varrho^1_Q$ define the three components of its orientation (magnetic dipole), and the $_{\alpha}\varrho^2_Q$ the five components of its alignment (electric quadrupole). The components $_{\alpha}\varrho^k_0$ are called longitudinal quantities and determine the "populations" of the Zeeman sublevels. The transverse quantities are $_{\alpha}\varrho^k_Q$, with $Q \neq 0$, related to the "Zeeman coherences". The three operators $_{ab}\varrho^1_Q$ are proportional to the optical electric dipole.

5.2.2. Excitation Matrix

The excitation matrix $\Lambda(v)$ is assumed to be isotropic and homogeneous and to be proportional to a Maxwellian velocity distribution

$$\begin{aligned} \Lambda(v) &= W_{\mathrm{M}}(v)\,[\lambda_{a\,a}T^0_0 + \lambda_{b\,b}T^0_0] , \\ W_{\mathrm{M}}(v) &= (u\sqrt{\pi})^{-1} \exp\left(-\frac{v^2}{u^2}\right) . \end{aligned} \tag{5.9}$$

The isotropy of Λ is not rigorous in all experimental cases, since the trapping of some fluorescence lines can introduce alignment along the axes of the capillary cell [5.56]. In that case, there is no theoretical difficulty in introducing $_{\alpha}T^2_Q$ tensor in (5.9), but the algebra becomes very complicated [5.14].

5.2.3. Hamiltonian

The Hamiltonian is expressed as the sum of the single atom part ($\mathscr{H}_0$), the Zeeman part ($\mathscr{H}_Z$), and the Hamiltonian of interaction with the laser beam $R(r, t)$,

$$\mathscr{H} = \mathscr{H}_0 + \mathscr{H}_Z + R(r, t) , \tag{5.10}$$

$$\mathscr{H}_0 = W_a\sqrt{2J_a+1}\,{}_aT^0_0 + W_b\sqrt{2J_b+1}\,{}_bT^0_0 , \tag{5.11}$$

[6] We use the normalization $\mathrm{Tr}({}_{\alpha\beta}T^k_Q \, {}_{\alpha'\beta'}T^{k'\dagger}_{Q'}) = \delta_{\alpha\alpha'}\delta_{\beta\beta'}\delta_{kk'}\delta_{QQ'}$ and the reduced matrix element is $\langle J_\alpha \| {}_{\alpha'\beta'}T^k_Q \| J_\beta \rangle = \delta_{\alpha\alpha'}\delta_{\beta\beta'}\sqrt{2k+1}$.

$$\mathscr{H}_Z = -M_Z H = \omega_a \sqrt{\frac{J_a(J_a+1)(2J_a+1)}{3}}\, {}_aT_0^1 + \omega_b \sqrt{\frac{J_b(J_b+1)(2J_b+1)}{3}}\, {}_bT_0^1 \,, \tag{5.12}$$

$$R(r,t) = -\boldsymbol{P}\cdot\boldsymbol{E}(r,t) = -\sum_q (-)^q P_q E_{-q} \,, \tag{5.13}$$

where W_a and W_b are the unperturbed level energies, ω_a and ω_b are the Zeeman splittings ($\omega_\alpha = g_\alpha \beta H$; g_α: Landé factor; β: Bohr magneton), $\boldsymbol{P}$ is the electric dipole operator, and $\boldsymbol{E}$ the laser-beam electric field. P_q and E_q are the standard components, i.e.

$$E_q = E_z \,, \qquad E_{\pm 1} = \frac{\mp 1}{\sqrt{2}}(E_x \pm iE_y) \,, \tag{5.14}$$

and $\boldsymbol{P}$ has only off-diagonal optical matrix elements and can be expressed in the form [$P_{ab} = (-)^{J_b - J_a} P^*_{ba}$, which is the reduced matrix element]

$$P_q = \left(\frac{P_{ab}}{\sqrt{3}}\right) {}_{ab}T_q^1 + \left(\frac{P_{ba}}{\sqrt{3}}\right) {}_{ba}T_q^1 \,. \tag{5.15}$$

The laser beam is composed of several modes

$$\boldsymbol{E}(r,t) = \sum_\mu \{\mathscr{E}^\mu \exp[-i(\omega_\mu t - k_\mu r)] + \mathscr{E}^{\mu *} \exp[i(\omega_\mu t - k_\mu r)]\} \,.$$

The complex vector $\mathscr{E}^\mu$ contains the relative phase of the modes and all the information about the polarization (relative phases of the three components). As all modes have the same polarization defined by the complex vector $\boldsymbol{e}$, we shall use the notation

$$\begin{aligned} \mathscr{E}^\mu &= \mathscr{E}^\mu \boldsymbol{e} = \sqrt{I_\mu}\exp(i\phi_\mu)\boldsymbol{e} \\ \boldsymbol{E}(r,t) &= \boldsymbol{e}\mathscr{E} + \mathrm{cc} = \boldsymbol{e}\sum_\mu \mathscr{E}^\mu \mathrm{e}^{-i(\omega_\mu t - k_\mu r)} + \mathrm{cc} \,. \end{aligned} \tag{5.16}$$

In the present discussion we consider a forward-traveling wave, therefore k_μ is positive. When the cell is inside the laser cavity (as in experiments on neon), one must take into account two waves of opposite $\boldsymbol{k}$ for each mode.

5.2.4. Relaxation

The relaxation phenomena include collisions and radiative effects. They are assumed to be isotropic [5.40, 41] and the "strong collision" model [5.34, 35] is used to describe the velocity changes due to collisions or trapping of fluorescence lines (no memory for the initial speed)

$$\left[\frac{d}{dt}\,{}_{\beta}\varrho_q^k(v,r,t)\right]_{\text{rel.}} = -\Gamma'_{\beta}(k)\cdot{}_{\beta}\varrho_q^k(v,r,t) + \gamma'_{\beta}(k)\cdot W_{\text{M}}(v)\cdot{}_{\beta}\overline{\varrho}_q^k(r,t)\,, \tag{5.17}$$

where ${}_{\beta}\overline{\varrho}$ is defined by (5.5) and $W_m(v)$ by (5.9). $\Gamma'_{\beta}(k)$ is the k-tensor relaxation rate, due either to the destruction of the tensorial quantity or to the velocity change of the atom. $\gamma'_{\beta}(k)$ is the restoration rate from any other velocity. The relaxation rates of the velocity-averaged density matrix ${}_{\beta}\overline{\varrho}$ are given by

$$\Gamma_{\beta}(k) = \Gamma'_{\beta}(k) - \gamma'_{\beta}(k)\,. \tag{5.18}$$

For the optical coherence, we shall neglect the velocity changes

$$\left[\frac{d}{dt}\,{}_{ab}\varrho_q^k(v,r,t)\right]_{\text{rel.}} = -\Gamma_{ab}(k)\cdot{}_{ab}\varrho_q^k(v,r,t)\,, \tag{5.19}$$

while, on the one hand, the velocity changes due to fluorescence trapping do not restore optical coherence, on the other hand, BERMAN and LAMB [5.36] have shown that there is no velocity-changing collision for the optical coherence if the scattering potentials are different in the a and b levels (a highly likely situation). In the case, it is not possible to define a classical trajectory for ${}_{ab}\varrho$, because the velocity changes are different in the a and b states.

5.2.5. Spontaneous Emission

The spontaneous emission from the upper level b to the lower level a produces a transfer of atomic quantities which is expressed by [5.27]

$$\left(\frac{d}{dt}\,{}_{a}\varrho_Q^k\right)_{tr} = \theta(b,a,k)_b\varrho_Q^k = -(-)^{J_a+J_b+k}\gamma_{ba}(2J_b+1)\begin{Bmatrix} k & J_b & J_b \\ 1 & J_a & J_a \end{Bmatrix}. \tag{5.20}$$

5.2.6. Developed Equations

When there is no laser beam, $R(r, t)$ cancels out in the Hamiltonian (5.10) and it is easy to see, from the form (5.9) of the source term Λ, that the only nonvanishing terms in the stationary solution of (5.6) are the populations of both levels ($\alpha = a$ or b)

$$ {}^{(0)}_{\alpha}\varrho^0_0(v) = {}^{(0)}_{\alpha}\bar{\varrho}^0_0 W_{\mathrm{M}}(v) = \frac{\lambda_\alpha}{\Gamma_\alpha(0)} W_{\mathrm{M}}(v) = \frac{n_\alpha W_{\mathrm{M}}(v)}{\sqrt{2J_\alpha + 1}}, \tag{5.21} $$

where λ_a is assumed to contain the spontaneous emission from b to a.

Projecting (5.6) onto the basis of irreducible tensors according to (5.8), and subtracting the laser-off equation, one obtains for the laser-induced modification of the density matrix

$$ \begin{aligned} &{}_a\dot{\varrho}^k_Q(v) + [iQ\omega_a + \Gamma'_a(k)]_a\varrho^k_Q(v) - \gamma'_a(k) W_{\mathrm{M}}(v)_a\bar{\varrho}^k_Q - \theta(b, a, k)_b\varrho^k_Q(v) \\ &= i \sum_{k'Q'q} (-)^{k+k'}\, {}^q_{ab}G^{k'k}_{Q'Q}[e_{-q}\mathscr{E}\, {}_{ab}\varrho^{k'}_{Q'}(v) P^*_{ab} \\ &+ (-)^{k+k'+Q} e^*_q \mathscr{E}^*\, {}_{ab}\varrho^{k'*}_{-Q'}(v) P_{ab}], \end{aligned} \tag{5.22a} $$

$$ \begin{aligned} &{}_b\dot{\varrho}^k_Q(v) + (iQ\omega_b + \Gamma'_b(k))\, {}_b\varrho^k_Q(v) - \gamma'_b(k) W_{\mathrm{M}}(v)\, {}_b\bar{\varrho}^k_Q \\ &= i \sum_{k'Q'q} {}^q_{ba}G^{k'k}_{Q'Q}[e_{-q}\mathscr{E}\, {}_{ab}\varrho^{k'}_{Q'}(v) P^*_{ab} + (-)^{k+k'+Q} e^*_q \mathscr{E}^*_{ab}\varrho^{k'*}_{-Q'}(v) P_{ab}], \end{aligned} \tag{5.22b} $$

$$ \begin{aligned} &{}_{ab}\dot{\varrho}^{k'}_{Q'}(v) + \left[i\left(\frac{\omega_a + \omega_b}{2} Q' - \omega\right) + \Gamma_{ab}(k')\right] {}_{ab}\varrho^{k'}_{Q'}(v) + i\,\frac{\omega_a - \omega_b}{2} \\ &\cdot \left\{Q' \frac{J_a(J_a+1) - J_b(J_b+1)}{k'(k'+1)}\, {}_{ab}\varrho^{k'}_{Q'}(v) + a(k')_{ab}\varrho^{k'-1}_{Q'}(v)\right. \\ &\left. + a(k'+1)_{ab}\varrho^{k'+1}_{Q'}(v)\right\} \\ &= iP_{ab}\mathscr{E}^* \sum_{kQq'} e^*_{-q'}[{}^{q'}_{ba}G^{k'k}_{Q'Q}\, {}_b\varrho^k_Q(v) + (-)^{k+k'}\, {}^{q'}_{ab}G^{k'k}_{Q'Q}\, {}_a\varrho^k_Q(v)] \\ &+ iP_{ab}\mathscr{E}^* e^*_{Q'} \delta_{k',1} \frac{n W_{\mathrm{M}}(v)}{\sqrt{3}}, \end{aligned} \tag{5.22c} $$

$$ {}_{ab}\varrho^{k'}_{Q'} = (-)^{J_a - J_b + Q'}\, {}_{ab}\varrho^{k'*}_{-Q'}, \tag{5.22d} $$

where $\omega = W_b - W_a$ is the atomic frequency, and the geometric factors are defined by

$$ {}_{\alpha\beta}^{\,q}G_{Q'Q}^{k'k} = (-)^{J_\alpha + J_\beta + Q'} \sqrt{(2k+1)(2k'+1)} \begin{pmatrix} k' & 1 & k \\ Q' & q & -Q \end{pmatrix} \begin{Bmatrix} k' & 1 & k \\ J_\alpha & J_\alpha & J_\beta \end{Bmatrix} \tag{5.23} $$

$$ a(k) = \frac{1}{k} \{[(J_a + J_b + 1)^2 - k^2][k^2 - (J_a - J_b)^2] \cdot [k^2 - Q^2][4k^2 - 1]^{-1}\}^{1/2} . \tag{5.24} $$

In (5.22c) the last source term contains the laser-off population inversion, see (5.21)

$$ n = \frac{n_b}{2J_b + 1} - \frac{n_a}{2J_a + 1} . \tag{5.25} $$

Rotating-Wave Approximation

In (5.22), the right-hand side of each equation represents the laser-induced coupling. In these terms the *rotating-wave approximation* has been made: for the atomic quantities (5.22a) and (5.22b) only source terms of low frequency ($\simeq \omega_\nu - \omega_\mu$) can be resonant. For ${}_{ab}\varrho$ (or ${}_{ba}\varrho$) only positive (negative) optical frequencies ($\simeq \omega_\nu$) are retained.

Normal Zeeman-Pattern Approximation

In the left-hand side of (5.22c) there is a coupling between different tensorial orders; this coupling, which disappears when $\omega_a = \omega_b$, is due to the fact that ${}_{ab}\varrho_Q^k$ is a linear superposition of all $\langle J_a M_a | \varrho | J_b M_b \rangle$ matrix elements with $M_a - M_b = Q$, which correspond to different optical frequencies when a magnetic field is applied (anomalous Zeeman effect). We shall neglect this coupling[7]. The approximation is then valid when $\omega_a - \omega_b \ll \Gamma_{ab}$ (unresolved anomalous Zeeman structure), but in many cases this range of validity may be considerable broadened (with the BLA, see Section 5.4, it is $\omega_a - \omega_b \ll N\Delta\omega$). Obviously, when J_a or $J_b = 0$, the coupling disappears without any approximation (as $k' = 1$).

To solve (5.22) an iterative procedure can be used [5.15, 37]. This method is valid only at low laser intensity as it is not easy to go up to high perturbation orders and the development does not converge with very high laser intensity. The advantage of this method is that no other approximations are needed concerning the laser spectrum and the relaxation processes. Furthermore, the perturbation method provides

[7] Using the $|J_a M_a\rangle\langle J_b M_b|$ basis (instead of the irreducible tensors) for the optical coherences alone, it is possible rigorously to avoid this coupling, since the Zeeman Hamiltonian becomes diagonal. Nevertheless, on this basis, it is necessary to assume all $\Gamma_{ab}(k')$ to be equal in order to avoid a new coupling by relaxations [5.14].

a good understanding of physical processes. It is summarized in Section 5.3, while Section 5.4 is devoted to nonperturbative solutions that are valid for high laser intensities.

5.3. Perturbation Method

5.3.1. Iteration

We are seeking for the stationary solution of (5.22) as a perturbation development according to the successive powers of the laser electric field

$$\varrho = {}^{(t)}\varrho - {}^{(0)}\varrho = {}^{(1)}\varrho + {}^{(2)}\varrho + {}^{(3)}\varrho \ldots . \tag{5.26}$$

Introducing this development into (5.22) we get a set of equations corresponding to each power of the laser field. They can be formally written as

$$ {}^{(n)}\mathring{\varrho} + i[\mathscr{H}_0 + \mathscr{H}_Z, {}^{(n)}\varrho] - \left(\frac{d\,{}^{(n)}\varrho}{dt}\right)_{\text{rel.}} = -i[R, {}^{(n-1)}\varrho] . \tag{5.27}$$

If we start with the first order and solve the equations, order by order, in their developed form (5.22), it is easy to see that:

a) At odd orders, only optical coherences ${}_{ab}\varrho^{k'}_{Q'}$ are obtained. They give the linear (first order) and the nonlinear (third, fifth, ... order) susceptibilities of the medium. They allow one to calculate the index of refraction and represent the basis of the laser theory [5.15].

b) At even orders, atomic quantities ${}_a\varrho^k_Q$ and ${}_b\varrho^k_Q$ appear alone. We shall focus our interest on them.

c) At every order, the source term on the right-hand side is the sum of several terms of the form $\exp[i(\Omega_n t - k_n r)]$, where $\Omega_n t - k_n r \approx \omega_\nu t - k_\nu r$ at odd orders, and $\Omega_n t - k_n r \approx (\omega_\nu - \omega_\mu)t - (k_\nu - k_\mu)r$ at even orders. Therefore the stationary solution is a sum of terms with the same space-time dependence. As

$$\left(\frac{\partial}{\partial t} + v\frac{\partial}{\partial r}\right)\exp[i(\Omega_n t - k_n r)] = i(\Omega_n - k_n v)\exp[i(\Omega_n t - k_n r)]$$

at each order, one gets a set of independent algebraic equations (one for each frequency [5.19]). Furthermore, it is easy to show that $(k_\nu - k_\mu)v$ can be neglected in even-order equations. This means that the Doppler effect is negligible at low frequencies such as $\omega_\nu - \omega_\mu$ (10^2 to 10^3 MHz with conventional gas lasers). This approximation may fail with very short lasers ($\omega_\nu - \omega_\mu$ large) and long-lived molecular levels (Γ_α small).

5.3.2. Linear Response of Atomic System

At second order, the atomic quantities are expressed as quadratic forms in the electric field; therefore, if the spectrum of the beam remains constant, they are proportional to the total laser intensity. One obtains

$$
\begin{aligned}
{}^{(2)}_{b}\varrho^{k}_{Q}(v) &= -n\frac{|P_{ab}|^2}{\sqrt{3}} \\
&\cdot \sum_{\substack{\nu\mu \\ q_1 q_2}} {}^{q_2}_{ba}G^{1k}_{q_1 Q}\, e^{*}_{q_1} e_{-q_2} \mathscr{E}^{\nu *} \mathscr{E}^{\mu} \mathrm{e}^{i[(\omega_\nu - \omega_\mu)t - k_\nu - k_\mu)r]} b^{k}_{Q}(v, \nu-\mu)\,,
\end{aligned}
\tag{5.28}
$$

$$
\begin{aligned}
b^{k}_{Q}(v, \nu-\mu) &= \frac{A^{\nu\mu}_{q_1 q_2}(v)}{\Gamma'_b(k) + i(\omega_\nu - \omega_\mu + Q\omega_b)} \\
&+ \left[\frac{1}{\Gamma_b(k) + i(\omega_\nu - \omega_\mu + Q\omega_b)}\right. \\
&- \left.\frac{1}{\Gamma'_b(k) + i(\omega_\nu - \omega_\mu + Q\omega_b)}\right] W_{\mathrm{M}}(v) \int A^{\nu\mu}_{q_1 q_2}(v')dv'\,,
\end{aligned}
\tag{5.29}
$$

$$
\begin{aligned}
A^{\nu\mu}_{q_1 q_2}(v) = W_{\mathrm{M}}(v) &\left[\frac{1}{\Gamma_{ab}(1) - i(\omega - q_1\omega_{\mathrm{Z}} - \omega_\nu + k_\nu v)}\right. \\
&+ \left.\frac{1}{\Gamma_{ab}(1) + i(\omega + q_2\omega_{\mathrm{Z}} - \omega_\mu + k_\mu v)}\right].
\end{aligned}
\tag{5.30}
$$

The geometrical factor G has been defined by (5.23), and $\omega_{\mathrm{Z}} = (\omega_a + \omega_b)/2$ is the mean Zeeman splitting (if $J_{a,b} = 0$, $\omega_{\mathrm{Z}} = \omega_{b,a}$).

In (5.29) the first term comes from atoms which have not changed velocity by collision (or trapping of fluorescence). The velocity distribution, defined by $A^{\nu\mu}_{q,q_2}(v)$, has a resonant behavior (for $K = Q = 0$ and $\nu = \mu$ it is nothing else but Bennett's holes [5.45]). The second term in (5.29) represents the contribution of atoms whose velocity has changed. Its velocity distribution is $W_{\mathrm{M}}(v)$, according to the strong-collisions model.

To get ${}^{(2)}_{a}\varrho^{k}_{Q}$, one has to exchange a and b indexes, to multiply (5.28) by $(-)^{k+1}$, and to take into account the spontaneous emission from the b level by replacing

$$[\Gamma'_b(k) + i(\omega_\nu - \omega_\mu + Q\omega_b)]^{-1} \quad \text{in (5.28) by}$$

$$
\begin{aligned}
&\frac{1}{[\Gamma'_a(k) + i(\omega_\nu - \omega_\mu + Q\omega_a)]_t} \\
&= \frac{1}{\Gamma'_a(k) + i(\omega_\nu - \omega_\mu + Q\omega_a)}\left[1 - \frac{\gamma_{ba}\mathscr{A}(b, a, 1, k)}{\Gamma'_b(k) + i(\omega_\nu - \omega_\mu + Q\omega_b)}\right]
\end{aligned}
\tag{5.31}
$$

with

$$\mathscr{A}(b,a,k',k) \tag{5.32}$$
$$=(-)^{J_a-J_b+k'}(2J_b+1)\begin{Bmatrix}k & J_b & J_b\\ 1 & J_a & J_a\end{Bmatrix}\begin{Bmatrix}k' & 1 & k\\ J_b & J_b & J_a\end{Bmatrix}\begin{Bmatrix}k' & 1 & k\\ J_a & J_a & J_b\end{Bmatrix}^{-1}.$$

Integrating (5.28) over v yields the final result (to obtain ${}^{(2)}_{a}\overline{\varrho}^{k}_{Q}$ the above substitutions are still valid)

$$\begin{aligned}{}^{(2)}_{b}\overline{\varrho}^{k}_{Q} = & -n|P_{ab}|^2 \sum_{\substack{q_1q_2\\ \nu,\mu}} {}^{q_2}_{ba}G^{1k}_{q_1Q}\, e^*_{q_1} e_{-q_2}\\ & \cdot \frac{\mathscr{E}^{\nu *}\mathscr{E}^{\mu}\,\mathrm{e}^{i[(\omega_\nu-\omega_\mu)t-(k_\nu-k_\mu)r]}}{\Gamma_b(k)+i(\omega_\nu-\omega_\mu+Q\omega_b)} \int A^{\nu\mu}_{q_1q_2}(v)dv\,.\end{aligned} \tag{5.33}$$

The last integral (optical line-shape factor) can be expressed with the help of the plasma dispersion function Z [5.46]

$$\begin{aligned}\int_0^\infty \frac{W_{\mathrm{M}}(v)dv}{\Gamma-i(\Omega+kv)} &= \frac{1}{u\sqrt{2}}\int_0^\infty \frac{\exp(-v^2/u^2)dv}{\Gamma-i(\Omega+kv)}\\ &= -\frac{i}{\Delta\nu}Z\left(\frac{\Omega}{\Delta\nu}+i\frac{\Gamma}{\Delta\nu}\right)=\frac{1}{\Delta\nu}[X(\Omega)+iY(\Omega)]\,,\end{aligned} \tag{5.34}$$

where $\Delta\nu = ku$ is the Doppler linewidth, and $X(\Omega)$ which is the imaginary part of Z is the Voight line shape. At the limit of the *Doppler approximation* ($\Delta\nu \gg \Gamma_{ab}$) $X(\Omega)$ become simply a gaussian curve

$$X(\Omega) \doteq \sqrt{\pi}\exp(-\Omega^2/\Delta\nu^2)\,. \tag{5.35}$$

dc *Terms: Hanle Effect*

If the fluorescence light emitted by the gas is detected with a dc detector, only the zero-frequency terms $(\omega_\nu-\omega_\mu)$ need be kept in (5.33). For the transverse quantities ($Q \neq 0$), the Z function varies very slowly when the magnetic field is scanned all over the resonance of the denominator ($\Delta\nu \gg \Gamma_b(k)$); therefore one can write

$$\begin{aligned}{}^{(2)}_{b}\overline{\varrho}^{k}_{Q}(\mathrm{dc}) = & -n\frac{|P_{ab}|^2}{\Gamma_b(k)+iQ\omega_b}\sum_{q_1q_2} {}^{q_2}_{ba}G^{1k}_{q_1Q}\, e^*_{q_1} e_{-q_2}\\ & \cdot \sum_\nu \frac{2I_\nu}{\Delta\nu} X(\omega-\omega_\nu)\,.\end{aligned} \tag{5.36}$$

This expression shows that the Hanle effect at very low laser intensity (no nonlinear effect) is identical to the Hanle effect obtained with conventional light sources; it is Lorentzian with the *usual width* $\Gamma_b(k)$, *which is insensitive to the velocity diffusion processes.* It can be shown that this property is valid for all velocity-change models if isotropy is assumed. If not, one needs ${}^{(2)}\varrho(v)$ to have the Maxwellian distribution $W_{\mathrm{M}}(v)$ (approximately valid when modes are numerous and close).

The only possibility of deviation from the ordinary Hanle-effect shape is when the hypothesis of isotropic relaxation fails (with monomode lasers this may occur; experiments are in progress to test this assumption).

For longitudinal quantities ($q_1+q_2=Q=0$) there is no resonant denominator. The optical line-shape factor, which becomes $\Sigma 2I_\nu(\Delta v)^{-1} \cdot X(\omega-\omega_\nu-q_1\omega_z)$, simply expresses the modification of the pumping rate due to the Zeeman shift of the line. This modification only occurs in high magnetic fields and is not sensitive in the Hanle-effect range (except for infrared lines for which $\Delta\nu$ is small).

The developed form of ${}^{(2)}_{\alpha}\varrho^k_Q$ is given in Appendix 2. We shall discuss here the most usual experimental case (cf. Section 5.1): the laser beam is σ linearly polarized ($e_0=0\, e_{\pm1}=-i/\sqrt{2}$) and the fluorescence light is detected perpendicular to the magnetic field with a σ linear analyzer ($\lambda_0=0, \lambda_{\pm1}=-(i/\sqrt{2})\exp(\pm i\phi)$; $\phi=0$ when the detected polarization is parallel to the laser polarization). The intensity of the $b\to g$ line is given by (Appendix 1):

$$l^\sigma_{bg}(\phi)=\frac{1}{3\sqrt{2J_b+1}}\,{}_b\varrho^0_0-(-)^{J_g+J_b}\begin{Bmatrix}2 & 1 & 1\\ J_g & J_b & J_b\end{Bmatrix} \cdot\left(\frac{1}{\sqrt{6}}\,{}_b\varrho^2_0+\mathrm{Re}\{{}_b\varrho^2_2\,\mathrm{e}^{2i\phi}\}\right).$$

The Hanle effect is easily obtained by the difference method. At low laser intensiv one gets

$$l^\sigma_{bg}(\phi=0)-l^\sigma_{bg}\left(\phi=\frac{\pi}{2}\right) =-nD|P_{ab}|^2(-)^{J_a+J_b+2J_g}\begin{Bmatrix}2 & 1 & 1\\ J_g & J_b & J_b\end{Bmatrix}\begin{Bmatrix}2 & 1 & 1\\ J_a & J_b & J_b\end{Bmatrix}\frac{\Gamma_b(2)}{\Gamma_b(2)^2+4\omega_b^2},$$

where D is the line-shape factor $\left(\sum_\nu\ldots\right)$ of (5.36). Similarly, the anisotropy of the laser-induced modifications in zero magnetic field appears to have exactly the form given in (5.4); it allows one to measure the ratio $\Gamma_b(2)/\Gamma_b(0)$.

Modulated Terms

The modes of the laser beam are almost equidistant, with a frequency spacing

$$\Delta\omega = \pi \frac{c}{L},$$

where L denotes the length of the laser. If the fluorescent-light detecting system is connected to a rf receiver tuned on the beat frequency $p\Delta\omega$ (this experiment was first performed by FORK et al. [5.2]), the measured signal is the modulus of a linear superposition (according to the geometry of the detection; cf. Appendix 1) of quantities of the general form

$$\begin{aligned} {}^{(2)}_{\;b}\overline{\varrho}{}^{k}_{Q}(p\Delta\omega) = &- \frac{n|P_{ab}|^2 \mathrm{e}^{ip\left(\Delta\omega t - \frac{\pi r}{L}\right)}}{\Gamma_b(k) + i(p\Delta\omega + Q\omega_b)} \sum_{q_1 q_2} {}^{q_2}_{ba}G^{1k}_{q_1 Q} e^{*}_{q_1} e_{-q_2} \\ &\cdot \sum_{\nu} \frac{2}{\Delta\nu} \mathscr{E}^{\nu *} \mathscr{E}^{\nu - p} X(\omega - \omega_\nu - q_1 \omega_{\mathrm{ZR}}) . \end{aligned} \tag{5.37}$$

Assuming again that $\Delta\nu \gg \Gamma_b(k)$, we have replaced the X function by its value when the first denominator is resonant ($\omega_Z = \omega_{\mathrm{ZR}} \simeq -p\Delta\omega/Q$). The longitudinal quantities ($Q=0$) are never resonant and they can be neglected if $\Gamma_b(k) \ll p\Delta\omega$. If this condition is not well fulfilled, the $Q=0$ terms distort the resonances due to the $Q \neq 0$ terms: indeed, the different terms have to be added vectorially. For instance, with the 6328 Å neon laser ($3s_2, J=1 \rightarrow 2p_4, J=2$) and the fluorescence line 6096 Å ($2p_4$, $J=2 \rightarrow 1s_4, J=1$), and with the usual geometry of the Hanle-effect experiments defined in the previous paragraph, the fluorescence light intensity is

$$l^\sigma = \frac{1}{3\sqrt{5}} {}^{(2)}_{\;a}\overline{\varrho}{}^{0}_{0} + \frac{\sqrt{21}}{30} \left[\frac{1}{\sqrt{6}} {}^{(2)}_{\;a}\overline{\varrho}{}^{2}_{0} \pm \frac{1}{2} \left({}^{(2)}_{\;a}\overline{\varrho}{}^{2}_{2} + {}^{(2)}_{\;a}\overline{\varrho}{}^{2}_{-2} \right) \right].$$

The $p=1$ modulated term is (neglecting spontaneous emission from b)

$$\begin{aligned} l^\sigma(\Delta\omega) \propto \mathrm{Re} \Bigg\{ \mathrm{e}^{i\Delta\omega t} \Bigg[&\frac{40/7}{\Gamma_a(0) + i\Delta\omega} + \frac{1}{\Gamma_a(2) + i\Delta\omega} \\ &\pm \frac{3/2}{\Gamma_a(2) + i(\Delta\omega + 2\omega_a)} \pm \frac{3/2}{\Gamma_a(2) + i(\Delta\omega - 2\omega_a)} \Bigg] \\ &\cdot \sum_{\nu} \mathscr{E}^{\nu *} \mathscr{E}^{\nu - 1} X(\omega - \omega_\nu) \Bigg\} . \end{aligned}$$

As the magnetic field is scanned, the amplitude of the $\Delta\omega$ modulation is given by the modulus of the bracket; with the experimental values [$\Gamma_a(0)=9$ MHz, $\Gamma_a(2)=16$ MHz, $\Delta\omega=82.5$ MHz] the curve obtained is very different from the square root of a Lorentzian curve that one would obtain if $\Gamma_a(0)$, $\Gamma_a(2) \ll \Delta\omega$. This shape is well verified experimentally [5.2, 14] (see Fig. 5.6).

The relative phase of modes appears to be of prime importance in determining the amplitude of modulated terms. If the mode frequencies are not exactly equidistant, the variations in the beat frequency $p\Delta\omega$ from one pair of modes to another are small enough to be negligible in the denominator of (5.37) which can be kept out of the sum over ν. However, these frequency differences appear as a time dependence of the relative phases of modes. If we choose any mode as the origin

$$\mathscr{E}^0 \, e^{i\omega_0 t} = \sqrt{I_0} \, e^{i(\omega_0 t + \phi_0)} ,$$

the ν mode can be written as

$$\mathscr{E}^\nu \, e^{i\omega_\nu t} = \sqrt{I_\nu} \, e^{i[(\omega_0 + n_\nu \, \Delta\omega)t + \phi_\nu(t)]} .$$

The amplitude of the $p\Delta\omega$ modulation is proportional to M_p such that

$$M_p^2 = \left| \sum_\nu \sqrt{I_\nu I_{\nu-p}} \, X(\omega - \omega_\nu - q_1 \omega_{ZR}) e^{i[\phi_\nu(t) - \phi_{\nu-p}(t)]} \right|^2_{av} \tag{5.38}$$

If the modes are free-running, all phase-dependent terms are cancelled out by averaging over the time constant of the apparatus. Therefore (5.38) is reduced to

$$M_p^2 = \sum_\nu X^2(\omega - \omega_\nu - q_1 \omega_{ZR}) I_\nu I_{\nu-p} , \tag{5.39}$$

which is of the order of $(N-p)I^2$ if there are N modes of approximate intensity I. In other words, the modulation intensity is simply the sum of the intensity of $N-p$ independent modulations.

On the other hand, if the modes are exactly equidistant, the interference terms in (5.38) do not disappear and they depend on the actual phases (time-independent). The most interesting case is the "phase-locked" operation of the laser for which all phases are equal (with a correct choice of the origin of time). In this case,

$$M_p^2 = \left[\sum_\nu X(\omega - \omega_\nu - q_1 \omega_{ZR}) \sqrt{I_\nu I_{\nu-p}} \right]^2 \tag{5.40}$$

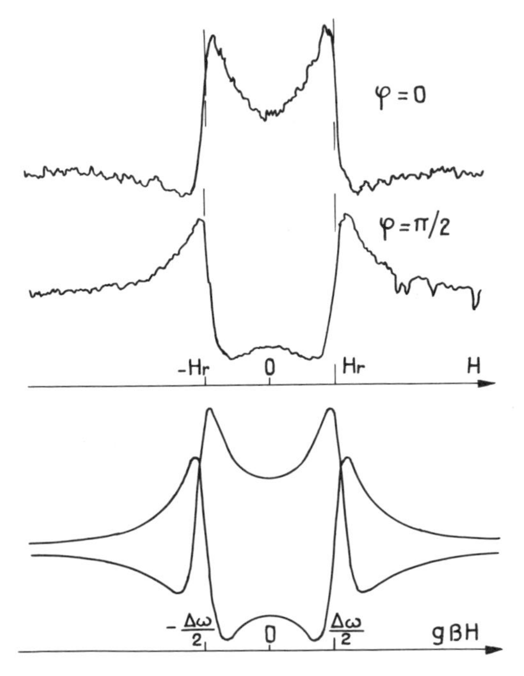

Fig. 5.6. Amplitude of the fluorescence modulation at the first beat frequency between modes (82.5 MHz). The experimental curves have been obtained on the 6096 Å line ($2p_4 \rightarrow 1s_4$) with the 6328 Å ($3s_2 - 2p_4$) laser line of neon. The theoretical curves have been obtained from (5.37) with approximately the actual values of relaxation parameters

is of the order of $(N-p)^2 I^2$. The modulation intensity is therefore much greater than in the free-running case. Experimentally, modulation of the fluorescence light has been observed only with phase-locked modes.

We point out that the phase-locked operation corresponds to a modulation of the laser intensity in short repetitive pulses. This fact shows that the resonances of the fluorescence-light modulations are very similar to those obtained in optical pumping experiments with *ordinary* light sources modulated in intensity [5.28].

5.3.3. Fourth-Order Effects: Saturation Resonances

Introducing the second-order solution (5.28) in the right-hand side of (5.22c)[8] one gets the third-order equation, the solution of which is inserted in (5.22a) and (5.22b) to provide the fourth-order equation.

[8] Let us recall that in (5.22c) $\omega_a - \omega_b$ was neglected and that the last term with $W_M(v)$ exists only in the first-order equation.

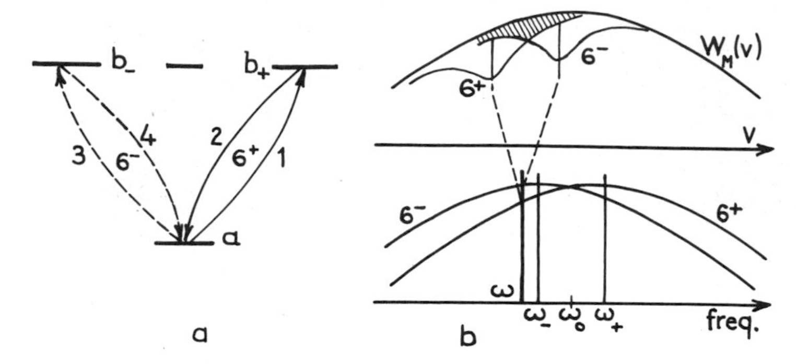

Fig. 5.7a and b. Illustration of the population effect. a) Diagram of the Zeeman sublevels ($J_a=0$, $J_b=1$). Numbers near the arrows refer to the successive perturbation orders. b) Distribution of the difference of atomic populations between a and b levels, according to velocity and Doppler-shifted frequency

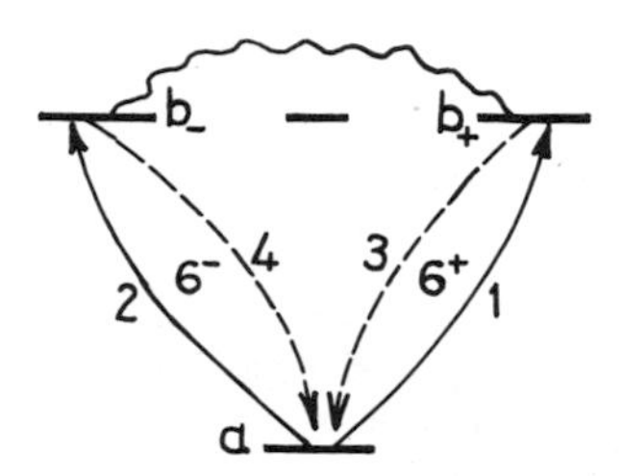

Fig. 5.8. Illustration of the Zeeman coherence effect. The wavy line represents the coherence between sublevels b_+ and b_-

Fourth order gives the first nonlinear correction to the linear atomic response obtained at second order. ${}^{(4)}_{\;b}\overline{\varrho}{}^{2}_{\pm 2}$ provides a correction to the Hanle-effect shape (radiative broadening) but, as explained in the introduction, fourth order is quite insufficient to give the curve shape for any laser intensity. Therefore, we shall focus our interest only on longitudinal quantities ${}^{(4)}_{\;b}\overline{\varrho}{}^{k}_{0}$, which will give us (at least qualitatively) an interpretation of the saturation phenomena observable with a π detection.

To avoid algebraic complications, we shall write formulas only for the experimental case studied so far for which the laser beam is σ linearly polarized.

$$e_0=0; \qquad |e_1|=|e_{-1}|=\frac{1}{\sqrt{2}}. \tag{5.41}$$

In that case there is no orientation (so long as the Zeeman splitting ω_b is small compared to the Doppler width $\Delta\nu$).

To show the most significant features of saturation resonances (population effect—Zeeman coherence effect—influence of phases and frequency spacing of modes), we treat first the very simplified cas $J_a=0$, $J_b=1$. Furthermore, we assume a simplified relaxation model with only three relaxation rates Γ_a, Γ_b, and Γ_{ab}, so that it is easy to project the density matrix on the usual basis $|J_\alpha M_\alpha\rangle\ \langle J_\beta M_\beta|$ instead of the basis of irreducible tensors. That formulation allows us to discuss the scheme of Zeeman sublevels (Fig. 5.7 and 5.8); for instance, the transverse alignment ${}_b\varrho^2_{-2}$ becomes simply coherence between b_+ and b_-, written as $\varrho_{b_+b_-}$.

In Subsection *Case* J_a *and* $J_b \neq 0$: *Landé g Factor Measurements* we shall study a more general transition.

Monomode Laser (*Simplified Model:* $J_a=0$, $J_b=1$)

Table 5.2 shows the chain of the matrix elements from step to step up to fourth order. Among the many possibilities we have chosen two paths, starting from the unperturbed population of the a level $[{}^{(0)}\varrho_a(v)]$, and making a contribution to the fourth-order correction to the same population $[{}^{(4)}\varrho_a(v)]$. They illustrate the two possible types of saturation resonances.

α) Population Effect (PE)

The first path (left-hand side of Table 5.2) begins with two interactions with the σ^+ component (arrows 1 and 2, Fig. 5.7a). At second order a dip is produced in the velocity distribution of the population of the a level (Fig. 5.7b). This dip is caused by the denominator $\Gamma_{ab}+i(\omega_+-\omega+kv)$, which expresses the resonance of the σ^+ component of the laser (ω) with a class of velocities centered on v_+, such as $\omega_+ + kv_+ = \omega$, and of width Γ_{ab}/k, Γ_{ab} being the homogeneous width. Taking into account the σ^- component, which produces a second dip ($\omega_- + kv_- = \omega$), one gets[9]

$$
{}^{(2)}\varrho_a(v) \propto \frac{W_{\mathrm{M}}(v)}{\Gamma_a} \cdot \left[\frac{|\mathscr{E}_+|^2}{\Gamma_{ab}+i(\omega_+-\omega+kv)} + \frac{|\mathscr{E}_-|^2}{\Gamma_{ab}+i(\omega_--\omega+kv)}\right] + cc\,. \tag{5.42}
$$

This formula is a simplified form of (5.28) for $k=Q=0$. Similar formulas can be written for ${}^{(2)}\varrho_{b_+}$ and ${}^{(2)}\varrho_{b_-}$.

[9] With the polarization defined in (5.41) $|\mathscr{E}_+|=|\mathscr{E}_-|$; nevertheless, the indices, referring to σ^+ and σ^- components are retained to clarify the discussion.

Table 5.2

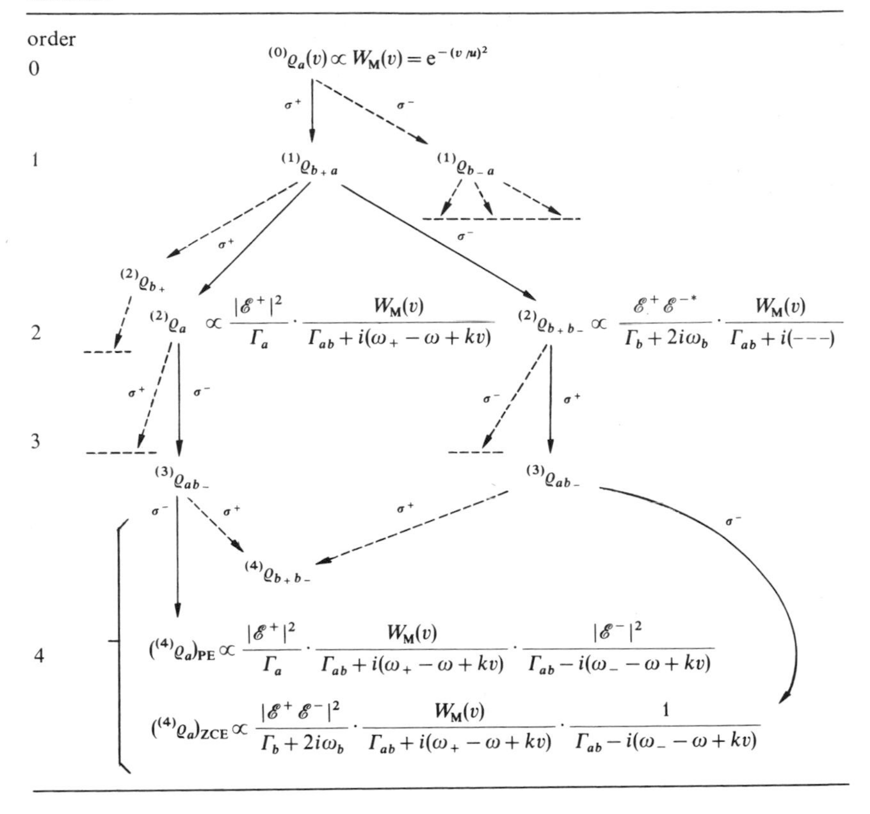

Now, following the first path in Table 5.2, we find two interactions with the σ^- component (arrows *3* and *4*, Fig. 5.7a). We get a contribution to $^{(4)}\varrho_a(v)$, depending on velocity, through two denominators that correspond, respectively, to the σ^+ and the σ^- holes in Fig. 5.7b. $^{(4)}\varrho_a(v)$ is important only when the two holes overlap; this occurs in a weak magnetic field ($\omega_+ - \omega_- = 2\omega_b \lesssim 2\Gamma_{ab}$). Due to this hole crossing, the total population exhibits a resonance in zero magnetic field

$$(^{(4)}\bar{\varrho}_a)_{PE} \propto \frac{1}{\Gamma_a} \frac{|\mathscr{E}_+|^2 |\mathscr{E}_-|^2}{2\Gamma_{ab} + 2i\omega_b} X(\omega_0 - \omega) + cc\,. \tag{5.43}$$

In other words, in zero magnetic field saturation increases because the σ^+ and σ^- components act on the same atoms; this is the *population effect* (PE) and is similar to the Lamb dip obtained with a standing wave. Its width is the homogeneous linewidth.

β) Zeeman Coherence Effect (ZCE)

Now let us study the second path in Table 5.2. A first interaction with σ^+ and a second with σ^- produce coherence between b_+ and b_- Zeeman sublevels (arrows *1* and *2* in Fig. 5.8). This coherence exhibits a resonance in zero magnetic field which is the *Hanle effect*. Its width is Γ_b [$\Gamma_b(2)$ in the general case—cf. (5.28) with $k=2$, $Q=\pm 2$ and (5.36)].

Following the second path with a σ^+ interaction at third order and a σ^- interaction at fourth order (Fig. 5.8), we get a new contribution to ${}^{(4)}\varrho_a(v)$, which contains a resonant denominator from the Hanle effect. The two other denominators express the optical resonance condition with σ^+ and σ^-. By integration over v they reduce to a simple denominator[10] (as in PE terms) which expresses the need for an optical coincidence between σ^+ and σ^- for the same atoms

$$({}^{(4)}\bar{\varrho}_a)_{\mathrm{ZCE}} \propto \frac{|\mathscr{E}_+\mathscr{E}_-|^2}{\Gamma_b+2i\omega_b} \frac{X(\omega_0-\omega)}{2\Gamma_{ab}+2i\omega_b} + \mathrm{c\cdot c} \tag{5.44}$$

This term, due to the Hanle effect at second order, is the so-called *Zeeman coherence effect* on saturation resonances.

γ) Shape of Saturation Resonance

The shape of the saturation resonance is given by the sum of both types of terms

$$\begin{aligned} \mathrm{Res} &\propto \frac{1}{2\Gamma_a}\cdot\frac{\Gamma_{ab}}{\Gamma_{ab}^2+\omega_b^2} + \mathrm{Re}\left\{\frac{1}{\Gamma_b+2i\omega_b}\,\frac{1}{2\Gamma_{ab}+2i\omega_b}\right\} \\ &= \left(\frac{1}{\Gamma_a} - \frac{1}{2\Gamma_{ab}-\Gamma_b}\right)\frac{1}{2}\cdot\frac{\Gamma_{ab}}{\Gamma_{ab}^2+\omega_b^2} \\ &+ \frac{1}{2\Gamma_{ab}-\Gamma_b}\cdot\frac{\Gamma_b}{\Gamma_b^2+4\omega_b^2}\,. \end{aligned} \tag{5.45}$$

At very low pressure, when $2\Gamma_{ab}=\Gamma_a+\Gamma_b$, the Lorentzian curve of width Γ_b alone remains. In that case, the two types of phenomena cannot be distinguished. On the other hand, when collisions occur, the homogeneous width Γ_{ab} generally increases faster than the Hanle effect width Γ_b. Therefore, the saturation resonance is the superposition of two Lorentzian curves: the broad one of width $2\Gamma_{ab}$ (= Lamb dip width) is due mainly to PE, while the narrow curve of width $\Gamma_b(2)$ (= Hanle effect) is due solely to ZCE [5.61].

[10] For this integration we assume the Doppler limit approximation ($\Delta v \gg \Gamma_{ab}$) to be valid (cf. Section 5.3.2).

Multimode Laser (Simplified Model)

In multimode lasers new resonances occur each time the beat frequency between two modes is equal to the Larmor frequency $(\omega_\mu - \omega_\nu = 2\omega_b)$ [5.2]. For these resonances it is also possible to distinguish a population effect and a Zeeman coherence effect.

α) The *population effect* is due to the crossing of σ^+ and σ^- holes in the velocity distribution of the population of the a level

$$({}^{(4)}\bar{\varrho}_a)_{\mathrm{PE}} \propto \frac{1}{\Gamma_a} \sum_{\nu,\mu} \frac{|\mathscr{E}_-^\nu|^2 |\mathscr{E}_+^\mu|^2 X\left(\omega - \dfrac{\omega_\nu + \omega_\mu}{2}\right)}{2\Gamma_{ab} + i(\omega_\nu - \omega_\mu + 2\omega_b)} + c\cdot c \tag{5.46}$$

$$= \frac{1}{\Gamma_a} \sum_{\nu,s} \frac{|\mathscr{E}_-^\nu|^2 |\mathscr{E}_+^{\nu-b}|^2 X\left(\omega - \omega_\nu + \dfrac{2\Delta\omega}{2}\right)}{2\Gamma_{ab} + i(s\Delta\omega + 2\omega_b)} + c\cdot c\,.$$

We neglect terms arising from second-order modulation of the populations [cf. (5.37) with $Q=0$]. These terms containing the off-resonance denominator $\Gamma_a + ip\Delta\omega$ are negligible if $\Delta\omega \gg \Gamma_a$ and Γ_b.

β) The *Zeeman coherence effect* is due to the second-order $b_+ b_-$ coherence, modulated at $\omega_\nu - \omega_\mu$ frequency, [cf. (5.37)] and demodulated at third and fourth order by interaction with a second pair of modes, σ^+ and σ^-, such that $\omega_\nu - \omega_\mu + \omega_\lambda - \omega_\kappa = 0$. The ZCE contribution is

$$({}^{(4)}\bar{\varrho}_a)_{\mathrm{ZCE}} \propto \sum_{\substack{\mu\nu \\ k\lambda}} \frac{\mathrm{Re}\{\mathscr{E}_-^{\nu*} \mathscr{E}_+^\mu \mathscr{E}_-^{\lambda*} \mathscr{E}_+^\kappa\}}{\Gamma_b + i(\omega_\nu - \omega_\mu + 2\omega_b)} \cdot \frac{X\left(\omega - \dfrac{\omega_\mu + \omega_k}{2}\right)}{2\Gamma_{ab} + i(\omega_\kappa - \omega_\mu + 2\omega_b)}$$

$$= \sum_{\nu,p,s} \frac{\mathrm{Re}\{\mathscr{E}_-^{\nu*} \mathscr{E}_+^{\nu-p} \mathscr{E}_-^{\nu-p-s*} \mathscr{E}_+^{\nu-s}\}}{\Gamma_b + i(p\Delta\omega + 2\omega_b)} \tag{5.47}$$

$$\cdot \frac{X\left(\omega - \omega_\nu + \dfrac{p+s}{2}\Delta\omega\right)}{2\Gamma_{ab} + i[(p-s)\Delta\omega - 2\omega_b]}\,,$$

where

$$\begin{cases} p\Delta\omega = \omega_\nu - \omega_\mu = \omega_\kappa - \omega_\lambda \\ s\Delta\omega = \omega_\nu - \omega_\kappa = \omega_\mu - \omega_\lambda\,. \end{cases} \tag{5.48}$$

Equation (5.47) can be interpreted exactly like (5.44). The first denominator expresses the resonances of second-order Zeeman coherences (zero frequency resonance expressed by (5.44) is the Hanle effect), while the second denominator requires the need for an optical coincidence, for the same class of atomic velocity, between the first pair of modes (ν, μ) and the second one (κ, λ).

Behavior of the Saturation Resonances with a Multimode Laser

The behavior of the above saturation resonances depends strongly on the relative magnitude of the mode spacing $\Delta\omega$ and of the relaxation rates, Γ_{ab}, Γ_a, and Γ_b.

α) Large Mode Spacing

$$\Gamma_a, \Gamma_b \lesssim \Gamma_{ab} \ll \Delta\omega \,. \tag{5.49}$$

Both denominators in (5.47) must be simultaneously resonant, therefore $s=0$, or in other words $\nu \equiv \lambda$, $\mu \equiv \kappa$. Only terms involving two modes (or one) are to be kept, therefore the relative phase of modes has no influence on the behavior of saturation resonances. Each resonance is the superposition of two Lorentzian curves, exactly as with the zero field resonance defined in (5.45). All resonances are well resolved from each other.

β) Mode Spacing of the Order of the Width of Holes

When the gas pressure increases, Γ_{ab} increases generally much faster than Γ_a or Γ_b [5.58, 59], therefore it is easy to get

$$\Gamma_a, \Gamma_b \ll \Delta\omega \lesssim \Gamma_{ab} \,. \tag{5.50}$$

Furthermore, we assume that the mode number N is large enough; typically $N\Delta\omega \simeq \Delta\nu$. These conditions are usual in optical pumping experiments with neon (pressure of the order of a few torr, $\Delta\omega \approx 80$ MHz).

With conditions as in (5.50), the width of PE resonances (5.46) is broader than their spacing, therefore *the* PE *resonances overlap so that they are no longer observable.* Hence, *the only observable resonances are the* ZCE *resonances*; indeed, resonances of the first denominator of (5.47) remain very narrow compared to $\Delta\omega$. On the other hand, resonances of the second denominator are broad and terms with $s \neq 0$ are no longer negligible.

With some algebra and some index permutations on (5.47) the p resonance can be written

$$\begin{aligned} \mathrm{Res}(p) \propto & \sum_{\nu s} X\left(\omega - \omega_\nu + \frac{p+s}{2}\Delta\omega\right) \mathrm{Re}(\mathscr{E}^{\nu *}\mathscr{E}^{\nu-p}\mathscr{E}^{\nu-p-s*}\mathscr{E}^{\nu-s}) \\ & \cdot \left\{ \frac{2\Gamma_{ab}-\Gamma_b}{(2\Gamma_{ab}-\Gamma_b)^2+(s\Delta\omega)^2} \frac{1}{\Gamma_b + i(p\Delta\omega + 2\omega_b)} \right. \\ & \left. + \frac{1}{2\Gamma_{ab}-\Gamma_b - is\Delta\omega} + \frac{1}{2\Gamma_{ab} + i[(p-s'\Delta\omega + 2\omega_b]} + c \cdot c \right\} . \end{aligned} \tag{5.51}$$

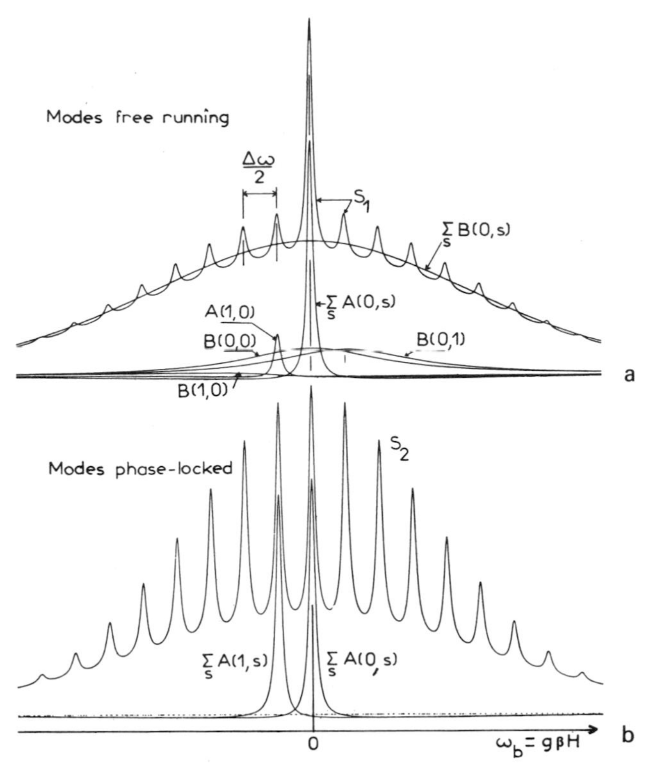

Fig. 5.9a and b. Computer calculation of saturation resonances produced by 11 modes ($I_\nu = \xi_\nu^2 = 3$ for $-3 < \nu > 3$, $I_\nu = 2$ for $\nu = \pm 4$, $I_\nu = 1$ for $\nu = \pm 5$). The Doppler width is $\Delta\nu = 800$ MHz. The mode spacing $\Delta\omega = 80$ MHz and the relaxation rates are in MHz: $\Gamma'_{ab} = 100$; $\Gamma'_b(0) = 9.6$; $\Gamma'_b(1) = 11$; $\Gamma'_b(2) = 11.5$; $\Gamma_b(2) = 7.5$; $\Gamma'_a(0) = 15$. These conditions correspond approximately to the laser line 7305 Å with 1.5 Torr of neon. $A(p, s)$ refers to the ZCE, see (5.56) and $B(p, s)$ to the PE terms (Eq. (5.46) gives the $B(0, s)$ term). a) *Modes free-running:* elementary resonances from ZCE and from PE are compared. The ZCE resonances are narrow; the zero-field one ($\Sigma_s A(0, s)$) is much higher than the others ($A(p, 0)$). The PE resonances are broad and unresolved, as shown by $\Sigma_s B(0, s)$. Off-resonance terms ($B(p, 0)$) are negligible. $S_1 = \Sigma_{s \text{ or } p=0} [A(p, s) + B(p, s)]$ is the resulting saturation signal. b) *Modes phase-locked:* ZCE resonance for $p = 1$ has the same order of magnitude as that in zero magnetic field. $S_2 = \Sigma_{sp} [A(p, s) + B(p, s)]$ is the total signal

It is easy to see that the second term produces broad resonances that all overlap and are not observable (like the PE resonances). Therefore, *the observed resonances have the shape of the Hanle effect* as they are due only to the first term in (5.51).

It is evident from (5.51) that the relative phase of modes is very important for determining the amplitude of resonances. Indeed, with the

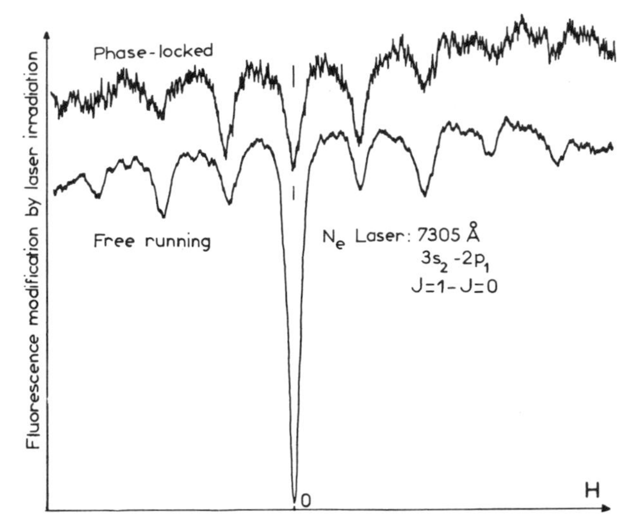

Fig. 5.10. Experimental records of the saturation signal on the fluorescence lines. As the detection is made with a π analyzer, the resonance in zero magnetic field is not the Hanle effect; like all lateral resonances, it can be explained only by nonlinear interactions (at least fourth order in the laser field). To obtain a spontaneously phase-locked oscillation we must decrease the laser intensity. Therefore the intensity of the phase-locked signal is much less than that of the free-running signal. Nevertheless, these curves clearly demonstrate the difference in behavior between the lateral resonances and the central one

notations of Subsection 5.3.2 (see *Modulated terms*)

$$\mathrm{Re}\{\mathscr{E}^{\nu*}\mathscr{E}^{\nu-p}\mathscr{E}^{\nu-p-s*}\mathscr{E}^{\nu-s}\} = \sqrt{I_\nu I_{\nu-p} I_{\nu-p-s} I_{\nu-s}} \cdot \cos(\phi_\nu - \phi_{\nu-p} + \phi_{\nu-p-s} - \phi_{\nu-s}). \tag{5.52}$$

If modes are phase-locked, the cosine is unity. All terms are important in the summation over ν and s. If the laser beam is composed of N modes with approximately the same intensity I, the resonance amplitude is of the order of ($\Gamma_b \ll 2\Gamma_{ab}$)

$$\mathrm{Res}(p)\,(\text{phase-locked}) \simeq \frac{I^2}{\Gamma_b \Gamma_{ab}} \cdot \left\{(N-p) + \sum_{s=1}^{N-p} 2(N-p-s)\left[1+\frac{s\Delta\omega}{2\Gamma_{ab}}\right]^{-1}\right\} \tag{5.53}$$

The amplitude slowly decreases from the central resonance ($p=0$) to the lateral ones in high magnetic field (Fig. 5.9b). *If modes are free-running* all phases are randomly varying (cf. Subsection 5.3.2) so that (5.52) cancels out when averaged (sum over ν; average over time), except for

$s=0$ or $p=0$. Therefore the resonance in zero magnetic field, which involves two modes ($p=0$; $\nu=\mu$; $\kappa=\lambda$), is not affected; its amplitude is the same as with phase-locked modes

$$\mathrm{Res}(p=0) \simeq \frac{I^2}{\Gamma_b \Gamma_{ab}} \left\{ N + \sum_{s=1}^{N} 2(N-s) \left[1 + \frac{s\Delta\omega}{2\Gamma_{ab}} \right]^{-1} \right\}. \tag{5.54}$$

On the other hand, resonances in nonzero magnetic field ($p \neq 0$) are strongly decreased by the vanishing of all terms with $s \neq 0$. Their amplitude is of the order of

$$\mathrm{Res}(p \neq 0)\,(\text{free-running}) = \frac{I^2}{\Gamma_b \Gamma_{ab}} (N-p)\,. \tag{5.55}$$

In conclusion, when modes are close compared to the homogeneous width [cf. (5.50)], the observed saturation resonances are due only to the ZCE; they are Lorentzian like the Hanle effect and are sensitive to the relative phases of modes (Fig. 5.9a). This conclusion is in good agreement with experimental observations on neon (Fig. 5.10) [5.14, 19].

Case J_a and $J_b \neq 0$; Landé g Factor Measurements

When $J_a=0$ and $J_b=1$, it is easy to measure the Landé factor of the b level by simultaneously measuring the position of resonances in the magnetic field scale, and the frequency of the beat nodes between modes. In the general case, it is not possible to say without calculation to what level the observed resonances must be attributed.

The general formalism [5.19] shows that the ZCE is now composed of two sets of resonances, one from the Zeeman coherence in the upper level (factor $[\Gamma_b(2) + i(p\Delta\omega + 2\omega_b)]^{-1}$) and the other from the Zeeman coherence in the lower level (factor $[\Gamma_a(2) + i(p\Delta\omega + 2\omega_a)]_t^{-1}$—cf. (5.31)). On the other hand, the optical coincidence factors of the ZCE terms, as well as the PE terms, are of the form $[\Gamma_{ab} + i\{(p-s)\Delta\omega + 2\omega_z\}]^{-1}$, where $\omega_z = (\omega_a + \omega_b)/2$.

When modes are closely spaced ($\Gamma'_{ab} \gtrsim \Delta\omega$), PE resonances disappear exactly as in the $J_a=0$ case, and similarly the Zeeman factors completely determine the position and shape of the ZCE resonances.

$$\begin{aligned}\mathrm{Res}(p)\,(\text{locked}) = 2n|P_{ab}|^4 \Big\{ & [C_b \mathscr{L}_b(p) + C_a \tilde{\mathscr{L}}_a(p)] \frac{\pi}{\Delta\omega\Delta\nu} \\ & \cdot \sum I_\nu I_{\nu-p} X\left(\omega - \omega_\nu + \frac{p}{2}\Delta\omega\right) \\ & + [C_b(\mathscr{L}_b(p) - \mathscr{L}'_b(p)) + C_a(\tilde{\mathscr{L}}_a(p) - \tilde{\mathscr{L}}'_a(p))] \\ & \cdot \left[\frac{1}{\Delta\nu} \sum_\nu \sqrt{I_\nu I_{\nu-p}}\, X\left(\omega - \omega_\nu + \frac{p}{2}\Delta\omega\right) \right]^2 \Big\},\end{aligned} \tag{5.56}$$

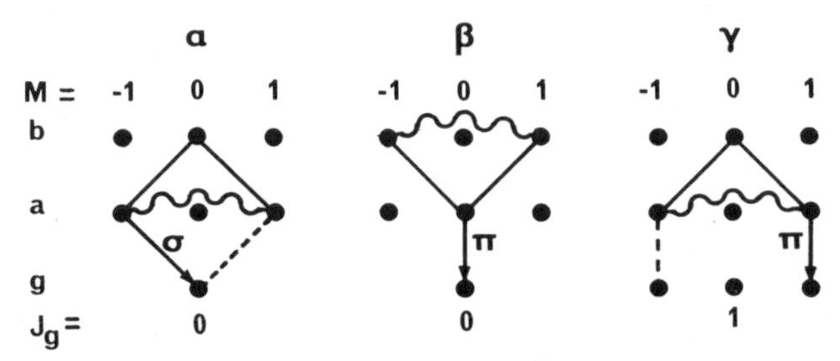

Fig. 5.11. $J_a = 1$, $J_b = 1$: graphical illustration of the three cases for which it is possible to attribute the observed saturation resonances to only one level, provided there is no coupling between the Zeeman sublevels by relaxation processes. Dots represent Zeeman sublevels, full lines: the laser interaction (σ^+ or σ^-), arrows: the fluorescence lines, and the wavy lines: the Zeeman coherence detected on the saturation signal

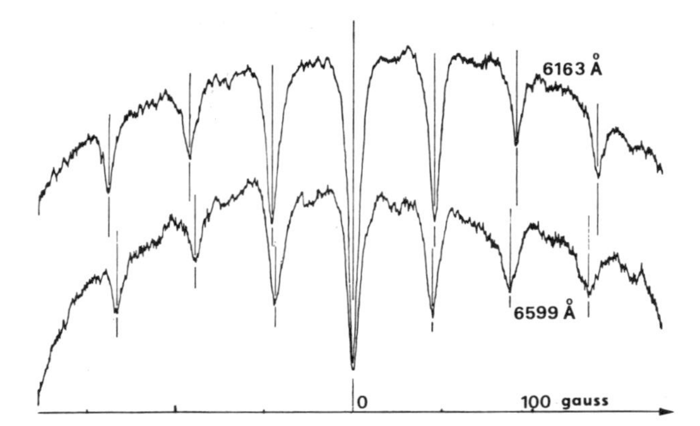

Fig. 5.12. Laser line 6401 Å ($3s_2, J=1 \leftrightarrow 2p_2, J=1$). The two curves are simultaneously recorded with two different fluorescence lines (π polarization). The 6163 Å line ($2p_2$, $J=1 \rightarrow 1s_3, J=0$) gives the Landé factor of the $3s_2$ level (case β in Fig. 5.11). The 6599 Å line ($2p_2, J=1 \rightarrow 1s_2, J=1$) gives the $2p_2$ Landé factor (case γ in Fig. 5.11). These records are obtained with $\Delta\omega = 166$ MHz and with experimental conditions for which $\Gamma_{2p_2}(0) = \Gamma_{2p_2}(2)$ (within 10%)

where $\mathscr{L}_b(p)$ and $\mathscr{L}'_b(p)$ are the Lorentzian resonances from the b level, and $\tilde{\mathscr{L}}_a(p)$ and $\tilde{\mathscr{L}}'_a(p)$ are the resonances from the a level distorted by spontaneous emission from the b level

$$\begin{aligned} \mathscr{L}_b^{(\prime)}(p) &= \Gamma_b^{(\prime)}(2)[\Gamma_b^{(\prime)}(2)^2 + (p\Delta\omega + 2\omega_b)^2]^{-1} \\ \tilde{\mathscr{L}}_a^{(\prime)}(p) &= \mathrm{Re}\{[\Gamma_a^{(\prime)}(2) + i(p\Delta\omega + 2\omega_a)]_t^{-1}\}\,. \end{aligned} \tag{5.57}$$

C_a and C_b are given in [5.19] as a function of the J values of the levels. It is important to notice that C_a and C_b depend on the fluorescence lines which are used to observe the resonances, and on their polarization.

This is due to the fact that the observed signal is a linear combination of the longitudinal quantities ${}^{(4)}_{a}\varrho^k_0$, which are not proportional to each other in the general case. The coefficients of the combination are determined by the method of detection.

For a $J_b=1$, $J_a=2$ laser line, the $J_a=2$ level appears to be the dominant level in all experimental cases. In fact, if $\Gamma_a(2)=\Gamma_b(2)$, C_a is approximately 21 times C_b [5.19]. Although it has been demonstrated in that case [5.20] that fourth-order perturbation theory rapidly diverges from the exact solution when the laser intensity is not very small (see Section 5.7.2) the above result has been clearly shown experimentally on neon laser lines 6328 Å and 1.15 μm [5.19].

For a $J_a=1$, $J_b=1$ laser line, the C_b values strongly depend on the polarization of the chosen fluorescence line and on the J_g value of the lower level g of this fluorescence line [5.19]. If there is only one relaxation rate for each level ($\Gamma_\alpha(0)=\Gamma_\alpha(2)$), and if the spontaneous emission on the laser line is negligible, it appears in some cases that either C_a or C_b vanishes, so that it is very easy to measure the Landé factors. These interesting cases can be easily found graphically (Fig. 5.11). Since there is only one relaxation rate for each level, there is no coupling between Zeeman sublevels and it is possible to discuss the case on the basis $|J_\alpha M_\alpha\rangle$ (instead of ${}_\alpha T^k_Q$). Consider, for instance, the case β in Fig. 5.11 (detection of the π component of an $a\to g$ line with $J_g=0$). We detect the population of the $M_a=0$ sublevel, which is coupled only to the coherence $M_b=-1\leftrightarrow M_b=+1$ by the σ polarized laser beam. Therefore saturation resonances observed with this configuration are due only to the b level. Note that this particular result is true at all orders of perturbation, as the three sublevels in question are never coupled to three others for any number of interactions with the laser. On the other hand, when $\Gamma_\alpha(2)\neq\Gamma_\alpha(0)$, the destruction of alignment by relaxation processes introduces a coupling between Zeeman sublevels and the previous discussion fails. This has been experimentally verified on neon laser line 6401 Å (Fig. 5.12).

5.4. Nonperturbative Theory

The fourth-order perturbation theory is well adapted to exhibit the principal features of the laser pumping of atoms and to provide a physical interpretation of the nonlinear processes, but a perturbation calculation is not appropriate to a quantitative study of these phenomena. The higher-order calculations, as performed in the formulation of the previous section, are too complicated on the one hand, and on the other hand the perturbation development is probably not convergent for high laser intensities. In order to obtain equations of motion of the atomic system that can be exactly solved and that are valid for arbitrary laser intensities, one must make some approximations. These approximations are discussed in the first part of this section; the second part presents the complete calculation for a $J=1-J=0$ laser transition.

5.4.1. Broad-Line Approximation

From the iteration procedure used in Section 5.3 it is clear that one can write at every order (assuming phase-locked operation)

$$ {}_{\beta}\varrho^{k}_{Q}(v,r,t)=\sum_{p}{}_{\beta}\varrho^{k}_{Q}(v,p)\mathrm{e}^{ip(\Delta\omega t-\Delta kr)}\,. \tag{5.58}$$

Here the discussion will be limited to the Hanle effect and to the zero-field saturation resonance (for more details, see [5.20]) that appears in magnetic fields weak enough to make the density matrix modulations negligible, due to the off-resonance factor: $[\Gamma_{\beta}(k)+i(Q\omega_{\beta}+p\Delta\omega)]^{-1}$, see (5.37). Thus, using only the $p=0$ term of (5.58), (5.22c) becomes [with normal Zeeman pattern approximation and assuming $\Gamma_{ab}(k')$ to be independent of k']

$$\begin{aligned} {}_{ab}\dot{\varrho}^{k'}_{Q'}(v)=&[i(\omega-Q'\omega_{z})-\Gamma_{ab}]\,{}_{ab}\varrho^{k'}_{Q'}(v)\\ &+iP_{ab}\sum_{\mu}\mathscr{E}^{\mu*}\zeta^{k'}_{Q'}(v)\mathrm{e}^{i(\omega_{\mu}t-k_{\mu}r)} \end{aligned} \tag{5.59}$$

with

$$\begin{aligned} \zeta^{k'}_{Q'}(v)=&\frac{1}{\sqrt{3}}e^{*}_{Q'}nW_{\mathrm{M}}(v)\delta_{k',1}+\sum_{kQq'}e^{*}_{-q'}\\ &\cdot[{}^{q'}_{ba}G^{k'k}_{Q'Q}\,{}_{b}\varrho^{k}_{Q}(v,0)+(-)^{k+k'}\,{}^{q'}_{ab}G^{k'k}_{Q'Q}\,{}_{a}\varrho^{k}_{Q}(v,0)]\,. \end{aligned} \tag{5.60}$$

Multiplying the stationary solution of (5.59) by $\mathscr{E}$, one gets the source term of (5.22a) and (5.22b) for the unmodulated part of ${}_{a}\varrho^{k}_{Q}$ and ${}_{b}\varrho^{k}_{Q}$

$$P^{*}_{ab}\mathscr{E}\cdot{}_{ab}\varrho^{k'}_{Q'}=iS(v)\zeta^{k'}_{Q'}(v) \tag{5.61}$$

with

$$S(v)=\sum_{\mu}\frac{|P_{ab}\mathscr{E}^{\mu}|^{2}}{\Gamma_{ab}+i(\omega_{\mu}-\omega-kv)}\,. \tag{5.62}$$

In (5.62), the Zeeman splitting $Q'\omega_Z$ has been neglected ($Q'\omega_Z \approx \Gamma_\beta(k) \ll \Gamma_{ab}$) and k_μ has been replaced by its mean value k. $|P_{ab}|^2 [\Gamma_{ab} + i(\omega_\mu - \omega - kv)]^{-1}$ is the atomic cross-section associated with the absorption or stimulated emission of a photon of frequency ω_μ. $S(v)$ can be understood as the probability that an atom of velocity v will interact with a laser photon so as to undergo a transition between levels a and b. The so-called broad-line approximation (BLA) then consists of two assumptions [5.20]:

a) The laser spectral width is larger than the Doppler width

$$N\Delta\omega \gg ku \tag{5.63}$$

and the modes nearly have the same intensity

$$\mathscr{E}^\mu \simeq \mathscr{E}\,. \tag{5.64}$$

Consequently, the series of (5.62) can be assumed to be infinite

$$S(v) = |P_{ab}\mathscr{E}|^2 \sum_{\mu=-\infty}^{+\infty} [\Gamma_{ab} + i(\omega_0 + \mu\Delta\omega - \omega - kv)]^{-1}\,. \tag{5.65}$$

This series is computable[11], yielding

$$S(v) = \gamma \frac{1 + i\tan\left(\pi \dfrac{\omega_0 - \omega - kv}{\Delta\omega}\right)\tanh\left(\pi \dfrac{\Gamma_{ab}}{\Delta\omega}\right)}{\tanh\left(\pi \dfrac{\Gamma_{ab}}{\Delta\omega}\right) + i\tan\left(\pi \dfrac{\omega_0 - \omega - kv}{\Delta\omega}\right)}\,, \tag{5.66}$$

where

$$\gamma = \pi \frac{|P_{ab}\mathscr{E}|^2}{\Delta\omega}\,. \tag{5.67}$$

b) When $\Gamma_{ab}/\Delta\omega$ is very large, $\tanh(\pi\Gamma_{ab}/\Delta\omega)$ goes to unity; $S(v)$ no longer depends on v and is equal to γ (laser-induced transition probability or laser pumping rate). Indeed, when $\Gamma_{ab} = \Delta\omega$, $\tanh(\pi\Gamma_{ab}/\Delta\omega) = 0.996$.

[11] Using the properties of the Γ function, it can be shown [5.29] that

$$\sum_{n=-\infty}^{+\infty} \frac{1}{n+z} = \frac{\pi}{\tan(\pi z)}\,,$$

where z is any complex number.

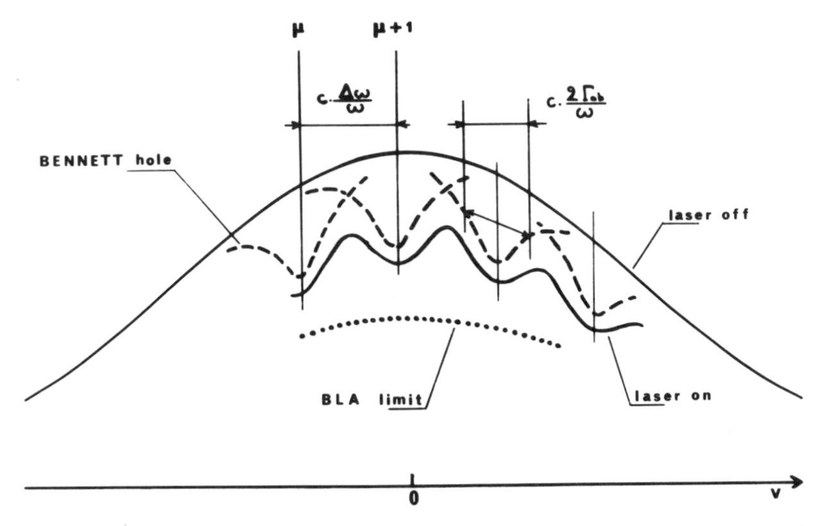

Fig. 5.13. Velocity distribution of a laser level. The laser-off distribution is proportional to the Maxwell partition function $W_M(v)$. When the laser is turned on, a set of equidistant "Bennett holes" appears, the position of which is determined by $\bar{k}v = \omega_\mu - \omega$ (dashed curves). When Γ_{ab} is greater than $\Delta\omega$, the hole structure disappears and the atomic response is proportional to $W_M(v)$ (BLA limit: dotted line) [5.20]

If we assume

$$\Gamma_{ab} \gtrsim \Delta\omega \,, \tag{5.68}$$

the relative variations of S with velocity are lower than 10^{-2}. This is the second condition of BLA, i.e., the *atomic response does not depend on the velocity*.

This property is also valid in nonzero magnetic fields, as long as the Larmor frequencies are much smaller than the laser spectral width,

$$\omega_a, \omega_b \ll N\Delta\omega \,. \tag{5.69}$$

With these conditions, the internal and external variables are uncorrelated and the density matrix factorizes ($\beta = a$ or b):

$${}_\beta\varrho^k_Q(v) = {}_\beta\bar{\varrho}^k_Q \, W_M(v) \,. \tag{5.70}$$

The first factor describes the internal spin state of the atoms and the second one gives the velocity dependence, which is Maxwellian. Relation (5.70) can be interpreted by pointing out that every term of (5.62) is associated with the hole [5.45] burnt by one laser mode in the atomic velocity distribution (Fig. 5.13). The hole burnt by the μ mode has a position determined by $\omega_\mu - kv = \omega$ and a width proportional to $2\Gamma_{ab}$. If $\Gamma_{ab} > \Delta\omega$, the holes are overlapping and the hole structure of the atomic response disappears. If, in addition, the modes cover the entire velocity distribution, the atomic response does not depend on the velocity; this leads to (5.70).

When (5.70) is valid, the internal state of the atoms is governed by equations of motion coupling the various density matrix components [5.20].

$$\begin{aligned}\frac{d}{dt}\,{}_b\bar{\varrho}^k_Q &= -(\Gamma_b(k) + iQ\omega_b)\,{}_b\bar{\varrho}^k_Q + \frac{n\gamma}{\sqrt{3}}\,{}_b g^k_Q \\ &\quad + \gamma \sum_{k''Q''} \left[{}_{ba}h^{kk''}_{QQ''}\,{}_a\bar{\varrho}^{k''}_{Q''} - {}_b h^{kk''}_{QQ''}\,{}_b\bar{\varrho}^{k''}_{Q''}\right].\end{aligned} \tag{5.71}$$

For level a, the equations are similar (with, in addition, the transfer by spontaneous emission $b \to a$). The geometrical coefficients, g and h, depend on the laser polarization and on the angular momenta [5.20]. The last term of the right-hand side of (5.71) expresses the laser-induced couplings between the elements of the density matrix. The ${}_a\bar{\varrho}$ term, which represents the effect of the laser on the atoms in the lower level, is due to the absorption of laser photons by atoms, while the ${}_b\bar{\varrho}$ term represents the contribution of stimulated emission.

In steady-state operation, (5.71) is exactly solvable for levels having small angular momenta. The solution has been obtained for $J=1-J=0$, $J=1-J=1$, and $J=1-J=2$ transitions when the laser is linearly σ polarized [5.20]. Here we shall study the case of a $J=1-J=0$ transition[12].

5.4.2. Exact Calculation for a $J=1$—$J=0$ Transition

As an example of exact solution, the detailed calculation for a $J_b=1 \leftrightarrow J_a=0$ transition is presented here for the case where the laser is linearly σ polarized along the Ox axis. This allows one to exhibit the principal features of the general solution.

[12] When the modes are phase-locked, the density matrix exhibits modulations resonant in strong magnetic fields (see *Modulated Terms* in Subsection 5.3). These modulations satisfy equations similar to (5.71) [5.20]. When the Landé factors of both levels are equal, these equations are strictly equivalent to (5.71). Then the Hanle effect and the modulated transverse alignment have the same behavior, as well as the zero-field saturation resonance and the lateral saturation resonances of the populations.

The equations of motions are deduced from (5.71)

$$
\begin{aligned}
{}_a\dot{\varrho}_0^0 = &- \Gamma_a(0)\,{}_a\varrho_0^0 + \frac{2n\gamma}{3} + \gamma_{ba}\sqrt{3}\,{}_b\varrho_0^0 \\
&+ \frac{2\gamma}{2}\left\{ -\,{}_a\varrho_0^0 + \frac{{}_b\varrho_0^0}{\sqrt{3}} + \frac{{}_b\varrho_0^2}{\sqrt{6}} - \frac{{}_b\varrho_2^2 + {}_b\varrho_{-2}^2}{2} \right\},
\end{aligned} \tag{5.72a}
$$

$$
\begin{aligned}
{}_b\dot{\varrho}_0^0 = &- \Gamma_b(0)\,{}_b\varrho_0^0 - \frac{2n\gamma}{3\sqrt{3}} \\
&- \frac{2\gamma}{3\sqrt{3}}\left\{ -\,{}_a\varrho_0^0 + \frac{{}_b\varrho_0^0}{\sqrt{3}} + \frac{{}_b\varrho_0^2}{\sqrt{6}} - \frac{{}_b\varrho_2^2 + {}_b\varrho_{-2}^2}{2} \right\},
\end{aligned} \tag{5.72b}
$$

$$
\begin{aligned}
{}_b\dot{\varrho}_0^2 = &- \Gamma_b(2)\,{}_b\varrho_0^2 - \frac{2n\gamma}{3\sqrt{6}} - \\
&- \frac{2\gamma}{3\sqrt{6}}\left\{ -\,{}_a\varrho_0^0 + \frac{{}_b\varrho_0^0}{\sqrt{3}} + \frac{{}_b\varrho_0^2}{\sqrt{6}} - \frac{{}_b\varrho_2^2 + {}_b\varrho_{-2}^2}{2} \right\},
\end{aligned} \tag{5.72c}
$$

$$
\begin{aligned}
{}_b\dot{\varrho}_2^2 = &- (\Gamma_b(2) + 2i\omega_b)\,{}_b\varrho_2^2 + \frac{n\gamma}{3} \\
&+ \frac{\gamma}{3}\left\{ -\,{}_a\varrho_0^0 + \frac{{}_b\varrho_0^0}{\sqrt{3}} + \frac{{}_b\varrho_0^2}{\sqrt{6}} - {}_b\varrho_2^2 \right\}.
\end{aligned} \tag{5.72d}
$$

In the steady-state operation ($\dot{\varrho} = 0$) the longitudinal components of the density matrix are proportional, independent of the laser intensity and of the magnetic field

$$
-\Gamma_a^*(0)\,{}_a\varrho_0^0 = \sqrt{3}\,\Gamma_b(0)\,{}_b\varrho_0^0 = \sqrt{6}\,\Gamma_b(2)\,{}_b\varrho_0^2\,, \tag{5.73}
$$

where

$$
\Gamma_a^*(0) = \Gamma_a(0)\left[1 - \frac{\gamma_{ba}}{\Gamma_b(0)}\right]^{-1}. \tag{5.74}
$$

Since the laser-induced population change of the β level is ${}_\beta\varrho_0^0\sqrt{2J_\beta + 1}$, the first relation of (5.73) expresses the conservation of the total population inside the a and b levels: all the atoms leaving one of the two levels due to the laser interaction must go into the other level. On the other hand, the proportionality between the population and the alignment of b (second relation) comes from the fact that the $m_b = \pm 1$ sublevels are equally excited by the laser, while the $m_b = 0$ sublevels is populated by the depolarization processes only. For instance, if these processes do not exist and $\Gamma_b(0) = \Gamma_b(2)$, relation ${}_b\varrho_0^0 = \sqrt{2}\,{}_b\varrho_0^2$ may be easily obtained from the fact there is no laser-induced change of the $m_b = 0$ population[13].

The equation satisfied by the transverse alignment is obtained by the elimination of the longitudinal components from (5.72).

$$[\Gamma_b(2) + 2i\omega_b]\,{}_b\varrho_2^2 = -\frac{\gamma}{3}\,{}_b\varrho_2^2 + \frac{\gamma}{3 + \eta\dfrac{\gamma}{\Gamma_b(2)}}\left[n + \eta\frac{\gamma}{6\Gamma_b(2)}({}_b\varrho_2^2 + {}_b\varrho_{-2}^2)\right], \tag{5.75}$$

where

$$\eta = \frac{1}{3} + \frac{2\Gamma_b(2)}{3\Gamma_b(0)} + \frac{2\Gamma_b(2)}{\Gamma_a^*(0)}. \tag{5.76}$$

The first term of the right-hand side of (5.75) describes the removal of the Zeeman coherence from level b, due to the stimulated emission. The second term in the bracket corresponds to the restitution of a part of this coherence into level b, such as appears after a transit inside the population of the sublevels. This restitution through the longitudinal components involves higher-order contributions in the laser field. Eliminating the imaginary part of ${}_b\varrho_2^2$ between relation (5.75) and its complex conjugate, one finds that the real part of ${}_b\varrho_2^2$ exhibits a Lorentzian dependence on the magnetic field, independent of the laser intensity,

$$\mathrm{Re}({}_b\varrho_2^2) \propto {}_b\mathscr{L}_2^2(H) = \left\{1 + \left[\frac{2\omega_b}{\gamma_b^I(2)}\right]^2\right\}^{-1}, \tag{5.77}$$

where

$$\gamma_b^I(2) = \Gamma_b(2)\left\{\frac{\left[1 + \dfrac{\gamma}{3\Gamma_b(2)}\right]\left[1 + (\eta+1)\dfrac{\gamma}{3\Gamma_b(2)}\right]}{1 + \eta\dfrac{\gamma}{3\Gamma_b(2)}}\right\}^{1/2}. \tag{5.78}$$

The theoretical variations of the Hanle-effect width with the laser pumping rate γ are shown in Fig. 5.14 for $\eta = 4.75$ and $\Gamma_b(2) = 11$ MHz.

[13] Relations (5.73), which are due to symmetry properties, are very general and valid independently of the laser spectral lineshape. The laser may be monomode or multimode, and BLA conditions are not necessary.

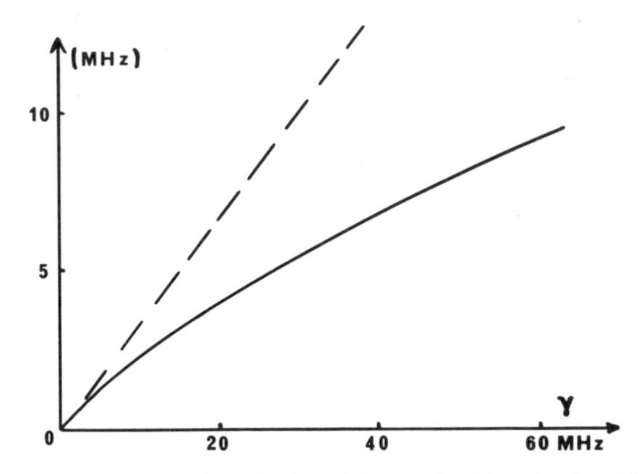

Fig. 5.14. Power-broadening of the Hanle effect of a $J_b=1$ level optically-pumped through a $J_b=1-J_a=0$ laser transition. The continuous curve gives the exact power-broadening $[\gamma_b^I(2)-\Gamma_b(2)]$, while the dashed line represents the fourth-order broadening $\gamma/3$. The numerical values of the relaxation rates are $\Gamma_a^*(0)=9$, $\Gamma_b(0)=3.7$ and $\Gamma_b(2)=11$, all in MHz [5.20].

For weak laser intensities, the power-broadening, equal to $\gamma/3$, is proportional to the pumping rate [this fourth-order broadening corresponds to the first term of the right-hand side of (5.75)]. For strong laser intensities, due to the restitution of a part of the Zeeman coherence through the sublevel populations [second term of (5.75)], the power broadening becomes lower than $\gamma/3$ and subsequently exhibits a nonlinear dependence on the laser pumping rate.

Equation (5.72) show that the longitudinal components are coupled to the real part of the transverse alignment. The populations exhibit a zero-field saturation resonance which has exactly the same shape as the Hanle effect[14]. The populations are proportional to

$$1-\mathscr{S}\cdot {}_b\mathscr{L}_2^2(H)\,, \tag{5.79}$$

where the relative amplitude of the saturation resonance is a homographic function of γ:

$$\mathscr{S}=\frac{\gamma}{3\Gamma_b(2)+(\eta+1)\gamma}\,. \tag{5.80}$$

[14] This saturation resonance corresponds to the "Zeeman coherence effect" of Subsection 5.3.3 only. In BLA ($\Gamma_{ab}>\Delta\omega$), the "population effect" (Zeeman-tuned holes crossing) vanishes because of the overlapping of the Bennett holes.

5.5. Saturation Experiments on $J=1$—$J=0$ Transitions

To demonstrate experimentally the results of the foregoing theoretical analysis, we detail the saturation experiments performed on the 1.52 μm ($2s_2 \rightarrow 2p_1$) and the 7305 Å ($3s_2 \rightarrow 2p_1$) neon laser lines [5.29]. This leads us to discuss some experimental problems.

The experimental set-up is the same as that described in Section 5.1. By putting the sample cell inside the laser cavity, one takes advantage of the high power density; the nonlinear effects will thus be important and their study will be made easier. However, unlike the case of an external cell, control of laser intensity is more difficult. The laser intensity is reduced by diaphragming the laser beam near the mirror at the opposite side of the cell (see Fig. 5.1). This method may involve changes in the spatial distribution of the laser power in the cell, and also changes in the mode number N. Hence, the principal difficulties of measurement lie in determining the variations in the pumping rate with the intensity i_λ of the beam coming out of the mirror situated near the cell. By introducing the power density $I=\Sigma_\mu |\mathscr{E}^\mu|^2$, γ can be written as

$$\gamma = \frac{\pi |P_{ab}|^2 I}{\hbar^2 N \Delta\omega} \tag{5.81}$$

($\hbar^2$ has been reintroduced for uniformity). Using the relation between $|P_{ab}|^2$ and the spontaneous transition probability γ_{ba} (Ref. [5.54], pp. 897–900), one obtains for $\gamma(i_\lambda)$ [5.29]

$$\frac{\gamma}{\gamma_{ba}} = \frac{3(2J_b+1)}{8\pi^2 \hbar c^2} \cdot \frac{L\lambda^3}{N} \cdot \frac{i_\lambda}{sT}, \tag{5.82}$$

where L is the cavity length, λ is the transition wavelength, s is the section of the laser beam, and T is the transmission factor of the laser mirror.

If s and N may be considered as constant, γ is a linear function of the experimental values of i_λ. In the opposite case, this is no longer true.

5.5.1. $2s_2$—$2p_1$ Neon Transition ($\lambda = 1.52$ μm)

The optical pumping of neon with the 1.52 μm line exhibits the propertics set out below.

The $2s_2$ Hanle Effect and the Saturation Resonance Show Similar Lorentzian Lineshapes

As predicted by the theory, the $2s_2$ Hanle effect always has a Lorentzian shape (Fig. 5.15, Curve *II*). Furthermore, the π fluorescence from level $2s_2$ and $2p_1$ exhibits a zero-field saturation resonance having exactly the shape of the Hanle effect (Fig. 5.15, Curve *I*). These properties are valid for any laser intensity and discharge conditions.

This equality of the widths of both effects represents an important point for verifying BLA validity. Indeed, in the neon pressure range used, the resonance line ($2s_2 \rightarrow 1p_0$, $\lambda = 627$ Å) is completely reabsorbed in the cell. The trapping on this line reduces the relaxation rate of the $2s_2$ alignment by 10 MHz [5.22]. According to the experimental results, this reduction of width occurs also in the saturation effect. Nonlinear effects necessitate several successive interactions of atoms with the laser. In order to have a saturation width reduced by trapping, the atoms that are re-excited to the $2s_2$ level by reabsorption of resonance photons must still interact with the laser beam. Since the trapping processes change the velocity of the excited atoms nearly randomly, one must conclude that, within the limit of the accuracy of the measurements, the atoms interact with the laser whatever their velocity. This shows the validity of BLA: nonlinear effects are not sensitive to the velocity changes.

Hanle-Effect Width and Saturation-Resonance Amplitude: Variations with Laser Intensity and Pressure [5.29]

Figure 5.16 shows the power dependence of the Hanle-effect width (upper curve) and of the saturation amplitude (lower curve). As this figure demonstrates, it is possible to fit the experimental points across the theoretical curves (5.78)–(5.80) by assuming the pumping rate to be proportional to the laser intensity. This derives from the fact that in the particular case of the 1.52 μm laser the beam section is not very dependent on the intensity. An experimental check of the theory is then obtained by determining $\Gamma_b(2)$, η, and the ratio γ/i_λ.

Figure 5.16 also shows that the saturation amplitude $\mathcal{S}$ is very far from the fourth-order perturbation value, which is given by the zero-intensity slope of the curve. In general, $\mathcal{S}$ is near to its asymptotic value of 0.175.

When γ is lower than

$$\gamma_0 = \frac{3\Gamma_b(2)}{\eta + 1}, \tag{5.83}$$

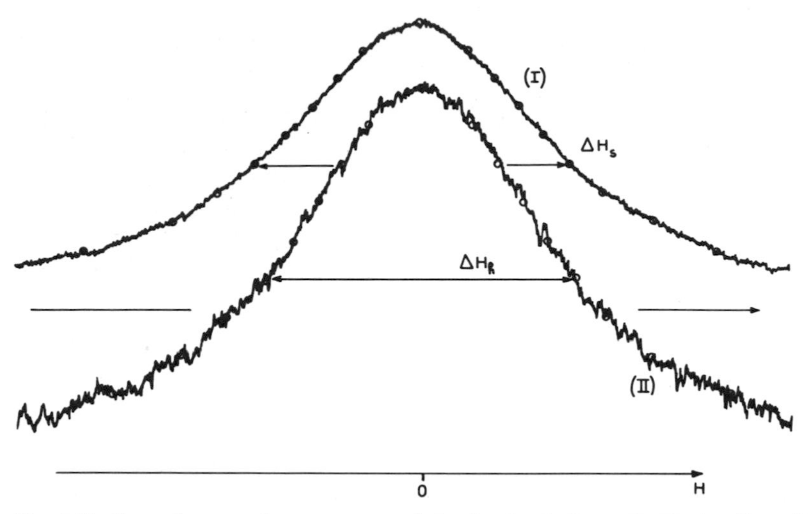

Fig. 5.15. *Curve I:* saturation resonance of the $2p_1$ level. *Curve II:* Hanle effect of $2s_2$ ($2s_2 - 2p_{10}$ line, $\lambda = 8868$ Å). Neon pressure: 0.9 Torr. The saturation width is $\Delta H_s = 10.8 \pm 0.3$ Gauss and the Hanle effect width is $\Delta H_h = 10.6 \pm 0.4$ Gauss. The circles are theoretical, using a Lorentzian lineshape [5.16]

$\mathscr{S}$ of (5.76) can be expressed as an infinite series

$$\mathscr{S} = \frac{1}{\eta + 1} \sum_p (-)^p \left(\frac{\gamma}{\gamma_0}\right)^{p+1}. \tag{5.84}$$

As a matter of fact, this expression gives the result of the perturbation theory when BLA is valid. Thus, the perturbation development is convergent only if $\gamma < \gamma_0$. In the case of Fig. 5.16 γ_0 is equal to 5.8 MHz, all the experimental points correspond to $\gamma > \gamma_0$, and γ can be as large as 90 MHz!

Finally two points are noteworthy:

i) $\Gamma_a^*(0)$ can be deduced from the experimental values of η and $\Gamma_b(2)$. [$\Gamma_b(2)/\Gamma_b(0)$ is measured from the anisotropy ratio of the $2s_2$ fluorescence, see Section 5.1.] This is an original method of measuring the relaxation time of a $J = 0$ level[15]. The $2p_1$ relaxation rate is found to be independent of the neon pressure and equal to 8.85 ± 1.25 MHz.

ii) An important point of the BLA theory is that the pumping rate γ depends only on the features of the laser field. Experiments have shown that γ is actually independent of pressure and of the discharge conditions in the cell [5.29].

[15] Strictly speaking, in $\Gamma_a^*(0)$ there is a correction factor, $1 - \gamma_{ba}/\Gamma_b(0)$, which is due to the $b \to a$ spontaneous emission, but this correction is often negligible.

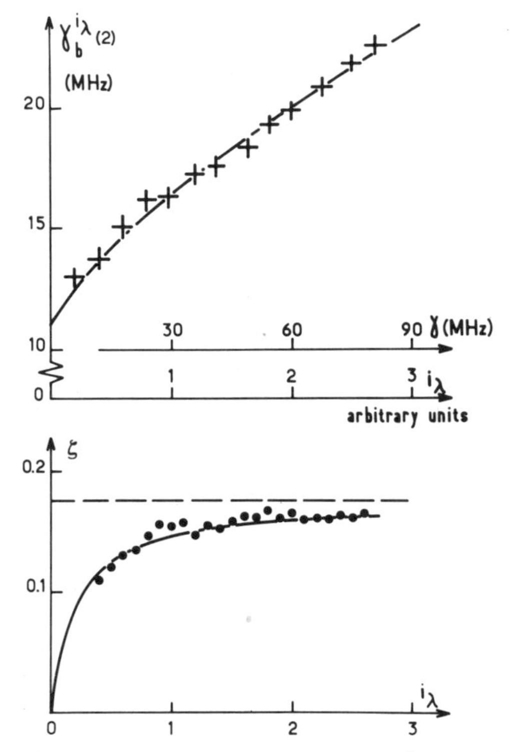

Fig. 5.16. Variations with laser intensity of the Hanle effect width (upper curve) and of the relative amplitude of the saturation resonance $\mathscr{S}$ (lower curve). ($\mathscr{S}$ is defined by the ratio between the height of the saturation resonance and the laser-induced change of the fluorescence in a strong magnetic field.) (Neon pressure: 1 Torr. The points and crosses are experimental, and the continuous curves are theoretical, using $\Gamma_b(2) = 11$ MHz, $\eta = 4.72$ and assuming ratio γ/i_λ to be constant) [5.29]

5.5.2. $3s_2$—$2p_1$ Neon Transition ($\lambda = 7305$ Å)

The purpose of this section is to present some interesting new features appearing in the experimental study of the $3s_2 - 2p_1$ transition. This study may be expected to provide an independent measurement of the $2p_1$ relaxation rate. This relaxation rate is obtained from η of (5.76). However, η does not appear in the fourth-order development of $\gamma_b^I(2)$ and $\mathscr{S}$, and subsequently may be measured at very high pumping rates only (in particular, from the asymptotic value of $\mathscr{S}$). We shall see that with the 7305 Å line, γ is always less than 4 MHz. In practice, the measurement of $\Gamma_a(0)$ is not possible. Hence, the previous measurement of $\Gamma_a(0)$ will be used for the interpretation of the experiments at 7305 Å.

Relaxation Rate of the $3s_2$ Level

Figure 5.17 shows the variations of the $3s_2$ Hanle-effect width (upper curve) and of the amplitude of the saturation resonance (lower curve) with the intensity of the 7305 Å laser line. In this case, it has not been possible to fit the experimental values across the theoretical curves by using a linear relation between γ and i_λ. This is because the cross section of the laser beam is not constant with varying laser intensities [5.29]. In order to analyze the experimental results, both the previous determination of the $2p_1$ decay rate and the $\Gamma_{3s_2}(2)/\Gamma_{3s_2}(0)$ ratio, as measured from the anisotropy degree of the $3s_2$ fluorescence (see Section 5.1) are used for calculating the η parameter. The comparison of simultaneous measurements of $\gamma_b^{i_\lambda}(2)$ and $\mathscr{S}$ then leads to the determination of the $3s_2$ alignment relaxation rate and of the pumping rate γ [using relations (5.78) and (5.80)]. The dashed line in Fig. 5.17 shows that the value obtained for $\Gamma_{3s_2}(2)$ does not depend on the laser intensity. This is a particular way of verifying the theory. The value obtained for $\Gamma_{3s_2}(2)$ is much more precise than the one obtained by extrapolation of the Hanle-effect width at vanishing laser intensities. Indeed, at low intensities, the signal-to-noise ratio is poor and the zero-intensity extrapolation is not easy to perform. In previous measurements [5.11, 22], $\gamma_b^{i_\lambda}(2)$ had been assumed to vary linearly with i_λ. This hypothesis had involved an underestimation of the Hanle-effect power-broadening. Figure 5.18 shows both the present measurements and the previous ones as functions of the neon pressure. $\Gamma_b(0)$ is deduced from $\Gamma_b(2)/\Gamma_b(0)$, which is measured by means of the zero-field anisotropy ratio of the $3s_2$ fluorescence[16]. Subsequently, the corrections of the $\Gamma_b(2)$ values affect $\Gamma_b(0)$. In pure neon, the 600 Å resonance line ($3s_2-1p_0$) is completely absorbed in the cell. This is why the zero-pressure values differ for $\Gamma_b(2)$ and $\Gamma_b(0)$. The theory of the coherent multiple scattering of the fluorescent lines [5.40, 47] shows that the $3s_2$ natural width and the $3s_2-1p_0$ transition probability may be deduced from these values:

$$\gamma_{3s_2} = 7.50 \pm 0.44 \text{ MHz}, \tag{5.85}$$

$$\gamma_{3s_2 \to 1p_0} = 5.29 \pm 0.43 \text{ MHz}. \tag{5.86}$$

These new values are in good agreement with measurements of the $3s_2$ orientation relaxation rate using a circular polarized 6328 Å laser line and an external cell (see Section 5.1).

---

[16] As this ratio does not depend on the laser intensity [5.11, 20] there is no problem of extrapolation at zero intensity.

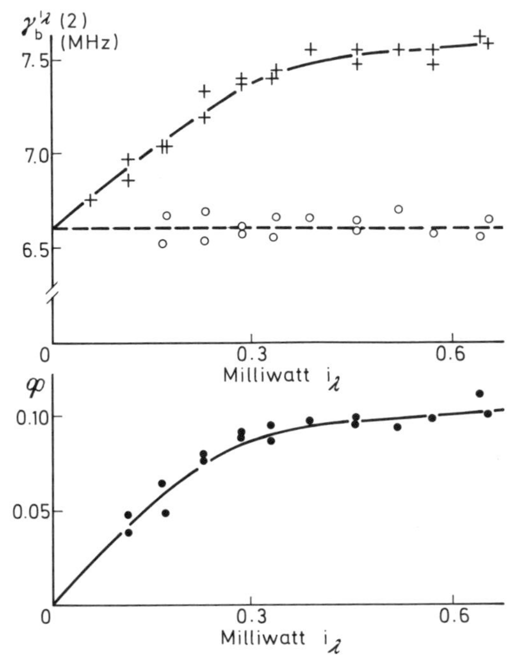

Fig. 5.17. $3s_2$ Hanle effect width (crosses, upper curve) and amplitude of the saturation resonance (points, lower curve) versus the 7305 Å laser intensity. The circles give the $3s_2$ alignment relaxation rate deduced from simultaneous *measurements* of $\gamma_b^{i\lambda}(2)$ and $\mathcal{S}$. (Neon pressure: 1.10 Torr; discharge intensity: 7 mA) [5.29]

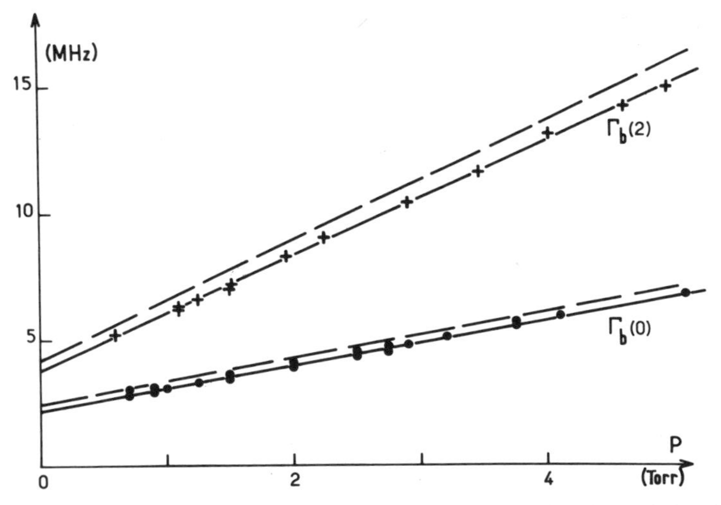

Fig. 5.18. Relaxation rates of the $3s_2$ alignment (crosses) and population (point;) versus neon pressure. The dashed lines give the erroneous previous measurements [5.21, 22]

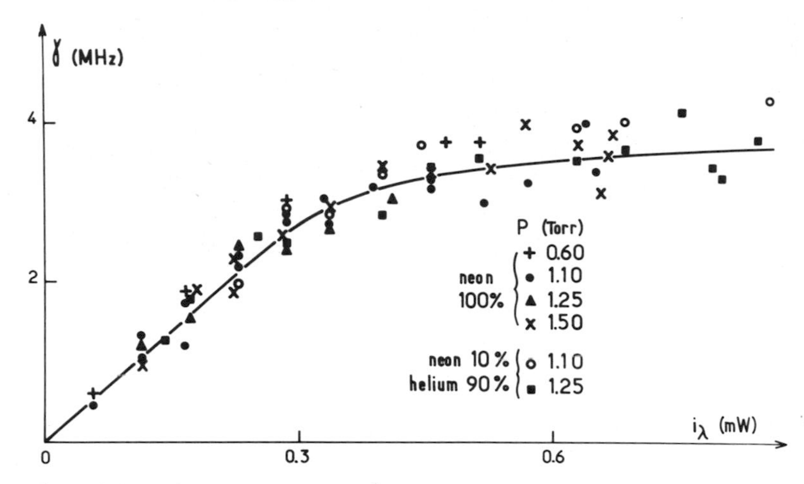

Fig. 5.19. Pumping rate versus 7305 Å laser intensity for different pressures and helium-neon mixtures. The continuous curve is theoretical, using $\gamma_{3s_2 \to 2p_1} = 0.05$ MHz [5.29]

Pumping Rate of the 7305 Å *Laser Line*

The pumping rate γ is the second quantity deduced from the simultaneous measurements of $\gamma_b^{i_\lambda}(2)$ and $\mathscr{S}$. Variations of γ with i_λ are shown in Fig. 5.19. As predicted by theory, γ does not depend on the pressure and the composition of the He–Ne mixture. Owing to the intensity-dependence of the beam section, γ is not proportional to i_λ. Indeed, there are two different operating conditions for the laser oscillation.

(i) When i_λ is lower than 0.3 mW, the cavity oscillates on axial modes only (TEM_{00} in the notation of Fox and Li [5.52]). In these conditions, the section is nearly constant and γ is a linear function of i_λ.

(ii) When i_λ is higher than 0.3 mW, the laser radiation usually fills the amplifier tube better. Transverse modes (TEM_{10}) appear, leading to a ring-shaped distribution of the intensity on the laser mirror. Because of this creation of transverse modes, the beam cross section increases in the same way as the intensity, and the power density—and subsequently the pumping rate—is no longer dependent on the laser intensity (see Fig. 5.19).

The number of modes can be obtained from a spectral analysis of the laser beam, and a direct measurement gives the transmission factor of the mirror. It is then sufficient to determine the $3s_2 - 2p_1$ transition probability (γ_{ba}) for obtaining γ as a function of the power density i_λ/s, see (5.82). The fitting of the theoretical curve to the experimental points of Fig. 5.19, is obtained with the following value [5.29]

$$\gamma_{ba} = (5 \pm 2)10^{-2} \text{ MHz}. \tag{5.87}$$

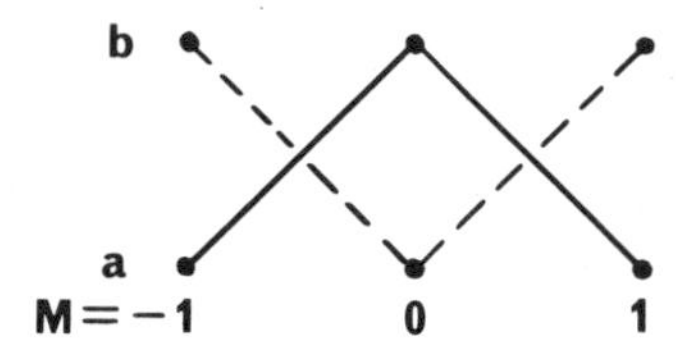

Fig. 5.20. Diagram of the σ transition for a $J=1, J=1$ line

This value is in good agreement with the theoretical one as determined from a spectroscopic study of neon by the parametric potential method [5.51].

5.6. $J_a = 1 - J_b = 1$ Laser Transition

This case has been experimentally studied on the 6401 Å neon line ($3s_2 - 2p_2$) (see [5.17, 29]). The main features are as follows:

a) The Hanle effects of both levels have a Lorentzian line shape. This can be easily understood from Fig. 5.20. If there is no depolarizing relaxation process inducing transitions between Zeeman sublevels, the set of optical transitions ($M_b = \pm 1 \leftrightarrow M_a = 0$) and the set ($M_a = \pm 1 \leftrightarrow M_b = 0$) are not coupled, and each set is equivalent to a $J = 1 \leftrightarrow J = 0$ transition. If depolarizing processes occur, they induce a coupling between the two sets. However, in the experimental case this coupling is too small to produce any observable departure from the Lorentzian shape.

b) The power broadening of the Hanle effect is the same in both levels. Theoretically, this is rigorous at fourth order and at higher order, too, if the relaxation rates are the same in both levels (due to the symmetry).

c) The Hanle effects of both levels contribute to the saturation resonance of the population, as observed in the π fluorescence. The relative contribution of the two Hanle effects depends on the fluorescence line (see Subsection 5.3.3).

5.7. Higher-Order Nonlinear Effects in $J = 1 - J = 2$ Transitions

For $J = 1 - J = 0$ and $J = 1 - J = 1$ laser transitions, population ($k = 0$) and alignment ($k = 2$) are the only tensorial components of the density matrix that may be induced in the laser levels, and these components appear from the second order in a perturbation development ("linear response").

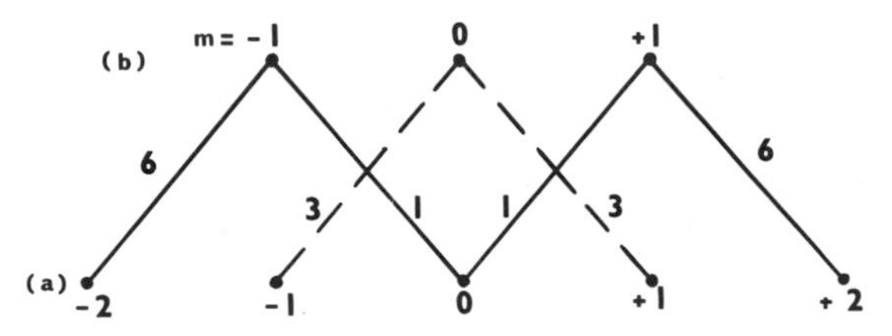

Fig. 5.21. Diagram of the transitions induced by a σ polarized laser between levels $b(J_b = 1)$ and a $(J_a = 2)$. The relative transition probabilities are indicated

For a $J_b = 1 - J_a = 2$ transition, the components of a multipole moment of order $k = 4$ may be created in the a level (electric hexadecapole moment ${}_a\varrho_Q^4$). But, in a perturbation development, this hexadecapole moment appears from the fourth order only. This is a typical feature of the non-linear atomic response. The first part of this section is devoted to the generation of this hexadecapole moment and its detection on the fluorescence; the second presents the observation of anomalous behavior in the saturation of $J = 1 - J = 2$ transitions.

5.7.1. Laser-Induced Hexadecapole Moment in the Neon $2p_4$ Level

Elementary arguments using the diagram of the transition (Fig. 5.21) allow one to explain the process of generating a hexadecapole moment into the $J_a = 2$ level [5.21]. To the second order in the laser field, the absorption of a σ-polarized photon by an atom in the $m_a = 0$ level creates coherence between the $m_b = \pm 1$ sublevels (transverse alignment of b). To the fourth order, stimulated emission of a photon induces $\Delta m = 4$ coherence between sublevels $m_a = \pm 2$: *this is a transverse hexadecapole moment* $({}_a\varrho_4^4)$ *which evolves at frequency* $4\omega_a$.

Owing to the dipolar character of the spontaneous emission, this multiple is not directly observable on the fluorescent lines from level a. However, to the sixth order in the field, this coherence is brought back in the $m_b = \pm 1$ sublevels by means of a photon absorption. Owing to this feedback, part of the second-order coherence of b is restored into level b at the sixth order. This restitution is maximum in zero magnetic field when Zeeman coherences do not precess, but is destroyed by very weak fields due to the successive evolution of coherence in b and a. On the Hanle effect of b, the coupling with the $\Delta m = 4$ coherence of a will appear as a very narrow peak in zero field.

This effect may be analyzed in a more precise way with the help of the theory presented in Section 5.4. For a $J_b = 1 - J_a = 2$ transition, there are four transverse components which precess under the influence of the magnetic field: alignment of levels $a({}_a\varrho_2^2)$ and $b({}_b\varrho_2^2)$, hexadecapole moment of level $a({}_a\varrho_2^4, {}_a\varrho_4^4)$. Equation (5.71) shows that these components are coupled both between themselves and to the populations of the sublevels. By eliminating the populations from (5.71) we find the following steady-state equations of motion for the transverse components [5.20]

$$\left(\Gamma_a(2)+\frac{5\gamma}{42}+2i\omega_a\right){}_a\varrho_2^2-\left(\frac{2\gamma}{5\sqrt{21}}+\frac{\sqrt{21}}{10}\gamma_{ba}\right){}_b\varrho_2^2+\frac{\gamma}{35\sqrt{3}}{}_a\varrho_2^4$$
$$-\frac{\gamma}{5\sqrt{21}}{}_a\varrho_4^4$$
$$=-\frac{n\gamma\sqrt{7}}{10\sqrt{3}}\chi_a^2+\frac{23\gamma}{210}(1-\chi_{aa}^{22}){}_aR_2^2 \tag{5.88a}$$
$$-\frac{\gamma}{5\sqrt{21}}(1-\chi_{ba}^{22}){}_bR_2^2+\frac{\gamma}{70\sqrt{3}}(1-\chi_{aa}^{42}){}_aR_2^4\,,$$

$$\left(\Gamma_b(2)+\frac{7\gamma}{30}+2i\omega_b\right){}_b\varrho_2^2-\frac{2\gamma}{5\sqrt{21}}{}_a\varrho_2^2-\frac{\gamma}{5\sqrt{7}}{}_a\varrho_2^4+\frac{\gamma}{5}{}_a\varrho_4^4$$
$$=\frac{n\gamma}{30}\chi_b^2+\frac{\gamma}{30}(1-\chi_{bb}^{22}){}_bR_2^2 \tag{5.88b}$$
$$-\frac{\gamma}{5\sqrt{21}}(1-\chi_{ab}^{22}){}_aR_2^2-\frac{\gamma}{10\sqrt{7}}(1-\chi_{ab}^{42}){}_aR_2^4\,,$$

$$\left(\Gamma_a(4)+\frac{4\gamma}{35}+2i\omega_a\right){}_a\varrho_2^4+\frac{\gamma}{35\sqrt{3}}{}_a\varrho_2^2-\frac{\gamma}{5\sqrt{7}}{}_b\varrho_2^2-\frac{\gamma}{10\sqrt{7}}{}_a\varrho_4^4$$
$$=\frac{n\gamma}{15\sqrt{7}}\chi_a^4+\frac{3\gamma}{28}(1-\chi_{aa}^{44}){}_aR_2^4 \tag{5.88c}$$
$$+\frac{\gamma}{70\sqrt{3}}(1-\chi_{aa}^{24}){}_aR_2^2-\frac{\gamma}{10\sqrt{7}}(1-\chi_{ba}^{24}){}_bR_2^2\,,$$

$$\left(\Gamma_a(4)+\frac{\gamma}{5}+4i\omega_a\right){}_a\varrho_4^4-\frac{\gamma}{5\sqrt{21}}{}_a\varrho_2^2$$
$$+\frac{\gamma}{5}{}_b\varrho_2^2-\frac{\gamma}{10\sqrt{7}}{}_a\varrho_2^4=0\,, \tag{5.88d}$$

where R is the real part of ϱ

$${}_\beta R_Q^k=\tfrac{1}{2}({}_\beta\varrho_Q^k+{}_\beta\varrho_Q^{k*})\,. \tag{5.89}$$

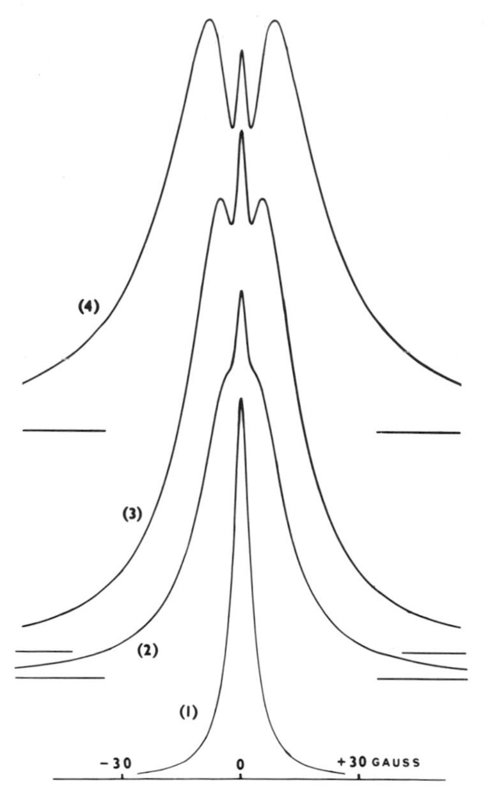

Fig. 5.22. Computer calculation of the lineshape of the Hanle effect from the $J=2$ level. The relaxation rates (MHz) are $\Gamma_a(0)=9.8$, $\Gamma_a(2)=\Gamma_a(4)=14.5$, $\Gamma_b(0)=2.75$, $\Gamma_b(2)=5.75$ and $\gamma_{ba}=0.5$ Landé factors: $g_a=1.299$ and $g_b=1.293$. Curves *1*, *2*, *3* and *4* correspond to pumping rates, respectively, equal to 10, 50, 70, and 115 MHz, [5.21]

Coefficients χ depend on γ and on the various relaxation rates, see [5.20]. In (5.88) the coupling between the Zeeman coherences appears according three processes:

a) The direct laser-induced coupling, which is proportional to the pumping rate γ. The corresponding terms have been put together on the left-hand side of (5.88). (The processes analyzed at the beginning of this section correspond to such a coupling between ${}_b\varrho_2^2$ and ${}_a\varrho_4^4$.)

b) An indirect coupling through the longitudinal components (populations of the sublevels)

$$ {}_\alpha\varrho_2^k \xrightarrow{(\gamma)} {}_{\alpha'}\varrho_0^{k'} \xrightarrow{(\gamma)} {}_{\alpha''}\varrho_2^{k''} \,. \tag{5.90} $$

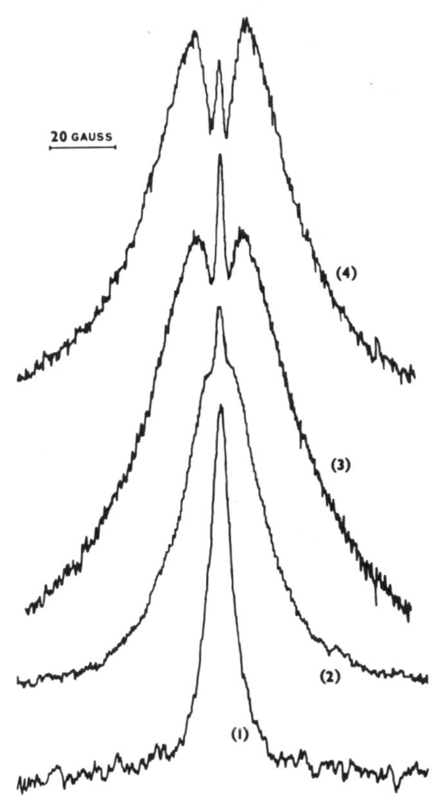

Fig. 5.23. $3s_2$ Hanle effect observed on the fluorescence at 7305 Å for laser irradiation at 6328 Å ($3s_2-2p_4$). (Neon pressure: 1.5 Torr; discharge intensity: 10 mA). Curves *1*, *2*, *3*, and *4* correspond to laser intensities, respectively, equal to 0.2, 1, 2, and 3.75 mW out of the laser cavity (from [5.21])

These terms, which come from the elimination of the longitudinal components on (5.71) have been put together on the right-hand side of (5.88) (R terms). Let us point out that

i) at the lowest order in the laser field, the coupling coefficients are proportional to $\gamma^2(1-\chi^{kk''}_{\alpha\alpha''}\alpha\gamma)$;

ii) this coupling involves the real part of ϱ^k_2 only, because the populations themselves are real.

c) The coupling induced by the spontaneous emission on the $b\rightarrow a$ (γ_{ba} term on the left-hand side of (5.88a)).

Equation (5.88) has been solved for any value of the magnetic field by a computer calculation. The theoretical lineshape obtained for the b Hanle effect is drawn in Fig. 5.22 for different values of the pumping rate.

α) At low laser intensities, the couplings are vanishing and the Hanle effect shape does not deviate from a Lorentzian one (Curve *1*).

β) When the intensity grows, as predicted by the physical arguments developed at the beginning of this section, the coupling between ${}_a\varrho_4^4$ and ${}_a\varrho_2^2$ induces a narrow central peak on the Hanle effect (Curve *2*).

γ) In addition, for high laser intensities, a dip of intermediate width appears (Curves *3* and *4*). This dip is due to the coupling between the $\Delta m=2$ coherences of levels b and a (${}_a\varrho_2^2, {}_a\varrho_2^4 \Rightarrow {}_b\varrho_2^2$). For instance, by absorption of a σ^+ photon, the coherence between sublevels $m_a=-2$ and $m_a=0$ is transferred into the $m_b=\pm 1$ sublevels (solid lines of Fig. 5.21). The resonance induced by this coupling is negative because the $\Delta m=2$ coherence of level a is opposite that of level b.

As is shown in Fig. 5.23, all the predicted effects have been observed on the Hanle effect of the $3s_2$ neon level optically pumped at the 6328 Å laser line. This presents experimental evidence for the generation of a hexadecapole moment in the $2p_4$ level through higher-order nonlinear processes. It would be interesting to measure the relaxation time of this moment in order to compare it with the relaxation time of the alignment. This measurement should be possible by fitting theoretical predictions and experimental results. This kind of work is in progress.

5.7.2. Anomalous Behavior of the Saturation Resonance of the $2p_4$—$1s_5$ Fluorescent Line ($\lambda=5944$ Å)

As explained previously (Subsections 5.3.3 and 5.4.2), the populations of the laser levels exhibit resonant variations in zero magnetic field, which are related to the Hanle effects of the levels. This saturation resonance may be observed on the π fluorescence emitted by these levels. For $J=1-J=0$ ($2s_2-2p_1, 3s_2-2p_1$) and $J=1-J=1$ ($3s_2-2p_2$) laser transitions, this resonance appears as a "saturation effect", i.e. it reduces the laser-induced fluorescence change in zero magnetic field. However, this is not always true for the $2s_2(J=1)-2p_4(J=2)$ transition ($\lambda=1.15$ µm): while the 6096 Å fluorescent line ($2p_4 \to 1s_4, J=1$) shows a resonance with the ordinary sign (Curve *1* of Fig. 5.24), the 5944 Å line ($2p_4 \to 1s_5, J=2$) exhibits a saturation resonance which *increases* the laser-induced fluorescence in zero field (Curve *2* of Fig. 5.24). As this effect is also observed for a laser irradiation at 6328 Å, this anomalous behavior appears to be connected with the following sequence of transitions

$$J_b=1 \xleftrightarrow{\text{laser}} J_a=2 \xrightarrow{\pi\,\text{fluorescence}} J_f=2\,.$$

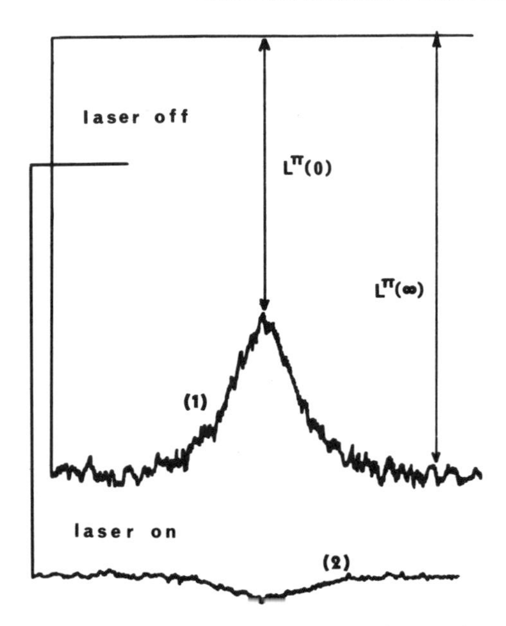

Fig. 5.24. Magnetic-field variations of the π fluorescent lines emitted on the $2p_4-1s_4$ ($\lambda=6095$ Å, Curve *1*) and $2p_4-1s_5$ ($\lambda=5944$ Å, Curve *2*) transitions. [Laser irradiation at 1.15 μm ($2s_2-2p_4$), neon pressure: 0.9 Torr] [5.18]

In order to explain it, we must calculate the saturation amplitude $\mathscr{S}$, see Fig. 5.24,

$$\mathscr{S}=\frac{L_\pi(\infty)-L_\pi(0)}{L_\pi(\infty)}. \tag{5.91}$$

By using (5.71) and assuming one relaxation time per level [$\Gamma_\alpha(k)=\gamma_\alpha$ and $\gamma_{ba}=0$], we obtain for the $J_a=2\rightarrow J_f=2$ fluorescence [5.20]

$$\mathscr{S}_{a2}=1-\frac{\Delta}{\Delta_0}\cdot\frac{1+\dfrac{32\gamma}{135}\left(\dfrac{1}{\gamma_a}+\dfrac{1}{\gamma_b}\right)}{1+\dfrac{\gamma}{270}\left(\dfrac{48}{\gamma_a}+\dfrac{55}{\gamma_b}\right)+\dfrac{\gamma^2}{1350\gamma_a}\left(\dfrac{10}{\gamma_a}+\dfrac{19}{\gamma_b}\right)}, \tag{5.92}$$

where

$$\Delta_0=\left[1+\frac{\gamma}{5}\left(\frac{1}{\gamma_a}+\frac{1}{\gamma_b}\right)\right]\left[1+\frac{4\gamma}{15}\left(\frac{1}{\gamma_a}+\frac{1}{\gamma_b}\right)\right] \tag{5.93}$$

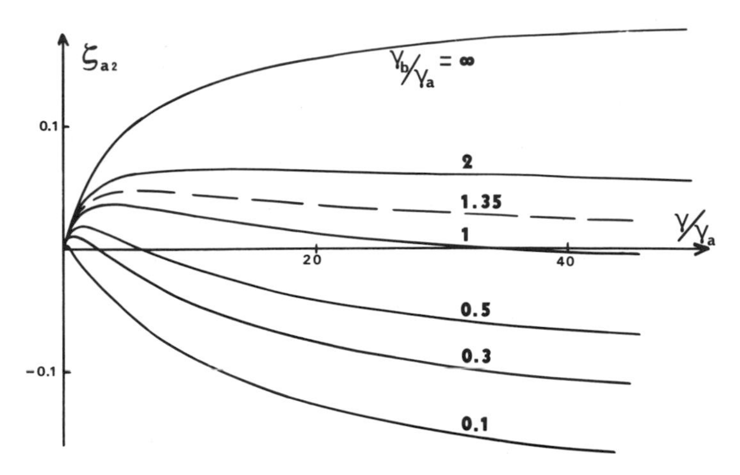

Fig. 5.25. Amplitude of the saturation resonance on the $J_a=2 \to J_f=2$ fluorescent line, for various ratios of the relaxation rates ($J_b=1-J_a=2$ laser transition. Model with two relaxation rates, γ_a and γ_b) [5.20]

and

$$\Delta=\left[1+\frac{\gamma}{10}\left(\frac{1}{\gamma_a}+\frac{2}{\gamma_b}\right)\right] \cdot\left[1+\frac{\gamma}{30}\left(\frac{8}{\gamma_a}+\frac{7}{\gamma_b}\right)+\frac{\gamma^2}{150\gamma_a}\left(\frac{2}{\gamma_a}+\frac{3}{\gamma_b}\right)\right]. \tag{5.94}$$

Figure 5.25 shows the variations of $\mathscr{S}_{a2}$ as a function of γ/γ_a for different values of γ_b/γ_a. The zero laser intensity slope of the curves $(11\gamma/270\gamma_a)$ represents a value like that obtained from the fourth-order perturbation calculation. For high laser intensities, the asymptotic value of ζ_{a2} is

$$\mathscr{S}_{a2}(\gamma\infty)=\frac{2\gamma_b^2+\gamma_a\gamma_b-5\gamma_a^2}{(\gamma_a+\gamma_b)(19\gamma_a+10\gamma_b)} \tag{5.95}$$

This asymptotic value is positive or negative according as γ_b/γ_a is higher or lower than $(\sqrt{41}-1)/4 \simeq 1.35$. Subsequently two cases are possible:

1) If γ_b is higher than $1.35\gamma_a$, $\mathscr{S}_{a2}$ keeps the sign of the fourth-order contribution, independently of the laser intensity. The zero-field saturation resonance always reduces the laser-induced change of the fluorescence intensity.

2) If γ_b is lower than $1.35\gamma_a$, $\mathscr{S}_{a2}$, which is positive for weak laser intensities, becomes negative when the laser intensity grows. This

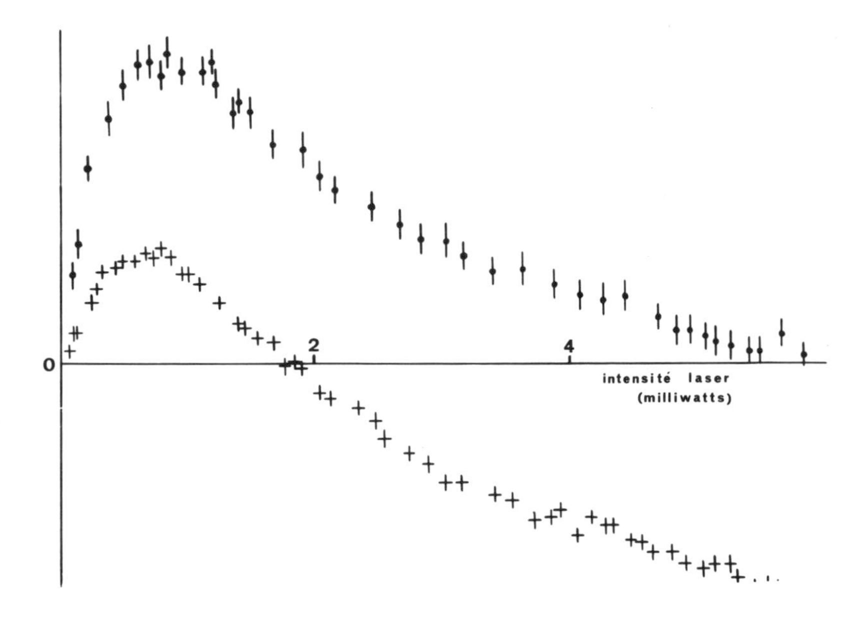

Fig. 5.26. Saturation resonance amplitude as measured by a technique of magnetic field modulation [5.18]. (Laser: 6328 Å, Π fluorescence: 5944 Å). (♦) neon traces in 1.5 Torr of helium. (+) 2.6 Torr of a 5–95% neon-helium mixture

reversal of the saturation resonance increases with decreasing γ_b/γ_a ratios. Subsequently, the anomalous behavior of the saturation seems to be closely connected with the mean lifetime of the b level, i.e. to the transit time of the atoms in this level.

This effect may be interpreted by using the diagram in Fig. 5.21 and pointing out that the population of the $m_a = \pm 2$ sublevels plays a leading part in the intensity of the $J_a = 2 \rightarrow J_f = 2$ fluorescence. Indeed, the $m_a = 2 \rightarrow m_f = 2$ transition is four times more probable than the $m_a = 1 \rightarrow m_f = 1$ transition, and the $m_a = 0 \rightarrow m_f = 0$ transition is forbidden.

Let us analyze the processes leading to the saturation resonance in the $m_a = 2$ sublevel. The stimulated emission of a σ-polarized photon by an atom in the $m_b = 1$ sublevel creates coherence between sublevels $m_a = 0$ and $m_a = 2$ (second-order contribution in a perturbation expansion). A new interaction with the laser (now a photon absorption) couples this coherence with the populations of sublevels $m_a = 0$, $m_b = 1$ and $m_a = 2$, and produces in these sublevels a saturation resonance which reduces, in zero magnetic field, the laser-induced population changes (fourth-order term). To the sixth order, the stimulated emission of a

σ^- laser photon brings back part of the $m_b = 1$ population into the $m = 2$ sublevel. Since the laser-induced population changes of these two sublevels are opposite, the latter process sets the saturation resonance of sublevel $m_b = 1$ over against that of $m_a = 2$ and *tends to increase the zero-field population change of* $m_a = 2$. This effect is important because the $m_b = 1 \rightarrow m_a = 2$ transition is six times more probable than the $m_b = 1 \rightarrow m_a = 0$ transition. On the other hand, this effect increases when the laser intensity is strong and when the population lifetime of level b becomes longer than that of level a (then the contribution of the $m_b = 1$ population grows relative to that of $m_a = 2$).

The effects predicted by Fig. 5.25 have been experimentally observed on the 5944 Å fluorescent line for a laser irradition at 6328 Å (Fig. 5.26). Changing the ratio $\gamma(3s_2)/\gamma(2p_4)$ has been produced by the use of the trapping processes on the 600 Å resonance line $(3s_2 - 1p_0)$.

In Fig. 5.26 points (♦) correspond to a very low partial pressure of neon (≈ 7 mTorr) and the relaxation rates are given by the natural widths of the levels. Then $\gamma(3s_2)$ and $\gamma(2p_4)$ are approximately equal and $\mathscr{S}$ is positive for any laser intensity.

Furthermore, crosses (+) correspond to a higher neon pressure (≈ 0.15 Torr). Because of a strong trapping of the 600 Å line, the $3s_2$ lifetime is lengthened and $\gamma(3s_2)$ becomes much smaller than $\gamma(2p_4)$ (the relaxation rate of the $3s_2$ population may be reduced by 70% for a complete trapping, see Subsection 5.5.2). Therefore a reversal of the saturation resonance is observed for high laser intensities.

Appendix 1

General Formula for the Fluorescence Light.
The Standard Components of the Polarization Vector

The fluorescence light intensity on the $b \rightarrow g$ line is given by

$$\begin{aligned} l_{bg}(\lambda) = & \frac{1}{3\sqrt{2J_b+1}}\, {}_b\varrho_0^0 + \frac{(-)^{J_b+J_g+1}}{\sqrt{2}} \begin{Bmatrix} 1 & 1 & 1 \\ J_g & J_b & J_b \end{Bmatrix} \\ & \cdot [(|\lambda_1|^2 - |\lambda_{-1}|^2)\, {}_b\varrho_0^2 - [(\lambda_1 \lambda_0^* + \lambda_0 \lambda_{-1}^*)\, {}_b\varrho_1^1 + c \cdot c]] \\ & + (-)^{J_b+J_g+1} \begin{Bmatrix} 2 & 1 & 1 \\ J_g & J_b & J_b \end{Bmatrix} \left[\frac{1 - 3|\lambda_0|^2}{\sqrt{6}}\, {}_b\varrho_0^2 \right. \\ & \left. - \left[\frac{(\lambda_1 \lambda_0^* - \lambda_0 \lambda_{-1}^*)}{\sqrt{2}}\, {}_b\varrho_1^2 + c \cdot c \right] + [\lambda_1 \lambda_{-1}^*\, {}_b\varrho_2^2 + c \cdot c] \right], \end{aligned}$$

where $\lambda_{0,\pm 1}$ are the standard components of the polarization vector $\boldsymbol{\lambda}$ of the detection. As well as the components of the polarization vector $\boldsymbol{e}$ of the laser beam, they can be defined with the help of Fig. 5.27. The direction of observation (or of the laser beam) is defined by θ and ϕ.

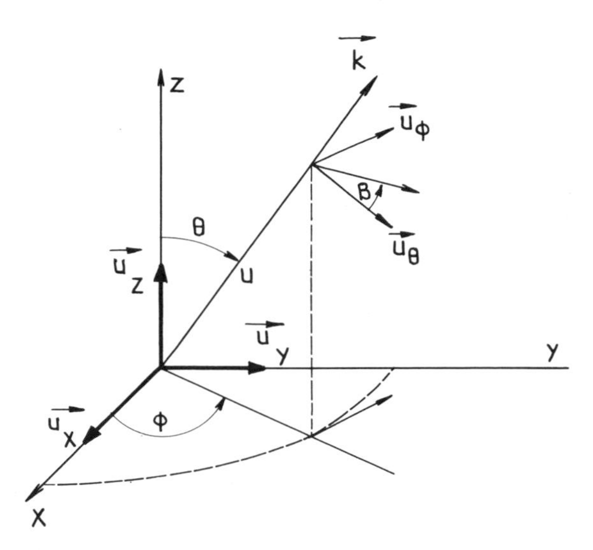

Fig. 5.27

The polarization vector is in the plane defined by $\boldsymbol{e}_\theta$ and $\boldsymbol{e}_\phi$. For a linear polarization one has

$$\boldsymbol{\lambda} = \boldsymbol{u}_\theta \cos\beta + \boldsymbol{u}_\phi \sin\beta \begin{cases} \lambda_0 = -\cos\beta\sin\theta \\ \lambda_{\pm 1} = \mp \dfrac{1}{\sqrt{2}}(\cos\beta\cos\theta \pm i\sin\beta)\mathrm{e}^{\pm i\phi} \end{cases}$$

and for circular polarization

$$\boldsymbol{\lambda}^r = -(1/\sqrt{2})(\boldsymbol{u}_\theta + i\boldsymbol{u}_\phi) \begin{cases} \lambda_0^2 = (1/\sqrt{2})\sin\theta \\ \lambda_{\pm 1}^2 = -(1/2)(1 \mp \cos\theta)\mathrm{e}^{\pm i\phi} \end{cases}$$

$$\boldsymbol{\lambda}^l = (1/\sqrt{2})(\boldsymbol{u}_\theta - i\boldsymbol{u}_\phi) \begin{cases} \lambda_0^l = -(1/\sqrt{2})\sin\theta \\ \lambda_{\pm 1}^1 = -(1/2)(1 \pm \cos\theta)\mathrm{e}^{\pm i\phi}\,. \end{cases}$$

Appendix 2

General Formula for the Linear Response

$$ {}^{(2)}_{\ b}\bar{\varrho}^{0}_{0} = -\frac{nD|P_{ab}|^2}{\Gamma_b(0)} $$

$$ {}^{(2)}_{\ b}\bar{\varrho}^{1}_{0} = -\frac{nD|P_{ab}|^2}{\Gamma_b(1)}(-)^{J_a+J_b+1}\begin{Bmatrix} 1 & 1 & 1 \\ J_a & J_b & J_b \end{Bmatrix}\frac{|e_1|^2-|e_{-1}|^2}{\sqrt{2}} $$

$$ {}^{(2)}_{\ b}\bar{\varrho}^{1}_{1} = \frac{nD|P_{ab}|^2}{\Gamma_b(1)+i\omega_b}(-)^{J_a+J_b+1}\begin{Bmatrix} 1 & 1 & 1 \\ J_a & J_b & J_b \end{Bmatrix}\frac{e_0^* e_{-1}+e_1^* e_0}{\sqrt{2}} $$

$$ {}^{(2)}_{\ b}\bar{\varrho}^{2}_{0} = \frac{nD|P_{ab}|^2}{\Gamma_b(2)}(-)^{J_a+J_b+1}\begin{Bmatrix} 2 & 1 & 1 \\ J_a & J_b & J_b \end{Bmatrix}\frac{1}{\sqrt{6}}[3|e_0|^2-1] $$

$$ {}^{(2)}_{\ b}\bar{\varrho}^{2}_{1} = \frac{nD|P_{ab}|^2}{\Gamma_b(2)+i\omega_b}(-)^{J_a+J_b+1}\begin{Bmatrix} 2 & 1 & 1 \\ J_a & J_b & J_b \end{Bmatrix}\frac{e_1^* e_0-e_0^* e_{-1}}{\sqrt{2}} $$

$$ {}^{(2)}_{\ b}\bar{\varrho}^{2}_{2} = -\frac{nD|P_{ab}|^2}{\Gamma_b(2)+2i\omega_b}(-)^{J_a+J_b+1}\begin{Bmatrix} 2 & 1 & 1 \\ J_a & J_b & J_b \end{Bmatrix}e_1^* e_{-1}, $$

where D is a constant optical lineshape factor, see Subsection 5.3.1.

For the lower level of the laser transition, the spontaneous emission from upper levels must be included according to (5.31).

References

5.1. A. JAVAN: Bull. Am. Phys. Soc. **9**, 489 (1964).

5.2. R. L. FORK, L. E. HARGROVE, M. A. POLLACK: Phys. Rev. Letters **12**, 705 (1964).

5.3. R. H. CORDOVER, J. PARKS, A. SZOKE, A. JAVAN: In *Physics of Quantum Electronics*, ed. by KELLEY, LAX, and TANNENWALD (McGraw-Hill, New York, 1966).

5.4. T. HÄNSCH, P. TOSCHEK: Phys. Letters **22**, 150 (1966).

5.5. B. DECOMPS, M. DUMONT: CRAS **262**B, 1004 (1966).

5.6. M. TSUKAKOSHI, K. SHIMODA: J. Phys. Soc. Jap. **26**, 758 (1969).

5.7. M. MCCLINTOCK, W. DEMTRÖDER, R. N. ZARE: J. Chem. Phys. **51**, 5509 (1969).

5.8. M. BROYER, J. VIGUE, J. C. LEHMANN: CRAS **273**B, 289 (1971).

5.9. F. P. SCHÄFER (ed.): *Dye Lasers* (Springer, Berlin, Heidelberg, New York, 1973).

5.10. S. SVANBERG, P. TSEKERIS, W. HAPPER: Phys. Rev. Letters **30**, 817, (1973).

5.11. B. DECOMPS, M. DUMONT: J. Phys. (Paris) **29**, 443 (1968); and IEEE J. Quant. Electr. QE-**4**, 916 (1968).

5.12. E. GIACOBINO: CRAS **276**B, 535 (1973).

5.13. E. GIACOBINO: Opt. Commun. **8**, 154 (1973).

5.14. M. DUMONT: Thesis, Paris (1971).

5.15. W. E. LAMB: Phys. Rev. **134**A, 1429 (1964).

5.16. M. DUCLOY: Opt. Commun. **3**, 205 (1971).

5.17. J. DATCHARY, M. DUCLOY: CRAS **274**B, 337 (1972); J. DATCHARY: Thèse de 3e cycle, Paris (1972).

5.18. M.P.GORZA, B.DECOMPS, M.DUCLOY: Opt. Commun. **8**, 323 (1973).
5.19. M.DUMONT: J. Phys. (Paris) **33**, 971 (1972).
5.20. M.DUCLOY: Phys. Rev. A**8**, 1844 (1973) and A**9**, 1319 (1974).
5.21. M.DUCLOY, M.P.GORZA, B.DECOMBS: Opt. Commun. **8**, 21 (1973).
5.22. B.DECOMPS: Thesis, Paris (1969).
5.23. M.DUMONT, B.DECOMPS: CRAS **269**, 191 (1969).
5.24. P.TOSCHEK: Private communication (1973).
5.25. M.DUCLOY, B.DECOMPS: CRAS **266**B, 412 (1968);
M.DUCLOY, E.GIACOBINO, B.DECOMPS: J. Phys. (Paris) **31**, 533 (1970);
E.FOURNIER: Thèse de 3e cycle, Paris (1969).
5.26. M.DUCLOY: Thèse de 3e cycle, Paris (1968).
5.27. M.DUCLOY, M.DUMONT: CRAS **266**B, 340 (1968); and J. Phys. (Paris) **31**, 419 (1970).
5.28. W.E.BELL, A.L.BLOOM: Phys. Rev. Letters **6**, 280 (1961).
5.29. M.DUCLOY: Thesis, Paris (1973);—Ann. Phys. Paris **8**, 403 (1973—1975).
5.30. R.H.CORDOVER, P.A.BONCZYK, A.JAVAN: Phys. Rev. Letters **18**, 730 and 1104 (1967).
5.31. H.K.HOLT: Phys. Rev. Letters **19**, 1275 (1967).
5.32. M.S.FELD, A.JAVAN: Phys. Rev. **177**, 540 (1968).
5.33. C.COHEN-TANNOUDJY: Thesis, Paris (1962): Ann. Phys. **7**, 423 and 469 (1962).
5.34. B.L.GYORFFY, M.BORENSTEIN, W.E.LAMB: Phys. Rev. **169**, 340 (1968).
5.35. S.G.RAUTIAN: JETP **51**, 1176 (1966) [Soviet. Phys. JETP **24**, 788 (1967)];
S.G.RAUTIAN, I.I.SOBEL'MAN: Soviet. Phys. Usp. **9**, 701 (1967).
5.36. P.R.BERMAN, W.E.LAMB: Phys. Rev. A **2**, 2435 (1970).
5.37. M.DUMONT, B.DECOMPS: J. Phys. (Paris) **29**, 181 (1968).
5.38. U.FANO: Rev. Mod. Phys. **29**, 74 (1957).
5.39. A.OMONT: J. Phys. (Paris) **26**, 26 (1965).
5.40. A.OMONT: J. Phys. (Paris) **26**, 576 (1965);—Thesis, Paris (1967).
5.41. M.I.D'YAKONOV, V.I.PEREL: Soviet. Phys. JETP **20**, 997 (1965).
5.42. M.LOMBARDI: Compt. Rend. Acad. Sci. **265**, 191 (1967);—J. Phys. (Paris) **30**, 631 (1969);—Thesis Grenoble (1970).
5.43. P.W.SMITH, T.HÄNSCH: Phys. Rev. Letters **26**, 740 (1971).
5.44. M.I.D'YAKONOV, V.I.PEREL: Soviet. Phys. JETP, **3**, 585 (1970).
5.45. W.R.BENNETT: Phys. Rev. **126**, 580 (1962).
5.46. B.FRIED, S.CONTE: *The Plasma Dispersion Function* (Academic Press, New York, London, 1961).
5.47. J.P.BARRAT: J. Phys. (Paris) **20**, 541, 633 and 657 (1959).
5.48. W.R.BENNETT, P.J.KINDLMANN: Phys. Rev. **149**, 38 (1966).
5.49. M.DUCLOY: IEEE J. Quant. Electron. QE-**8**, 560 (1972).
5.50. M.DUCLOY: Opt. Commun. **8**, 17 (1973).
5.51. S.FENEUILLE, M.KLAPISCH, E.KOENIG, S.LIBERMANN: Physica **48**, 571 (1970).
5.52. A.G.FOX, T.LI: Bell Syst. Techn. J. **40**, 453 (1961).
5.53. J.Z.KLOSE: Rev. **141**, 181 (1966).
5.54. A.MESSIAH: *Mécanique Quantique* (Dunod, Paris, 1964).
5.55. P.W.SMITH: IEEE J. Quant. Electron. QE-**8**, 704 (1972).
5.56. C.G.CARRINGTON, A.CORNEY: Opt. Commun. **1**, 115 (1969); and J. Phys. (London) B **4**, 849 (1971).
5.57. E.B.SALOMAN, W.HAPPER: Phys. Rev. **144**, 7 (1966).
5.58. P.W.SMITH: J. Appl. Phys. **37**, 2089 (1966).
5.59. T.HÄNSCH, P.TOSCHER: IEEE J. Quant. Electron. QE-**4**, 530 (1968) and QE-**5**, 61 (1969).
5.60. E.GIACOBINO: J. Phys. Letters **36** L65 (1975).
5.61. M.GORLICKI, M.DUMONT: Opt. Commun. **11**, 166 (1974).

6. Laser Frequency Measurements, the Speed of Light, and the Meter*

K. M. EVENSON and F. R. PETERSEN

With 5 Figures

The spectral characteristics of electromagnetic radiation are determined by either its vacuum wavelength or its frequency (and, of course, the speed of light is the product of the two). Before the advent of lasers, infrared and visible spectra were measured using wavelength techniques. Lasers have fractional linewidths approaching those of radio and microwave oscillators and have many orders of magnitude more spectral radiance than incoherent sources, as shown in Fig. 6.1; thus, direct frequency measuring techniques as well as wavelength techniques can be used and the frequency techniques provide much more resolution and accuracy in measuring the spectral characteristics of the radiation. Frequency measuring techniques are limited only by the accuracy of the fundamental standard and the stabilities of the sources available and, as a result, are better by several orders of magnitude. Hence, lasers which provide intense coherent sources of electromagnetic radiation extending from the microwave through the visible to the ultraviolet portion of the electromagnetic spectrum (see Fig. 6.2) can be considered as either wavelength or frequency sources.

In spite of the fact that lasers provided coherent frequency sources in the infrared and visible, frequencies could not immediately be measured because no device capable of generating frequencies of a few THz from harmonics of cw sources was known to exist. Secondly, the laser frequency was not very stable; that is, in the case of the gas laser although its short term linewidth was a few hundred hertz, over a long period, its frequency could vary within the Doppler and pressure broadened gain curve of the laser. This gain curve might vary from less than a hundred to several thousand megahertz depending on the laser. Therefore, even though the instantaneous frequency of the laser could be measured, there was no reference point other than the broad gain curve whose center could not be located very precisely. Happily, solutions to both of these problems were found: for the first, the extension of frequency measurements into the infrared by means of the metal-on-metal point contact diode [6.1], and for the second, the development of the technique of

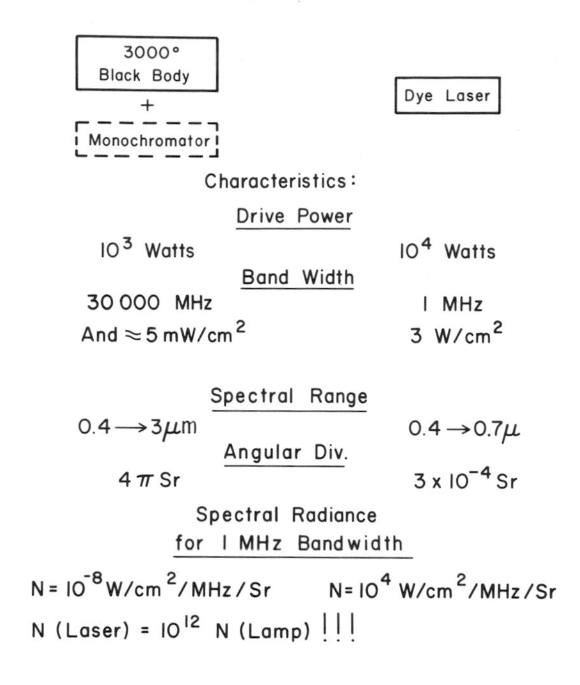

Fig. 6.1. Comparison of tunable light sources. The laser is characterized by its small spectral band width and angular divergence resulting in a large spectral radiance per unit band width. The narrow tuning-range disadvantage of a typical gas laser is to a large extent being removed by the development of the dye laser

saturated absorption [6.2] (see e.g. Subsect. 1.4.2). These two developments meant that coherent, highly stable, short wavelength sources of radiation existed whose frequency and wavelength could be directly related to the primary frequency and wavelength standards.

With the perfection of highly reproducible and stable lasers, their wavelength-frequency duality becomes of wider interest. We begin to think of lasers as frequency references for certain kinds of problems such as high resolution optical heterodyne spectroscopy [6.3]. The extension of absolute frequency measurements, linking the cesium standard (accurate to about 2×10^{-13}) [6.4] to these lasers, provides accuracy as well as resolution to the absolute frequencies involved. At the same time, we use the wavelength aspect of the radiation, for example, in precision long-path interferometry [6.5]. Indeed, the increasing resolution which these new spectroscopic techniques provide may very well usher in an era of precision and accuracy in frequency and length measurements undreamed of a few years ago.

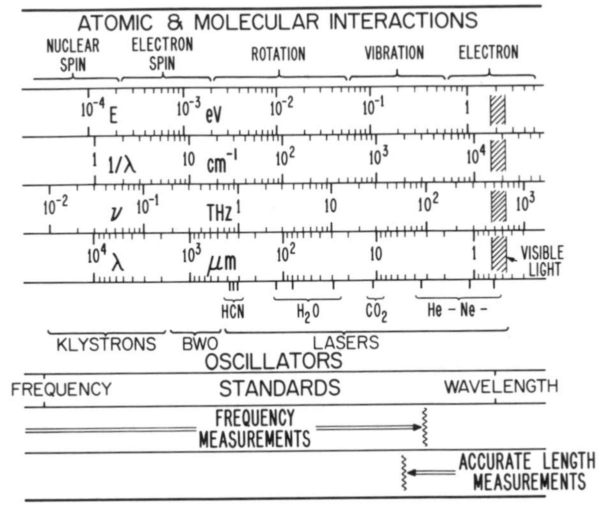

Fig. 6.2. The electromagnetic spectrum showing oscillators, frequency and wavelength standards, and measurement regions. The extension of frequency measurements into the infrared has produced an overlap of the accurate ($<4\times10^{-9}$) wavelength measurement region and the frequency measurement region for the first time. The frequency standard is the zero field hyperfine structure separation in the ground state of ^{133}Cs which is defined to be 9192631770 Hz. The wavelength standard is the transition between the $2p_{10}$ and $5d_5$ levels of ^{86}Kr with the vacuum wavelength of this radiation defined to be 1/1650763.73 m.

Recent advances in stabilizations with saturated absorption techniques (see Chapter 1) have produced lasers with one-second fractional frequency instabilities [6.2] as small as 5×10^{-13}. Although the fractional uncertainty in frequency reproducibility is somewhat larger, stabilized lasers are already excellent secondary frequency standards. With additional development in laser stabilization and infrared frequency synthesis some of these devices can be considered as contenders for the primary frequency standard role; the 3.39 μm He–Ne laser, for example, currently has a fractional frequency uncertainty in reproducibility [6.2] of less than one part in 10^{11}. The duality implies that the radiation must have a similar wavelength characteristics, i.e., $\Delta\lambda/\lambda\approx10^{-11}$ which is more than 100 times better than the current length standard [6.6]. Hence, the stabilized laser must be considered to have tremendous potential in wavelength as well as frequency standards applications and perhaps in both [6.7].

It has been clear since the early days of lasers that this wavelength-frequency duality could form the basis of a powerful method to measure the speed of light. However, the laser's optical frequency was much too high for conventional frequency measurement methods. This fact led to the invention of a variety of modulation or differential schemes, basically conceived to preserve the small interferometric errors associated with the short optical wavelength, while utilizing microwave frequencies which were still readily manipulated and measured. These microwave frequencies were to be modulated onto the laser output or realized as a difference frequency between two separate laser transitions [6.8]. Indeed, a proposed major long-path interferometric experiment [6.9] based on the latter idea has been made obsolete by the recent high-precision direct frequency measurement [6.10]. An ingenious modulation scheme, generally applicable to any laser transition, has recently produced successfully an improved value for the speed of light [6.11]. While this method can undoubtedly be perfected further, its differential nature leads to limitations which are not operative in direct frequency measurements.

Recent ultrahigh resolution measurements of both the frequency and wavelength of the methane stabilized He–Ne laser yielded a value of the speed of light 100 times more accurate [6.12] than that of the previously accepted value [6.13]. This significant increase in accuracy was made possible by the extension of frequency measurements to the region of the electromagnetic spectrum where wavelength measurements can be made with high accuracy and by the use of very stable lasers.

It is the purpose of this chapter to describe the stabilization of lasers by saturated absorption, laser frequency measuring techniques, experimental details of the speed of light measurement, and some possibilities for a new standard of length.

6.1. Stabilization of Lasers by Saturated Absorption

Homogeneous as well as inhomogeneous broadening in gas lasers results in gain curves which are many MHz wide while cavity linewidths are of the order of only 1 MHz. Therefore, the frequency of a gas laser is determined within the confines of the gain curve largely by the optical path length between the mirrors. The fractional frequency instability is equal to the fractional change in optical path length, and since it is difficult to keep this parameter less than one part in 10^{-7}, passively stabilized lasers have in general been free to drift many MHz over long periods of time.

The best short term stability is obtained by taking steps to control the optical path length between the mirrors. These steps have included use of cavity materials with near zero expansion coefficients, cavity

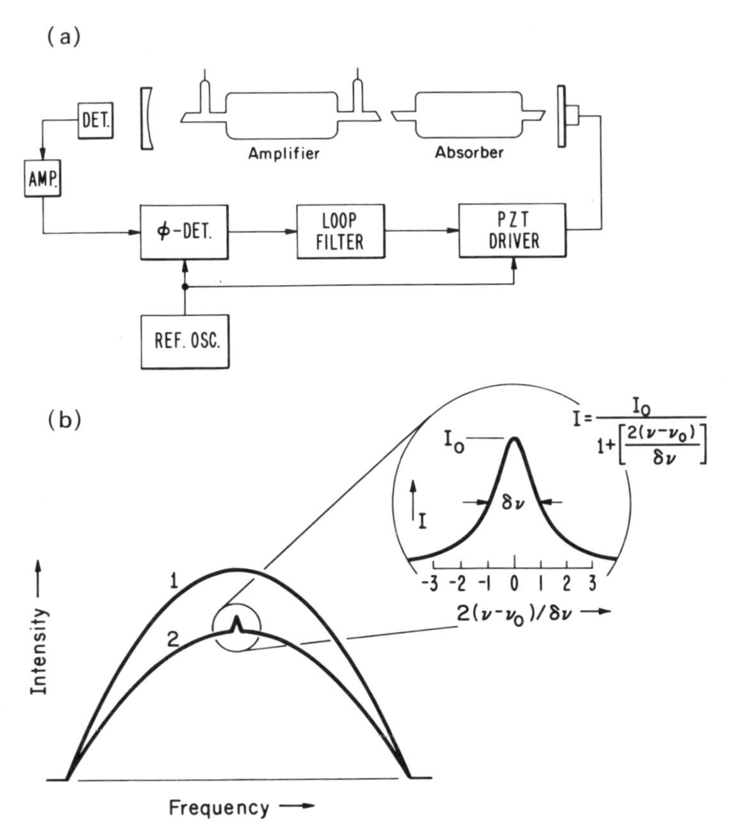

Fig. 6.3a and b. Schematic for laser stabilization by the technique of molecular saturated absorption. Diagram (a) shows the servo-controlled laser with an internal molecular absorption cell. Diagram (b) shows 1) the gain curve of the laser without absorber, and 2) the gain curve of the amplifier-absorber combination. The frequency of the laser is locked to the peak of the narrow "emission-type" feature shown on curve *2*

temperature control, isolation from mechanical and acoustic disturbances, as well as other ingenious ideas and devices. Also in a discharge excited gas laser, it is necessary to use a highly stabilized power supply in order to maintain a constant index of refraction. Some of the most sophisticated passively stabilized gas lasers have been designed by FREED and the interested reader is referred there [6.14] for more details. The observed spectral width is always controlled by environmental perturbations giving rise to frequency modulation. However, with care, lasers with short term spectral linewidths of the order of a kHz can be constructed.

Long term stability of laser oscillators can be attained by servo controlling the frequency to some more stable reference device. For example, long term stability may be provided by locking the laser to the resonant frequency of a temperature controlled, mechanically and thermally stable, passive cavity. However, for good resetability a molecular reference which is less subject to environmental perturbations is more suitable, provided that Doppler and pressure broadening can be eliminated or at least greatly reduced. Molecular beam type devices suggest themselves. However, the molecular saturated absorption technique attains nearly the same goal and yet avoids most of the experimental complications of beam devices. With lasers, the technique was first demonstrated by LEE and SKOLNICK [6.15] with the 6328 Å laser and a ^{20}Ne absorption cell excited by a dc gas discharge. Later a similar technique was developed by BARGER and HALL [6.16] with the 3.39 μm He–Ne laser to the point where one second fractional frequency instabilities are now only a few parts in 10^{13}. A schematic for laser stabilization by the technique of saturated absorption is shown in Fig. 6.3a + b. Since the linewidths are inversely proportional to the molecular transit time across the laser beam, HALL and BORDE [6.3], by increasing the diameter of the laser beam and cooling the absorption cell, have observed saturated absorption lines which are only 7 kHz wide. This refinement represents a resolution greater than 10^{10} and is narrow enough for observation of the methane hyperfine structure. Other lasers of interest here wich have been stabilized by saturated absorption include the 6328 Å laser which has been locked to various iodine absorption lines [6.17] and all of the lines in both the 9 and 10 μm bands of the CO_2 laser which have been stabilized to CO_2 itself [6.18, 19].

Since the wavelength of the radiation coming from an oscillator is simply related to the frequency through the velocity of propagation, it is reasonable to assume that the wavelength stability will be the same as the frequency stability. Moreover, since the coherence length is inversely proportional to the spectral linewidth of the radiation source, lasers are excellent wavelength standards. Thus we have laser oscillators which are excellent secondary frequency and wavelength standards at a number of convenient points in the electromagnetic spectrum. The frequencies of all the aforementioned molecular stabilizing transitions except those at 6328 Å have been measured absolutely [6.10, 20] to an accuracy exceeding that of the length standard.

6.2. Frequency Measurement Techniques

Direct frequency measurements require counting the number of cycles of electromagnetic radiation in a given period. Modern electronic circuitry permits direct cycle counting up to about 500 MHz with the timing generally provided by a quartz crystal oscillator, or for the highest accuracy, directly by a cesium clock. A measurement of a higher frequency requires a heterodyne technique in which two oscillators (one whose frequency is known) are mixed together so that a countable frequency difference is generated. The known frequency may be synthesized as an exact harmonic of a directly countable lower frequency by irradiating an appropriate nonlinear device. To synthesize microwave frequencies, oscillators of about a hundred MHz irradiating a silicon diode will generate useful harmonics as high as 40 GHz. The desired harmonic is generally mixed with the unknown frequency in the same diode so that the heterodyne beat note of a few MHz is finally generated and can be directly counted.

The above process may be illustrated in the following equation

$$\nu_x = l\nu_a \pm \nu_b , \tag{6.1}$$

where ν_x is the unknown frequency, ν_a the reference frequency, and ν_b the countable beat note frequency. The harmonic number l is generally determined by making an approximate frequency determination from a wavelength measurement. At frequencies from about 40 GHz to 1000 GHz (1 THz), the silicon diode is still operable; however, several steps are usually required in a frequency measurement chain extending from some countable reference frequency. For this purpose, klystrons between 60 and 80 GHz are often phase locked to a lower frequency klystron and these are used to cover the range between 200 and 1000 GHz with less than 15 harmonics in each step. The harmonic number is kept below 15 because at frequencies above a few hundred gigahertz the overall harmonic generation-mixing process is less efficient and consequently, one must use a lower harmonic number.

6.3. High-Frequency Diodes

In addition to the silicon diode already mentioned, another type of harmonic generator mixer, the point contact Josephson junction, is a much more efficient diode in the generation of high harmonics than is the silicon diode. Its use has been extended a little higher in frequency than [6.21] the silicon diode to 3.8 THz by the direct generation of the 401*st* harmonic from an X-band source. Work is continuing on this device to determine its high frequency limit.

A third type of harmonic generator-mixer, the metal-metal diode, is the only device which has been used at frequencies above 3.8 THz. This metal-metal point contact diode has been the most useful at laser frequencies, and has, thus far [6.22], operated to 88 THz (3.39 μm). High frequency applications of this diode were first demonstrated in Javan's laboratory [6.1] by generating and mixing the third harmonic of a pulsed water vapor laser with a CO_2 laser. Specifically, the tungsten-nickel diode has been more useful than many other metal-metal combinations which were tested. At these laser frequencies, harmonic generation, sideband generation, and mixing occur in the same diode, that is, the harmonics or sidebands are not actually propagated from diode to diode, but a single diode is radiated with all of the necessary frequencies. As an example, one of the steps in measuring the frequency of a CO_2 laser is to compare the frequency of its 9.3 μm $R(10)$ line with the third harmonic of a water vapor laser.

In this case,

$$\nu_{CO_2, R(10)} = 3\nu_{H_2O} - \nu_{kly} + \nu_{beat}, \tag{6.2}$$

where $\nu_{CO_2, R(10)}$ is the unknown frequency (32.134267 THz), and $\nu_{H_2O} = 10.718069$ THz, $\nu_{kly} = 0.019958$ THz, $\nu_{beat} = 0.000020$ THz. In this heterodyning, the radiation from all three sources simultaneously irradiates the diode and the resulting 20 MHz beat note is amplified and measured in a counter or spectrum analyzer.

Some characteristics of this type of diode have been measured [6.23], and so far, attempts at extending its range to the visible have shown that it mixes at frequencies as high as 583 THz (5145 Å). However, harmonic generation beyond 88 THz has not been demonstrated. Although the exact physical mechanism by which the harmonic generation-mixing occurs is not yet understood, there are at least three different possible explanations for the phenomena. One is a tunneling process through an oxide layer [6.24], another is a field emission process [6.25], and a third is a quantum mechanical scattering model [6.26].

To illustrate the way in which the diode is used in laser frequency measurement, a direct measurement of the methane transition used to stabilize the 3.39 μm He–Ne laser will be described. It is this frequency which is multiplied by the precisely determined methane wavelength to yield a definitive value of the speed of light.

6.4. The Speed of Light

The speed of light, c, is possibly the most important of all the fundamental constants [6.27]. It enters into the conversion between electrostatic and electromagnetic units; it relates the mass of a particle to its energy in the well known equation $E = mc^2$; and it is used as well in many relationships connecting other physical constants. In ranging measurements, very accurately measured transit times for electromagnetic waves are converted to distance by multiplying by the speed of light. Examples are geophysical distance measurements which use modulated electromagnetic radiation [6.28], and astronomical measurements such as microwave planetary radar and laser lunar ranging [6.29]. Recent experiments have set very restrictive limits on any possible speed dependence on direction [6.30] or frequency [6.31].

Because of its importance, more time and effort have been devoted to the measurement of c than any other fundamental constant. In spite of this fact, a highly accurate result has remained elusive even in recent times—mainly because of ever present systematic errors which have been difficult to properly evaluate. Indeed, the probable value of c has changed, and the suggestion has been made that the real value may be changing linearly or even periodically with time [6.32]. However, in view of recent electro-optical measurements, microwave interferometer measurements, spectroscopic measurements, and the laser measurement to be described here, it appears that the probable value of c is converging to a constant within the stated errors and that the changing nature of c in the past was related to unaccounted inaccuracy in previous experiments.

Three different techniques have been used to measure the speed of light: 1) time of flight techniques; 2) ratio of electrostatic to electromagnetic units; and 3) frequency and wavelength measurements ($\lambda \nu = c$). The first quantitative measurement of c was an astronomical one in which the time of flight of light across the earth's orbit around the sun was measured by ROEMER in 1676. Early time-of-flight terrestrial measurements utilized long accurately-measured base lines and either rotating toothed wheels or mirrors for measuring the time interval. One of the most accurate early measurements of c was an electrostatic to electromagnetic ratio experiment by ROSA and DORSEY [6.33] in 1906.

Measurements of c in which both the wavelength and frequency of an oscillator were measured played a significant role in proving that radio and light radiation were electromagnetic in nature. This method was used by FROOME [6.13] to obtain the accepted value of c in use since 1958. His method used a moving reflector type of microwave interferometer operating at 72 GHz which was used to measure the microwave wavelength in terms of the length standard. The vacuum wavelength thus obtained when corrected for diffraction and other systematic effects was multiplied by the measured frequency to give a value for the speed of light. The frequency could be related to the primary standard with good accuracy, but the major estimated experimental error was connected with uncertainties in the wavelength measurement. At these long wavelengths the diffraction problem forced the use of a long air path which resulted in uncertainties in the measurement of the path length and in the index of refraction.

Speed of light determinations from wavelength and frequency measurements have traditionally suffered from the problems illustrated by FROOME's experiment. To measure the frequency, it is best to do the experiment at a frequency not too far removed from the primary Cs frequency where extremely stable oscillators can be made and where frequencies are easily measurable. However, to measure the wavelength, it is best to do the experiment close to the visible ^{86}Kr wavelength standard where wavelengths can be more easily compared and where diffraction problems are not as severe. The extension of frequency measurements into the infrared portion of the electromagnetic spectrum has in a sense solved this dilemma, and has been responsible for the 100 fold increase in the accuracy in the value of c. The frequency of the methane stabilized helium-neon laser is over 1000 times higher in frequency than that of the oscillator used in FROOME's measurement of c. Direct frequency measurements were recently extended to this frequency [6.22] and subsequently refined [6.10] to the present accuracy of 6 parts in 10^{10}. The wavelength of this stabilized laser has been compared [6.34–37] with the krypton-86 length standard to the limit of the usefulness of the length standard (approximately 4 parts in 10^9) [6.6]. The product of the measured frequency and the wavelength yields a new, definitive value for the speed of light, c. That measurement will now be described.

6.5. Methane Frequency Measurement

To measure the frequency of methane, a chain of oscillators extending from the cesium frequency standard to the methane stabilized helium-neon laser was used. The chain is shown in Fig. 6.4. The three saturated-absorption-stabilized lasers are exhibited in the upper right-hand section, the transfer chain oscillators are in the center column, and the cesium frequency standard is in the lower right-hand corner. The He–Ne and CO_2 lasers in the transfer chain were offset locked [6.16], that is, they were locked at a frequency a few megahertz different from the stabilized lasers. This offset-locking procedure produced He–Ne and CO_2 transfer oscillators without the frequency modulation used in the molecular-stabilized lasers. The measurements of the frequencies in the entire chain were made in three steps shown on the right-hand side, by using standard heterodyne techniques [6.22, 38–40].

Conventional silicon point-contact harmonic generator-mixers were used up to the frequency of the HCN laser. Above this frequency, tungsten-

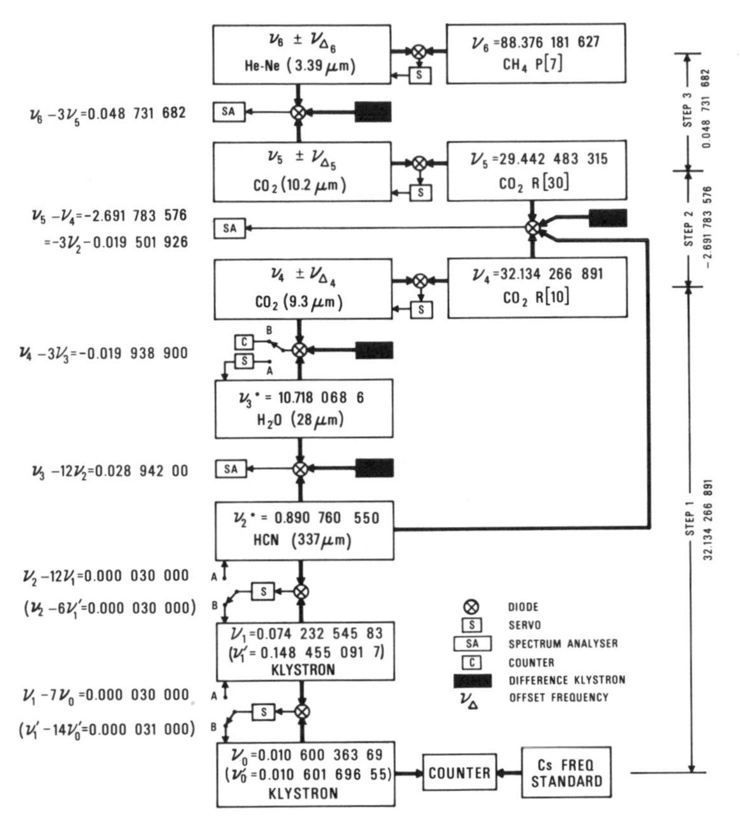

Fig. 6.4. Stabilized laser frequency synthesis chain. All frequencies are given in THz; those marked with an asterisk were measured with a transfer laser oscillator tuned to approximate line center

on-nickel diodes were used as harmonic generator-mixers. These metal-metal diodes required 50 or more mW of power from the lasers to obtain optimum signals. The 2-mm-long, 25 μm-diam. tungsten antenna, with a sharpened tip which lightly contacted the nickel surface, seemed to couple to the radiation in two separate manners. At 0.89 and 10.7 THz it acted like a long wire antenna [6.41, 42], while at 29–88 THz its concial tip behaved like one-half of a biconical antenna [6.42]. Conventional detectors were used in the offset-locking steps.

The methane-stabilized He–Ne laser used in these experiments is quite similar in size and construction to the device described by HALL

[6.2]. The gain tube was dc excited, and slightly higher reflectivity mirrors were employed. The latter resulted in a higher energy density inside the resonator and consequently a somewhat broader saturated absorption. Pressure in the internal methane absorption cell was about 0.01 Torr (1 Torr = 133.3 N/m^2).

The two 1.2-m-long CO_2 lasers used in the experiments contained internal absorption cells and dc-excited sealed gain tubes. A grating was employed on one end for line selection, and frequency modulation was achieved by dithering the 4-m-radius-of-curvature mirror on the opposite end. CO_2 pressure in the internal absorption cell was 0.020 Torr. The laser frequency was locked to the zero-slope point on the dip in the 4.3 μm fluorescent radiation [6.18]. The 0.89-, 10.7-, and 88-THz transfer lasers were 8-m-long linearly polarized cw oscillators with single-mode output power greater than 50 mW. The Michelson HCN laser has been described [6.43]. The H_2O laser used a double-silicon-disk partially transmitting end mirror, and a 0.5-mil polyethylene internal Brewster-angle membrane polarized the laser beam. The 8-m He–Ne laser oscillated in a single mode without any mode selectors because of a 4-Torr pressure with a 7:1 ratio of helium to neon. The resultant pressure width was approximately equal to the Doppler width, and a high degree of saturation allowed only one mode to oscillate.

Conventional klystrons used to generate the four difference frequencies between the lasers were all stabilized by standard phase-lock-techniques, and their frequencies were determined by cycle counting at X-band.

An interpolating counter controlled by a cesium frequency standard of the NBS Atomic Time Scale [6.44, 45] counted the 10.6-GHz klystron in the transfer chain. This same standard was used to calibrate the other counters and the spectrum-analyzer tracking-generator.

In step 1, a frequency synthesis chain was completed from the cesium standard to the stabilized $R(10)$ CO_2 laser. All difference frequencies in this chain were either measured simultaneously or held constant. Each main chain oscillator had its radiation divided so that all beat notes in the chain could be measured simultaneously. For example, a silicon-disk beam splitter divided the 10.7-THz beam into two parts: one part was focused on the diode which generated the 12th harmonic of the HCN laser frequency, the remaining part irradiated another diode which mixed the third harmonic of 10.7-THz with the output from the 9.3 μm CO_2 laser and the 20-GHz klystron.

Figure 6.4 shows the two different ways in which the experiment was carried out. In the first scheme (output from mixers in position A), the HCN laser was frequency locked to a quartz crystal oscillator via the 148- and 10.6-GHz klystrons, and the frequency of the 10.6-GHz klystron was counted. The H_2O laser was frequency locked to the stabilized CO_2 laser, and the beat frequency between the H_2O and HCN lasers was measured on the spectrum analyzer. In the second scheme (output from mixers in position B), the 10.6-GHz klystron was phase locked to the 74-GHz klystron, which in turn was phase locked to the free-running HCN laser. The 10.6-GHz klystron frequency was again counted. The free-running H_2O laser frequency was monitored relative to the stabilized CO_2 laser frequency, and the beat frequency between the H_2O and HCN lasers was measured as before on the spectrum analyzer.

In step 2, the difference between the two CO_2 lines was measured. The HCN laser remained focused on the diode used in step 1, which now also had two CO_2 laser beams focused on it. The sum of the third harmonic of the HCN frequency, plus a microwave frequency, plus the measured rf beat signal is the difference frequency between these two CO_2 lines. The two molecular-absorption-stabilized CO_2 lasers were used directly, and the relative phase and amplitudes of the modulating voltages were adjusted to minimize the width of the beat note. The beat note was again measured on a combination spectrum analyzer and tracking-generator-counter. The roles of the CO_2 lasers were interchanged to detect possible systematic differences in the two laser-stabilization systems.

In step 3, the frequency of the $P(7)$ line in methane was measured relative to the 10.18 μm $R(30)$ line of CO_2. Both the 8-m 3.39 μm laser and the CO_2 laser were offset locked from saturated-absorption-stabilized lasers and thereby not modulated. The 10- to 100-MHz beat note was again measured either on a spectrum analyzer and tracking generator, or in the final measurement when the S/N ratio of the beat note was large enough (about 100), directly on a counter.

The measurements were chronologically divided into four runs, and values for each of the steps and for ν_4, ν_5, and ν_6 were obtained by weighting the results of all runs inversely proportional to the square of the standard deviations. The largest uncertainty came in step 1; however, a recent measurement by NPL [6.46] gave a value of the $R(12)$ line which was only 2×10^{-10} (7 kHz) different from the number obtained by adding the $R(12)$–$R(10)$ difference [6.20] to the present $R(10)$ value. Thus, the first step of the experiment has been verified.

The final result is:

Molecule	Line	λ [μm]	Frequency [THz]
$^{12}C^{16}O_2$	$R(30)$	10.18	29.442483315(25)
$^{12}C^{16}O_2$	$R(10)$	9.3	32.134266891(24)
$^{12}CH_4$	$P(7)$	3.39	88.376181627(50)

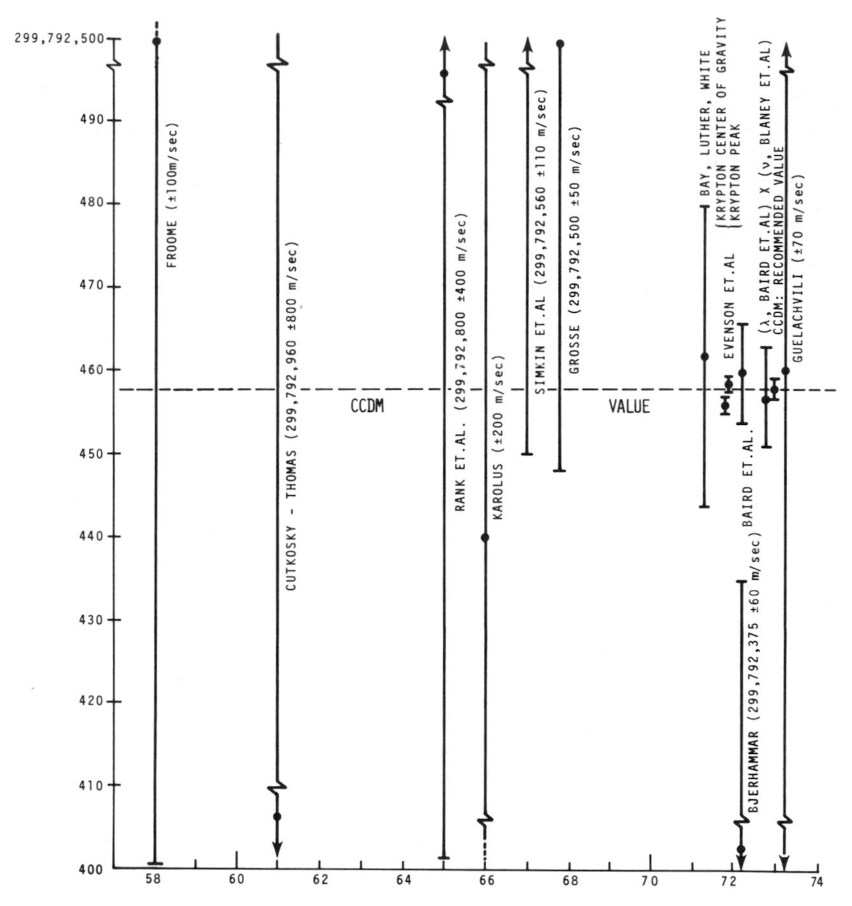

Fig. 6.5. Accurate speed of light values in meters per second since 1958. The value "recommended" by CCDM is 299792.458 m/s with an uncertainty of about 4 parts in 10^9. This value maintains continuity in the meter and may be used when the meter is redefined

The numbers in parentheses of the right column are 1-standard-deviation-type errors indicating uncertainties in the last two digits.

In a coordinated effort, the wavelength of the 3.39 μm line of methane was measured with respect to the ^{86}Kr 6058 Å primary standard of length by BARGER and HALL [6.34]. Using a frequency-controlled Fabry-Perot interferometer with a pointing precision of about 2×10^{-5} orders, a detailed search for systematic offsets inherent in the experiment, including effects due to the asymmetry of the Kr standard line, was made. Offsets due to various experimental effects (such as beam misalignments, mirror curvatures and phase shifts, phase shift over the exit aperture,

Table 6.1. Speed of light measurements since 1958

Year	Author	Ref.	Method	c [km/s]	δc [km/s]
1958	FROOME	[6.13]	Radio interferometer	299792.5	0.1
1961	CUTKOSKY and THOMAS	[6.48]	Ratio of units	299791.96	0.8
1965	KOLIBAYEV	[6.49]	Electro-optical	299792.6	0.06
1965	RANK et al.	[6.50]	Spectroscopic	299792.8	0.4
1966	KAROLUS	[6.51]	Electro-optical	299792.44	0.2
1967	GROSSE	[6.52]	Electro-optical	299792.5	0.05
1967	SIMKIN et al.	[6.53]	Radio interferometer	299792.56	0.11
1971	BJERHAMMAR	[6.54]	Electro-optical	299792.375	0.060
1972	BAY et al.	[6.11]	He–Ne $\lambda\nu$ (0.633 μm)	299792.462	0.018
1972	BAIRD et al.	[6.47]	CO_2 $\lambda\nu$ (9 and 10 μm, avg.)	299792.460	0.006
1972	EVENSON et al.	[6.12]	He–Ne $\lambda\nu$ (3.39 μm) c.g.	299792.4562	0.0011
			peak	299792.4587	0.0011
1973	GUELACHVILI	[6.55]	Spectroscopic	299792.46	0.07
1973	BAIRD and BLANEY et al.	[6.47, 46]	CO_2 $\lambda\nu$ (9.32 μm)	299792.457	0.006
1973	CCDM	[6.6, 10]	He–Ne $\lambda\nu$ (3.39 μm)	299792.458	0.0012

diffraction, etc.) were carefully measured and then removed from the data with an uncertainty of about 2 parts in 10^9. Their results indicate an asymmetry in the krypton line which resulted in two different values of λ_{CH_4}. One value arises if the center of gravity of the krypton line is used to define the meter, and another if the peak value is used. This reproducibility for a single wavelength measurement illustrates the high precision which is available using the frequency-controlled interferometer.

At the 5th session of the consultative committee on the definition of the meter (CCDM) [6.6], results of BARGER and HALL as well as measurements made at the International Bureau of Weights and Measures [6.36], the National Research Council [6.37], and the National Bureau of Standards at Gaithersburg [6.35] were all combined to give a "recommended" value for the wavelength of the transition of methane used to stabilize the He–Ne laser. The recommended value is

$$\lambda_{CH_4[P(7),\,\text{band}\,\nu_3]} = 3\,392\,231.40 \times 10^{-2}\ \text{m}\,. \tag{6.3}$$

Multiplying this recommended wavelength of methane by the measured frequency yields the value for the speed of light:

$$c = 299\,792\,458\ \text{m/sec}\ (\Delta c/c = \pm 4 \times 10^{-9})\,, \tag{6.4}$$

which was also recommended by the CCDM [6.6] to be used in distance measurements where time-of-flight is converted to length and for converting frequency to wavelength and vice versa.

This result is in agreement with the previously accepted value of $c = 299\,792\,500(100)$ m/sec and is about 100 times more accurate. A recent differential measurement of the speed of light has been made by BAY et al. [6.11]; their value is 299792462(18) m/sec, which is also in agreement with the presently determined value. A third recent, highly accurate value may be obtained by multiplying the frequency of the $R(12)$, 9.3 μm line of CO_2 measured by BLANEY et al. [6.46] by the wavelength measured by BAIRD et al. [6.47] which yields the value of 299 792 458 (6) m/sec for c. Also shown is another value by BAIRD [6.47], however, it used the CO_2 frequencies which were measured in this experiment, and hence, does not represent an independent value.

The fractional uncertainty in this value for the speed of light, $\pm 4 \times 10^{-9}$, arises from the interferometric measurements with the incoherent krypton radiation which defines the international meter. This limitation is indicative of the remarkable growth in optical physics in recent years; the present krypton-based length definition was adopted only in 1960.

Various measurements of c since FROOME's work are listed in Table 6.1 and are plotted in Fig. 6.5. One sees a remarkable convergence of the values of c for the first time in history!

6.6. Possible New Standard of Length

In addition to the "recommended" values of the wavelength of methane and the value of the speed of light made at the 5th CCDM meeting [6.6], a value for the wavelength of iodine was also recommended:

$$\lambda_{^{127}I_2[R(127),\text{band }11-5, i\text{ component}]} = 632991.399 \times 10^{-12}\ \text{m}\,.$$

These "recommended" values are in agreement with wavelength measurements to the limits possible with the krypton length standard (that is, about $\pm 4 \times 10^{-9}$). It is "recommended" that either of these values be used to make length measurements using these stabilized lasers in the interim before the meter is redefined.

It is also significant that no further work was recommended on the present length standard, the krypton lamp, which is far inferior to a laser and will probably soon be replaced by one.

As a result of the recommendations made by CCDM, two different definitions of a new length standard must be considered. First, we can continue as before with separate standards for the second and meter, but with the meter defined as the length equal to $1/\lambda$ wavelengths in vacuum of the radiation from a stabilized laser instead of from a ^{86}Kr lamp. Either the methane-stabilized [6.2, 16] He–Ne laser at 3.39 μm (88 THz) or the I_2-stabilized [6.17] He–Ne laser at 0.633 μm (474 THz) appear to be suitable candidates. The 3.39 μm laser is already a secondary frequency standard in the infrared, and hopefully, direct measurements of the frequency of the 0.633 μm radiation will give the latter laser the same status in the visible. The 3.39 μm laser frequency is presently known to within 6 parts in 10^{10}, and the reproducibility and long term stability have been demonstrated to be better by more than two orders of magnitude. Hence, frequency measurements with improved apparatus in the next year or two are expected to reduce this uncertainty to a few parts in 10^{11}. A new value of the speed of light with this accuracy would thus be achievable if the standard of length were redefined in terms of the wavelength of this laser.

Alternately, one can consider defining the meter as a specified fraction of the distance light travels in one second in vacuum (that is, one can fix the value of the speed of light). The meter would thus be defined in terms of the second and, hence, a single unified standard would be used for frequency, time, and length. What at first sounds like a rather radical and new approach to defining the meter is actually nearly one hundred years old. It was first proposed by Lord Kelvin in 1879 [6.56]. With this definition, the wavelength of all stabilized lasers would be known to the same accuracy with which their frequencies can be measured. Stabilized lasers would thus provide secondary standards of both frequency and length for laboratory measurements, with the accuracy being limited only by the reproducibility, measurability, and long term stability. It should be noted that an adopted nominal value for the speed of light is already in use for high-accuracy astronomical measurements [6.57], thus, there are currently two different standards of length in existence: one for terrestrial measurement and one for astronomical measurements. A definition which fixes c and unites these two values of c would certainly be desirable from a philosophical point of view.

Independent of which type of definition is chosen we believe that research on simplified frequency synthesis chains bridging the microwave-optical gap will be of great interest, as will refined experiments directed toward an understanding of the factors that limit optical frequency reproducibility. No matter how such research may turn out, it is clear that ultraprecise physical measurements made in the interim can be preserved through wavelength or frequency comparison with a suitably stabilized laser such as the 3.39 μm methane device.

Frequencies are currently measurable to parts in 10^{13}, and hence the over-all error of about six parts in 10^{10} for the frequency measurement can be reduced. This measurement was performed fairly quickly to obtain a frequency of better accuracy than the wavelength. It should be possible to obtain considerably more accuracy by using tighter locks on the lasers. For example, the 8-m HCN laser has recently been phase locked [6.58, 59] to a multiplied microwave reference which currently determines the HCN laser linewidths. An improved microwave reference could be a superconducting cavity stabilized oscillator [6.60] for best stability in short term (narrowest linewidth) coupled with a primary cesium beam standard for good long term stability.

The relative ease with which these laser harmonic signals were obtained in these frequency measurements indicates that the measurement of the frequencies of visible radiation now appears very near at hand. Even if the point contact metal-on-metal diode is inoperable above 88 THz, conventional nonlinear optical techniques (i.e., 2nd harmonic generation in crystals) could still be used to extend direct frequency measurements to the visible. Such measurements should greatly facilitate one's ability to accurately utilize the visible and infrared portion of the electromagnetic spectrum.

Acknowledgement

The authors which to acknowledge J. S. WELLS, B. L. DANIELSON, and G. W. DAY who participated in the measurement of the frequency of the methane stabilized He Ne laser and R. L. BARGER and J. L. HALL who first stabilized the laser and measured its wavelength. We also express our gratitude to D. G. MCDONALD and J. D. CUPP whose work with the Josephson junction was a parallel effort to ours in trying to achieve near-infrared frequency measurements. We note here that experiments using the Josephson junction to measure laser frequencies are continuing, and may well lead to better methods for near-infrared frequency synthesis in the future.

References

6.1. V. DANEU, D. SOKOLOFF, A. SANCHEZ, A. JAVAN: Appl. Phys. Letters **15**, 398 (1969).
6.2. J. L. HALL: In *Esfahan Symposium on Fundamental and Applied Laser Physics*, ed. by M. FELD and A. JAVAN (Wiley, New York). To be published.
6.3. J. L. HALL, CH. BORDE: Phys. Rev. Letters **30**, 1101 (1973).
6.4. D. J. GLAZE, H. HELLWIG, S. JARVIS, Jr., A. E. WAINWRIGHT, D. W. ALLAN: 27th Annual Symposium on Frequency Control, Fort Monmouth, N. J., USA (1973).
6.5. J. LEVINE, J. L. HALL: J. Geophys. Res. **77**, 2592 (1972).
6.6. Comite Consultatif pour la Definition du Metre, 5th session, Rapport (Bureau International des Poids et Mesures, Sevres, France, 1973).
6.7. DONALD HALFORD, H. HELLWIG, J. S. WELLS: Proc. IEEE **60**, 623 (1972).

6.8. J. L. HALL, W. W. MOREY: Appl. Phys. Letters **10**, 152 (1967).
6.9. J. HALL, R. L. BARGER, P. L. BENDER, H. S. BOYNE, J. E. FALLER, J. WARD: Electron. Technol. **2**, 53 (1969).
6.10. K. M. EVENSON, J. S. WELLS, F. R. PETERSEN, B. L. DANIELSON, G. W. DAY: Appl. Phys. Letters **22**, 192 (1973).
6.11. Z. BAY, G. G. LUTHER, J. A. WHITE: Phys. Rev. Letters **29**, 189 (1972).
6.12. K. M. EVENSON, J. S. WELLS, F. R. PETERSEN, B. L. DANIELSON, G. W. DAY, R. L. BARGER, J. L. HALL: Phys. Rev. Letters **29**, 1346 (1972).
6.13. K. D. FROOME: Proc. Roy. Soc. Ser. A **247**, 109 (1958).
6.14. C. FREED: IEEE J. Quant. Electron. QE **4**, 404 (1968).
6.15. P. H. LEE, M. L. SKOLNICK: Appl. Phys. Letters **10**, 303 (1967).
6.16. R. L. BARGER, J. L. HALL: Phys. Rev. Letters **22**, 4 (1969).
6.17. G. R. HAINES, C. E. DAHLSTROM: Appl. Phys. Letters **14**, 362 (1969); G. R. HAINES, K. M. BAIRD: Metrologia **5**, 32 (1969).
6.18. C. FREED, A. JAVAN: Appl. Phys. Letters **17**, 53 (1970).
6.19. F. R. PETERSEN, D. G. MCDONALD, J. D. CUPP, B. L. DANIELSON: Phys. Rev. Letters **31**, 573 (1973).
6.20. F. R. PETERSEN, D. G. MCDONALD, J. D. CUPP, B. L. DANIELSON: "Accurate Rotational Constants, Frequencies, and Wavelengths from $^{12}C^{16}O_2$ Lasers Stabilized by Saturated Absorption". Laser Spectroscopy, R. G. BREWER and A. MOORADIAN, Ed. (Plenum, New York, 1974).
6.21. D. G. MCDONALD, A. S. RISLEY, J. D. CUPP, K. M. EVENSON, J. R. ASHLEY: Appl. Phys. Letters **20**, 296 (1972).
6.22. K. M. EVENSON, G. W. DAY, J. S. WELLS, L. O. MULLEN: Appl. Phys. Letters **20**, 133 (1972).
6.23. E. SAKUMA, K. M. EVENSON: IEEE J. Quant. Electron. QE **10**, 599 (1974).
6.24. S. M. PARIS, T. K. GUSTAFSON, J. C. WIESNER: IEEE J. Quant. Electron. QE **9**, 737 (1973).
6.25. A. A. LUCAS, P. H. CUTLER: 1st European Conf. Condensed Matter Summaries, Florence, Italy (September 1971).
6.26. ERIC G. JOHNSON, JR.: Paper in preparation.
6.27. The interested reader will find a useful, critical discussion of the speed of light in K. D. FROOME, L. ESSEN: *The Velocity of Light and Radio Waves* (Academic Press, New York, 1969).
6.28. E. BERGSTRAND: Ark. Mat. Astr. Fys. **36**A, 1 (1949).
6.29. P. L. BENDER, D. G. CURRIE, R. H. DICKE, D. H. ECKHARDT, J. E. FALLER, W. M. KAULA, J. D. MULHOLLAND, H. H. PLOTKIN, S. K. POULTNEY, E. C. SILVERBERG, D. T. WILKINSON, J. G. WILLIAMS, C. O. ALLEY: Science **182**, 229 (1973).
6.30. T. S. JASEJA, A. JAVAN, J. MURRAY, C. H. TOWNES: Phys. Rev. 133A, 1221 (1964), using infrared masers; D. C. CHAMPENEY, G. R. ISAAK, A. M. KHAN: Phys. Letters **7**, 241 (1963), using Mössbauer effect.
6.31. B. WARNER, R. E. NATHER: Nature (London) **222**, 157 (1969), from dispersion in the light flash from pulsar NP 0532, obtain $\Delta c/c \leq 5 \times 10^{-18}$ over the range $\lambda = 0.25$ to 0.55 μm.
6.32. For example, see M. E. J. GHEURY DEBRAY: Nature **133**, 464 and 948 (1934).
6.33. E. B. ROSA, N. E. DORSEY: Bull. Nat. Bureau of Standards **3**, 433 (1907).
6.34. R. L. BARGER, J. L. HALL: Appl. Phys. Letters **22**, 196 (1973).
6.35. R. D. DESLATTES, H. P. LAYER, W. G. SCHWEITZER: Paper in preparation.
6.36. P. GIACOMO: 5th session of the Comite Consultatif pour la Definition du Metre, BIPM, Sevres, France, 1973.
6.37. K. M. BAIRD, D. S. SMITH, W. E. BERGER: Opt. Commun. **7**, 107 (1973).

6.38. L. O. HOCKER, A. JAVAN, D. RAMACHANDRA RAO, L. FRENKEL, T. SULLIVAN: Appl. Phys. Letters **10**, 5 (1967).

6.39. K. M. EVENSON, J. S. WELLS, L. M. MATARRESE, L. B. ELWELL: Appl. Phys. Letters **16**, 159 (1970).

6.40. K. M. EVENSON, J. S. WELLS, L. M. MATARRESE: Appl. Phys. Letters **16**, 251 (1970).

6.41. L. M. MATARRESSEE, K. M. EVENSON: Appl. Phys. Letters **17**, 8 (1970).

6.42. *Antenna Engineering Handbook*, ed. by HENRY JASIK (McGraw-Hill, New York 1961) Chapters 4 and 10.

6.43. K. M. EVENSON, J. S. WELLS, L. M. MATARRESE, D. A. JENNINGS: J. Appl. Phys. **42**, 1233 (1971).

6.44. D. W. ALLAN, J. E. GRAY, H. E. MACHLAN: IEEE Trans. Instr. Meas. IM **21**, 388 (1972).

6.45. H. HELLWIG, R. F. C. VESSOT, M. W. LEVINE, P. W. ZITZWITZ, D. W. ALLAN, D. GLAZE: IEEE Trans. Instr. Meas. IM **19**, 200 (1970).

6.46. T. G. BLANEY, C. C. BRADLEY, G. J. EDWARDS, D. J. E. KNIGHT, P. T. WOODS, B. W. JOLLIFFE: Nature **244**, 504 (1973).

6.47. K. M. BAIRD, H. D. RICCIUS, K. J. SIEMSEN: Opt. Commun. **6**, 91 (1972).

6.48. a) J. L. THOMAS, C. PETERSON, I. L. COOTER, F. R. KOTLER: J. Res. Nat. Bur. Std. **43**, 291 (1949);
b) R. D. CUTKOSKY: J. Res. Nat. Bur. Std. **65**A, 147 (1961);
c) B. N. TAYLOR, W. H. PARKER, D. N. LANGENBERG: Rev. Mod. Phys. **41**, 375 (1969).
Note: Original work was reported in a) and b). The calculated value for the speed of light includes corrections which were reported in b) and c).

6.49. V. A. KOLIBAYEV: 1965, Geodesy and Aerophotography, No. 3, p. 228 (translated for the American Geophysical Union).

6.50. D. H. RANK: J. Mol. Spectrosc. **17**, 50 (1965).

6.51. A. KAROLUS: 5th Intern. Conf. Geodetic Measurement, 1965, Deutsche Geodetische Kommission, München (1966), p. 1.

6.52. H. GROSSE: Nachr. Karten- und Vermessungswesen (Ser. I) **35**, 93 (1967).

6.53. G. S. SIMKIN, I. V. LUKIN, S. V. SIKORA, V. E. STRELENKII: Izmeritel, Tekhn, **8**, 92 (1967); [Translation: Meas. Tech. **1967**, 1018].

6.54. A. BJERHAMMAR: Tellus **XXIV**, 481 (1972).

6.55. G. GVELACHVILI: Ph. D. Thesis, Université de Paris-Sud, Centre D'Orsay (1973).

6.56. W. F. SNYDER: IEEE Trans. Instr. Meas. IM **22**, 99 (1973).

6.57. P. BENDER: Science **168**, 1012 (1970).

6.58. J. S. WELLS, DONALD HALFORD: NBS Techn. Note No. 620 (May 1973).

6.59. J. S. WELLS, D. G. MCDONALD, A. S. RISLEY, S. JARVIS, J. D. CUPP: Rev. Appl. Phys. **9**, 285 (1974).

6.60. S. R. STEIN, J. P. TURNEAURE: Electron. Letters **8**, 431 (1972).

Additional References with Titles

1.1. Doppler-Limited Spectroscopy

R. V. Ambartzumyan, G. I. Bekov, V. S. Letokhov, V. I. Mishin: Excitation of high-lying states of the sodium atom by dye laser radiation and their autoionization in an electric field. JETP Lett. **21**, 279 (1975).

A. M. Angus, E. E. Marinero, M. J. Colles: Opto-acoustic spectroscopy with a visible cw dye laser. Opt. Commun. **14**, 223 (1975).

E. N. Anlonov, V. G. Koloshnikov, V. R. Mironenko: Quantitative measurement of small absorption coefficients in intracavity absorption spectroscopy using a cw dye laser. Opt. Commun. **15**, 99 (1975).

M. Arditi, J.-L. Picqué: Application of the light-shift effect to laser frequency stabilization with reference to a microwave frequency standard. Opt. Commun. **15**, 317 (1975).

L. Armstrong, Jr., B. L. Beers, S. Feneuille: Resonant multiphoton ionization via the Fano autoionizing formalism. Phys. Rev. A **12**, 1903 (1975).

V. M. Arutyunyan, T. A. Papazyan, Yu. S. Chilingaryan, A. V. Karmenyan, S. M. Sarkisyan: Investigation of resonance polarization phenomena during the passage of laser radiation through potassium vapor. Sov. Phys. JETP **39**, 243 (1975).

V. N. Bagratashvili, I. N. Knyazev, V. S. Letokhov, V. V. Lobko: Resonance excitation of C_2H_4-molecule luminescence by a pulsed high-pressure continuous tunable CO_2 laser. Opt. Commun. **14**, 426 (1975).

H. A. Baldis, H. Pépin, B. Grek: Third harmonic generation from laser-produced plasma. Appl. Phys. Lett. **27**, 291 (1975).

A. Bambini, G. J. Troup: A new approach to the resonant interaction between atoms and strong electromagnetic field. Opt. Commun. **14**, 176 (1975).

N. G. Basov, V. T. Galochkin, S. I. Zavorotnyi, V. N. Kosinov, A. A. Ovchinnikov, A. N. Oraevskii, A. V. Pankratov, A. N. Skachkov, G. V. Shmerling: Threshold character of the visible fluorescence excited by IR laser radiation. JETP Lett. **21**, 32 (1975).

P. Bensoussan: Experimental determination of highly excited *F* levels in potassium by three-photon absorption spectroscopy. Phys. Rev. A **11**, 1787 (1975).

D. M. Bloom, G. W. Bekkers, J. F. Young, S. E. Harris: Third harmonic generation in phase-matched alkali metal vapors. Appl. Phys. Lett. **26**, 687 (1975).

D. M. Bloom, J. F. Young, S. E. Harris: Mixed metal vapor phase matching for third harmonic generation. Appl. Phys. Lett. **27**, 390 (1975).

M. Boesl, H. J. Neusser, E. W. Schlag: High resolution lifetime measurements of isolated vibronic levels in naphtalene $C_{10}H_8$. Chem. Phys. Lett. **31**, 1 (1975).

M. Boesl, H. J. Neusser, E. W. Schlag: Deuterium isotope effect of single vibronic levels in naphtalene. Chem. Phys. Lett. **31**, 7 (1975).

P. A. Braun, A. N. Petelin: The dynamical Stark effect in diatomic molecules. Sov. Phys. JETP **39**, 775 (1975).

R. G. Bray, R. M. Hochstrasser, H. N. Sung: Two-photon excitation spectra of molecular gases: new results for benzene and nitric oxide. Chem. Phys. Lett. **33**, 1 (1975).

R. D. H. BROWN, G. P. GLASS, I. W. M. SMITH: The relaxation of HCl ($v=1$) and DCl ($v=1$) by O atoms between 196 and 400 K. Chem. Phys. Lett. **32**, 517 (1975).

D. H. BURDE, R. A. MCFARLANE, J. R. WIESENFELD: Collisional quenching of excited iodine atoms I($5p^5\ ^2P_{1/2}$) by I_2. Chem. Phys. Lett. **32**, 296 (1975).

J. L. CARLSTEN, T. J. MCILRATH, W. H. PARKINSON: Absorption spectrum of the laser-populated ^3D metastable levels in barium. J. Phys. B **8**, 38 (1975).

G. M. CARTER, D. E. PRITCHARD, T. W. DUCAS: Steady-state excitation of a sodium beam using a cw laser. Appl. Phys. Lett. **27**, 498 (1975).

M. R. CERVENAN, N. R. ISENOR: Multiphoton ionization yield curves for Gaussian laser beams. Opt. Commun. **13**, 175 (1975).

C. CHACKERIAN, JR., L. P. GIVER: Measurement of weak IR absorption with a tunable laser: the hydrogen $S_2(1)$ line strength. Appl. Opt. **14**, 1993 (1975).

F. C. M. COOLEN, H. L. HAGEDOORN: Detection of ^{20}Na atoms and measurement of sodium-vapour densities by means of atomic resonance-fluorescence. J. Opt. Soc. Am. **65**, 952 (1975).

D. COTTER, D. C. HANNE, R. WYATT: Infrared stimulated Raman generation: effects of gain focussing on threshold and tuning behaviour. Appl. Phys. **8**, 333 (1975).

M. CRANCE, P. JUNCAR, J. PINARD: A new method for measuring relative oscillator strengths using a cw dye laser. J. Phys. B **8**, 2461 (1975).

T. F. DEATON, D. A. DEPATIE, T. W. WALKER: Absorption coefficient measurements of nitrous oxide and methane at DF laser wavelength. Appl. Phys. Lett. **26**, 300 (1975).

L. S. DITMAN, JR., R. W. GAMMON, T. D. WILKERSON: High resolution emission spectra of laser-excited molecular iodine as the excitation frequency is tuned through and away from resonance. Opt. Commun. **13**, 154 (1975).

T. W. DUCAS, M. G. LITTMANN, R. R. FREEMAN, D. KLEPPNER: Stark ionization of high-lying states of sodium. Phys. Rev. Lett. **35**, 366 (1975).

C. FABRE, S. HAROCHE: Observation of giant polarizabilities in atomic sodium Rydberg states. Opt. Commun. **15**, 254 (1975).

H. S. FREEDHOFF, M. E. SMITHERS: Radiative transitions at the doublet splitting frequency in the resonant Stark effect. J. Phys. B **8**, L 209 (1975).

D. M. FRIEDRICH, W. M. MCCLAIN: Polarization and assignment of the two-photon excitation spectrum of benzene vapor. Chem. Phys. Lett. **32**, 541 (1975).

T. F. GALLAGHER, S. A. EDELSTEIN, R. M. HILL: Radiative lifetimes of the S and D Rydberg levels of Na. Phys. Rev. A **11**, 1504 (1975).

H. J. GERRITSEN, G. NIENHUIS: Multidirectional Doppler pumping: a new method to prepare an atomic beam having a large fraction of excited atoms. Appl. Phys. Lett. **26**, 347 (1975).

CH. R. GOLDSCHMIDT, G. STEIN, E. WÜRZBERG: Nanoseconds uv laser study of radiationless deactivation from upper electronic excited states in solution: Tb^{3+}. Chem. Phys. Lett. **34**, 408 (1975).

E. H. A. GRANNEMAN, M. J. VAN DER WIEL: Two photon ionization measurements on atomic caesium by means of an argon ion laser. J. Phys. B **8**, 1617 (1975).

H. GREENSTEIN, C. W. BATES, JR.: Line width and tuning effects in resonant excitation. J. Opt. Soc. Am. **65**, 33 (1975).

K. C. HARVEY, R. T. HAUKINS, G. MEISEL, A. L. SCHAWLOW: Measurement of the Stark effect in sodium by two-photon spectroscopy. Phys. Rev. Lett. **34**, 1073 (1975).

S. S. HASSAN, R. K. BULLOGH: Theory of the dynamical Stark effect. J. Phys. B **8**, L 147 (1975).

R. B. HIGGINS: Three level atomic systems as models for the dynamical Stark effect in the Sodium D_2 Line. J. Phys. B **8**, L 321 (1975).

H. K. HOLT: Laser intracavity absorption. Phys. Rev. A **11**, 625 (1975).

K. K. HUI, D. I. ROSEN, T. A. COOL: Intermode energy transfer in vibrationally excited O_3. Chem. Phys. Lett. **32**, 141 (1975).

H. INABA, T. KOBAYASI: Infrared laser radar technique using heterodyne detection for range resolved sensing of air pollutants. Opt. Commun. **14**, 119 (1975).

A. KASDAN, E. HERBST, W. C. LINEBERGER: Laser photonelectron spectroscopy of CH^-. Chem. Phys. Lett. **31**, 78 (1975).

A. P. KAZANTSEV: Radiation from an atom in an external electromagnetic field. Sov. Phys. JETP **39**, 601 (1975).

F. KEILMANN, R. L. SHEFFIELD, J. R. R. LEITE, M. S. FELD, A. JAVAN: Optically pumping and tunable laser spectroscopy of the ν_2 band of D_2O. Appl. Phys. Lett. **26**, 19 (1975).

H. J. KIMBLE, L. MANDEL: Time development of the light intensity in resonance fluorescence. Opt. Commun. **14**, 167 (1975).

H. J. KIMBLE, L. MANDEL: Problem of resonance fluorescence and the inadequacy of spontaneous emission as a test of quantum electrodynamics. Phys. Rev. Lett. **34**, 1485 (1975).

M. LAMBROPOULOS, S. E. MOODY, S. J. SMITH, W. C. LINEBERGER: Observation of electric quadrupole transitions in multiphoton ionization. Phys. Rev. Lett. **35**, 159 (1975).

S. A. LEE, R. WALLENSTEIN, T. W. HÄNSCH: Hydrogen $1S$–$2S$ isotope shift and $1S$ lamb shift measured by laser spectroscopy. Phys. Rev. Lett. **35**, 1262 (1975).

R. K. LENGEL, D. R. CROSLEY: Rotational dependence of vibrational relaxation in $A\,^2\Sigma^+$ OH. Chem. Phys. Lett. **32**, 261 (1975).

C. LEOMPTE, G. MAINFRAY, C. MANUS, F. SANCHEZ: Laser temporal-coherence effects on multiphoton ionization processes. Phys. Rev. A **11**, 1009 (1975).

M. MAEDA, F. ISHITSUKA, Y. MIYAZOE: Dye-laser amplified atomic absorption flame spectroscopy. Opt. Commun. **13**, 314 (1975).

F. METZ: Theory of the $S_0 \rightarrow S_1$ two-photon transition in benzene. Chem. Phys. Lett. **34**, 109 (1975).

V. P. OLEINIK, M. M. CHUMACHKOVA: Multiphoton transitions in a hydrogenlike atom. Sov. Phys. JETP **40**, 460 (1975).

V. P. OLEINIK, V. A. SINYAK: On the Compton effect in a strong electromagnetic field. Opt. Commun. **14**, 179 (1975).

G. PETTY, C. TAI, F. W. DALBY: Nonlinear resonant photoionization in molecular iodine. Phys. Rev. Lett. **34**, 1207 (1975).

A. S. PINE, N. MENYUK: Optically pumped cw InSb lasers for NO spectroscopy. Appl. Phys. Lett. **26**, 231 (1975).

M. REDON, H. GUREL, M. FOURIER: Infrared microwave double resonance in NH_3: studies of collision-induced transitions in the presence of a high field Stark effect. Chem. Phys. Lett. **30**, 99 (1975).

D. E. ROBERTS, E. N. FORTSON: Rubidium isotope shifts and hyperfine structure by two-photon spectroscopy with a multimode laser. Opt. Commun. **14**, 332 (1975).

R. D. RUNDEL, F. B. DUNNING, H. C. GOLDWIRE, JR., R. F. STEBBINGS: Near-threshold photoionization of xenon metastable atoms. J. Opt. Soc. Am. **65**, 628 (1975).

H. SCHRÖDER, H. J. NEUSSER, E. W. SCHLAG: The temporal development of intracavity absorption in a dye laser. Opt. Commun. **14**, 395 (1975).

S. E. SCHWARTZ, G. I. SENUM: Laser fluorescence spectroscopy and lifetime of the $^2B_1(K'>0)$ electronic state of NO_2. Chem. Phys. Lett. **32**, 569 (1975).

R. C. SEPUCHA: Vibrational relaxation of $CO_2(00^01)$ in pure CO_2. Chem. Phys. Lett. **31**, 75 (1975).

C. Y. SHE, K. W. BILLMAN: Infrared-pumped third-harmonic and sum frequency generation in diatomic molecules. Appl. Phys. Lett. **27**, 76 (1975).

H. K. SHIN: Vibrational deactivation of HF ($v=1$) in the H_2O+HF dimer: HF ($v=1$)+ H_2O (000) HF ($v=0$)+H_2O (001)+ΔE. Chem. Phys. Lett. **32**, 218 (1975).

R. F. Stebbings, C. J. Latimer, W. P. West, F. B. Dunning, T. B. Cook: Studies of xenon atoms in high Rydberg states. Phys. Rev. A **12**, 1453 (1975).
W. H. Stevenson, R. D. Santos, S. C. Mettler: A laser velocimeter utilizing laser-induced fluorescence. Appl. Phys. Lett. **27**, 395 (1975).
R. L. Swofford, W. M. McClain: The effect of spatial and temporal laser beam characteristics on two-photon absorption. Chem. Phys. Lett. **34**, 455 (1975).
K. Tohma: A simple model for intracavity absorption. Opt. Commun. **15**, 17 (1975).
A. A. Varfolomeev: Nonlinear processes in a system of resonantly interacting atoms. Sov. Phys. JETP **39**, 985 (1975).
R. Vetter, R. Damaschini, J. Cahen, J. Brochard: Study of excitation transfers by dye laser pumping. J. Phys. B **8**, L 275 (1975).
D. F. Zaretskii, V. P. Krainov: Resonance excitation of atomic levels in a strong electromagnetic field. Sov. Phys. JETP **39**, 257 (1975).
D. F. Zaretskii, V. P. Krainov: Resonance multiphoton excitation of atomic levels in a strong electromagnetic field. Sov. Phys. JETP **40**, 647 (1975).

1.2. Application to Chemistry and Isotope Separation

R. V. Ambartzumyan, V. S. Letokhov, E. A. Ryabov, N. V. Chekalin: Isotopic selective chemical reaction of BCl_3 molecules in a strong infrared laser field. JETP Lett. **20**, 237 (1975).
D. Arnoldi, K. Kaufmann, J. Wolfrum: Chemical-laser-induced isotopically selective reaction of HCl. Phys. Rev. Lett. **34**, 1597 (1975).
H. R. Bachmann, H. Nöth, R. Rinck, K. L. Kompa: Infrared laser specific reactions of boron compounds. CO_2 laser control of the exchange reactions $B(CH_3)_nBr_m+HBr \rightarrow B(CH_3)_{n-1}Br_{m+1}+CH_4$. Chem. Phys. Lett. **33**, 261 (1975).
R. J. Butcher, R. J. Danovan, C. Fortakis, D. Fernie, A. G. A. Rae: Photodissociation laser isotope effects. Chem. Phys. Lett. **30**, 398 (1975).
S. M. Freund, J. J. Ritter: CO_2 TEA laser-induced photochemical enrichment of boron isotopes. Chem. Phys. Lett. **32**, 255 (1975).
S. E. Harris, D. B. Lidow: Isotope separation by optically pumped ionizing collisions. Appl. Phys. Lett. **26**, 104 (1975).
K. S. Kochelashvili, N. V. Karlov, A. N. Orlov, R. P. Petrov, Y. N. Petrov, A. M. Prokhorov: Selective heterogeneous separation of vibrationally excited molecules. JETP Lett. **21**, 302 (1975).
J. L. Lyman, R. J. Jensen, J. Rink, C. P. Robinson, S. D. Rockwood: Isotropic enrichment of SF_6 in S^{34} by multiple absorption of CO_2 laser radiation. Appl. Phys. Lett. **27**, 87 (1975).
J. B. Marling: Laser isotope separation. Chem. Phys. Lett. **34**, 84 (1975).
H. Nakamura: Branching ratio for preionization and predissociation of a superexcited state of a diatomic molecule. Chem. Phys. Lett. **33**, 151 (1975).
L. J. Radziemski, Jr., S. Gerstenkorn, P. Luc: Uranium transitions and energy levels which may be useful in atomic-photoionization schemes for separating ^{238}U and ^{235}U. Opt. Commun. **15**, 273 (1975).
S. D. Rockwood, J. W. Hudson: Laser driven synthesis of $BHCl_2$ from BCl_3 and H_2. Chem. Phys. Lett. **34**, 542 (1975).

1.3. Coherent Transient Effects

S. S. Alimpiev, N. V. Karlov: Nutation effect in the molecular gases BCl_3 and SF_6. JETP **39**, 260 (1975)
P. R. Berman, J. M. Levy, R. G. Brewer: Cohorent optical transient study of molecular collisions: theory and observations. Phys. Rev. A **11**, 1668 (1975).

R. G. BREWER, E. L. HAHN: Coherent two-photon processes: transient and steady-state cases. Phys. Rev. A **11**, 1641 (1975).
J. P. HERITAGE, T. G. GUSTAFSON, C. H. LIN: Observation of coherent transient birefringence in CS_2 vapor. Phys. Rev. Lett. **34**, 1299 (1975).
R. JODOIN, L. MANDEL: Optimum conditions for superradiant damping. Opt. Commun. **15**, 284 (1975).
M. MATSUOKA: Doppler-free, two-photon induced coherence and emission. Opt. Commun. **15**, 84 (1975).
M. MATSUOKA, H. NAKATSUKA, J. OKADA: Free precession decay of two-photon-induced coherence in Ca vapor. Phys. Rev. A **12**, 1062 (1975).

1.4. Natural Linewidth-Limited Spectroscopy

E. V. BAKLANOV: Influence of field on the Lamb dip center shift with account of recoil effect. Opt. Commun. **13**, 54 (1975).
K. BERGMANN, W. DEMTRÖDER, P. HERING: Laser diagnostic in molecular beams. Appl. Phys. **8**, 65 (1975).
W. K. BISCHEL, P. J. KELLY, CH. K. RHODES: Observation of Doppler-free two-photon absorption in the ν_3 bands of CH_3F. Phys. Rev. Lett. **34**, 300 (1975).
P. A. BONCZYK: Saturated absorption by HF gas within a HF laser cavity. Appl. Phys. **6**, 125 (1975).
PH. CAHUZAC, R. VETTER: Sub-Doppler spectroscopy of xenon laser lines with a saturated-amplification method: resolution up to the natural width. Phys. Rev. Lett. **34**, 1070 (1975).
W. W. CHOW, M. O. SCULLY, J. O. STONE, JR: Quantum-beat phenomena described by quantum electrodynamics and neoclassical theory. Phys. Rev. A **11**, 1380 (1975).
B. COUILLARD, A. DUCASSE: Refractive index saturation effects in saturated absorption experiments. Phys. Rev. Lett. **35**, 1267 (1975).
B. COUILLARD, A. DUCASSE: Saturated absorption experiments using a free running cw dye laser. Opt. Commun. **13**, 398 (1975).
C. DELSART, J.-C. KELLER: Absorption line narrowing in a three-level system of Ne under interaction with two quasi-resonant lasers. Opt. Commun. **15**, 91 (1975).
L. A. HACKEL, K. H. CASLETON, S. G. KUKOLICH, S. EZEKIEL: Observation of magnetic octupole and scalar spin-spin interactions in I_2 using laser spectroscopy. Phys. Rev. Lett. **35**, 568 (1975).
R. C. HILBORN: Theory of the time and frequency dependence of near resonant Raman scattering and quantum beats. Chem. Phys. Lett. **32**, 76 (1975).
G. S. HURST, M. G. PAYNE, M. H. NAYFEH, J. P. JUDISH, E. B. WAGNER: Saturated two-photon resonance ionization of He (2^1S). Phys. Rev. Lett. **35**, 82 (1975).
H. W. KUGEL, M. LEVENTHAL, D. E. MURNICK, C. K. N. PATEL, O. R. WOOD, II: Infrared-X-ray double-resonance study of $2P_{1/2}$-$2S_{1/2}$ splitting in hydrogenic fluorine. Phys. Rev. Lett. **35**, 647 (1975).
V. S. LETOKHOV, V. P. CHEBOTAYEV: Resonance phenomena in saturation of absorption by laser radiation. Sov. Phys. Usp. **17**, 467 (1975).
V. S. LETOKHOV: Emission and absorption of γ-radiation by nuclei in molecules interacting with a light field. Phys. Rev. A **12**, 1954 (1975).
T. W. MEYER, CH. K. RHODES, H. A. HAUS: High-resolution line broadening and collision studies in CO_2 using nonlinear spectroscopic techniques. Phys. Rev. A **12**, 1993 (1975).
W. RADLOFF, E. BELOW: Lamb-dip spectroscopy of some lower carbon-hydrogen compounds. Opt. Commun. **13**, 160 (1975).
S. D. ROSNER, R. A. HOLT, T. D. GAILY: Measurement of the zero-field hyperfine structure of a single vibration-rotation level of Na_2 by a laser-fluorescence molecular-beam-resonance technique. Phys. Rev. Lett. **35**, 785 (1975).

R. SALOMAA, S. STENHOLM: Two-photon spectroscopy: effects of a resonant intermediate state. J. Phys. B **8**, 1795 (1975).

IM THEK-DE, O. P. PODAVALOVA, A. K. POPOV, G. KH. TARTAKOVSKII: Resonant four-photon parametric processes in Ne in the field of a single-frequency He-Ne-laser and their use in non-linear spectroscopy. JETP Lett. **21**, 195 (1975).

P. TSCHERIS, R. GUPTA: Measurement of hyperfine structure of the $8^2S_{1/2}$ and $9^2S_{1/2}$ states of rubidium, and $12^2S_{1/2}$ state of cesium by stepwise dye-laser spectroscopy. Phys. Rev. A **11**, 455 (1975).

CH. C. WANG, L. I. DAVIS, JR.: Saturation of resonant two-photon transitions in thallium-vapor. Phys. Rev. Lett. **35**, 650 (1975).

J. F. WARD, A. V. SMITH: Saturation of two-photon-resonant optical processes in cesium vapor. Phys. Rev. Lett. **35**, 653 (1975).

B. G. WHITEFORD, K. J. SIEMSEN, H. D. RICCIUS, G. R. HANES: Absolute frequency measurements of N_2O laser transitions. Opt. Commun. **14**, 70 (1975).

J. P. WOERDMAN, M. F. H. SCHNURMANS: Spectral narrowing of selective reflection from sodium vapor. Opt. Commun. **14**, 248 (1975)

Chapter 2

I. Infrared Laser Spectroscopy

J. P. ALDRIDGE, R. F. HOLLAND, H. F. FLICKER, K. W. NILL, T. C. HARMAN: High resolution Q-branch Spectrum of CO_2 at 618 cm^{-1}. J. Mol. Spectrosc. **54**, 328 (1975).

F. ALLARIO, C. H. BAIR, J. F. BUTLER: High-resolution spectral measurements of SO_2 from 1176.0 to 1256.8 cm^{-1} using a single PbSe laser with magnetic and current tuning. IEEE J. Quant. Electron. QE-**11**, 205 (1975).

J. R. ARONSON, P. C. VON THÜNA, J. F. BUTLER: Tunable diode laser high resolution spectroscopic measurements of the ν_2 vibration of carbon dioxide. Appl. Opt. **14**, 1120 (1975).

E. N. BAZAROV, G. A. GERASIMOV, YU. I. POSUDIN: Absorption of carbon dioxide laser lines in OsO_4 vapor. Sov. J. Quant. Electron. **4**, 106 (1974).

R. S. ENG, P. L. KELLEY, A. R. CALAWA, T. C. HARMAN, K. W. NILL: Tunable diode laser measurements of water absorption line parameters. Molec. Phys. **28**, 653 (1974).

R. S. ENG, A. MOORADIAN, H. FETTERMAN: InAs spin-flip laser operation at 3 μm. Appl. Phys. Lett. **25**, 453 (1974).

P. L. HOUSTON, J. I. STEINFELD: Low-temperature absorption contour of the ν_3 band of SF_6. J. Mol. Spectrosc. **54**, 335 (1975).

G. P. MONTGOMERY, JR., J. C. HILL: High-resolution diode-laser spectroscopy of the 949.2 cm^{-1} band of ethylene. J. Opt. Soc. Am. **65**, 579 (1975).

A. S. PINE: Doppler-limited molecular spectroscopy by difference-frequency mixing. J. Opt. Soc. Am. **64**, 1683 (1974).

A. S. PINE: Doppler-limited spectra of the ν_3 vibration of $^{12}CH_4$ and $^{13}CH_4$. J. Mol. Spectrosc. **54**, 132 (1975).

W. G. PLANET, J. R. ARONSON, J. F. BUTLER: Measurements of the widths and strengths of low-J lines of the ν_2 Q branch of CO_2. J. Mol. Spectrosc. **54**, 331 (1975).

H. PRIER, W. RIEDEL: NO spectroscopy by pulsed PbSSe diode lasers. J. Appl. Phys. **45**, 3955 (1974).

R. T. THOMPSON, JR., J. M. HOELL, JR., W. R. WADE: Measurements of SO_2 absorption coefficients using a tunable dye laser. J. Appl. Phys. **46**, 3040 (1975).

II. Tunable Lasers

S. P. ANOKHOV, V. I. KRAVCHENKO, M. S. SOSKIN: Rapid spectroscopy using lasers with swept lasing frequency. Opt. Spectrosc. **36**, 106 (1974).

M. J. COLLES, C. R. PIDGEON: Tunable lasers. Rept. Progr. Phys. **38**, 329 (1975).

J. T. GANLEY, F. B. HARRISON, W. T. LELAND: The spin-flip Raman laser as a tunable infrared source. J. Appl. Phys. **45**, 4980 (1974).
J. P. GOLDSBOROUGH: Scanning single frequency cw dye laser techniques for high-resolution spectroscopy. Opt. Engng. **13**, 523 (1974).
S. H. GROVES, K. W. NILL, A. J. STRAUSS: Double heterostructure $Pb_{1-x}Sn_xTe$-PbTe lasers with cw operation at 77 K. Appl. Phys. Lett. **25**, 331 (1974).
D. C. HANNA, B. LUTHER-DAVIES, R. C. SMITH, R. WYATT: CdSe down-converter tuned from 9.5 to 24 μm. Appl. Phys. Lett. **25**, 142 (1974).
G. K. KLAUMINZER: Etalon/grating synchronized scanning of a narrowband pulsed dye laser. Opt. Engng. **13**, 528 (1974).
K. W. NILL: Tunable infrared lasers. Opt. Engng. **13**, 516 (1974).

III. Laser Pollution Monitoring

K. ASAI, T. IGARASHI: Detection of ozone by differential absorption using CO_2 laser. Opt. Quant. Electron. **7**, 211 (1975).
E. G. BURKHARDT, C. A. LAMBERT, C. K. N. PATEL: Stratospheric nitric oxide measurements during daytime and sunset. Science **188**, 1111 (1975).
R. L. BYER: Remote air pollution measurement. Opt. Quant. Electron. **7**, 147 (1975).
W. M. FAIRBANK, JR., T. W. HÄNSCH, A. L. SCHAWLOW: Absolute measurement of very low sodium-vapor densities using laser resonance fluorescence. J. Opt. Soc. Am. **65**, 199 (1975).
W. B. GRANT, R. D. HAKE: Calibrated remote measurements of SO_2 and O_3 using atmospheric backscatter. J. Appl. Phys. **46**, 3019 **(1975)**.
E. D. HINKLEY, A. R. CALAWA: Diode lasers for pollutant monitoring, in: *Analytical Methods Applied to Air Pollution Measurement*, ed. by R. K. STEVENS and W. F. HERGET (Ann Arbor Science Publishers, Inc., Ann Arbor 1974), Chapt. 3.
T. KOBAYASI, H. INABA: Infra-red heterodyne laser radar for remote sensing of air pollutants by range-resolved differential absorption. Opt. Quant. Electron. **7**, 319 (1975).
R. T. KU, E. D. HINKLEY, J. O. SAMPLE: Long-path monitoring of atmospheric carbon monoxide with a tunable diode laser system. Appl. Opt. **14**, 854 (1975).
R. M. MEASURES, W. R. HOUSTON, D. G. STEPHENSON: Laser-induced fluorescent decay spectra – a new form of environmental signature. Opt. Engng. **13**, 494 (1974).
D. C. O'SHEA, L. G. DODGE: NO_2 concentration measurements in an urban atmosphere using differential absorption techniques. Appl. Opt. **13**, 1481 (1974).
C. K. N. PATEL, E. G. BURKHARDT, C. A. LAMBERT: Spectroscopic measurements of stratospheric nitric oxide and water vapor. Science **184**, 1173 (1974).

Chapter 3

T. BERGEMAN: Intensities, line shift, and resonance interference in zero-field optical double resonance. J. Chem. Phys. **61**, 4515 (1974).
T. BERGEMAN, R. N. ZARE: Fine structure, hyperfine structure, and Stark effect in the NO $A\,^2\Sigma^+$ state by optical radio-frequency double resonance. J. Chem. Phys. **61**, 4500 (1974).
R. G. BREWER, R. L. SHOEMAKER, S. STENHOLM: Collision-induced optical double resonance. Phys. Rev. Lett. **33**, 63 (1974).
R. W. FIELD, R. S. BRADFORD, H. P. BROIDA, D. O. HARRIS: Excited state microwave spectroscopy on the A $^1\Sigma$ state of BaO. J. Chem. Phys. **57**, 2209 (1972).
M. REDON, H. GUREL, M. FOURIER: Infrared-microwave double resonance in NH_3: studies of collision-induced transitions in the presence of high field Stark effect. Chem. Phys. Lett. **30**, 99 (1975).
R. L. SHOEMAKER, S. STENHOLM, R. G. BREWER: Collision-induced optical double resonance. Phys. Rev. A **10**, 2037 (1974).

R. Solarz, D. H. Levy: Microwave optical double resonance in electronically excited NO_2. J. Chem. Phys. **58**, 4026 (1973).
R. Solarz, D. H. Levy, R. F. Curl, Jr: High resolution observation of microwave-optical double resonance in NO_2. J. Chem. Phys. **60**, 1158 (1974).
M. Takami, K. Shimoda: Microwave spectrum of HCOOH in the $\nu_{CH}=1$ vibrational state observed by laser-microwave double and triple resonance. Japan. J. Appl. Phys. **13**, 1699 (1974).
T. Tanaka, K. Abe, R. F. Curl, Jr.: Microwave optical double resonance and fluorescence spectrum of NO_2 excited by the 4579 Å line of the argon-ion laser. J. Mol. Spectrosc. **49**, 310 (1974).
T. Tanaka, A. D. English, R. W. Field, D. A. Jennings, D. O. Harris: Microwave optical double resonance of N_2O with a tunable cw dye laser, J. Chem. Phys. **59**, 5217 (1973).
T. Tanaka, R. W. Field, D. O. Harris: Microwave optical double resonance of NO_2 with a tunable cw dye laser. II. Excited state microwave transitions. J. Chem. Phys. **61**, 3401 (1974).
A. Yogov, Y. Haas: Vibrational relaxation of biacetyl studied by a visible-infrared double resonance technique. Chem. Phys. Lett. **21**, 544 (1973).
R. C. L. Yuan, J. M. Preses, G. W. Flynn, A. M. Ronn: V—V and V—T/R energy transfer studies of C_2H_4 by infrared laser double resonance. J. Chem. Phys. **59**, 6128 (1973).

Chapter 4

D. Baierl, W. Kiefer: Hot band and isotropic structure in the resonance Raman spectrum of bromine vapor. J. Chem. Phys. **62**, 306 (1975).
J. J. Barrett, A. B. Harvey: Vibrational and rotational-translational temperatures in N_2 by interferometric measurement of the pure rotational Raman effect. J. Opt. Soc. Am. **65**, 392 (1975).
M. Berjot, L. Bernard, T. Theophanides: Variation in the depolarization factor of the fundamental band of iodine during the continuous transition of the resonance fluorescence to the resonance Raman effect. Canad. J. Spectrosc. **18**, 128 (1973).
G. Black, R. Sharpless, T. Slanger: Measurements of vibrationally excited molecules by Raman scattering. I. The yield of vibrationally excited nitrogen in the reaction $N+NO \rightarrow N_2+O$. J. Chem. Phys. **58**, 4792 (1973).
G. Black, H. Wise, S. Schechter, R. L. Sharpless: Measurements of vibrationally excited molecules by Raman scattering. II. Surface deactivation of vibrationally excited N_2. J. Chem. Phys. **60**, 3526 (1974).
R. Butcher, W. Jones: Study of the rotational Raman spectra of $^{14}N^{15}N$ and $^{15}N_2$ using a Fabry-Perot etalon. J. Chem. Soc. Faraday Trans. II **70**, 560 (1974).
A. Chapput, B. Reussel, G. Fleury: Vapor phase Raman spectroscopy of three deuterated forms of formaldehyde. J. Raman Spectrosc. **1**, 507 (1973).
J. Cherlow, H. Hyatt, S. Porto: Raman scattering in hydrogen halide gases. J. Chem. Phys. (to be published, 1975).
A. Compaan, W. Langor, D. Eden, H. Swinney: Collisional excitation of carbon monoxide by H_2. Astrophys. J. **185**, L105 (1973).
F. de Martini: High resolution nonlinear spectroscopy of molecular vibrational resonances in gases, in: *Proceedings of the Esfahan Symposium on Fundamental and Applied Laser Physics* (Wiley-Interscience, New York, 1973), p. 549.
S. Dworetsky, R. Hozack: Comment on "Resonance Raman scattering in I_2 below the dissociation limit". Phys. Rev. A **8**, 3257 (1973).
W. Fletcher, J. Rayside: High resolution vibrational Raman spectrum of oxygen. J. Raman Spectrosc. **2**, 3 (1974).

D. G. FOUCHE, R. K. CHANG: Observation of resonance Raman scattering below the dissociation limit in I_2 vapor. Phys. Rev. Lett. **29**, 536 (1972).
L. FROMMHOLD: Interpretation of Raman spectra of van der Waals dimers in argon. J. Chem. Phys. **61**, 2996 (1974).
M. GARVEY, G. S. KENT: Raman backscatter of laser radiation from the stratosphere. Nature **248**, 124 (1974).
R. GAUFRES: On some possibilities offered by separation of scattering spectra traces in Raman spectroscopy of gases. Compt. Rend. Hebd. Seances Acad. Sci. B **277**, 297 (1973).
C. GRAY: Calculation of the pressure-broadening of HCl and DCl infrared and Raman lines. J. Chem. Phys. **61**, 418 (1974).
K. HAMADA: Rotational wings of Raman bands in gaseous and liquid carbon disulfide. J. Phys. Soc. Japan **35**, 1564 (1973).
R. HILL, D. HARTLEY: Focused multiple-pass cell for Raman scattering, Appl. Opt. **13**, 186 (1974). See also A. STURTEVANT: Focussed multiple cell for Raman scattering: Comment, Appl. Opt. **13**, 1739 (1974); and R. HILL, D. HARTLEY: Authors reply . . . , Appl. Opt. **13**, 1739 (1974).
T. HIRSCHFELD, W. MUELLER, S. KLAINER: A simple very high temperature cell for Raman spectroscopy. Appl. Spectrosc. **29**, 88 (1975).
W. HOLZER, Y. LEDUFF: Collision-induced light scattering observed at the frequency region of vibrational Raman bands. Phys. Rev. Lett. **32**, 205 (1974).
W. HOLZER, R. OULLION: Forbidden Raman bands of SF_6: Collision induced Raman scattering. Chem. Phys. Lett. **24**, 589 (1974).
W. KIEFER: Laser-excited resonance Raman spectra of small molecules and ions – A review. Appl. Spectrosc. **28**, 115 (1974).
R. KEIJSER, J. LOMBARDI, K. VAN DER HOUT, H. DEGREOT, B. SANCTUARY, H. KNAAP: Evidence for resonance collisions in rotational energy transfer; Raman lines of H_2. Phys. Lett. A **45**, 3 (1973).
R. KEIJSER, K. VAN DER HOUT, H. KNAAP: Pressure broadening of the rotational Raman lines of hydrogen isotopes. Physica **76**, 577 (1974).
M. LAPP, C. PENNEY, L. GOLDMAN: Vibrational Raman scattering temperature measurement. Opt. Commun. **9**, 195 (1974).
M. LAPP, C. PENNEY: *Laser Raman Gas Diagnostics* (Plenum Press, New York, 1974).
Y. LEDUFF, W. HOLZER: Raman scattering of HF in the gas state and in liquid solution. J. Chem. Phys. **60**, 2175 (1974).
J. LEVATTER, R. SANDSTROM, S. CHITIN: Raman cross-sections measured by short-pulse laser scattering and photon counting. J. Appl. Phys. **44**, 3273 (1973).
M. MARSDEN, G. BIRD: Resonance Raman spectrum of gaseous nitrogen dioxide. J. Chem. Phys. **59**, 2766 (1973).
H. MORAAL: Kinetic theory of rotational Raman line width. Physica **73**, 379 (1974).
F. MOYA, S. DRUET, J. TARAN: Gas spectroscopy and temperature measurement by coherent Raman anti-stokes scattering. Opt. Commun. **13**, 169 (1975).
J, NESTOR, E. LIPPINCOTT: The effect of the internal field on Raman scattering cross-sections. J. Raman Spectrosc. **1**, 305 (1973).
N. OHASHI, H. WATANABE, S. MATSUOKA: Three mirror Raman cell for gases. Jap. J. Appl. Phys. **12**, 1103 (1973).
C. PENNEY, R. ST. PETERS, M. LAPP: Absolute rotational Raman cross-sections for N_2, O_2 and CO_2. J. Opt. Soc. Am. **64**, 712 (1974).
V. PODOBEDEV, A. PYNDYK, KH. STERIN: High-speed recording of Raman spectra of liquids and gases. Opt. Spectrosc. **34**, 478 (1973).
A. PYNDYK, V. PODOBEDEV: Apparatus for the recording of Raman spectra of gases. Opt. Spectrosc. **35**, 31 (1973).

R. St. Peters, S. Silverstein, M. Lapp, C. Penney: Resonant Raman scattering or resonance fluorescence in I_2 vapor? Phys. Rev. Lett. **30**, 191 (1973).

R. St. Peters, S. Silverstein: Manifestations of pressure broadening on tuned resonance Raman fluorescence. Opt. Commun. **7**, 193 (1973).

R. Schwiesow, W. Abshire: Relative Raman cross-section of O_3 for four Ar^+ laser frequencies. J. Appl. Phys. **44**, 3708 (1973).

S. Silverstein, R. St. Peters: Explanation of high-foreign gas pressure effects on resonant light scattering in I_2 vapor. Phys. Rev. A **9**, 2720 (1974).

S. Silverstein, R. St. Peters: A new effect in the depolarization of resonant light scattering from molecules in the vapor phase: I_2. Chem. Phys. Lett. **23**, 140 (1973).

D. Stephenson: Raman cross-sections of selected hydrocarbons and freons. J. Quant. Spectrosc. Radia. Transf. **14**, 1291 (1974).

W. Stricker, J. Hochenbleicher: Laser excited Raman spectrum of gaseous fluorine. Z. Naturforsch, A **28** A, 27 (1973).

C. Wang, R. Wright: Raman studies of the effect of density on the Fermi resonance in CO_2. Chem. Phys. Lett. **23**, 241 (1973).

C. Wang, R. Wright: Effect of density on the Raman scattering of molecular fluids. I ... Gaseous N_2. J. Chem. Phys. **59**, 1706 (1973).

P. Williams, D. Rousseau, S. Dworetsky: Resonance fluorescence and resonance Raman scattering lifetimes in molecular iodine. Phys. Rev. Lett. **32**, 196 (1974).

Some Very Recent Papers

R.V. Ambartzumian, N. P. Furzikov, V. S. Letokhov, A. A. Puretsky: Measuring photoionization cross-sections of excited atomic states. Appl. Phys. **9** (1976).

Y. V. Baklanov, B. Y. Dubetsky, V. P. Chebotayev: Nonlinear Ramsey resonance in optical region. Appl. Phys. **9**, 177 (1976)

A. F. Bernhardt: Isotope separation by laser deflection of an atomic beam. Appl. Phys. **9**, 19 (1976).

J. W. Eerkens: Spectral Considerations in the Laser Isotope Separation of Uranium Hexafluoride. Appl. Phys. **10** (1976).

P. R. Hammond, A. N. Fletcher, D. E. Bliss, R. A. Henry, R. L. Atkins: Search for efficient, near uv lasing dyes. III. Monocyclic and miscellaneous dyes. Appl. Phys. **9**, 67 (1976).

J. Häger, W. Hinz, H. Walther, G. Strey: High resolution spectroscopy of ethylene by means of spin-flip-Raman laser. Appl. Phys. **9**, 35 (1976).

P. R. Hammond, A. N. Fletcher, R. A. Henry, R. L. Atkins: Search for efficient, near uv lasing dyes. I. Substituent effects of bicyclic dyes. Appl. Phys. **8**, 311 (1975); – II. AZA substitution in bicyclic dyes. Appl. Phys. **8**, 315 (1975).

V. S. Letokhov, B. D. Pavlik: Spectral line narrowing in a gas by atoms trapped in a standing light wave. Appl. Phys. **9** (March 1976).

G. Marowsky, F. P. Schäfer, J. W. Keto, F. K. Tittel: Fluorescence studies of electron-beam pumped POPOP dye vapor. Appl. Phys. **9**, 143 (1976).

M. Sargent: Laser saturation grating phenomena. Appl. Phys. **9**, 127 (1976).

K. Shimoda: Doppler-free resonant two-photon transition for isotope separation. Appl. Phys. **9** (March 1976).

C. O. Weiss, G. Kramer: Vibrational relaxation in the HCOOH-far-infrared laser. Appl. Phys. **9**, 175 (1976).

Subject Index

W. Demtröder **Laser Spectroscopy**

2nd., enlarged edition. 16 figures. 3 tables. III, 106 pages. 1973 (Topics in Current Chemistry, Vol. 17)

Laser Spectroscopy

Proceedings of the 2nd International Conference, Mégève, France, June 23–27, 1975
Editors: *S. Haroche; J. C. Pebay-Peyroula; T. W. Hänsch; S. E. Harris*
230 figures. 30 tables. X, 468 pages (5 pages in French). 1975 (Lecture Notes in Physics, Vol. 43)

Forthcoming titles

Topics in Applied Physics

High-Resolution Laser Spectroscopy K. Shimoda (editor)

K. Shimoda: Introduction. – *K. Shimoda:* Line Broadening and Narrowing Effects. – *P. Jacquinot:* Atomic Beam Spectroscopy. – *V. S. Letokhov:* Saturation Spectroscopy. – *J. L. Hall:* Recent Studies on Very High Resolution Spectroscopy. – *V. P. Chebotayev:* Three-Level Laser Spectroscopy. – *N. Bloembergen, M. D. Levenson:* Doppler-Free Two-Photon Spectroscopy

Laser Monitoring of the Atmosphere E. D. Hinkley (editor)

S. H. Melfi: Remote Sensing for Air Quality Management. – *V. E. Zuev:* Laser Transmission of the Atmosphere. – *R. H. T. Collis, P. B. Russell:* Lidar Measurement of Particles and Gases by Elastic Backscattering and Differential Absorption. – *H. Inaba:* Detection of Atoms and Molecules by Raman Scattering and Resonance Fluorescence. – *E. D. Hinkley, R. T. Ku, P. L. Kelley:* Molecular Pollutant Detection by Differential Absorption. – *R. T. Menzies:* Laser Heterodyne Detection Techniques

Topics in Modern Physics

Beam-Foil Spectroscopy S. Bashkin (editor)

S. Bashkin: Introduction. – *S. Bashkin:* Instrumentation. – *I. Martinson:* Wavelengths Measurements and Level Analysis. – *L. Curtis:* Lifetime Measurements. – *I. Sellin:* Autoionizing Levels. – *H. Marrus:* Studies of H-like and He-like Ions of High Z. – *W. Whaling, L. Heroux:* Applications to Astrophysics. – *O. Sinanoglu:* Fundamental Calculation of Level Lifetimes. – *W. Wiese:* Systematic Effects in Z-Dependence of Oscillator Strengths. – *J. Macek, D. J. Burns:* Coherence, Alignment, and Orientation Phenomena

COLLECTING POTTERY & PORCELAIN

The Facts At Your Fingertips

MIVA CITO

COLLECTING POTTERY & PORCELAIN

The Facts At Your Fingertips

General Editor:
Janet Gleeson

Special Consultants:
Keith Baker,
Gordon Lang,
John Sandon &
Lita Solis-Cohen

Collecting Pottery & Porcelain
The Facts At Your Fingertips

First published in Great Britain in 1997 by Miller's,
an imprint of Mitchell Beazley, both divisions of
Octopus Publishing Group Ltd.,
2–4 Heron Quays,
London E14 4JP

Reprinted 1998 (twice), reissued 2001

Executive Editor Alison Starling
Executive Art Editor Vivienne Brar
Project Editor Nina Sharman
Text editor Wendy Lee
Designer Louise Griffiths
Picture Research Jenny Faithfull
Production Jilly Sitford
Special photography Chris Halton, Tim Ridley
Indexer Hilary Bird

A CIP catalogue for this book is available from the British Library

ISBN 185732 898 1

Printed and bound by Toppan Printing Co., China

Front cover *(Top left)* Meissen group of Messetin and Columbine, c.1740, £1,800–2,500. *(Top right)* Coffe cup and saucer, c.1930, £30–40. *(Centre)* Small Wemyss pig, c.1910, £300–400. *(Bottom left)* Blue and white transfer-printed plate by Joseph Stubbs, c.1822–35, £180–240. *(Bottom right)* Clarice Cliff conical jug, c.1930, £600–800.
Front flap William de Morgan "Persianware" vase, c.1900, £10,000 or more.
Back cover Böttger Kakiemon flared beaker, c.1725, £1,000–1,500
Back flap Staffordshire spaniel (one of a pair), mid-19thC, £1,100–1,500.
Page 2 *(Top left)* Deruta wet-drug jar, c.1507, £15,000–20,000. *(Top right)* English delft bowl, c.1740, £1,000–1,400. *(Centre left)* Plate from a Meissen dinner service, c.1763–1764, £400–600. *(Centre)* Chinese blue and white vase, early 18thC, £1,300–3,000 depending on size. *(Centre right)* Westerwald blue-glazed jug, late 17thC, £1,200–1,800. *(Bottom left)* Susie Cooper Kestrel-shape Teapot, 1930s, £60–90.
(Bottom right) Staffordshire creamware equestrian group, c.1785, £6,000–10,000.
Page 3 *(Top left)* "Rhodian ware dish late 16thC, £10,000–20,000. *(Top right)* Meissen teapot, c.1745, £3,000–4,000. *(Centre)* Salt-glazed agateware seated cat, c.1745, £800–1,200. *(Bottom left)* Red Nicolson vase, late 19th/early 20thC. *(Bottom right)* Staffordshire Toby jug, c.1830, £280–400.

CONTENTS

INTRODUCTION

Browse through any auction or antiques fair catalogue and ceramics of one sort or another will invariably feature prominently. Visually appealing, remarkably diverse, available at an incomparably wide spectrum of prices, pottery and porcelain have long captured the attentions of collectors; and as the world of antiques and collectables in general attracts an ever-growing number of enthusiasts, ceramics seem set to remain as popular as ever.

But what is it about pottery and porcelain that has made them so fascinating to so many people throughout the centuries? In bridging the gap between the useful and the decorative, ceramics provide a fascinating insight into the lifestyle of our forebears, combined with huge aesthetic appeal. Today we may prefer to keep a bleeding bowl or an *albarello* on the dresser rather than collect blood or store chemicals in it, but our enjoyment of such objects is greatly enhanced by an understanding of their origins. Learning how such objects were made and what inspired their decoration is as much a part of collecting as being able to spot the bargains, an aspect on which this book focuses wherever possible.

Clay was used from the earliest times to create functional objects in which man could store and serve food and other essentials. As potters became increasingly adept at using the raw materials available they began adorning their primitive vessels with increasing sophistication, developing coloured pigments to decorate them and glazes to make them water-resistant. The potter's blossoming technical prowess evolved hand in hand with an increasingly sophisticated society, in which there was a demand not only for useful objects but for decorative, prestigious ones as well. So the humble clay pots fashioned by primitive man gave rise to refined clay, stoneware and porcelain objects that were beautiful enough to be regarded as works of art.

The most dramatic development in the history of ceramics came with the development of porcelain in China in the Tang dynasty (618–906AD). When, in the 16th century, Portuguese traders began exporting Chinese porcelain in quantity to the Western world, porcelain objects quickly became highly prized among the affluent few who were able to afford them, and potters in Europe vied continuously with one another for the secret of how to make this magical, translucent substance. Kings and emperors throughout Europe squandered fortunes on endeavouring to discover the secret of making porcelain, and the royal cachet associated with the leading factories that were eventually established at Meissen, Vienna, Sèvres, Copenhagen and St Petersburg helped fuel the public passion for other sophisticated products. Today, as we handle and enjoy these amazingly fragile survivors of the past, their technical and artistic skill remains compelling – little wonder they are as keenly collected as ever.

But with such a dauntingly varied array of subjects from which to choose, how does the new collector know where to begin? Most enthusiasts focus on a particular area, and there are many ways in which the field of ceramics can be broken down into smaller specialities. You could concentrate on ceramics of a particular period and type – such as 17th-century French faience, 18th-century salt-glazed stoneware; transfer-printed ware or 19th-century majolica. Alternatively, you could choose an interesting factory, perhaps Meissen, Wedgwood or Spode. Many other collectors form attractive collections by concentrating on a particular type of ware – say teapots, or coffee cups or saucers.

Traditionally the most popular ceramic collecting areas have been early Oriental ceramics, 17th- and 18th-century European pottery, and 18th-century European porcelain made by famous factories. As prices for these highly prized wares have risen and good examples have become less readily available, collectors have gradually turned their attentions to other less exclusive collecting areas. In recent decades subjects such as blue and white transfer-printed pottery, majolica, or the products of some of the most innovative 20th-century factories and designers, including Carlton, Clarice Cliff, Susie Cooper and Shelley, have seen a huge surge in public interest. The good news for collectors is that, whatever your budget, taste and interests, you are almost certain to find some area of ceramics that fits the bill.

Taking time to learn about your chosen subject is of course one of the best ways to avoid making expensive mistakes, and learning to spot the bargains. This book aims to offer a general introduction to some of the major collecting areas, both traditional and new, and provide you with some historical background along with practical information on what to look for and what prices to expect to pay. Once you have chosen a subject there are a wealth of detailed specialist books on almost every topic covered in this book from which you can expand your knowledge – many are listed on pp168–9. You should also spend time visiting public collections, salerooms and dealers' galleries to examine similar pieces closely before you begin to buy. Remember that simply reading books or looking at pieces in a museum is no substitute for hands-on experience. Ceramics is a three-dimensional art; therefore if you take time to familiarize yourself with the feel and look of the genuine, spotting fakes will be far less difficult. When you visit a dealer or saleroom do not be afraid to pick pieces up and handle them. Look at the body, the appearance of the glaze, the style and colours used in the decoration; consider the weight and any distinctive imperfections. Take the marks into account but do not rely on them entirely as a sign of authenticity. If you do make the odd mistake, bear in mind that you are in good company – most experts have occasionally been fooled, and making mistakes is part of the learning process that all collectors go through.

Whatever collecting area you decide to focus on, value is largely determined by common factors: rarity, size, decorative appeal, popularity and condition are almost invariably critical in establishing the value of any piece of pottery or porcelain. Pieces of unusual form, shape, style or decoration are always appealing to serious collectors and therefore likely to command far higher prices than more readily available wares. Visually appealing pieces are also in keen demand, and the value of pieces made in the 19th and 20th centuries is especially reliant on their decorative appeal. Large pieces are usually more striking than smaller ones and therefore tend to be more valuable. Similarly, pairs or sets of decorative objects, such as vases, appeal to the decorative market and will often cost more than twice the price of a single object. Both pottery and porcelain are inherently fragile, and the value of all types of ceramics is enormously affected by damage. Provenance is yet another factor that can add dramatically to value. Even fairly run-of-the-mill pieces are given extra cachet if they are known to have belonged to a famous person, and labels from well-known collections will always boost value. Fashion and popularity also have a bearing on price. As certain areas enjoy a surge of interest and prices go up others may fall into relative decline.

The categories and objects we have chosen to illustrate in this book are intended to provide an overview of an extraordinarily complex and wide-ranging subject and also to reflect the types of ceramics you are likely to see on the market today. The objects are grouped mainly by type of ware and factory; but in some cases they highlight some of the more popular growing collecting areas such as commemorative ceramics, blue and white, lustre and ironstone. We hope that the varied assortment of information we have included will fire your appetite to learn more about this enthralling subject and that the collection you build as a result will provide you with a source of lasting pleasure – and who knows, perhaps you will even make a profit too.

The values given in this book for featured objects reflect the sort of prices you might expect to pay for similar pieces at an auction house or from a dealer. As there are so many variable factors involved in the pricing of antiques, the values should be used only as a general guide.

MAIN AREAS OF PRODUCTION – EUROPE & THE UNITED STATES

European and American pottery and porcelain are generally classified according to the type of ware for example, earthenware, maiolica, faience and stoneware, and the area in which the piece was produced. The map below shows many of the most famous centres of European and American production each of which developed their own distinctive style often drawing on influences from other European centres or from imported Oriental ceramics. In some instances the areas marked refer to a single prominent factory, such as Meissen, Sèvres or Coalport, in others names such as Delft, Moustier or Marseilles denote a region or town in which small manufacturers were grouped, rather than an individual maker.

MAIN AREAS OF PRODUCTION – CHINA & JAPAN

Pottery and porcelain was produced in China probably earlier than anywhere else. Chinese ceramics are often classified by the area, dynasty and reign in which they were produced. Porcelain was first produced in the north of China during the Tang dynasty. The famous porcelain of the northern Song dynasty (960–1127) comes from Hebei and Henan and is known as Dingyao. Jingdezhen became a centre of ceramic production in the Yuan period and became paramount in the Ming era and remains so today.

Ceramics have been made in Japan since the Neolithic period. The main centre of pottery production was at Seto and the term *Seto-mono* became equivalent to pottery itself. Porcelain was made in Japan from 1616 when a source of china clay was discovered at Arita.

559
LOT

PART 1

BUYING & SELLING POTTERY & PORCELAIN

ABOVE A MEISSEN FIGURE OF HARLEQUIN FROM THE *COMMEDIA DELL'ARTE*, C.1740. £8,000–10,000

LEFT A COLLECTION OF STAFFORDSHIRE FIGURES, POTTERY AND PORCELAIN FROM THE 19TH CENTURY.

AUCTIONS

Auctions can be one of the most exciting and fruitful ways to buy and sell pottery and porcelain. There is a huge variety of salerooms from which to choose. On the one hand you have the larger establishments which divide ceramic sales into categories according to the age and country of origin or type of ware. Specialist categories may include English ceramics, European porcelain, 19th-century ceramics, Art Nouveau, Clarice Cliff, Wedgwood, Staffordshire figures and studio ceramics. On the other hand there are numerous smaller firms where ceramics may well be sold alongside furniture and other works of art, or where ceramics of all types are sold together. If you are a novice collector, buying at auction, whether you decide to visit a grand London saleroom or a small country auction house, is not nearly as daunting as it might seem provided you follow a few guidelines.

BUYING AT AUCTION CHECKLIST

- STUDY THE CATALOGUE
- VIEW THOROUGHLY
- STICK TO YOUR LIMIT

CATALOGUES

Most salerooms provide a catalogue of the sale that contains a list of the items in the order in which they will be sold. Catalogues vary from glossy publications with colour illustrations and lengthy descriptions, to simple typed sheets. They are available from one month to a week prior to the sale and are worth buying and studying before you visit the saleroom to give you an idea of what is on offer. Most catalogue descriptions will state the name of the manufacturer if known, the type of body (materials used), the sort of object, a description of the decoration, a date or approximate date of manufacture as well as details of any marks. In some catalogues they also provide information on any damage.

VIEWING

In the days before the sale takes place there will be a "view" at which you can examine the ceramics on offer. It is important always to try and attend the view; do not be tempted to turn up on the morning of the sale and expect to be able to see everything because more often than not the porters will be preparing the lots for the auction and it will be impossible to view properly. Examine anything that catches your eye very, very carefully and make

The highly charged atomosphere of the saleroom brings added excitement to collecting.

up your own mind as to its authenticity and age. Bear in mind that although most auction house catalogues include a disclaimer stating that their descriptions express only the "opinion" of the cataloguer, they are legally bound by the Trade Descriptions Act. Don't be afraid to handle the objects to look at the marks and for signs of damage or restoration; ask for assistance if need be. Remember to sift through boxes of multiple lots, as this is often where some of the best bargains are to be found. Most of the larger auction houses include estimates in the catalogue to give you some idea of what you can expect to pay for a lot. If there are no estimates in the catalogue they are often pinned up at the view; if not, ask the valuer in charge of the sale what he or she expects the lot will fetch and they will give you some idea based on the reserve price and the level of interest shown during the view. The valuer might also be able to shed more light on the piece you are interested in and give you some idea as to its provenance.

PAYMENT

While you are at the view, find out about procedures for payment and collection if you intend to bid. Some auction houses do not accept credit cards, others will expect cheques to clear before they let you take your purchases away and most will expect you to pay for and collect anything you buy within five working days; if you do not you will often be charged interest and storage. If you cannot attend the sale you can leave a bid with the commissions clerk. The clerk is bound to buy for you as cheaply as possible, but even so it is always best to attend the sale if you can.

BIDDING

Some sales are day-long affairs, and if the lot in which you are interested comes towards the end of the sale you may not want to sit through the whole event. You can find out from a member of staff roughly how quickly the auctioneer sells (usually between 100–120 lots per hour) and work out approximately when you need to arrive.

On the day of the sale you may need to register before you are able to bid. This usually means filling in a form giving details of your name, address and bank. Some salerooms will then issue you with a piece of card with a bidding number on it (sometimes called a paddle) which you hold up if you are successful in your bidding. Before the sale starts decide on your limit and try to stick to it rather than getting carried away in the heat of the moment and regretting it later. Don't forget that on top of the hammer price you will have to add the buyer's commission of between 10 and 15 percent and VAT on the commission.

When the lot on which you want to bid goes under the hammer do not panic if the auctioneer seems to be ignoring you. Many auctioneers only take bids from two people at once; when one drops out that will be your chance to join in.

SELLING

Selling at auction is no more complicated than buying, although you might find taking your prized dinner service or tea set in for inspection and valuation rather nerve-wracking at first. If you intend visiting one of the larger auction houses it is worth phoning ahead to make sure the relevant expert will be available. There is no charge for a verbal over-the-counter valuation (for written valuations see p20) and the expert will often be able to give you a fairly accurate idea of what the piece might make at auction on the spot, although they might need to take it in for further research before deciding finally on the estimate. Remember that the figure they quote is the hammer price; from this the auction house will deduct a commission of around 10 percent, insurance, and photographic charges if the piece is illustrated in the catalogue. If you are selling ceramics of specialist interest you may be advised to wait for a suitable sale. The expert will also advise you on a reserve price (usually on or just below the low estimate), which prevents the lot being sold below market value should the sale turn out to be poorly attended.

DEALERS

Buying pottery and porcelain from a reputable dealer is one of the easiest and user-friendly ways to begin collecting art. Unlike buying at auction you do not have to make up your mind in an instant as to how much you are prepared to pay; in fact some dealers may even let you take things home on approval so you can decide at your leisure whether or not to buy. But how do you know if a particular dealer is "reputable" or not? A good indication of an honest and knowledgeable dealer is one who belongs to a trade association such as the BADA (British Antique Dealers Association) or LAPADA (London and Provincial Antique Dealers Association). If you see the logo of one of these organizations in a dealer's window it is a sign that he has agreed to adhere to a strict code of practice and that his stock and expertise have been assessed and approved by the association's committee.

BUYING FROM A DEALER CHECKLIST

- **CHOOSE A REPUTABLE DEALER**
- **ASK QUESTIONS ABOUT THE OBJECTS**
- **GET A DETAILED RECEIPT**

Once you find a friendly dealer it is worthwhile establishing a rapport with him, since he will often look out for special pieces for your collection and share the benefits of his experience with you. Learning from someone with first-hand practical knowledge of a particular subject can save you from making costly mistakes, and a dealer will often give you an insight into the workings of the art market that would be impossible to learn from books and other more conventional sources.

Many general antique shops have odd pieces of pottery or porcelain in stock. Other dealers are more highly specialized and may deal only in ceramics of a certain kind – there are well-known dealers whose stock only includes 18th-century English porcelain, Japanese ceramics, Staffordshire figures, Art Nouveau ceramics or early English pottery. If you are a regular buyer or ask to be put on the mailing list of interesting dealers, you may be invited to exhibition previews, which give you the chance to buy before the public at large have seen the goods on offer, and in the case of a contemporary potter this may give you the opportunity to meet the artist in person and discuss his or her work firsthand.

Although buying from specialist dealers has advantages, don't ignore the less specialist sources. The more recently popular ceramics can still be found for very modest sums. Hidden in your local house clearance emporium you may discover a highly desirable Doulton figure or a stylish piece of Susie Cooper, and you will probably pay less for it than you would in a specialist's shop. However, if you buy from such sources you cannot expect the same level of guidance that you would receive in a specialist shop. Not everything bought from a junk shop turns out to be a bargain.

If you are buying the work of a relatively well-known pottery do try and price pieces in more than one source before you make an expensive purchase. Useful sources of reference include the annual *Miller's Antiques Price Guide* which publishes prices of ceramics sold at auction and in dealers' galleries along with a brief description. Bear in mind that although the price of a work of art at auction may well be lower than if you buy from a dealer, the dealer may well have paid to restore the piece to a high standard, and unlike buying at auction, when you buy from a dealer there are no other costs such as buyer's commission.

BUYING FROM A DEALER OR GALLERY

When you have some idea of the market value of the ceramics of your choice, don't haggle aggressively but do ask the dealer for his "best price" – you may be pleasantly surprised by a small discount. Before you buy ask the dealer as many questions as you can about the piece that has caught your eye. If possible find out where he bought it, because a good provenance adds to value. Has it been heavily restored? Are the marks "right"? Once you have paid make sure you are given a receipt with the dealer's name, trading address and a full description of the piece, including the name of the manufacturer if known, type of ware, date or approximate date of manufacturer and the price paid. Keep the receipt in a safe place as it will be useful for insurance purposes, and in the unlikely event as to there being a later dispute over authenticity.

"KNOCKERS"

Avoid selling to people who call unsolicited at your home, or put a note through your door asking whether you have any "antiques" for sale. Many of these so-called "knockers" are highly unscrupulous people whose aim is to trick you into parting with your property for much less than it is worth.

SELLING TO A DEALER

Eye-catching headlines about works of art that sell for hundreds of thousands of pounds at auction can give the misleading impression that selling in the saleroom is more profitable than selling direct to a dealer. Contrary to popular belief this is not always the case, and as with buying there can be significant advantages to selling to a dealer. In the first place, the sum you agree with the dealer is the sum you will actually receive; there are no other deductions, such as seller's commission, photographic charges and insurance. If you agreed, say, to sell a vase for £500 to a dealer, you would receive exactly that amount. Sell a vase at auction for the same hammer price and you will receive (after several weeks) a cheque for around £420 (assuming commission at 12½ percent, 1 percent insurance and VAT on the commission); photographic charges would reduce the figure still further.

Before you sell to a dealer it is always worth checking the value of your property with more than one person. The public nature of an auction room means that even if the cataloguer has under-valued your property the chances are that it will still realize its market value. When you sell to a dealer you have no such safeguard so it is wise to seek more than one opinion before you agree a final figure. Always try to sell to a specialist in the type of ware you are offering. An expert in Chinese ceramics will have a better knowledge of the current values of, say, a Canton jardinière than a general antiques dealer and so is more likely to offer you a realistic price.

Ceramics on display at a specialist dealer's gallery.

If the collection you want to sell is large, or the piece fragile and difficult to transport, showing dealers a photograph is a good way of gauging interest and avoiding inadvertent damage. Most dealers will not charge for a verbal valuation and many will visit your home to save you from moving the pieces.

FAIRS, MARKETS & BOOT SALES

Apart from buying at auction or a dealer's gallery there are numerous other places in which you can find pottery and porcelain to add to your collection. There are street markets such as Portobello Road, Camden Lock and Bermondsey in London and the famous Marché aux Puces in Paris which take place on one or two days every week and where dealers in a wide variety of art and antiques congregate to sell their wares. At these large markets you will find a combination of shops, covered stalls and some stalls that are no more than a trestle table on the street. Not surprisingly you are likely to come across an equally wide cross-section of dealers at a street fair; some of them will be members of trade associations, others will not, and while the unpopular English law of *marché ouvert*, which facilitated the legal sale of stolen property at markets, has now been changed, it is still up to you to satisfy yourself as to the dealer's integrity.

BUYING CHECKLIST AT FAIRS & MARKETS

- **COMPARE PRICES AND STOCK AT SEVERAL STALLS**
- **BE THERE EARLY TO FIND THE BEST BARGAINS**
- **TRY TO GET A RECEIPT**

Large annual antiques fairs have enjoyed a huge rise in popularity in recent years and can be one of the most appealing and safe ways for novice collectors to buy. You can buy a wide range of antiques and collectables at some of the large regular fairs. There is also a well-known ceramics fair (see p164) which focuses solely on pottery and porcelain. Most of the larger fairs are advertised in the national press and in specialist publications.

Fairs and boot sales are often a fertile source of bargains for keen collectors.

If you are unsure of what type of ceramics you want to collect a large fair is an ideal starting place since you have a concentration of dealers all assembled under one roof, offering a wide selection of pottery and porcelain. In addition, you often find unusual and rare pieces are exhibited for the first time at major fairs. Many dealers invest considerable sums of money in taking a stand at one of the large fairs so they will often "hold back" prize objects to make their stand more appealing.

The wide choice of dealers at specialist fairs also gives you a golden opportunity to compare prices of similar types of ware without wearing out your shoe-leather. Another advantage that most of the larger prestigious fairs offer is the added safeguard of a vetting committee. "Vetting" means that every piece of pottery or porcelain on a dealer's stand will have been examined and approved as authentic by a panel of independent experts. Some vetted fairs also operate a "dateline", which means that only pieces before a given date, say 1890, are eligible to be sold.

Most large fairs charge an entrance fee which may include the cost of a catalogue of exhibitors. These publications contain the business addresses and telephone numbers of

Hopeful buyers at a local car boot sale.

all the exhibitors and are a useful future source of reference to locate reputable dealers even if you don't buy from them at the fair. Provided a dealer lives within easy travelling distance of your home, don't forget that if you spot something that catches your eye but can't quite make up your mind, you can always visit the dealer after the fair at his or her premises to view the stock in a more leisurely fashion. Many dealers regard fairs as an essential method of contacting new collectors and are only too happy for you to visit them later, although there is of course no guarantee that they won't have sold the object you liked.

SMALLER FAIRS

There is a world of difference between these large, professionally organized events and the plethora of smaller antiques and collectors' fairs that take place in such diverse venues as church halls, schools and even car parks. Fairs of this type are usually advertised in the local press; you may pay an entrance fee to get in to them but there is generally no catalogue, no vetting committee and no guarantee of the authenticity of the goods on offer. It is therefore up to you to decide whether what you see is worth the money you are asked for it.

CAR BOOT SALES

Boot fairs have attracted huge attention recently as without fail each year a handful of hitherto unrecognized heirlooms are bought for a few pence and later sold for tens of thousands of pounds. Like smaller fairs, boot sales are held in a wide variety of venues, including car parks and farmers' fields, and are usually advertised in the local press. Although the chances of finding a very valuable object are, realistically, pretty remote, modest bargains are not unusual, and they can be good places to buy 20th-century ceramics for reasonable sums. The best way to find a bargain at a boot sale is to turn up early and to remember to take a torch if it's winter and the light is poor. Dealers are always first on the scene, sifting through the contents of cardboard boxes and tea chests as they are unpacked. If you can beat them to it you stand a good chance of finding something interesting and unusual that might just turn out to be more valuable than you expected.

PART 2

CARING FOR YOUR COLLECTION

ABOVE CLARICE CLIFF TEAPOT C.1935.
£400–600

LEFT AN ASSORTMENT OF BLUE AND WHITE PORCELAIN ATTRACTIVELY DISPLAYED ON WALL BRACKETS ABOVE A CHIMNEYPIECE.

SECURITY

One of the less pleasant aspects of collecting any type of antique is the increasing risk of theft. As the incidence of theft and burglary steadily rises it has never been more important to protect and insure your collection properly. Theft is the biggest single risk to antique collections and currently accounts for around 60 percent of all insurance claims for art; fire, the next single biggest risk, is responsible for around 25 percent of art claims.

Nevertheless the news is not all bad. If you are unlucky enough to be burgled a thief is far less likely to help himself to your antiques than your CD player or video, which are much easier for the average burglar to dispose of. Among the various antique collecting fields, silver, jewellery and small clocks and watches are the most portable and obviously valuable, hence are most vulnerable to theft. Although ceramics are more vulnerable to accidental damage than many other types of antique, thefts are still relatively rare. For this reason, provided you are prepared to shop around, you should be able to insure your collection at reasonable cost.

SECURITY CHECKLIST

- MAKE YOUR HOME SECURE
- PHOTOGRAPH VALUABLES BUT DO NOT KEEP THE PHOTORAPHS IN YOUR HOME
- KEEP AN INVENTORY OF YOUR COLLECTION

INSURING

If you have only a handful of relatively inexpensive pieces you may find it is most practical to include them on your general household contents. If on the other hand the value of your collection amounts to more than about £50,000 it would almost certainly be worth your while to insure with a specialist firm. (If they are worth £25,000-50,000 you should investigate both options.) The advantage of dealing with a specialist firm is that the various risks associated with different types of antique are reflected in lower premiums. Specialist firms may have an annual charge of as little as 30p per £100 insured for antiques whereas if you include art in a general policy you may pay at least twice as much. The disadvantage of dealing with a specialist is that they are not really interested in smaller collections and many have minimum premiums of around £500 per year. If you need help in choosing the best way of insuring and protecting your collection the Council for the Prevention of Art Theft (see p164) will be able to advise you further.

VALUATIONS

Once you have shopped around to find the best insurance policy, you will probably find the insurance company requires a written valuation of the ceramics (and any other objects) in your collection. Most auction houses have departments that undertake insurance valuations; there are also specialist valuers (many of them auction house-trained) or you can also try a reputable local dealer.

Valuations are well worth having because in the event of a claim a recent valuation by a reputable source will make disputes with the insurance company over the value of an item less likely. When you take an object to an auction house or dealer for an informal valuation there is no charge, but you can expect to pay for a written insurance valuation. Once again

Over-the-counter valuations are normally free, but a fee will usually be charged for written insurance valuations.

costs vary widely and it is worth shopping around to compare prices.

As a general guide there are three ways in which valuations can be priced: firstly, as a percentage of the sum insured, usually between ½ and 1½ percent. Unless you only need a valuation for one or two items this is usually the most expensive way to pay. Secondly, the charges may be based on how long it takes a valuer to assess your property. As a rule a well-qualified valuer will be able to catalogue and value around 100–300 objects a day. Finally, you may find you can agree a fixed rate for the valuation.

Whichever method you opt for, remember to negotiate before the valuation is made. Make sure whoever you select for the job is properly qualified and has sufficient expertise in the subject or you may find, in the event of a claim, it is impossible to replace something for the specified amount. If, say, you have a highly specialized collection of early English pottery you would be well advised to use a company with an expert in the field (such as a larger auction house) rather than consulting a general ceramics dealer who may have little knowledge of current market prices.

Whoever you choose for the job, make sure that the valuation is acceptable to the insurer. The valuation document should include a full description of the object. In the case of ceramics this would involve the maker, the type of object, the material (earthenware, porcelain, stoneware etc.), the date or approximate date, the size (height before width) and the value for insurance purposes.

Before the valuer begins you should agree with him the level of the valuation. This depends largely on where you would go to replace the objects in the event of their loss. Would you go shopping in an expensive city gallery or at a local auction room? Obviously the higher the sum placed on a piece, the more expensive it will be to insure, but at the very least the figure should be about 20 percent more than what you would expect to get should you sell at auction, in order to allow for auction house commissions. Don't forget that if you under-insure you could find yourself unable to replace lost items satisfactorily.

SECURITY

The vast majority of household burglaries are opportunistic and you can reduce the likelihood of a break-in by making your home as uninviting to burglars as possible. Local crime prevention officers will give you free advice on the best security measures to take. Good locks on doors, windows and security lights are all useful deterrents and if an alarm is necessary don't forget that the cost of installing it can often be offset by a reduction in insurance premiums – so make sure you tell your insurance company if you decide to install one.

STOLEN PROPERTY

One of the fundamental ways in which art and antiques differ from electronic equipment, however valuable, is that for many collectors art holds not only commercial but also sentimental value and very often a cheque is a poor substitute for the loss of a much-loved piece of pottery or porcelain. In the unfortunate event of theft there are, however, several measures you can take to help recover your stolen property; all of them depend on having photographs of the missing objects.

The police have a computer database run by the Art and Antiques Squad into which details of stolen objects can be entered to give police throughout the country instant information on such property. The Art Loss Register is another computerized database containing details of stolen property to which dealers, auctioneers and insurance companies subscribe. Dealers and auction houses use the system to check that objects they are offered are not stolen. The cost of including a stolen object on the register is around £20 and the success rate is impressive. There are also specialist publications which can help you trace stolen property. *Trace*, a British magazine in which you can place photographs of stolen property, is distributed nationally and internationally to auction houses and dealers. *IFAR Reports* is a similar publication in the United States.

DISPLAY, CARE & RESTORATION

Of all types of antiques ceramics are among the most vulnerable to damage. If you have recently bought an old piece of pottery or porcelain, it may well have cracks or chips and you may wonder how much, if any, restoration it requires. Modern technology has resulted in the development of restoration techniques of extraordinary effectiveness; not only is it now possible to repair cracked and broken objects with the use of new adhesives, it is also possible to replace even large missing areas with a body that will almost match that of the original.

In the case of serious damage the extent of the restoration is, however, a matter of opinion and something that is worth discussing at length with a skilled restorer. In general the modern approach to restoration favours keeping work to a minimum and avoiding respraying the surface to disguise repairs at the expense of the original decoration.

RESTORATION, HANDLING & DISPLAY CHECKLIST

- USE A PROFESSIONAL RESTORER
- AGREE ON THE COST AND HOW MUCH RESTORATION YOU WANT BEFORE THE WORK BEGINS
- REMOVE LIDS BEFORE HANDLING
- DO NOT LIFT PIECES BY HANDLES
- MAKE SURE HOLDERS AND STANDS ARE THE CORRECT SIZE AND SECURELY FITTED

Always use professional restorers. Restoration is a highly skilled art; an unqualified restorer can cause untold damage to a precious piece of pottery or porcelain and greatly reduce its value. You can find a local qualified restorer through one of the various trade associations (see p164). The restorers that are registered with one of these organizations are professionally qualified and will advise you on the necessary conservation and restoration of your purchase as well as help you to care for it correctly once you take it home.

CLEANING CERAMICS

Any type of ceramic on open display is bound to build up a residue of dirt and dust over time; and while this may not threaten the well-being of the object it is obviously unsightly and needs to be dealt with periodically. Different types of ceramic bodies need to be cleaned in various ways. Although hard-paste porcelain and

Restoring damaged ceramics using modern resin and pigments.

Modern restoration techniques and materials mean that ceramics can be restored increasingly effectively.

high-fired stonewares, despite their apparent fragility, may safely be washed by hand in mild soapy water (unless they have been restored), unglazed ceramics, however, should not normally be exposed to water and should only be dusted. The body of most unglazed ceramics is porous and if you wash them the dirt will soak into the body and cause discoloration. Low-fired earthenwares and other soft-paste types of porcelain and bone china may also be discoloured and damaged by washing, but may usually be wiped gently with a damp cloth. If in doubt about how to deal with a certain type of ceramic do not be afraid to consult a specialist dealer or restorer for advice.

HANDLING CERAMICS

There is no doubt that damage is most likely to occur while ceramics are being handled. However, examining pieces "hands on" is an essential part of learning about them and enjoying them and provided you observe a few simple guidelines there should be no need to feel nervous. When picking up objects with lids, for example, teapots or covered vases, always remove the lid first, preferably without picking it up by the knop which is vulnerable to damage and may be insecure. Lift hollow objects by supporting the body gently using two hands; do not pick up pieces by their handles – again these are often weakened by use. When replacing an object after examining it put it down gently, well away from any other objects on the surface. Try to avoid touching gilding which is easily rubbed and worn through handling.

DISPLAY

However beautiful a piece of pottery or porcelain is, if it is poorly displayed you will considerably diminish its impact. In order to display ceramics effectively they usually need to be securely supported, and placed somewhere they can be enjoyed – out of the way of danger from accidental knocks.

Old plates are often bought with rusty wire hangers still in position; these are an extremely undesirable method of display – as well as being unreliable, they can exert excessive pressure on the body, causing cracks and chips. Remove wire hangers by cutting through the metal on the reverse. Do not attempt to bend the wire from the front as this can chip the rim. If you wish to hang plates or dishes on a wall the best method is to use an acrylic hanger; these are transparent, adjustable and do not have any sharp points that could scratch the surface of the dish. Modern plastic-covered wire hangers in the correct size are suitable for more robust types of ceramics since they have spring adjustments which mean they will not press too tightly on the body. If you wish to display ceramics such as bowls, cups and saucers, dishes or plates on a flat surface like a cabinet shelf or a table, there is also a range of extremely well-designed acrylic stands; again it is imperative that the stand should be exactly the right size to give proper support. Whichever method of display you choose make sure that the ceramics are well protected from vibrations which can cause them to move and fall down or knock into one another.

COLOUR

POTTERY COLOURS

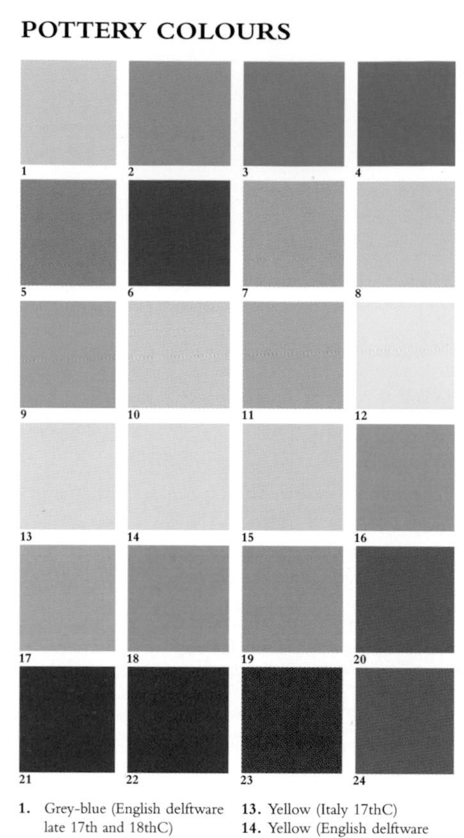

1. Grey-blue (English delftware late 17th and 18thC)
2. Turquoise (Iznik, Turkey mid-16th–end 17thC)
3. Mid-blue (English delftware late 17th and 18thC)
4. Mid-range cobalt (Venice and Urbino *istoriati*, Italy 16thC)
5. Deep blue (Faenza, Italy c.1525–50
6. Dark blue (Nevers, France 1650–1700)
7. Greyish-cobalt (Deruta, Italy c.1520–50)
8. Turquoise (Italy late 15thC, English delftware 17thC)
9. Green-blue (Urbino *istoriato*, Italy c.1525–75)
10. Grey-green (Castelli, Siena, Italy 18thC)
11. Grey-green (English delftware 18thC)
12. Yellow-green (Urbino, Italy c.1520–50)
13. Yellow (Italy 17thC)
14. Yellow (English delftware late 17th and 18thC)
15. Ochre (Deruta, Italy c.1500–25)
16. Ochre (Holland early 17thC)
17. Sienna (Deruta, Siena, Italy early 16thC)
18. "Sealing wax red" (Iznik, Turkey mid-16th–end 17thC)
19. Lustre (Spain 15th–16thC)
20. Ruby lustre (Gubbio, Italy c.1500–55)
21. Red (France, Holland, London 18thC)
22. Puce "purple manganese" (Strasbourg, France c.1750–1800)
23. Manganese brown (Italy, Spain 15th–16thC)
24. Stone (Urbino, Italy c.1525–75)

PORCELAIN COLOURS

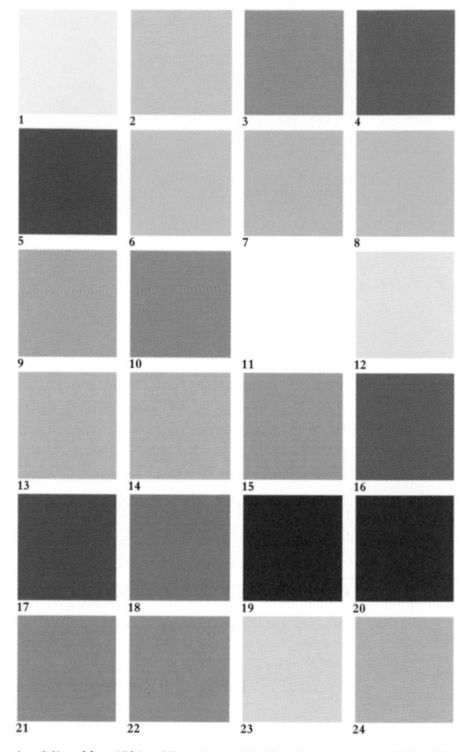

1. Misty blue (China Yingqing 12th–14thC)
2. Turquoise (Meissen c.1740–70, Derby c.1770)
3. *Bleu céleste* (Sèvres, from 1753)
4. Persian blue (Chinese 16thC)
5. Dark cobalt (Ming & Qing 14th–18thC)
6. Turquoise (Chelsea Gold Anchor period)
7. Emerald green (Bow 1660–1770)
8. Apple green (Worcester c.1770, Sèvres c.1756)
9. Early Meissen green (1720–40)
10. Brunswick green (English factories, early 19thC)
11. Lemon-yellow (Meissen c.1730–50)
12. Egg yolk (16thC Chinese)
13. Brownish yellow (Arita, c.1650–1700, late Ming *wucai*)
14. Tan (German, Swiss 1760–70)
15. Rich terracotta (English Neo-classical wares)
16. Russet (Fürstenberg c.1760–70)
17. Dark brown (German 18thC)
18. Iron-red (All factories)
19. Purple (Meissen, 1740s)
20. Claret (Vienna 19thC)
21. English puce (Longton Hall 1749–60)
22. Puce (German factories, mid-18thC)
23. *Rose Pompadour* (from c.1757)
24. Lilac (Meissen, 1750s)

The colours or enamels with which pieces of pottery or porcelain are decorated provide an important clue as to their date and place of origin. Certain colours are associated with particular factories; others were enormously fashionable at a particular time. Rare colours can add greatly to the value of a particular piece. The colours shown above are an approximation of the most common colours used by certain manufacturers and the key gives an indication of when and where they were mostly used.

PART 3

IDENTIFICATION

ABOVE A BERLIN GOLD GROUND CABINET CUP AND SAUCER C.1820. £3,000–4,000

CLASSIFICATION

Ceramics are primarily classified by the type of body – the materials – from which they are made, and secondly by the type of glaze used. The body of a glazed piece can usually be seen on the base; on most pieces the glaze does not entirely cover the footrim but leaves the underlying body partly visible.

BODY

The materials from which ceramics are made can be broadly divided into three families:

Pottery Made from an earthenware body. Depending on the minerals the clay contains, pottery may be white, buff, brown, red, grey or black in colour; it is coarser in texture than porcelain and the body is opaque.

Stoneware A fine water-resistant body created by firing clays at a high temperature. Some stonewares are slightly translucent and can sometimes resemble porcelain.

Porcelain A mixture of china clay and petuntse (hard paste), or crushed glass or quartz and white clay (soft paste). It is a fine material, that gives a musical note if gently tapped, and may be semi-translucent if held to the light.

POTTERY

The earliest ceramics were made from earthenware, that is, clay fired at a relatively low

English delft flower tub c.1690 – an example of English tin-glazed earthenware pottery. £2,500–3,500

Doulton stoneware jugs made c.1870. £150–250 each

temperature of under 2,200°F (1,200°C). Unglazed earthenware is porous and in order to hold liquids has to be glazed. The types of earthenware commonly seen today include Islamic pottery, Italian and Spanish maiolica, French and German faience, delftware, slipware, Whieldon ware, Staffordshire pottery, creamware, Prattware, pearlware and majolica.

STONEWARE

First used by ancient Chinese potters of the Shang dynasty, stoneware was made from a type of clay that could withstand firing to temperatures of 2,500°F (1,400°C). At these high temperatures components of the clay melted, forming a water-resistant body that was extremely robust. Western stoneware was independently developed in Germany in the Middle Ages. There are two distinct types of stoneware: the first is a fine, white-bodied variety like Sieburg and later Staffordshire. The second, a grey-bodied ware sometimes covered in brown salt glaze which looks pitted like an orange skin, was produced in Cologne, Westerwald and Fulham, London.

PORCELAIN

Porcelain bodies are broadly divided into two categories: hard paste and soft paste. Familiarity with the different types of body can help identify unmarked pieces.

Hard-paste porcelain Has a smooth texture resembling icing sugar. First discovered by ancient Chinese potters hard-paste porcelain was subsequently made by Meissen in Germany and in other European factories throughout the 18th century.

Soft-paste porcelain Is slightly more granular and chips may have a rougher appearance compared to hard paste. Developed in Italy from the 16th century in an attempt to replicate true Oriental porcelain, the colour of the soft-paste body varies between pure white and grey, depending on the minerals used. Colour can also help with identification since certain factories are associated with bodies of a particular shade.

BONE CHINA

Bone china was supposedly developed by Josiah Spode at the end of the 18th century. It is a less expensive method of porcelain making containing calcified bone and is the most commonly found type of porcelain made in England after c.1820; it was produced by Spode, Derby, Rockingham, Coalport, Minton and others. The body of objects made from bone china is pure white in appearance.

GLAZES

Glazes are used to make a porous body watertight and also to decorate a piece. They may be translucent, opaque or coloured. A variety of different glazes were used, each identifiable by its distinctive characteristics.

POTTERY GLAZES

On pottery there are three main types of glaze: lead, tin and salt.

Lead glaze Used on most European earthenwares including creamware and pearlware as well as Chinese Tang ware. Lead glaze is transparent and shiny in appearance and may be coloured with metal oxides.

Tin glaze A glaze containing tin oxide which gives an opaque white appearance. Tin glaze is found on maiolica, faience and Delftware where it was used to give the earthenware body a pure white finish resembling porcelain. Tin glaze was applied to the fired body, which was then decorated with enamels, and finally coated with clear lead glaze before the piece was returned to the kiln for refiring.

Salt glaze A method commonly used for stoneware where salt was thrown into the kiln during firing at a high temperature. The sodium in the salt fused with silicates in the clay to form a glassy surface. Salt-glazed wares were made in Germany, England as well as the United States. Some salt glaze has a distinctive pitted surface (see p26), which can be a help with identification.

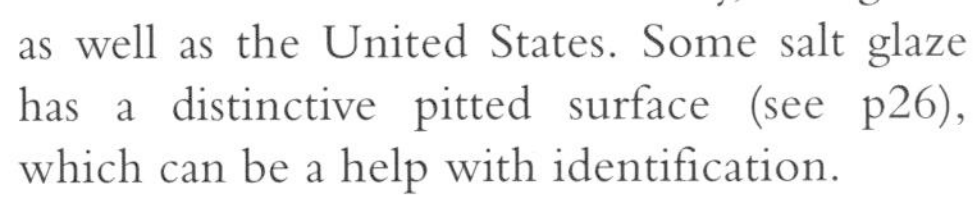

Bow soft-paste porcelain figures of New Dancers 1760–1765. £2,500–3,500

PORCELAIN GLAZES

Glazes on porcelain vary according to the factory concerned and often have a highly distinctive appearance ranging from matt to glassy, opaque to translucent. Glazes can also give porcelain a characteristic tinge of colour; some glazes appear to be pure white while others can seem bluish, yellowish, greyish or greenish in appearance.

Hard paste The glaze typically has a glassy thin appearance.

Soft paste The glaze typically is thicker, softening detailed modelling and pooling in the curves and crevices of a piece.

DECORATION & COLOURS

Ceramics have throughout the centuries been decorated in an extraordinary variety of ways. The technique, subjects and style of decoration all give useful clues as to the origins and age of a piece. Decoration adds considerably to value: certain subjects and styles are particularly sought after and command a premium.

TECHNIQUES

Underglaze blue One of the most popular decorative techniques used on both pottery and porcelain. The method involved painting the design using cobalt oxide on to the biscuit body before it was glazed. First used on 13th-century Persian pottery, the method was later adopted by the Chinese to decorate porcelain. Underglaze blue decoration was a continuing tradition in European pottery and from the late 18th century in England factories such as Caughley, Liverpool and Worcester decorated wares with underglaze blue transfer prints.

High-fired colours A range of colours that were able to withstand firing to a high temperature of 2,200–2,400°F (1,200–1,300°C) and could be applied on the surface or given a final coat of pure lead glaze in the same way as underglaze blue. These colours were only available in a limited palette: cobalt blue, manganese/purple, copper green, antimony yellow and ochre. They are often known as *grand feu* enamels. They were used in Spain, Italy, France, Holland, Germany and England.

Lidded stoneware bowl by Bernard Leach, decorated with an incised sgraffito design. £4,000–6,000

Overglaze enamels Painted decoration, also known as *petit feu* enamels, made from a mixture of metallic oxides and molten glass, applied after the piece had been glazed, which was then refired at a lower temperature in a muffle kiln, causing the glass to fuse with the glaze. Overglaze enamels may be less hard-wearing than underglaze colours but were available in a far wider range of colours.

Slip decoration A technique for decorating pottery using solutions of variously coloured clays. The piece was typically dipped in a slip (see p62) and then designs were created using slips of contrasting colours which were trailed, combed or scratched to create a decorative effect. The technique was much used in England from the 17th century and slips were also used by Doulton and Moorcroft in the 19th century. The tradition was revived by studio potters such as Bernard Leach in the 20th century.

Sgraffito This technique involves scratching into the body of the piece before firing to create an incised design. Sgraffito has been used on earthenware and stoneware, and remains popular with studio potters such as Lucie Rie. Scratch blue was a type of incised decoration where sgraffito designs were rubbed with cobalt oxide or other coloured pigments before the piece was fired. The technique was used on early English stoneware and by the Doulton factory in the late 19th century.

Lustre decoration An ancient decorative glaze made from metallic oxides that creates a distinctive metallic sheen. Lustre glazes were applied to pieces after glazing and firing, and were then refired at a lower temperature. The technique was used in Spanish pottery probably from the 15th century onwards and it later became a popular finish with English potters in the 19th century.

Gilding Gold may be applied in a variety of different ways: by mixing it with oil, mercury or honey and then refiring the piece at a lower temperature. Alternatively, liquid gold can be applied cold. Mercury gilding was popular on the Continent and looks distinctively bright and shiny; honey gilding was

20th-century Wedgwood lustre vase designed by Daisy Makeig-Jones. £2,000–3,000

used in England from c.1750 and looks slightly dull in appearance; oil gilding was used on Staffordshire pottery and early English porcelain, and pieces finished in this way are especially vulnerable to wear.

APPLIED DECORATION

Sprigging A technique involving making relief decorations from separate pieces of porcelain, clay or moulded slip and applying it to the body of the piece. Meissen and many other leading factories applied flowers to ornamental objects. In the case of Wedgwood, applied decoration was done in slips made in colours which contrasted with the body creating a cameo effect.

Pâte sur pâte A refined form of applied decoration, *pâte sur pâte* is the term that is given to the cameo-like decoration created by building up layers of translucent white porcelain slip on a contrasting body and carving back to create a low-relief design; the technique was used by the Sèvres factory in the 19th century and also at the Minton and Royal Worcester factories.

Jewelling A decorative technique first used by Sèvres in the late 18th century but popular with English factories such as Copeland, Minton and Worcester in the 19th century. The method involved decorating porcelain with tiny raised dots of coloured enamels over gilding, creating a rich gem-like effect.

PRINTED DECORATION

Transfer printing: A method first used in the mid 18th century to make decorating ceramics less costly. Transfer-printed decoration involved taking an impression from an engraved copper plate on to transfer paper and applying it to the ceramic body. Transfer printing may be overglaze or underglaze, and was used on both porcelain and pottery. Early examples were printed in monochrome, sometimes with hand enamelling. Colour printed designs were developed in the 1840s by F. & R. Pratt. Patterns had to be specially amended for different shaped wares and interesting variations are sometimes seen.

Bat printing A technique used in the early 19th century as an alternative to transfer printing. The method involved using flexible glue bats instead of transfer paper as a way of transferring a design from an engraved copper plate to the pottery or porcelain body.

Staffordshire pottery meat dish decorated with a coloured transfer print of Windsor Castle, c.1840. £400–600

MARKS

Throughout the centuries ceramics have been marked by their manufacturers and these marks can provide useful information about the maker. Many factories such as Sèvres, Chelsea and Derby changed their marks throughout their long histories and these changes can also help with dating a piece. There are numerous specialist publications available (see pp168–9) that will provide you with invaluable detailed information about the marks in your particular collecting field.

SPURIOUS MARKS

Marks cannot always however be taken at face value. In Chinese ceramics pieces were frequently marked with earlier reign marks as a sign of respect for the skill of earlier potters. The famous Meissen crossed swords mark was imitated by numerous factories throughout Europe; Worcester's crescent mark was also used by Bow; and numerous other factories capitalized on the popularity of successful factories by copying designs and imitating marks.

TYPES OF FACTORY MARK

Marks found on ceramics commonly include a factory mark. This may be a symbol, such as the anchor of Chelsea; initials, monograms and letters, such as the letter "Z" used by the Zurich factory; or the name of the factory written as a word or signature: Minton for example is marked in this way. Marks may be applied to a piece in a variety of different ways: they may be painted by hand in underglaze or overglaze enamels, incised by hand, impressed with a stamp, or printed.

OTHER MARKS

Other marks on ceramics may refer to the various individuals involved:

- the designer
- the decorator
- the modeller (in the case of figures)
- the gilder.

There may also be a pattern serial number which provides useful information when the factory's records have survived.

PRINTED MARKS

Printed marks were used by most major English factories of the 19th century. Small changes in the wording can help with dating:

- From c.1820 – marks with a royal coat of arms began to be seen
- From c.1896 – marks with the word "England" or other country's names, following the McKinley Tariff Act of 1891
- 20th century – marks with the words "Made in England", or "Bone china".

REGISTRATION MARKS

A diamond registration mark was used on English ceramics between 1842 and 1883; letters and numbers in the corners of the diamond refer to the date of manufacture. After 1883 "Rd No" (the registered number) followed by the number itself was used.

FAKES AND COPIES

Copies and imitations of earlier styles have been made throughout the centuries. Most early imitations were not intended to deceive but show admiration for earlier designs. Deliberate faking with the intention to mislead started in the 18th century with the decoration of Sèvres porcelain blanks by independent

A Samson rabbit tureen c.1900, imitating a Chelsea Red Anchor original. £1,000–1,500

decorators who marked them as though they were genuine Sèvres. During the 19th century many imitations of earlier pieces were made, often honestly marked by their maker.

EDMÉ SAMSON

The most prolific imitator of early ceramics, Samson made copies of Oriental, German, French and English ceramics from the mid-19th century. Much of Samson's output was marked with an entwined or squared "S", crossed batons or the "Worcester" seal mark. As values have risen in the 20th century these imitations have become collectable in their own right. Problems arise when unscrupulous dealers remove the Samson mark.

ORIENTAL CERAMICS

- Imitation "Tang" pieces were produced in the early 20th century after the genuine pieces became more widely available following excavations in China. The honest marks on reproductions were often removed by dealers at a later date.
- Imitation pieces were made in the Far East from c.1920; these are sometimes identifiable by heavy potting and poor-quality calligraphy.
- Emil Samson, son of Edmé, made good-quality copies of later Chinese and Japanese porcelain, usually marked with a distinctive square-shaped "S".

ITALIAN POTTERY

- Many copies and fakes of Italian maiolica were made in the 19th century; these are sometimes identifiable by their loosely painted margins, or by overly important borders painted in colours that are too bright.
- The Cantagalli factory made numerous imitations of early Italian pottery and marked them with a cockerel.

GERMAN CERAMICS

- Earlier stoneware styles were revived in the 19th century – copies are discernible by their flat rather than concave bases.
- Copies of 18th-century Meissen were made throughout the 19th century by Samson and others; wares were also imitated by minor factories in the 18th century.
- Samson's glaze is often too glassy and smooth and the modelling lacks definition.

The Cantagalli factory mark found on late 19th- and 20th-century imitation maiolica and Iznik ware.

FRENCH CERAMICS

- Copies of French faience were made in the 19th century – sometimes identifiable by their stiffly painted decoration.
- Nantgarw produced imitations of French-style porcelain in the early 19th century, but these are highly collectable.
- Coalport and Minton made imitations, mainly of larger Sèvres pieces.
- Many small Paris makers made imitation Sèvres in the 19th century, often with spurious marks.
- Copies of Chantilly and St Cloud were made in hard paste by the Samson factory and others in the late 19th century.

ENGLISH CERAMICS

- Elaborate slipware has been faked at least since the 1920s. Stoneware owl jugs have recently been faked. Look for signs of wear.
- Copies of early stoneware and agateware were produced in the 1920s.
- 19th-century animal tureens imitating Chelsea were made by Samson.
- Worcester's 18th-century wares were copied in Paris in the 19th century.
- Some of Derby's ornamental wares were copied by the Samson factory.

PART 4

CERAMIC FILE

ABOVE A 19TH-CENTURY JARDINIÈRE BY WILLIAM MOORCROFT IN THE CORNFLOWER PATTERN. £4,000–6,000

LEFT A COLOURFUL ASSORTMENT OF 19TH CENTURY ENGLISH POTTERY AND PORCELAIN CAN MAKE AN EYE-CATCHING DISPLAY.

Isolated from the rest of the world, China was responsible for almost every major development in pottery and porcelain, and throughout the centuries Chinese ceramics have exerted a profound influence over those made in the Far East, Middle East and Europe.

The earliest wares evolved between c.3000 and 1500BC in the Neolithic cultures of the Shandong Peninsula, the Gansu corridor and elsewhere. The pieces that have survived were made as funerary objects and were unglazed and simply decorated with pigments derived from the earth.

During the Shang dynasty (1700–1027BC, China's first historic period, pottery became more and more refined and shapes were based on those of metal artefacts made at the same time. The first lead and feldspathic glazes were not developed until the Han dynasty (206BC–220AD), when high-fired stonewares were made. By the Tang dynasty (618–906AD), commerce and culture began to make themselves felt in the development of ceramics. Outside influences from central Asia gradually infiltrated China, altering the appearance of ceramics, and Chinese wares began to spread throughout the Middle East carried by Arab traders. Potters rediscovered white clay which led ultimately to the development of porcelain.

Considering the great age and historical importance of these early ceramics they are still available at relatively modest prices. Yue wares from the Han dynasty are available from £300–500 and celadon from the Yuan dynasty can also be comparatively inexpensive.

EARLY CHINESE POTTERY

There are few extant examples of pottery made before the Han dynasty (206BC–220AD). A surprising number of Han funerary pottery wares have survived. These were made to accompany the deceased person to the after-life. They are highly stylized and primitively modelled and include figures, animals, models of buildings and vessels. During the Tang dynasty figures became increasingly refined and naturalistically modelled; some objects were unglazed, others were decorated with pigment or straw-coloured glaze, or polychrome glaze known as *sancai* (three-coloured glaze). Horses and camels, are among the most sought-after objects made at this time. Unglazed or straw-glazed figures tend to be less expensive than those with three-coloured glaze.

▲ **FUNERARY JARS**
This jar made in the Neolithic Yangshao culture is decorated with simple black and red pigment and the swirling pattern with chequered roundels is typical of this type. The lower part was left undecorated because the jar would have been partly sunk in earth in the tomb. Incredibly, jars of this type, made nearly 4,000 years ago, are not particularly rare and the value would only be £1,200–1,800.

WHAT TO LOOK FOR

Examine the glaze carefully for signs of authenticity. Points to note are:

- translucent glaze that shows imperfections in the body beneath
- crackling in the glaze
- unevenness – the glaze usually pulls to the base
- an element of iridescence caused by being buried in the soil.

▶ MODEL BUILDINGS

Models of contemporary buildings were usually included in tombs; this is a watchtower made in the Han dynasty. Other buildings you might find from this period include: storehouses, barns, well-heads, gatehouses and granaries. This is quite large (13½in [34cm] high) and therefore worth £2,000–3,000; smaller objects cost from £300.

◀ TANG POTTERY

A heavy baluster jar from the Tang dynasty (618–906); the heavy form and small rim are typical characteristics of this period and the decoration is created in a similar way to tie-dye in a standard palette of greens, brown and white. £8,000–10,000

BEWARE

Fakes of horses of this type abound and have been made since the 1920s. The best fakes can be difficult for novice collectors to identify – handling the genuine article is the best way to learn to spot them.

▶ HORSES

Figures such as this unglazed horse were made in the Tang dynasty from different sections: the neck, legs and rump were all probably separately moulded and then joined together. Value depends primarily on the impact of the object. Figures should be crisply modelled and powerful; unusual poses can also greatly add value. This is a conventional pose and unglazed, but still worth £2,500–3,500; if the horse had a foot raised or was slightly rearing it would be far more valuable.

CHINESE STONEWARES

Stoneware is made by firing clay at very high temperatures (2,500°F (1,400°C) and above) during which process part of the clay melts and hardens creating a robust, water-tight and often slightly translucent body. Stoneware began to be produced in China in the Shang dynasty (1700–1027BC) and by the Han dynasty (206BC–220AD) had become refined enough to be considered more than just a utilitarian material. A range of small objects for the educated élite, such as water-droppers (used in calligraphy), oil lamps and vases, began to be made, often in the form of animals like frogs and rams. These pieces were often decorated using a roulette wheel with incised or carved repeating patterns.

Celadon, the most famous type of Chinese stoneware, was produced in two main centres. A distinctive olive-green coloured celadon was produced in the north-west region of China in Henan and Shaanxi during the Song dynasty (960–1279). After this time the southern celadons of Longquan (in Zhejiang) became the main centres of production and wares were made in a soft blue-green colour.

Among the other important types of stoneware made in China was Cizhou ware, which was produced mainly in the northern provinces of Henan, Hebei and Shaanxi. Pieces are often highly decorative, characteristically adorned with painted, incised or punched designs which proved to be inspirational in the 20th century for modern studio potters such as Bernard Leach (see p152). Jun wares are another highly sought-after type of Chinese stoneware. Produced in Linru, central Henan, these pieces were typically heavily potted and thickly glazed in various shades of blue, sometimes with splashes of purple and, more rarely, celadon-like green.

Although larger examples of Chinese stoneware are keenly collected and often attract very high prices, small, less rare items such as teabowls are readily available for modest sums and can form an interesting and visually appealing collection.

◀ **YUE WARE**
This oval jar dates from 265–317AD and was probably inspired by metal vessels. It is typically glazed in an olive-green colour; often (as pictured here) the glaze does not cover the entire outer surface and where the base has been left exposed the body has fired to a reddish colour in the kiln. £1,500–2,500

▶ **NORTHERN CELADON**
Bowls are the most commonly seen examples of northern Chinese celadon, and this one (12thC) is of typical conical form with an olive-green glaze and moulded floral decoration. If the decoration is carved or includes figures the value increases. £1,500–2,000

◀ **JUN WARE**
This small shallow dish (12th–13thC) is one of the most common Jun shapes. Pieces such as this were fairly robustly potted, with a thick wedge-shaped footrim measuring about ¼in (0.5cm) across. Colours range from this pale lavender glaze, to a greener blue and dark blue and purple. The glaze was generously applied and dribbles on the underside are also typical of this type of ware. £1,500–2,000

▶ **BLACKWARE TEABOWLS**
This late Song teabowl made in the Fujian province has a typical form with a shallow ridge beneath the rim so it could be easily held. Bowls like this were first used in monasteries by Buddhist monks who introduced tea drinking to China. Glazes on blackware vary from black to this streaked dark brown glaze known as "hare's fur". £1,500–2,000

◀ **STEM BOWLS**
Stem bowls were used for ritual purposes and the earliest date from the Tang dynasty. This 15th-century Longquan example is typically coated in a thick glaze that softens the contours of the body, although around the rim you can just see the white body where the glaze has run during firing. £1,500–2,000

BEWARE
Celadon has long been coveted by collectors and has never really been out of production. Modern imitations of southern celadon pieces are in most respects, apart from their slightly mechanical and glassy appearance, very similar to the originals.

CHINESE PORCELAIN

Porcelain evolved gradually from the high-fired white wares made in China in the Tang dynasty (c.618–906), the earliest pieces were greyish or creamy and slightly gritty in appearance. By the Song dynasty (960–1279) forms were increasingly sophisticated, their shapes and decoration often inspired by flowers such as the lotus, chrysanthemum and peony. In the northern region of Dingyao potters began producing carved and moulded porcelain bowls and dishes, fired in coal kilns producing an ivory or buff-coloured appearance. Underglaze blue decoration on porcelain first began to be used c.1330 and most early wares were decorated in bands or registers of pattern. The overall effect is crowded and complex by comparison with the more balanced and spacious designs of early Ming. By the 15th century the first polychrome wares began to be produced although some of these later wares are not as skilfully painted as the renowned *doucai* porcelain of the Chenghua period (1465–87). Ding ware remains rare and expensive; imperial Ming porcelain is among the most valuable of all early Chinese ceramics. By contrast in southern Qingbai huge quantities of bluish and greenish bowls survive and these are still modestly priced.

◀ DINGYAO

The warm ivory tone and moulded floral decoration are characteristic of Dingyao. This porcellaneous stoneware dish dates from 1115–1234 and was glazed and fired upside down hence the metal-bound rim to protect the unglazed edge. The double gourd ewer characterizes the elegance of northern Song and was made 960–1127.
Dish £3,000–5,000;
ewer £4,000–6,000

MARKS

大明宣德年製

Marks first became used regularly during the Ming dynasty in the reign of Xuande c.1426 and usually comprise four or six characters. By c.1500 marks on Chinese porcelain are usually those of earlier reigns.

▶ CARVED DECORATION

Superior carved decoration increases the value of early Chinese porcelain. This 13th-century Qingbai (Yinqing) ewer for oil or wine is typically thinly potted and made of a chalk-white sugary body covered in a thin semi-translucent bluish glaze. The name Yingqing in fact means shadowy or misty blue.
£1,000–1,500

◀ **EARLY YUAN UNDERGLAZE DECORATION**
The decoration on early blue and white porcelain was usually divided into separate registers or panels, each of which could be completed by a different hand. The decoration on this Yuan *kendi* (left) and ewer (right) has several registers decorated with classic scrolls, plantain leaves and squared scroll bands.
£3,000–5,000 each

► **MING BLUE AND WHITE**
A late Ming blue and white saucer dish of the second half of the 16th century; the traditional twin fish design denotes marital fidelity and harmony. Note that the blue decoration has characteristic flecks which have been caused by oxidization through the glaze.
£3,000–5,000

◀ **MING POLYCHROME DECORATION**
The typical palette of blue, iron red, dirty yellow, green and manganese-brown was created by first painting the design in underglaze blue, firing, enamelling with the other colours and refiring. Although designs such as those used on this *wucai* square dish (1573–1619) are rather confused, they have great charm and are highly sought after by collectors.
£2,000–3,000

CHINESE EXPORT

The large-scale export of Chinese ceramics to Europe began shortly after the arrival of the Portuguese in China in 1517. Thousands of pieces went to Lisbon as ballast in ships laden with spices, lacquer, silks and other luxury items. Throughout the 17th century the Dutch monopolized the trade with China and Japan, while the English dominated the export of porcelain throughout the 18th century. Early export goods are sometimes indistinguishable from those made for the home market, and it was not until the 17th century that the Dutch began requesting European shapes such as mustard pots, cuspidors (spittoons), coffee pots and narrow-necked jugs. By the 18th century there was also a thriving demand for armorial porcelain. The trade in Chinese export porcelain waned as European porcelain factories and creamware became increasingly competitive in the late 1700s. After 1735, instead of potting and enamelling wares in the potteries and shipping them to Canton (Guangzhou) for export, porcelain was dispatched from the potteries in the white (when it was less fragile) and enamelling shops were set up in the Canton area; vast quantities of decorative "Canton" wares were made throughout the 19th and 20th centuries.

Much export porcelain is still readily available at modest sums and you can find pairs of polychrome or blue and white plates for between £300 and 400.

▶ ***KRAAK-PORSELEIN***
The early export wares made in the Ming period were often decorated with panels filled with repeating flowers, animals, birds or figures and this distinctive type of decoration is known as *kraak-porselein*. (*Kraak* is a Dutch word for the type of Portuguese ship that transported this sort of porcelain.) This bottle is a good example of the characteristic simple style. £1,000–1,500

◀ ***FAMILLE VERTE***
Another popular palette for export wares was formed of translucent colours known as *famille verte* ("green family"); this replaced the old *wucai* palette and used overglaze blue and slightly raised enamelling. The decorative figure in a garden was a favourite theme in the Qing dynasty. £400–600

► **SHAPES**
The influence of French silver designs is reflected in this mid-18th-century hexagonal dish. The decoration is fairly dense and detailed and the blue underglaze decoration is very pure – later it became greyer and the decoration was less accomplished. £200–250

MARKS
Marks on Chinese export porcelain include:
- symbols such as a tripod, lozenges, lotus and artemisia leaf
- earlier reign marks.

▼ **CANTON *FAMILLE ROSE***
Canton porcelain of the 19th century is usually crammed with figures and detail, and pink, turquoise and gilding are predominant. This large vase is typically colourful. £1,200–1,600

▲ ***FAMILLE ROSE***
From c.1718 onwards porcelain in the *famille rose* palette was decorated using colours to which white had been added to make them opaque. Despite the slightly misleading term *famille rose* (meaning "pink family"), pink is not necessarily the dominant shade. The scattered decoration of this 18th-century dish is typical of the export ware of that period. £100–200

KOREAN CERAMICS

Among the earliest Korean ceramics to survive are unglazed stonewares produced from about 100AD. These were simply decorated with repeating motifs incised or punched into the surface. Celadon stonewares reflecting the influence of Chinese pottery rank among Korean potters' most beautiful achievements, and enjoyed a golden age from 1150–1250. Another important type of stoneware, Punch'ong ware, was similar to celadon but more elaborately decorated with stamped designs decorated with white slip (see p62). Porcelain was produced by Korean potters from the 11th century onwards, but although Korean porcelain often reflects the inspiration of Chinese design, it is usually identifiable by its idiosyncratic imperfections. Unfortunately, the paste tended to sag during firing and you will frequently find pieces of irregular shape or with splits and flaws.

Nevertheless Korean ceramics are highly collectable and even 19th-century examples can attract high prices. Surprisingly, the more affordable collecting areas include very early stonewares, which are often relatively modestly priced considering their age.

◀ KOREAN CELADON
Korean potters were the only manufacturers of early celadon to use inlaid decoration. Designs cut out of the clay body were filled with white or black slips; this 13th century pear-shaped bottle has a white floral and geometric design. Other celadon wares were decorated with moulded, stamped or carved designs. £2,000–4,000

DECORATION
A wide variety of techniques were used on Korean celadon:
- moulding or stamping
- carving
- incising
- inlaying incised designs with coloured slip.

▲ WARES
Porcelain objects for the scholar's table, such as brush droppers, brush holders and censers, were produced in Korea in the 18th and 19th centuries. Some pieces, such as this two-handled censer, are adorned somewhat casually with some sort of pierced decoration, which was very popular in the late Yi dynasty (16th–19thC). £1,500–2,000

▶ UNDERGLAZE BLUE DECORATION
This 19th-century blue and white globular bottle with a tall neck is painted with a typical soft greyish underglaze blue design of a coiled dragon. Other favourite motifs include bamboo, peony flowers, birds, landscapes and roundels.
£1,000–1,500

▲ KOREAN STONEWARES
This Korean stoneware bowl is decorated in olive-green on a white ground. The slightly angular bulbous shape was popular in the late 15th and 16th centuries, both in Korea and Japan.
£5,000–10,000

IDENTIFICATION

THE UNDERSIDE OF THIS 13TH CENTURY INCISED CELADON LOTUS BOWL HAS MANY OF THE IMPERFECTIONS YOU WILL FIND ON KOREAN CERAMICS.
£1,500–2,000

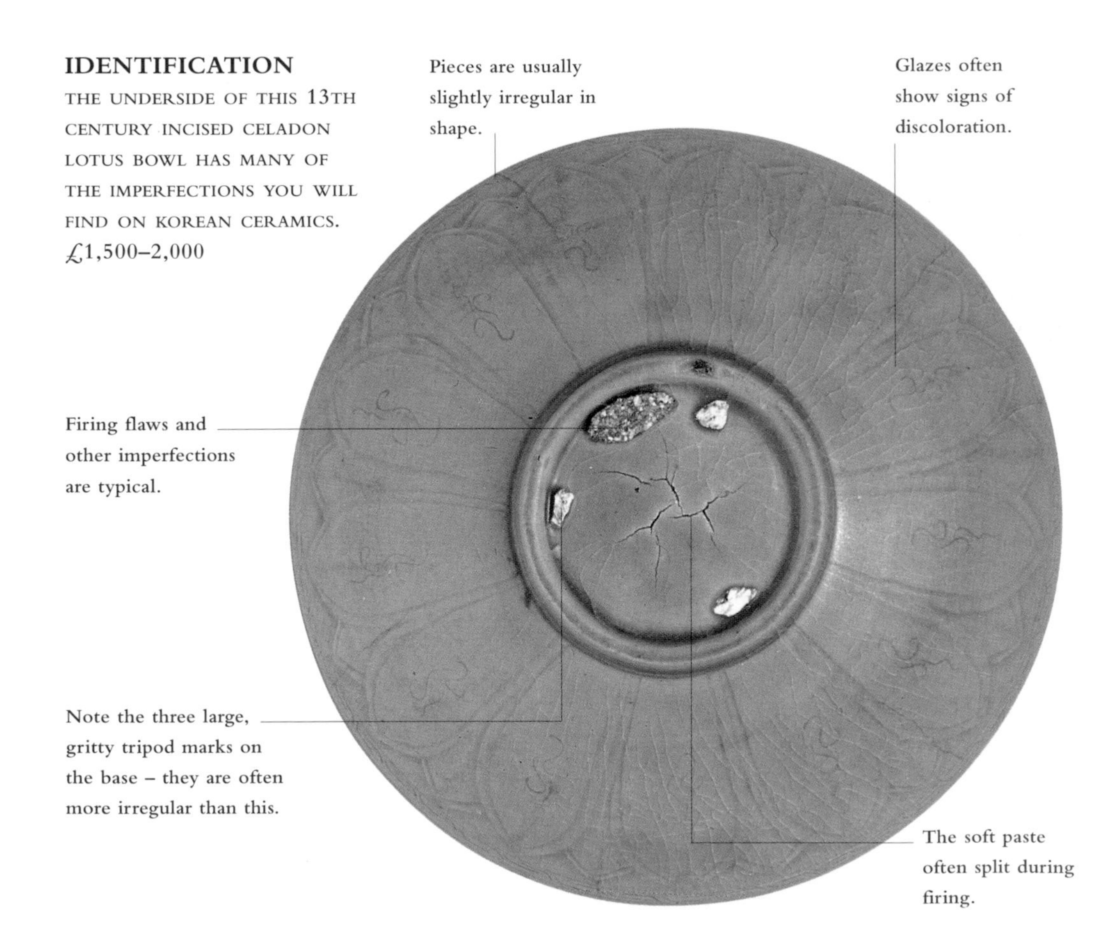
Pieces are usually slightly irregular in shape.
Glazes often show signs of discoloration.
Firing flaws and other imperfections are typical.
Note the three large, gritty tripod marks on the base – they are often more irregular than this.
The soft paste often split during firing.

EARLY JAPANESE CERAMICS

Stoneware has been produced in Japan from Neolithic times but unlike their Chinese counterparts the Japanese did not begin producing porcelain until the early 17th century. Many Japanese ceramics reflect the cross-fertilization of ideas from China mingled with distinctively Japanese elements. Stonewares show the influence of blackware and celadon in characteristic regional styles. Early porcelain wares were made in white or blue and white, and by 1650 the Japanese had established a market in export wares with Dutch traders. Kakiemon and Imari, the two most famous types of Jápanese coloured enamel decoration, were developed in the second half of the century. After the civil wars in China (c.1680) the Chinese began to produce much less expensive porcelain than the Japanese. By 1757 the Dutch had virtually ceased trading porcelain with Japan. The Japanese export market dwindled until its revival in the late 19th century. The vast majority of Japanese ceramics made for Western markets therefore predate c.1760 or were made after c.1860.

MARKS

With a few exceptions, the only marks on Japanese porcelain are borrowed from the Chinese reign marks of the Ming dynasty.

▶ ARITA PORCELAIN

Arita porcelain takes its name from the region of northern Kyushu, Japan where many kilns were centred. Vast quantities of blue and white wares, such as this *Enghalskrug* or narrow-necked jug, were produced for the export market. This example dates from c.1660–80 and was probably intended to be metal-mounted. £2,000–3,000

◀ DECORATION

Designs on late 17th- and early 18th-century porcelain are often stiffly painted because the decorators worked from crudely copied wooden patterns given to them by the Dutch. The shape of this *kendi* (c.1665) is also based on Chinese forms; although intended originally to hold purified water, export *kendi* were often used as hookah bases. £1,500–2,000

COPIES
Kakiemon style, colours and designs were copied throughout Europe in the 18th century by Meissen, Chelsea, Chantilly and many other factories.

◄ KAKIEMON
Named after a celebrated potter, Kakiemon porcelain is recognizable by its distinctive palette and often sparse decoration. Colours used include iron red, deep sky blue, turquoise, yellow, aubergine/manganese and black. Decoration is delicate and allows much of the white porcelain to show. The floral pattern on this lobed dish (c.1700) is asymmetrical – another typical Kakiemon feature. £3,000–5,000

▼ VASES
Vases like this (early 18thC) were popular export items and have a distinctive shape with straight conical sides on a ridged foot and a deep cover with flanges that protrude at a sharp angle. Covers may have finials in the shape of a tall flame, an animal or a figure. Most were originally from a garniture which included three vases and matching tall beakers that would have been displayed on chimneypieces or cabinets. £6,000–8,000

▲ IMARI
Elaborately decorated wares in a characteristic palette of blue, iron red and gilding are classified as Imari wares – the name comes from the port where such pieces were shipped. Imari porcelain may also include shades of greenish yellow, turquoise and manganese. This large dish (c.1700) is typically decorated with an elaborate design known as "three friends", showing figures among rock work. £3,000–5,000

LATER JAPANESE CERAMICS

The Japanese ceramics industry enjoyed a revival of popularity in the late 19th century when trading links with the Western world were reestablished. Most 19th-century ceramics repeat earlier styles often with complex shapes. As production increased, underglaze blue was often applied by stencil and decoration appears stiff by comparison with earlier wares. Satsuma stonewares enjoyed a heyday of popularity in the late 19th and early 20th centuries, and these highly decorative pieces decorated with a crackled creamy glaze, heavy gilding and subdued colours are widely available. Prices depend mainly on size and visual appeal of the decoration. From c.1900, another prolific and successful manufacturer of porcelain aimed at the European market was Noritake. This factory is still in production today but early pieces reflecting the influence of Art Nouveau and Art Deco are becoming increasingly sought after, and there is a specialist club for Noritake collectors.

◀ SATSUMA
This high-shouldered jar decorated with women and children gathering blossom has many features that are typical of 19th-century Satsuma stoneware including: a creamy wash ground; elaborate figurative subject – geisha or holy men are also favourite subjects; lavish gilding overlaid with a geometrical floral design. By this time many better-quality pieces of pottery were signed, as this is. A signature may add slightly to value. £2,500–3,500

▶ FUKAGAWA
Naturalism is a dominant theme in later decoration and although this heavy baluster jar is painted in traditional colours with a favourite theme – carp among a fishing net – the sense of movement is typical of the better-quality wares of the late 19th century. £1,000–1,500

MATERIALS

- Earthenwares produced for the home and export markets: Satsuma, Kyoto, Awara
- Porcelain for home and export: Arita, Imari, Fukagawa, Hirado, Seto, Noritake.

► **STYLE**
The decoration on later Imari porcelain, such as this large charger (c.1880) tends to be very dense. The colour balance changes subtly so that red rather than blue is generally the dominant shade. The vessel in front is a traditional 17th-century form, based on a wooden pail. Dish £600–1,000; charger £500–700

▼ **VALUE**
Later Japanese ceramics vary enormously in quality. These Satsuma censers (c.1900–10) are crowded with scenes of holy men painted predominantly in shades of gold and russet. Pieces such as this were exported in vast quantity to the West as decorative items rather than refined works of art. £400–500

MARKS
- Imari is unmarked or occasionally has an impressed stamp.
- Noritake has a fleur-de-lis motif.
- Good Kyoto and Satsuma often carry the potter's name.
- Fukagawa has a Mount Fuji mark.

► **NORITAKE**
In the early 20th century Noritake produced hand-painted wares at a price comparable to printed European ceramics. This plate, with its stylized floral decoration, reflects the influence of Art Deco. Noritake exported large quantities of wares to America. Many of its 1920s and 1930s designs were conceived in America by eminent designers such as Frank Lloyd Wright. £40–60

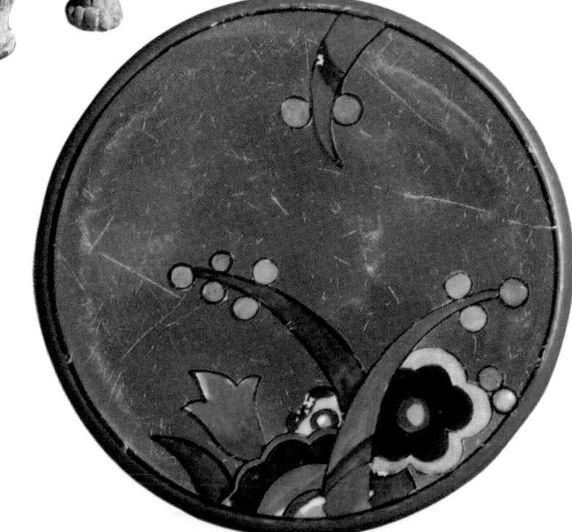

Islamic Ceramics

Richly diverse Islamic pottery has throughout the centuries exerted a powerful influence over the development of ceramics in the West. The Islamic world spanned from Spain in the West, to the Indus valley in the East and this vast area was responsible for a similarly huge range of decorative traditions and potting techniques.

The first main centres of pottery production included Mesopotamia, the fertile plain between the rivers Tigris and Euphrates, where tin-glazed earthenware and lustre decoration were developed and refined between the 8th and 13th centuries. From the Middle Ages until the 15th century Persian centres in Kashan and Rayy (in what is now Iran) were preeminent, producing elaborately decorated lustre pottery decorated with intricate calligraphic designs or naturalistic patterns. In the 16th and 17th centuries the southern and eastern areas of Persia rose to prominence producing blue and white wares intended to rival Chinese porcelain, and were sometimes even sold to unwary buyers as the genuine article.

Among the most famous examples of Islamic pottery are wares made by Turkish potters around Iznik and later at Kutahya. These colourful pieces were made from the 15th century onwards and their distinctive patterns include many of the flowers that grew wild in the surrounding area such as tulips and carnations. Turkish potters produced a wide range of wares including tiles for mosques and other buildings as well as objects such as candlesticks and lamps, and tablewares including such items as bowls, plates and cups.

ISLAMIC POTTERY

Few pieces of Islamic pottery are marked and attributions are reliant on knowledge of the decoration and pottery bodies produced in each area. Although Islamic pottery is internationally collected, the field is particularly popular with Middle Eastern buyers. With European collectors, Iznik pottery is among the most popular of all types of Islamic ceramics and public interest has been boosted by recent research. Early wares fetch relatively high prices, but later pieces decorated with traditional patterns are available for modest sums.

▲ **LUSTREWARES** This elegant pear-shaped 17th-century bottle, probably made in Kashan, is an example of lustre decoration. The technique involved decorating the pottery with metal oxides such as gold and silver and firing the piece in a reducing atmosphere so that the oxygen is taken from the compound to leave pure metal behind. £2,000–3,000

► **PERSIAN BLUE & WHITE**
Some of the finest Middle Eastern blue and white pottery, such as this 17th-century *kendi*, was produced in Meshed and Kirman in eastern and southern Persia and designs rely heavily on those of Chinese late Ming and transitional porcelain. Unlike Chinese porcelain, some Persian examples are outlined in black. £1,500–2,000

▲ **TILES**
Iznik potters produced large numbers of tiles and this is a fine example, typically brilliantly coloured with bright cobalt, turquoise and Armenian bole (red). The tulip and carnation motifs are among the most popular of all on Iznik pottery. £800–1,200

► **PERSIAN FRITWARE**
A fine example of turquoise glazed fritware, this elegant vase was probably produced in Persia in the 13th or 14th century. Fritware was made from a silica-rich clay which when fired formed a semi-translucent body with a glassy appearance. £1,200–1,800

◄ **IZNIK DESIGNS**
Many Iznik designs were loosely derived from Chinese porcelain. Both the lotus (left) and the whorl (right) designs are from early Ming (16thC–17thC) border patterns. Left £800–1,200; right £700–1,000

European Pottery

Pottery had been made in Europe from Neolithic times but the Arab invasion of Spain in the 8th century marked an important turning point in its evolution. The Moors introduced lustre and other sophisticated glazes, which had been evolved in Egypt and Mesopotamia. These new techniques enabled potters to make objects of unrivalled sophistication and pieces of pottery achieved the status of decorative works of art rather than being merely regarded as utilitarian objects.

Gradually during the centuries that followed, potters in Italy, France, Germany and Holland evolved their own distinctive decorative styles and potting techniques. The earliest European pottery that collectors are likely to see is made from an earthenware body which has been lead-glazed to make it watertight. One of the largest pottery "families" was tin-glazed earthenware. Known as maiolica in Italy, faience in France, Delft in the Netherlands and delftware in England, tin glaze was so opaque that it allowed potters to simulate the effect of white porcelain. Other early wares include stoneware discovered by potters in the Rhineland who also developed salt glazes. As more refined bodies such as creamware were developed in the 18th century, and new glazing techniques evolved, so too did new styles of decoration. The patterns with which pottery is adorned reflect the prevailing fashion for Baroque, Rococo and Classical ornament.

For collectors European pottery offers a hugely diverse and fascinating collecting area at a similarly varied range of prices.

SPAIN & PORTUGAL

Spain is especially noted for its fine lustrewares that were made from at least the 14th century onwards in the centres of Málaga and Manises. Production of Spanish lustre continued until the 18th century and beyond but the finest and most valuable pieces are those of the 14th and 15th centuries. Early wares are dominated by Islamic motifs; later on, European designs prevail. Other leading Spanish pottery centres where polychrome wares were produced include Talavera de la Reina, Puente del Arzobispo, and later Alcora. In the 19th century primitively decorated earthenware replaced the sophisticated pottery of earlier times. Early examples of Spanish pottery are rare and fetch extremely high prices; non-armorial pieces from the 16th and 17th centuries sell for between £1,000 and £2,000; prices for small bowls and dishes from the 17th century begin at around £100.

▲ **EARLY DESIGNS** This dish (c.1400) is skilfully painted with a repeated tree of life motif and symbols representing benevolence. The colours, cobalt and lustre, have been applied in two stages. Early pieces of this quality are rare and values high. £10,000–12,000

◄ PORTUGUESE POTTERY
Loose, flowing brushwork is a characteristic feature of much of 17th-century Portuguese pottery, which was often based on Chinese Ming designs. Here the figure of the deer is derived from Chinese ceramics; it has, however, been westernized by the addition of antlers. £1,000–1,500

MARKS
It is rare to find Spanish pottery that has been marked by its factory; however, some later pieces made in the centres of Talavera de la Reina and Alcora may be found to have factory marks.

► LATER LUSTRE
Reproductions of early designs were made in Spain and Italy from the last quarter of the 19th century onwards. The decoration of these blue and copper lustre chargers has a crowded mechanical appearance that points to a date in the late 19th century. £200–300

▲ DECORATION
Repeating patterns were much used on Spanish and Portuguese pieces after the 15th century. The spiral moulded border on this early 16th century dish was inspired by silver designs and the intricate knot patterns and geometric designs were very popular. £1,500–2,000

◄ TALAVERA POTTERY
Inkwells such as this were made in the 17th century at Talavera de la Reina and Puente del Arzobispo in Spain. The crude feathery decoration in cobalt, manganese and ochre is a characteristic feature of these pieces. £150–250

ITALY & SICILY

The technique of making tin-glazed earthenwares called maiolica developed in Italy in the 15th century, in the important centres of Deruta and Gubbio. The earliest pieces were adorned with primitive designs and a limited palette but gradually decoration became increasingly sophisticated – often based on engravings of ornamental subjects or narrative scenes – and an increasingly wide range of colours were used. By the 16th century Castel Durante, Urbino and Siena emerged as leading areas of production and a fully developed style evolved, known as "beautiful style" Shapes became increasingly ambitious and many pieces were based on Mannerist silver forms. In the 17th and 18th centuries high-quality pottery was produced in Venice and Castelli and the influence of Chinese porcelain and an interest in classicism are reflected in many of the subjects chosen for decoration. Pottery was also made in several centres in Sicily, most notably in Palermo.

▶ LUSTREWARE
Lustre was probably introduced to Italy from Spain in the late 15th century and Deruta, where this charger (c.1520) was made, emerged as an important pottery centre. The Cupid which decorates the centre was probably copied from a woodblock print and the piece is painted using lustre blue for the outline of the design. £8,000–12,000

▲ DRUG JARS
Jars made for storing drugs are divided into two main types: this one has a spout and was meant to hold wet drugs; those of straighter form and with no spout, known as *albarelli* (see right), were for dry drugs. The jar above dates from the late 17th century and is simply decorated in blue and white, in a style typical of Savona where it was made. £700–1,000

◀ ALBARELLI
Used in pharmacies and hospitals from late medieval times, the *albarello* was derived from Islamic pottery. This one was made in Sicily c.1600 and is decorated with a portrait of a saint surrounded with trophies. Sicilian pottery commands a premium and this is considerably more valuable than Italian *albarelli* of similar date and decoration. £2,000–3,000

▶ **LATER ITALIAN MAIOLICA**
Ulysse Cantagalli was one of the most prolific manufacturers of late 19th-century reproductions. He marked his wares with a cockerel. This Cantagalli two-handled oval bowl painted in the Urbino style with a landscape is typical of late 19th-early 20th-century imitations of earlier Italian maiolica. Even so it would be worth £300–500 for decorative appeal alone.

ISTORIATO DISHES

FROM C.1520 POTTERS IN URBINO BEGAN PRODUCING HIGHLY DECORATIVE DISHES WHICH WERE TREATED LIKE A CANVAS AND ENTIRELY COVERED WITH A NARRATIVE SCENE. THIS DISH DATES FROM C.1550, AND IS WORTH £5,000–7,000, BUT SOME *ISTORIATO* DISHES CAN FETCH MUCH MORE. FACTORS AFFECTING VALUE INCLUDE:

- DATE
- QUALITY OF THE PAINTING
- SIGNATURE
- CONDITION.

MARKS

- Early Italian and Sicilian pottery is only sporadically marked. Later wares are more frequently found to be marked – often with the words "Faenza" and "Made in Italy".
- Marks often refer to the individual potter and the town where he was working.
- Rare marks include: Caffaggiolo, Urbino (Castel Durante), Faenza, Gubbio, Savona.

AUTHENTICITY
Copies of *istoriato* dishes were made in the 19th century (see above) and these can even confuse the experts. This dish from an important collection was sold with a thermoluminescence certificate stating it was fired between 350 and 550 years ago.

SUBJECTS
Most scenes are based on engravings illustrating stories by Ovid; this one shows Meleager and the Calydonian Boar.

COLOURS
The predominant shades of ochre and yellow are typical of early *istoriato* pieces; later in the following decades the palette became brighter and more varied.

FRENCH FAIENCE

Early French faience looks very similar to pottery made in Italy because Italian potters migrated to Rouen and Lyons early in the 16th century, bringing with them their native style. Identifiably French faience was being produced in the leading centre of Nevers by the 17th century and styles show the dual influences of Baroque and Chinese design. As the century waned Rouen superseded Nevers as the most important centre and throughout the 18th century potters moved away from the use of hot, high-fired colours to *petit feu*, a more refined, delicate palette. Other important centres of French pottery at this time included the towns of Strasbourg, Sceaux, Marseilles and Moustiers. During the 19th century French pottery returned to earlier styles and there were few new developments although French creamware called *faïence fine* was made in response to competition from the English company Wedgwood.

The market for French pottery remains largely dominated by collectors in France. Among the most valuable pieces are large early objects such as cisterns and vases. Pieces made in the 19th century in earlier styles are often highly decorative and available to the collector for modest sums.

◀ **ROUEN**
The blue and red palette and formally painted style of decoration are typical of pottery that was produced in Rouen during the late Baroque period. This large dish dates from c.1720–30 and the central panel showing a vase of flowers on a table and a deer with rockwork is clearly inspired from designs that were frequently used to decorate earlier Chinese porcelain. £2,000–3,000

▶ **FORMS**
Silver shapes provided the inspiration for many domestic pieces made at the Rouen factories from around the beginning of the 18th century. This helmet-shaped ewer form was popular in both England and France and the combination of blue and red is inspired by the Japanese Imari palette. £3,000–4,000

► **STRASBOURG** Milk-white, near-perfect glazes and botanical painting of the highest order are characteristic of Strasbourg pottery made in the 18th century. This detail of a delicately painted plate, dominated by puce shows how refined decoration became with the use of typical *petit feu* colours. Some plates of this type are signed on the back and this can add to their value. £2,000–3,000

▲ **LATER POTTERY** French pottery of the 17th and 18th centuries enjoyed a revival of popularity towards the end of the 19th century. The present-day value of pieces of faience such as this blue and white Nevers-style vase (c.1880–1900) is based mainly on the size and decorative appeal of the individual item. £1,000–1,500

MARKS

Many pieces of French faience are marked with the initials of the maker or with a cipher which is characteristically painted by hand.

▼ **MARSEILLES** Eleven factories flourished in the Marseilles region and one of the most famous was Veuve Perrin, where this dish was made c.1760. Pottery decorators were encouraged to attend drawing academies, and this free, spontaneous style of decoration is typical of much Marseilles pottery. £1,200–1,800

VEUVE PERRIN

Pottery made by the Veuve Perrin factory is marked with the initials "VP".

DUTCH DELFT

The technique of making tin-glazed earthenware was brought to the Netherlands at the beginning of the 16th century by immigrant Italian potters. By the mid-17th century, as Dutch Delft pottery flourished, the brewing industry declined, and many potteries were set up in disused breweries. This is why there are potteries with names like The Three Barrels, The Two Ships and The Claw. Around 33 factories flourished in the region around Delft, most producing a wide range of ornamental and everyday objects. Blue and white was the dominant palette and the vast majority of wares were heavily influenced by Chinese porcelain imported in huge quantity to Dutch ports. Although Dutch Delft bears a strong resemblance to English wares of similar date collectors should remember that at present prices for Dutch Delft are usually somewhat lower than for English.

▶ TILES & PLAQUES The Dutch were prolific makers of tiles and plaques and some factories were solely devoted to their output. Most were painted in blue and white or polychrome, and price depends on the quality of decoration and detail. These early 17th-century examples are attractively delineated with slightly unusual subjects and this increases their value. £200–500 each; later versions £50 or less

▲ EXOTIC OBJECTS Objects inspired by Oriental imports were popular in the late 17th and 18th centuries. This metal-mounted ewer, in the style of Chinese export porcelain from the start of the 18th century, shows how Dutch potters responded to the fashion for chinoiserie in their own designs. £1,500–2,500

ENGLISH OR DUTCH?

The following points can help with identification:

Dutch

- Ambitious forms
- Pinholes in greyish body
- Ambitious brush work
- Often marked

English

- Simpler forms
- Fine body
- Simple painted decoration work
- Rarely marked

◀ **COWS**
Another favourite of Delft potters, cows were made in considerable numbers, usually in white, but sometimes, as in this case, painted. This piece dates from the middle of the 18th century and was probably made to mark an agricultural festival. £1,000–2,000

▶ **CONDITION**
As with all ceramics condition can have an enormous bearing on price. Because tin-glaze earthenware is a fairly soft material some chipping is to be expected but it is often possible to buy pieces with more serious imperfections at greatly reduced prices. This plate, however, is by the most sought-after of all Dutch pottery painters, Frederich van Frytom, so the value remains extremely high. £10,000–15,000

BEWARE
Later copies of earlier, more valuable styles often carry the phoney marks of famous factories such as APK and PAK, and these marks should therefore not be relied upon for identification.

◀ **GARNITURES**
Inspired by Chinese porcelain versions that were imported into the country, the Dutch Delft potters first produced garnitures in the 17th century. This 19th-century vase is part of a five-piece garniture in the Cachemire style made originally in the early 18th century. The sides are ribbed and decorated with elaborate panels of flowers. Even though it is a later copy it is highly decorative and the set would still be worth £4,000–6,000.

GERMANY

Stonewares were made in the Sieburg and Cologne areas of the Rhineland from the late Middle Ages and evolved independently of those made in China. Regional specialities became gradually more pronounced and by the 17th century Westerwald, Cologne, Frechen, Raeren and Kreussen were all important centres of stoneware production.

German potters also made faience and the industry was well established by the second half of the 17th century in Frankfurt, Hanau and later Berlin. German faience is easily confused with Dutch Delft and heavily influenced by Chinese designs. Early wares were mainly confined to bulbous jugs known as *Birnkrug* and *Enghalskrug*, and deep dishes known as *Buckelplatte,* while in the 18th century cylindrical tankards called *Walzenkrug* were made. These usually have hinged pewter lids and footrims and are available for modest sums.

◄ WESTERWALD
This stoneware *Kugelbauchkrug,* or fat-bellied tankard, was made in the Westerwald region c.1710 for the English market. It is decorated with a portrait of Queen Anne and has several features typical of Westerwald pottery: moulded and applied, and incised motifs; a blue wash ground; detailing in manganese. Similar jugs with portraits of George I are more common and therefore less valuable. £500–800

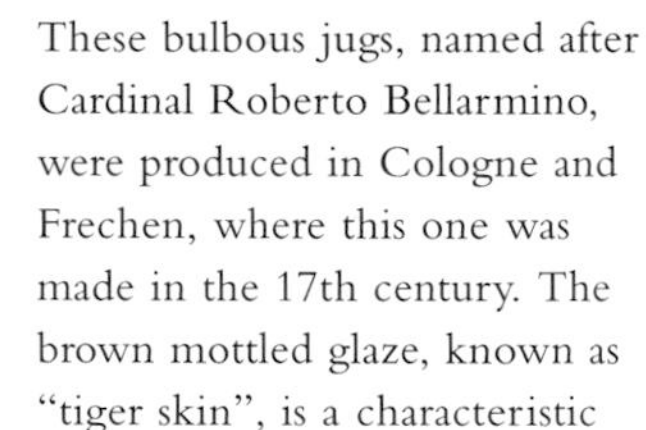

► BELLARMINES
These bulbous jugs, named after Cardinal Roberto Bellarmino, were produced in Cologne and Frechen, where this one was made in the 17th century. The brown mottled glaze, known as "tiger skin", is a characteristic feature of these jugs. £500–700

BEWARE
The base of genuine bellarmines should have an impressed oval mark like a thumb print where the piece was removed from the wheel. Later copies are flat on the base.

▲ TANKARDS
Faience tankards were produced in large quantities in Germany throughout the 18th century and this particular example is decorated with a favourite German motif – a bird among sponged foliage. Tankards are sometimes marked on the base but most are not. £1,000–1,500

► **FRANKFURT**
The blue and white figures in landscapes which decorate these Frankfurt faience bottles (made c.1700) are based on Chinese designs in the "transitional" style, one of the most important influences on European pottery in the latter 17th century. The brilliant shade of blue is a sign of quality and adds to their decorative appeal.
£1,000–2,000 (for the pair)

▼ **WIESBADEN**
One of several minor producers of German faience in the late 18th century, the Wiesbaden factory specialized in scattered floral decoration painted in green, a form of decoration known as *en camaïeu vert*. This bullet-shaped teapot (c.1775) is slightly damaged, hence the relatively low value. Pieces are sometimes marked on the base, as this one is, "WD".
£600–900

▲ **STYLES & DECORATION**
Spirally moulded forms, derived from silver shapes, were produced in the southern regions of Frankfurt, Hanau, Ansbach and Nuremberg; the scattered floral decoration is typical of Frankfurt faience made c.1700. £2,000–3,000

British Pottery

Most collectable British pottery dates from c.1650 onwards; earlier pieces are scarce and usually highly priced. Throughout the 17th and early 18th centuries lead-glazed wares dominated production and today their naive charm and colourful decorative appeal make them highly sought after by collectors. As the 18th century progressed new materials and glazes evolved and certain makers, such as Whieldon and Wedgwood, became associated with certain types of ware. English pottery is therefore categorized in a variety of ways: by the type of body (creamware, basalt, agateware, redware); its glaze or decoration (salt glaze, lead glaze, lustre); or by the name of the potter with whom it is associated (Whieldon ware, Prattware).

The production of all ceramics depends on the ready availability of the necessary raw materials and for this reason the Staffordshire region, where there were rich deposits of clay and readily available fuel, dominated pottery production from the late 18th century onwards. Among the potters who flourished in the region were Whieldon, Spode and Wedgwood.

Potteries became increasingly mechanized during the 19th and 20th century; output increased and these pieces are available for modest sums today. In the late 19th century the reaction against industrialization inspired a revival of hand-crafted pottery, and studio potteries such as the Martin Brothers and William de Morgan flourished. In the 20th century modern designs became an important element in industrial pottery. Today designs by Clarice Cliff, Eric Ravilious and Keith Murray are highly sought after.

ENGLISH DELFTWARE

Rather confusingly tinglaze earthenware made in England is known as "English delftware", while tin glaze made in the Netherlands is called "Delftware". London was the most important early centre of English delftware and after c.1650 centres were established in Bristol, Brislington, Liverpool, Glasgow and Dublin. English delftware is characterized by its simple forms and colourful decoration, much of which shows the influence of Chinese porcelain. Among the most sought-after pieces are the "blue dash chargers": large dishes so called because of the blue strokes that decorate the rims. Rare examples can fetch tens of thousands of pounds although damaged pieces are far less valuable. Unlike most ceramics, prices for early delft are hardly affected by damage, as pieces are so scarce.

▲ **COMMEMORATIVE CHARGERS**
Chargers decorated with portraits of monarchs, popular members of the royal family or military leaders feature Charles I and II, William and Mary, Queen Anne and George I, II and III. Value depends on rarity and on the complexity of the background. The sponged trees seen on this Bristol example with a portrait of William and Mary (c.1690) are among the simplest of backgrounds to be found on chargers.
£2,500–3,500

◀ **SUBJECT MATTER**
Among the most sought-after delftware subjects are the "farmyard" series – plates decorated with various birds and domestic animals, including peacocks, cockerels and hounds, or as here an egret with a worm in its beak. Blue and white is less valuable than polychrome so this plate (c.1920–40) is worth £2,000–2,500.

MARKS
Although often dated, English delftware is not usually marked and can therefore be confused with Dutch Delft which is usually less valuable (see p56 for differences).

▲ **DECORATION**
Painted decoration tended to become increasingly sensitive and sophisticated as the 18th century progressed. The delicately painted landscape with figures that decorated this vase are typical of the mid-18th century and similar decoration is also seen on plates, dishes and other wares. £1,500–2,000

▲ **POSSET JUGS**
Posset jugs like this were used for a slightly alcoholic gruel or porridge which was drunk from the spout, and were made from the mid-17th century to the mid-18th century. The bulbous form with two handles and curving spout is the most commonly seen. £1,500–2,500

WHAT TO LOOK FOR
Among other forms of unusual and decorative English delftware, you can find the following:
- shaving and bleeding bowls
- puzzle jugs
- flower bricks
- shoes
- apothecaries' pill slabs.

SLIPWARE

Slipware is so called because the earthenware body was decorated with "slip": a creamy solution of different coloured clays and water. Pieces were usually dipped into white slip and then extra details were applied by trailing, dotting and combing other slips of varying russet and chocolate colours onto the surface of the body. Decorated pieces were then coated in a thick lead glaze that turned the white clay yellow when the piece was fired.

The overall effect has a primitive charm that collectors in Britain and the United States seem to find highly appealing. Early slipware, which was made from the beginning of the 17th century in Staffordshire and elsewhere in England, can fetch extremely high prices.

▶ MAKERS
Rare early pieces marked with the name of their makers are generally the most sought-after slipwares. This armorial charger (c.1685) is marked by William Talor. Other names to keep a lookout for include: Thomas Toft, George Talor, John Wright, John Simpson and Samuel Malkin.
£20,000–30,000

◀ DECORATION
This handsome large circular dish (c.1690–1710) is based on silver shapes from Holland and Germany. The simple decoration was created by dipping the piece into cream slip and then trailing and dotting the dark brown, chestnut and cream slip to form this particularly striking decorative effect.
£4,000–5,000

▶ VALUE
Makers of slipware did not generally mark their pieces; most names or initials refer to the owner rather than the maker. Because this posset cup, made c.1690, has been inscribed with the name of its original owner, the value is enormously boosted. Even though the piece has fairly serious damage it would still be worth
£5,000–8,000

OWL JUGS

AMONG THE MOST DECORATIVE AND VALUABLE PIECES OF SLIPWARE, OWL JUGS SUCH AS THIS ONE (MADE C.1700) WERE MOST PROBABLY USED FOR DRINKING TOASTS TO WELCOME GUESTS. JUGS LIKE THIS HAVE BEEN WIDELY FAKED BUT THIS PARTICULAR EXAMPLE HAS MANY OF THE SIGNS OF AUTHENTICITY YOU SHOULD LOOK FOR ON A GENUINE PIECE. £15,000–20,000

The glaze is beginning to break and crackle.

There are signs of wear on the studs.

The handle is irregularly shaped and heavy – a fake would probably be more even and finer.

This marbled effect, a favourite technique with slipware potters, was created by dipping the piece into brown and cream slip and then combing the two to mix them slightly.

The base is also thickly potted and uneven – again, fakes tend to be thinner and rather regular in finish.

LATER SLIPWARES

Slipware continued to be made in Britain throughout the 19th century. Affordably priced baking dishes and other utilitarian wares invariably have great charm and can form a highly decorative collection.

▶ WARES

Slipware was also popular for humble domestic wares such as this late 17th-century baluster mug. The robust form is typical of the less elaborate pieces of the time but the date adds value. Such pieces were made to be used and may be damaged – this is slightly cracked and chipped. £5,000–7,000

SALT-GLAZED STONEWARE

Salt-glazed stoneware is made by firing pottery in the kiln to a high temperature (2,500°F (1,400°C)) and then throwing in salt. Sodium from the salt combines with silicates in the clay to form a glassy glaze with a distinctive granular surface – often described as looking like the skin of an orange. The technique was developed in Germany and spread to England in the late 17th century, when it was patented by John Dwight of Fulham, London. Throughout the 18th century salt-glazed wares were made in large quantities in Staffordshire and Nottinghamshire. Teawares and tablewares were the commonest products but various decorative items were also made, including pew groups, cats and other animals. Later pieces were sometimes decorated with transfer prints. Early sculptural pieces tend to attract extremely high prices but you can find moulded plates and other smaller objects from £100. Among the more unusual forms of stoneware is agateware, in which differently coloured clays were rolled together to produce an attractive marbled effect. Popular in the 18th century, agateware was often astonishingly sharply cast and forms often reflect the shapes of silver of the same date. Pieces such as pecten shell teapots and caddies can fetch £1,000.

▶ DECORATION
Most salt glaze was finished with moulded decoration and then covered in a cream-coloured glaze. Pieces decorated with coloured enamels such as this green teapot decorated with large pink roses (c.1760) are much less common and therefore highly collectable. Other typically coloured grounds include blue and aubergine. £2,000–4,000

◀ RARITY
Although made in the 18th century at a similar date to the teapot above, this example is rather more ambitious in form and the raised moulded cartouche or decorative frame around the painted scene is a rare feature. The pink ground colour is also unusual. £4,000–6,000

► LOVING CUPS
Two-handled loving cups originally symbolized friendship and trust; later on they became associated with christenings and weddings. It is also possible to find three-handled cups and these are known as "tygs". This typical mid-18th century example is of the two-handled variety and is inscribed with its original owner's initials and is dated 1756. £2,000–3,000

◄ NOTTINGHAM STONEWARE
Brown stoneware was a speciality of potters in the Nottingham region during the 18th century. This pierced tankard has a double shell so it still holds liquid. £1,500–2,000

► SGRAFFITO DECORATION
Incised or scratched decoration, known as sgraffito, was used on stoneware from the third quarter of the 18th century. Simple foliate subjects and inscriptions typically painted in blue, as seen on this small tea caddy, are characteristic. £2,000–3,000

▲ STYLES
The influence of contemporary silver design can be seen in the form of this candlestick (c.1750); the heavily scrolled, foliate base reflects the Rococo style and the form of the piece is more ambitious than most. £1,000–1,500

WHIELDON & OTHER COLOURED LEAD-GLAZED WARES

Thomas Whieldon, one of the most famous of the early Staffordshire potters, has given his name to a type of ware decorated with distinctive mottled glazes. The effect was created using coloured lead glazes which mingled together during firing. Whieldon was probably the first potter to develop the coloured lead-glaze technique but numerous other potters in the region also made similar objects; since wares of this type are not generally marked, they are usually called "Whieldon-type". A large proportion of Whieldon ware consists of domestic items such as plates and teawares, or decorative objects such as figurative groups and animals; large hollow wares such as coffee pots are far rarer and therefore tend to command the highest prices. Plates and small jugs will cost under £1,000.

◀ DECORATION
This Whieldon-type cornucopia is typically decorated with applied moulded decoration. Cornucopias (the word means "horn of plenty") were made to attach to a wall and used to hold flowers. You also find salt-glaze versions and both types are highly sought after. £2,500–3,500

▲ FRUIT AND VEGETABLES
Naturalistic moulds of cauliflowers, cabbages, lettuces, melons and, as here, pineapples were used by all the Staffordshire potters from the mid-18th century for teapots, coffee pots, tureens and various other objects; this one dates from c.1760. The Victorians later made copies of many of these shapes but they tend to be much more heavily potted and larger, and are far less desirable. £4,000–6,000

WHAT TO LOOK FOR

Several characteristic features can help the collector identify Whieldon ware:

- irregular glaze
- blurred colours that typically run together
- limited palette
- thinly potted body
- slightly iridescent surface covered with regular craquelure, a fine network of cracks running through the glaze.

► **COLOURS**
The coloured glazes used for Whieldon ware are restricted to brown, green, grey, yellow and blue. In this case the plate (c.1760) is glazed in brown and green creating a distinctive blotchy "tortoiseshell" effect. £200–300

TOBY JUGS

AMONG THE AMUSING NOVELTY OBJECTS PRODUCED AFTER C.1750 IN THE STAFFORDSHIRE REGION USING COLOURED LEAD GLAZES ARE TOBY JUGS. THESE RESEMBLE A COMICAL MAN WEARING A TRICORN HAT AND WERE BASED ON HARRY ELWES, A WELL-KNOWN CHARACTER NICKNAMED TOBY PHILPOT BECAUSE OF HIS LEGENDARY CAPACITY FOR DRINK.

Toby jugs originally had a cover in the hat. Most are now missing, but an original cover can considerably increase the value of the piece.

Many different forms of toby jug were made and each is known by a special name; this is the most commonly seen "traditional" type. Other forms of the period include: the Thin Man, the Collier, the Sharp Face, Rodney's Sailor, Admiral Lord Howe, the Coachman, the Fiddler and the Stepped. £1,500–2,000

This is one of the most sought-after toby jugs – made by Ralph Wood II, the colours are well controlled and they have not drifted.

LATER TOBY JUGS

During the late 18th and early 19th centuries brightly coloured Prattware toby jugs were made (see p72) – these are usually worth under £500. Numerous poor-quality copies were also produced in France and elsewhere; these are available for £100 or less.

CREAMWARE & PEARLWARE

During the 1740s a new cream-coloured earthenware with a transparent lead glaze began to make its appearance. Known as creamware, this pottery was refined enough to be considered a substitute for porcelain and from the mid-century was being produced by numerous potters in Staffordshire, Leeds, Liverpool, Bristol and Swansea. Creamware was decorated with moulded decoration, painted enamels or transfer prints and its most successful manufacturer was Josiah Wedgwood who, as a clever marketing ploy, christened his products "Queensware", ostensibly in deference to Queen Charlotte. Pearlware was a variation of creamware with a slightly blue tinge. Developed by Wedgwood in 1779 it remained popular with several manufacturers throughout much of the 19th century. A huge range of highly decorative objects was made in creamware and pearlware and these remain widely available for modest sums, although early rarities can fetch over £1,000.

▶ PATTERNS

Chintz patterns, as used to decorate this globular teapot, were popular decoration for creamwares made c.1770; other popular hand-painted designs include bouquets of flowers, ruined castles and picturesque cottages. £1,500–2,000

▲ TERMINOLOGY

There is an overlap between pearlware and Prattware (see p72). Pearlware figures in high-fired colours are often called Prattware. This figure dates from c.1820. £1,200–1,800

▶ COMMEMORATIVES

Creamware and pearlware decorated with transfer prints or decorated in low relief with political personalities of the day were popular from the late 18th to mid-19th century. Value depends largely on the rarity of the subject. This particular example of a commemorative plate depicts Admiral Earl Howe and dates from c.1795; it would be worth £500–700.

◄ **BOTANICAL SUBJECTS**
Botanical subjects enjoyed a heyday of popularity during the early 19th century; this service dates from c.1810. Most plant and flower specimens were copied from contemporary prints, and pieces such as these were made in both England and Wales.
£3,000–4,000 for the set

► **SERVICES**
Vast transfer-printed services were produced in creamware by Wedgwood and other manufacturers. Compared with modern-day equivalents these can offer surprisingly good value. This extensive Wedgwood service (c.1830), transfer-printed in brown in the Indian Temple pattern, contains 93 dinner plates, 20 dessert plates, 23 soup plates and numerous serving dishes. £4,500–6,500

◄ **PEARLWARE**
Pearlware was used to produce a huge variety of decorative figures and novelty wares, and was left in the white or decorated with coloured enamels. Late 18th or early 19th century models of sheep like these were a particular favourite, as were figurative groups. £1,200–1,500 for a pair

WEDGWOOD

Josiah Wedgwood occupies a pivotal position in the development of ceramics. His factories at Burslem and Etruria were the first to introduce industrial production techniques to pottery. Wedgwood's name is synonymous with several different types of body such as black basalt, *rosso antico*, caneware, jasper ware, agate ware and creamware (see p68), all of which became hugely fashionable and were much copied. A master self-publicist, Wedgwood made high-quality wares that achieved renown throughout England, Europe and America and his fame has endured two centuries. The factory's 20th-century products are of similarly high quality and remain keenly sought after by collectors today (see p146). Wedgwood was also the first potter to commission leading artists to create designs for his products. John Flaxman was responsible for many of the jasper ware designs of the 18th and 19th centuries. The range of Wedgwood available is so enormous that many collectors tend to concentrate on particular objects, on specific types of body or on certain designers. Early marked pieces tend to attract the highest prices.

◀ BLACK BASALT WARE
Basalt ware is among Wedgwood's most popular products, and was produced at the Etruria factory which specialized in ornamental wares. Wedgwood produced a range of library busts of both classical and contemporary figures. This portrait of Sir Walter Scott dates from the 1820s and is quite small (7–8in [18–20cm] high). Some busts are of much larger proportions and were made to stand on top of library bookcases. £300–500

BEWARE
Later in the 19th century a rival factory called Wedgewood (note the additional "e") also marked its products with its name. Wares by this manufacturer are generally of inferior quality and therefore far less valuable. Wedgwood wares were also imitated on the Continent by factories such as Sèvres, but few fakes are marked.

MARKS

WEDGWOOD

wedgwood

Although early pieces were unmarked, the vast majority of wares made after 1768 were marked "Wedgwood & Bentley", Wedgwood or with initials. From 1860 a system of date coding was used.

▶ LATER WARES
During the 19th century Wedgwood continued to employ some of the leading decorators and designers of the day. This is one of a set of six plates which were painted by A. Walker. The set of plates features fish, which typifies the late 19th-century predilection for naturalistic themes. £100–200 (for a single plate)

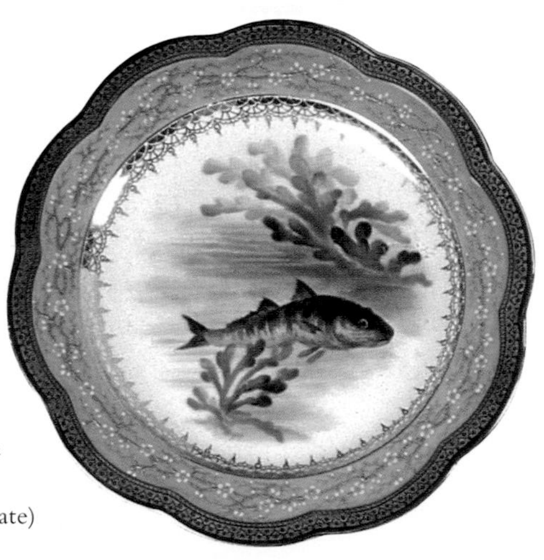

JASPER WARE

INSPIRED BY ANTIQUE CAMEOS AND ROMAN GLASS AND TOMBS, JASPER WARE WAS A FORM OF UNGLAZED STONEWARE INTRODUCED C.1767. IT BECAME ONE OF THE COMPANY'S MOST POPULAR PRODUCTS. THIS PEGASUS VASE WAS MADE IN THE 19TH CENTURY. £5,000-10,000

The simple cylindrical shape is characteristic of the Neo-classical period.

Jasper ware was made in a variety of colours: green (as here), yellow, lilac, claret, black and white and most commonly blue. Early blue jasper ware was far more distinctive in colour than 20th-century versions: either a very deep purplish blue, or a strong slate blue – not the pale shade nowadays called "Wedgwood" blue.

Many designs were based on drawings by artists of the day such as John Flaxman and inspired by antiquity. These figures illustrate domestic employment and were designed by Elizabeth, Lady Templetown.

Cutting of the applied cameo design should be crisp and very refined with a slight translucence in the shallower areas.

QUEENSWARE

Wedgwood's Queensware, a refined type of creamware (see p68), was enormously popular from c.1765 and was generally used for tea and coffee services. Its fame soon spread to Europe, where it proved equally popular and was widely copied. Queensware was named in honour of Queen Charlotte, the wife of King George III. Wares were decorated with moulded designs, transfer prints or hand-painted decoration.

► IMITATIONS

Wedgwood's success spawned numerous imitations. Among his copyists the most successful was Turner of Lane End; Adam and Spode also made many Wedgwood imitations. This unmarked belt buckle (late 18th early 19th century) may be by Wedgwood but is probably by an imitator such as Turner; if it were marked it would be worth four times as much. £200–300

PRATTWARE

Prattware takes its name from the Pratt family whose pottery in Fenton, Staffordshire developed a distinctive type of lead-glazed earthenware mainly decorated with relief moulding and high-fired colours. Wares of this type were also made in vast quantities by numerous other factories elsewhere in Staffordshire, Yorkshire and Scotland from c.1775–1835. The name Prattware is nowadays a generic term used to describe pearlware earthenwares decorated in a distinctive palette of blue, green, yellow, orange, brown and purple. Themes for the hand-decorated earthenwares included rustic subjects such as milk maids, classical figures, political and royal subjects and animals. Wares include hollow domestic objects and novelty and decorative items such as money boxes and figural groups, and are rarely marked. Value relies heavily on the elaborateness of the decoration and the rarity of the form. Commemorative Prattware is especially popular with today's collectors.

PRATTWARE JUGS
Inky blue, olive-green and ochre, the palette characteristic of most Prattware, was typically used to highlight the low relief-moulded decoration. This baluster jug was made c.1800.

CONDITION
As always damage reduces value. Even with this crack the jug is worth £300–400, but in top condition it would probably fetch 50 percent more. If the subject matter were royalty, value would be even higher.

The jug is adorned with a well-nourished toper sitting in a landscape – a favourite theme c.1800.

HANDLES
The shapes of handles vary; this is very plain but some are far more elaborate in form.

◄ COMMEMORATIVE SUBJECTS
Admiral Rodney is featured on this late 18th century jug, he is typical of the sort of naval and military hero who appear on Prattware jugs. Other figures include Admiral Nelson, Queen Caroline, King George IV, the Duke of York and even Napoleon Bonaparte. £600–900

BEWARE

Fake commemorative Prattware jugs have been seen on the market, these are usually recognizable by their softer or weaker colours and over-regular craquelure. Look for signs of genuine wear and tear on the base and only buy from a reputable dealer if you are in doubt.

► FIGURATIVE GROUPS
Prattware figures such as this dairyman with his cow and dog are more scarce than hollow wares and therefore can command a premium price. This figure probably dates from c.1820; the sponged base is typical of this type of ware, as is the high-temperature palette – blue, orange, green and muddy brown. £1,000–1,200

PRINTED DECORATION

"Prattware" is also the term used to describe a type of stippled colour printing developed by Felix Pratt in the middle of the 19th century. The technique was frequently used to decorate pot lids.

◄ ORNAMENTS
This "Babes in the Wood" group (c.1820–30) is unusually simply moulded in the manner of late flatbacks. It depicts children against a foliate background and constitutes one of a pair. Such pieces were intended as mantel ornaments and were made by pressing clay into a mould and adding extraneous details by hand. Another popular set of classical figures is composed of "Faith, Hope and Charity". £100–150

LUSTRE

Lustre or metallic decoration, first used in European pottery probably as early as the 13th century in Spain, enjoyed a revival of interest in Britain in the 19th century. Lustre decoration was applied in a variety of ways: either limited to bands or reserves, applied all over or streaked using a "splash" technique. The Wedgwood factory introduced a mottled range called "Moonlight Lustre". Lustre was often used in combination with stencil-printed decoration and is most commonly seen adorning jugs, teapots, mugs and chamber pots. Such objects were produced in numerous potteries in Staffordshire, Sunderland, Wales and Scotland. Lustre is generally available for modest sums although the price increases if the piece is unusually large or decorated with an unusual print.

▶ COMMEMORATIVE LUSTRE
Pink lustre rims were often applied to the rims of white-bodied commemorative wares in the early 19th century. This plate (c.1816) is decorated with a print showing Princess Charlotte and her husband Prince Leopold with Goody Bewley, to whom they later presented a large copy of the Bible – the Princess's kindness was legendary and this subject was a popular one. £250–300

◀ DECORATION
Wares were frequently adorned with a combination of lustre and enamelling. The body of this 19th-century moulded Staffordshire jug is decorated with a landscape in the background, hounds and a colourful trailing vine motif, all of which are highlighted in lustre. Bands of lustre have also been applied on the inside rim, base and handle of the jug. £100–150

▶ COLOURS
Lustre colours depended on the metal oxides which were used and on the colour of body beneath. Pink or purple lustre, which has been used to decorate this chalice (c.1830), was created using a mixture of gold and copper oxides on top of a white body. If gold lustre was applied to a dark body, however, it created a rich bronze colour, while silver was made from platinum oxides. £100–150

BEWARE
Reproductions and fakes of more elaborate pieces of lustre are occasionally seen – look for signs of genuine age.

MARKS
Lustre was produced in many small potteries and few pieces are marked.

IRONSTONE

Ironstone, a type of robust earthenware still produced today, was patented in 1813 by Charles James Mason, a Staffordshire potter. "Mason's ironstone" was a spectacularly successful material for dinner services because it held the heat well and was particularly hardwearing. Mason's was taken over by Ashworth's in 1859 but continued to produce ironstone. Other factories also produced their own versions of the material to which they gave names such as granite china, opaque china and stone china.

Ironstone can form an attractive theme for a collection since pieces are usually decorated in colourful pseudo-Oriental patterns and shapes are often extremely inventive. Jugs cost from £50 while large services can fetch £1,000 or more, depending on the number of pieces.

► SERVICES
Typically decorated in a colourful Oriental style, this sauce tureen (with cover and stand) and side plate is part of a dinner service (c.1820) containing 95 pieces. The value of the service would be £7,000–8,000

MARKS
Mason's ironstone has an impressed mark; changes in details can help with dating. Spode's impressed mark usually includes the words "New Stone".

◄ WARES
Apart from services, ironstone was also used for a range of items large and small, including ornamental vases, jardinières, wine coolers, jugs, teapots, miniature wares, fireplaces, garden seats and card racks such as this example, made c.1820. Rare shapes command a premium and this rack would fetch £450–550.

► SPODE STONE CHINA
Prior to the introduction of Mason's ironstone, Spode produced a similar material called "stone china" (c.1805). This jug (c.1820) is part of a tea and coffee set containing over 35 pieces; it is decorated with an imitation of a Chinese *famille rose* design. Other makers of similar materials are: Davenport, Minton, Wedgwood, Ridgway, E. & C. Challinor. £250–350

STAFFORDSHIRE FIGURES

In the early 19th century dozens of small potteries in the North Staffordshire region produced a range of decorative pottery figures. The earliest represented idyllic themes: shepherds and shepherdesses, musicians and courting couples were typically brightly enamelled and crisply detailed, with modelling in the round even though they were meant mainly to be seen from one side only. From c.1840 potters began making simpler "flatback" figures, which as their name suggests were not decorated on the reverse. Flatbacks were usually press-moulded and produced in vast numbers in a huge range of subjects, ranging from royal personalities, naval and military heroes, and politicians to criminals, theatrical personalities and animals. Despite their primitive qualities Staffordshire figures have great decorative appeal and enjoy huge popularity with present-day collectors. Prices range from under £100 to £1,000 or more and are affected by the quality of the modelling and enamelling, as well as the subject matter: collectors' favourites include sporting and theatrical subjects and rare and exotic animals like elephants and lions.

◀ EARLY FIGURES
A good example of an early Staffordshire group (c.1825). This piece, representing an amorous couple, is well modelled in the round and painted with considerable attention to fashionable detail – note the woman's plumed hat, the height of post-Regency fashion. £1,000–1,500

▲ COLOURS
Brilliant enamelling was a feature of early Staffordshire groups and such well-detailed pieces as this are always in demand. The subject of this piece (c.1825–30) depicts a couple being married by a blacksmith – a reference to the New Marriage Act of 1823, which attempted to stop secret marriages. £2,000–3,000

BEWARE

Numerous fakes and copies of Staffordshire figures have been made, some using original 19th-century moulds. Signs which should arouse suspicion include:

- lighter weight – copies are usually slip-cast from pale chalky clay
- absence of irregularities on the hollow inside made from tools and fingers
- over-regular craquelure – cracking in the glaze should be erratic, if present at all
- deliberately blackened bases.

▲ SUBJECTS
Royal personalities were endlessly popular with the Staffordshire potters and these groups of children with dogs (c.1850–60) probably represent the eldest children of Queen Victoria: the Prince of Wales (later Edward VII) and the Princess Royal. There are numerous different figures of the royal children, some in Scottish dress. £200–300 for the pair

▼ LATER FIGURES
This figure showing Samson wrestling with a lion is probably based on a contemporary circus act like that of Isaac van Ambrugh. In common with most later figures, it is less well modelled and the back is not detailed. The colours are typical of c.1850–60 with strong use of red and ochres. £80–120

◀ DOGS
Spaniels, poodles and greyhounds were among a variety of dogs modelled by the Staffordshire potters and in this example (1860–80) three dogs surround a fake clock face. The russet colours of the spaniel on the right are instantly recognizable and repeated on hundreds of similar Staffordshire dogs. £120–180

BLUE & WHITE TRANSFER-PRINTED WARE

Although today blue and white transfer-printed wares are avidly collected, when they were first made they were considered little more than utilitarian domestic pieces. Dinner and tea services were made in large numbers and for this reason plates far outnumber other pieces. The decoration was inexpensively produced using underglaze blue transfer prints taken from engraved copper plates and needed only one firing. Different copper plates had to be created for each different size and form produced, and sometimes designs are oddly adapted to fit more unusual shapes. The earliest pieces were inspired by Chinese ceramics; the famous Willow pattern was a pastiche of Oriental design produced by numerous English manufacturers. The heyday of blue and white transfer wares was the late 18th to the mid-19th centuries and pieces made within this period tend to be most sought after. Prices range from £30–50 for a simple plate to £1,000 or more for a large rarity.

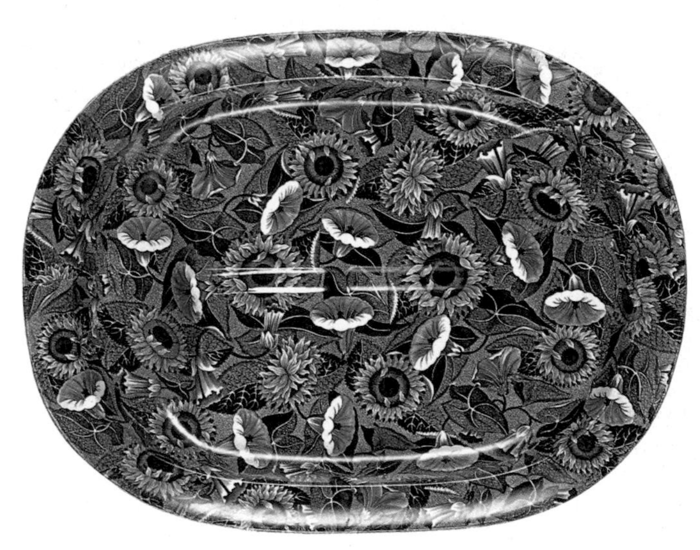

◀ MANUFACTURERS
Numerous makers throughout the country produced blue and white wares and while most collectors do not collect by factory, pieces by Spode and Rogers, the main makers, are particularly sought after. This Convolvulus-pattern Spode platter (c.1830) is both decorative and by a top maker so would be worth £800–900.

▼ FORMS
One of the appealing aspects of collecting these wares is the huge range of shapes available – from *bourdaloues* (chamberpots for females) and foot baths, to washstand sets, platters, drainers and tureens. Larger unusual pieces such as this dish (c.1820–25) from a Caramanian sandwich set by Spode (see p79) are always popular. £250–300

WHAT TO LOOK FOR

Other subjects to look out for include:

- eye-catching animals – the Durham Ox is a favourite
- Arctic scenery
- views of Oxford and Cambridge colleges.

REPRODUCTIONS

Fakes are not generally a problem although many of the most popular patterns were produced until the late 19th century; later versions tend to be of inferior quality and are less sought after. Modern reproductions are usually easily identifiable, but beware of a Dr Syntax series made in the 1920s based on 18th-century Rowlandson prints – these are not as old as they may look.

▲ PATTERNS
Unusual patterns can have a considerable bearing on value. Many were taken from fashionable engravings such as topographical views of foreign scenes. Among the most popular are Spode's Indian Sporting, the design used on this service (c.1820–30), which was based on engravings after Daniel. Caramanian – a series of views of Turkey (see p78) – is also sought after. £150–1,500 per piece

▼ OTHER DESIGNS
This plate (1822–35) features a view of Fair Mount near Philadelphia on a plate by Joseph Stubbs. Many Staffordshire earthenwares intended for export showed American views. £180–240

MARKS

Marks do add to value, especially if marked by small makers, but unless made by one of the larger factories with a retail shop most pieces are not marked.

MAJOLICA

Inspired loosely by the Italian low reliefs of della Robbia and the old French pottery of Bernard Palissy, which enjoyed huge popularity during the 19th century, majolica ware was produced in England, Europe and America from the mid-19th century. Majolica was modelled in relief and decorated either with pigments added to the body itself or, more usually, to the glaze. Different parts of the design were painted with coloured pigments which when fired melded together with the glaze. The manufacturer's ability to control the glaze during firing was obviously critical and on poor-quality examples the colours bled together with disastrous effect. Until the 1980s majolica ware was a relatively undervalued market but a prominent exhibition brought its decorative charms to the attentions of interior decorators and prices rose dramatically as a result. At present recent market fluctuations appear to have steadied and the market for good-quality pieces seems as strong and buoyant as ever although buyers are becoming increasingly discerning.

▶ MANUFACTURERS
Top factories were Minton, George Jones and Wedgwood, but many other large factories turned their hand to majolica including Royal Worcester and Copeland, as did small manufacturers such as Holdcroft and Brownfield. This c.1875 humorous punchbowl supported by Mr Punch is by George Jones, one of the best 19th-century majolica manufacturers. £3,500–4,500

◀ DECORATIVE WARES
Novelty domestic wares such as teapots, game tureens and, as here, strawberry sets were typical of the ornamental tablewares made in the 19th century. This double bird set is by George Jones. Majolica was also used for architectural objects such as fountains and wall tiling, jardinières and garden seats. £2,500–3,500

▶ VALUE
Big names command a premium as do pieces made in America, even though their quality is inferior to English pieces of similar value. Small unusual pieces are also highly coveted by collectors – a very unusual game tureen can make £20,000 or more, although less exceptional pieces fetch under £1,000. This 1880 Minton hare and duck tureen is of a remarkable design and is worth £20,000–24,000.

► **FORMS**
This Wedgwood leaf pickle stand (c.1880) combines plant forms with majolica glazes. Small pieces such as this are available for modest sums. £100–150

CONDITION

Majolica is particularly prone to chipping and flaking but because pieces are bought primarily for their decorative appeal value is less dramatically affected by damage than for other types of ceramics and restoration is fairly common.

▲ **SMALLER MAKERS**
Unmarked majolica or pieces by less well-known makers can still attract high prices if the design is strong. This majolica teapot (c.1878) is by a smaller maker but the design reflects the influence of fashionable Japanesque styles. £600–800

► **DECORATION**
High Victorian taste favoured extraordinarily elaborate decoration, as seen on this typical vase of c.1890. Although rather large it is unmarked and would therefore be worth far less than a piece by Minton. £150–200

REPRODUCTIONS & FAKES

Although fakes are not a major problem some unmarked pieces have had fake marks added with an etched "Minton" to boost their value. Modern reproductions are usually easily identifiable by their lighter potting and their modelling which is less well-defined.

DOULTON

One of the most prolific and innovative manufacturers of the 19th and 20th centuries, the Doulton factory's success was largely due to the entrepreneurial spirit of Henry Doulton who took over the sole running of the company in 1854. The factory based in Lambeth, London produced utilitarian stonewares such as water filters until 1862 when decorative items, designed by students from the nearby Lambeth School of Art, began to be produced. Over the following decades Doulton established a line in decorated stonewares and employed numerous designers. In 1882 Doulton took over a second factory in Burslem, Staffordshire, where porcelain and earthenware tablewares were produced. The factory became "Royal Doulton" in 1901 and the Burslem factory is still in production although the Lambeth factory closed in 1956. Products are so diverse that collectors often concentrate on wares of a particular type, such as flambé wares, stonewares, faience wares or kingswares. Prices start at around £50 for ornamental wares; the highest prices are commanded by pieces signed by their leading artists, which can fetch over £10,000.

▶ CHINÉ DECORATION
Another popular method of decoration, known as "chiné manner" was patented by Slater for Doulton's. The technique involved pressing lace on to the wet surface of the clay and finishing with hand colouring and gilding. Vast quantities of these wares were made from the 1880s up to 1939 and they remain relatively inexpensive. £50–70

▲ DECORATION
Stonewares were influenced by 16th- and 17th-century designs and often feature incised or applied and hand-carved decoration. Motifs such as beaded borders are typical characteristics as is the restrained palette in shades of blue and brown. This 1879 jug by Florence Barlow and Emma Martin is typically decorated with beaded roundels and paler scrolling foliage. £120–180

◀ HANNAH BARLOW
Deer provide a recurring decorative theme in Hannah Barlow's work, but the *pâte sur pâte* decoration on this vase (made c.1885 at the Lambeth factory) is unusual and was added by William Baron, who later set up his own pottery in Devon (see p86). £300–400

MARKS
Wares are marked with the name of the factory, impressed, printed or painted; they are also marked with the initials of the artist and the date.

DOULTON DESIGNERS		
NAME	**SPECIALITY**	**MARKS**
George Tinworth	Speciality modelling, religious plaques, frog and mice groups	GT + Doulton
Hannah Barlow	Incised decoration, animal subjects	BHB + Doulton
Florence Barlow	Incised bird subjects	FEB + Doulton
Frank Butler	Bold shapes, decorated with natural forms	FAB + Doulton
Harry Nixon	Song and Chang, flambé wares	H Nixon (on side of piece generally)
Mark V Marshall	Well-modelled stoneware - worked with Martin Brothers	MVM
Eliza Simmance	High-quality Art Nouveau designs	ES

▲ MICE GROUPS
Among the most popular novelties for which George Tinworth is famed are novelty mice groups (c.1890) such as this which shows mice playing instruments. Tinworth also produced frog groups and imp musicians which are equally popular. £700–900

▼ SONG WARE
Inspired by Chinese ceramics Royal Doulton's Song wares came in a choice of subjects and colourways. This "Song" vase (c.1920) by Arthur Charles Eaton is painted with a typically exotic bird. £1,200–1,500

MARTIN WARE

One of the most original "studio" potteries of the 19th century was established by the Martin brothers in a studio in Fulham and later in Southall, London. There were four brothers, Robert Wallace, Charles Douglas, Walter Frazer and Edwin Bruce. Robert Wallace, a sculptor and the firm's leading creative force, was fascinated by the grotesque gargoyles he encountered while working as an apprentice on the Houses of Parliament, and was responsible for the introduction of the extraordinary bird jars with movable heads, known as "Wally" birds, that are among the company's most distinctive and individual wares. The company also made jugs and mugs with faces on each side, vases, some in the form of naturalistic gourds, ordinary jugs and other ornamental pieces.

▶ BIRDS
Martin Brothers' birds reflect the Victorian fascination with the grotesque and their quirky features were often based on people the brothers knew personally. These examples were made in the late 19th or early 20th century, and like most others of their type have removable heads. Value is affected by size; large pieces are worth £8,000–10,000, while smaller examples would fetch around £2,000–3,000.

MARKS
Pieces are usually incised "R W Martin Bros, London & Southall" and dated. Early pieces may just be signed "Martin".

▲ TILES
Tiles are among the Martin Brothers' more unusual products; the fish with which this panel (c.1890) is decorated are also seen on vases. £3,000–4,000

BEWARE

Martinware has enjoyed huge popularity and fakes do appear on the market, but these are usually identifiable by their lack of subtlety in decoration and modelling. Although it has a Martin "signature" this spoon warmer in the shape of a tortoise is far inferior in its modelling and clearly a fake. Fakes have sometimes been made by taking a mould from an original piece; hence the marks are also moulded. This should indicate to the prospective buyer that the piece has been faked.

WILLIAM de MORGAN

One of the most important figures in the Arts and Craft movement, de Morgan was employed as a designer by William Morris before he set up his own business. De Morgan specialized in producing tiles and lustre wares, inspired by Islamic and Hispano-Moresque ceramics, and in "Persian" wares based on Turkish pottery designs. Among the leading decorators he employed were Fred and Charles Passenger and Joe Juster, all of whom often signed pieces with their initials rather than with the de Morgan mark. De Morgan's highly decorative wares have a ready following and can fetch very high sums. "Persian" pieces fetch £2,000 or more according to size and choice of decoration. Tiles start at around £50 for 6in (15cm) sizes in the most usual patterns, many of which were produced for fireplace surrounds; 8in (20cm) tiles with an unusual pattern can fetch £500 or more.

▶ "PERSIAN" WARES
Among the most distinctive and valuable pieces of de Morgan pottery are Persian-inspired wares such as this vase (c.1880). It has been painted by leading decorator Fred Passenger with peacocks and formalized flowers. The vivid turquoise and green shades and stylized floral decoration are based on those found on Persian and Iznik pottery.
£10,000–12,000

▶ DECORATION
Full-masted galleons and stylized fish were favourite de Morgan motifs; here dolphins and a galleon combine on a vase (c.1890) made in ruby lustre, for which de Morgan is also famed. A range of triple lustre glazes known as "Moonlight" and "Sunset suite" are the most complex and valuable of all de Morgan's lustre wares.
£5,000–6,000

MARKS
An impressed or painted mark with the name of the company is usually found, frequently in conjunction with a motif such as an abbey, a tulip or a rose. The motifs relate to the period and location of the various factory sites de Morgan established.

▶ TILES
De Morgan produced tiles (c.1890) in a wide range of styles and patterns and prices depend on decorative appeal and size. Although elaborately decorated these 6in (15cm) tiles are less expensive than if they formed part of a large decorative panel.
£100–400 each

DEVON POTTERIES

The discovery of deposits of high-quality red clay suitable for making fine terracotta wares together with a burgeoning tourist industry and a growing demand for souvenirs, encouraged a handful of small potteries to set up in Devon in the mid- to late 19th century. In 1869 the Watcombe Terracotta Company began producing painted wares and unglazed pieces decorated with turquoise borders and glazed interiors. The Torquay Terracotta Company made figurative subjects, painted wares and ceramics in the style of the Aesthetic movement. Another South Devon pottery was Aller Vale which made wares decorated with incised mottoes such as jugs and teapots as well as now obsolete objects such as plaques for resting curling tongs. In North Devon near Fremington Edwin Beer Fishley produced pottery simply decorated with coloured slips or incised designs. Another important pottery located at Barnstaple was C. H. Brannam, which became well known for its "Royal Barum" ware. The company's reputation grew in the 1880s when it began selling through Liberty's, London, and later pieces are in general more elaborately coloured and more varied. Decorators include James Dewdney and William Leonard Baron; the latter eventually set up his own pottery in Barnstaple too.

◀ **VASES**
Among the most expensive pieces of Devon pottery are those decorated in the style of the Aesthetic movement – such as this pair of vases made c.1878 by the Torquay Terracotta Company. The vases, designed by Louis F. Day and painted by Alexander Fisher (senior) to represent "morning" and "night", are worth £2,000–3,000 but less spectacular examples sell from £100.

VALUES
Prices vary considerably but these features are desirable:
- pieces in artistic styles
- large sculptural pieces
- large slip-decorated vases.

▶ **ALLER VALE**
Pixies proved to be a popular motif with tourists; this pixie-adorned jug (late 19thC) was made by the Aller Vale pottery, one of the most prolific of the Devon potteries that later merged with Watcombe. £80–120

► C. H. BRANNAM
Although the birds and floral roundels that decorate this jug (c.1910) are heavily influenced by Oriental ceramics, the use of incised decoration in conjunction with coloured slips is typical of the Brannam pottery. £150–200

▲ TERRACOTTA FIGURES
Unglazed terracotta figures were a popular product of the Watcombe and Torquay Terracotta companies in the late 19th century. This angel made by the Torquay Terracotta Company is typical of the religious subjects produced there. Portrait busts, classical subjects and figural groups were also produced. £120–180

▼ SOUVENIRS
Among the less expensive tourist souvenirs produced in the Devon studios some surprising novelty items can be found. This Baron of Barnstaple pottery bell (c.1920) is modelled in the form of a female head. It carries a rather dubious inscription – "The perfect woman speaks only when tolled." £80–120

MARKS
The marks most commonly found on Devon pottery include:
- **"Aller Vale"** impressed name or incised hand-written signature mark
- **"Torquay Terracotta Co"** initials or name in full
- **"Watcombe"** usually a printed mark.

A millennium after the development of porcelain in the Far East the tantalizing secret of how to make it was still undiscovered by European potters. By the late 16th century, as Portuguese traders began to export Oriental porcelain to Europe, demand outstripped supply, prompting European potters to try to discover the elusive formula. Although various factories in France and England managed to make soft-paste porcelain in the 17th century, it was not until 1708 that a German alchemist, Johann Friedrich Böttger, found a formula for hard-paste porcelain. In 1710 the Meissen factory opened near Dresden; it dominated porcelain production for the next 50 years or so, but as the secret leaked out, factories such as Vienna, Fürstenberg and Nymphenburg began hard-paste production too.

In due course, Meissen's preeminence was challenged by the French royal porcelain manufactory of Sèvres, which began making wares with spectacular coloured grounds and raised gilding of unrivalled quality. Many of the most famous European factories continued to produce porcelain throughout the 19th century and beyond, although from the mid-19th century few new ideas and styles evolved and most production remained heavily retrospective until after World War I, when some factories began producing refreshingly innovative designs.

For collectors European porcelain offers a vast range of prices and objects, from modestly priced cups and saucers to elaborate figurative groups.

EARLY MEISSEN

Meissen's earliest wares included a type of red stoneware called jasper ware, which is extremely rare and valuable, and porcelain decorated in Oriental style. In 1720 the factory was joined by J. G. Herold, a skilful chemist who developed a new range of colours, and in the 1730s by J. J. Kaendler who was a brilliant modeller. During this period painted decoration became more European in style and modelling became an increasingly important part of the overall decoration.

▲ **TEAWARES**
Tea and coffee drinking were fashionable among the wealthy and aristocratic classes in the 18th century, and large quantities of wares survive. This teabowl (c.1725) is a good example of Meissen's early fascination with chinoiserie decoration.
£2,000–3,000

MARKS

K.P.M

Beware The Meissen mark was copied by numerous other leading factories in Europe and England and is certainly no guarantee of authenticity. The crossed swords are more usually alone. The "KPM" mark is only used on grand early pieces pre-1740.

MEISSEN SAUCERS

THIS TYPICAL MEISSEN SAUCER MADE C.1760 WAS ORIGINALLY PART OF A TEA AND COFFEE SERVICE ALL DECORATED IN THE SAME PATTERN. SAUCERS ON THEIR OWN ARE GENERALLY INEXPENSIVE AND POPULAR WITH COLLECTORS AS THEY ARE EASY TO DISPLAY. £100–150

The modest gilding is a characteristic of much early German porcelain.

The idyllic landscape subject and the absence of a frame around the painting are also typical Rococo features.

The reverse side would be marked in the usual way with the crossed swords and possibly with an impressed number.

Monochrome decoration was fashionable in the mid-18th century; puce was particularly popular in the Rococo period.

▲ COLOURED GROUNDS

Coloured grounds were introduced in the latter 1720s and always command a premium with collectors. The turquoise ground on this faceted ovoid teapot (c.1735) reflects the colour of Chinese celadon; other colours you might come across include yellow, claret and dark blue. £3,000–4,000

▼ BEWARE

Meissen copies are quite common and can be difficult to spot; this one by Samson (see p31) is based on a group of c.1740. Value if it were Meissen, £4,000–6,000; the Samson version is worth £450–600.

19TH- & 20TH-CENTURY MEISSEN

During the early 19th century Meissen, in common with many other leading European porcelain manufacturers, produced elaborately decorated and gilded pieces in the prevailing Empire style. Painted decoration became increasingly elaborate, with the surface treated as a canvas and adorned with romantic topographical views or Classical subjects. As the century progressed the factory concentrated on repetitions of 18th-century designs and this revivalist trend continued until the 1920s when designers such as Paul Scheurich and von Lowenich began to produce imaginative modern pieces. Although less sought after by serious collectors 19th-century Meissen has obvious decorative appeal.

▶ STYLES
Pierced work had been made from the mid-18th century but became increasingly popular from the 1820s as part of the fashion for Rococo style. The deeply modelled contours of trailing flowers with small floral cartouches that adorn this ornamental basket (c.1840) are typical of the 19th century's love of opulent decoration. £400–600

◀ HUNTING THEMES
The hunt proved a popular theme in the 1760s and continued to inspire many German factories in the following century. This base of this figure (c.1880) is perhaps more elaborate than an 18th-century example, and it is larger in scale. £2,000–3,000

▶ SUBJECTS
Scenes based on classical antiquity were popular in the late 19th century and this scene showing Europa and the Bull is after a painting by Boucher. The blue and gilded border reflects the style made popular by Sèvres. Wares of this type were copied by many other European factories in the late 19th century. It is their decorative appeal rather than their rarity that makes them sought after. £400–600

MARKS

19th- and 20th-century marks on figures are usually bigger than 18th-century ones. Between World Wars I and II the blades are curved but still hand-painted.

◄ FIGURES
This figurative group contains a typical 19th-century cocktail of decorative styles: elements of Neo-classicism in the border and urn; and the rustic naturalism of the Rococo period in the base and costumes. The slightly fiddly, detailed costumes are also characteristic of the 19th century. £1,000–1,500

► FORMS
Forms became more complex at the end of the 19th century, and decoration often features a confusion of different styles. In this example Watteauesque scenes, first made popular between c.1740 and 1745, are teamed with stylized flowers typical of a later period and the vase's shape is clearly derived from Classical prototypes. £2,000–3,000

◄ 20TH-CENTURY MEISSEN
One of the most innovative designers working at Meissen in the 1930s, Professor Paul Scheurich specialized in figures with elongated limbs and decadent poses in the Art Deco style and his work commands high prices. This group of a woman and Moor is modelled and decorated with typical refinement and delicacy. £2,500–3,500

OTHER GERMAN FACTORIES

Within a few decades of Meissen's discovery of the secret of porcelain-making, a potent mixture of greed and treachery helped to spread the formula throughout Europe and by the mid-18th century a handful of other German factories had been established. Among the most important of these were Höchst, Frankenthal, Ludwigsburg and Nymphenburg. Other less prolific factories included Fulda and Fürstenberg. Although

	FRANKENTHAL	HÖCHST
	 A Frankenthal group of shepherd and flock, 1773. £4,000–5,000	 Children playing by J. P. Melchior, c.1767 £1,500–2,000
CHARACTERISTICS	• Shallow rounded base with shallow Rococo scrolls broadly washed in gilding • Mound encrusted with mossy shavings of porcelain • Fine white body	• Very pale yellow, pink and blue details • Simple grassy mound base with brown streaks • Sweet plump childish faces • Creamy body
KEY MODELLERS	• Johann Friedrich Lück • J. W. Lanz • Conrad Linck	• J. P. Melchior • Johann Friedrich Lück • Laurentius Russinger
MARKS	• Initials "CT" in underglaze blue • Sometimes numbers to show year of production 	• Wheel mark, usually in underglaze blue, also in purple or blue • Incised numbers and letters sometimes 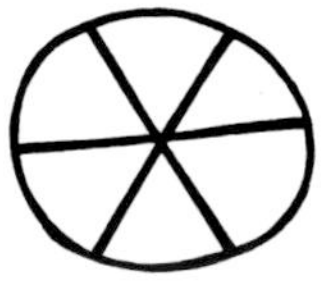

the factories produced a similar range of products, many based on designs first developed by Meissen, each developed its characteristic style and used distinctive paste, and some knowledge of both can help with identification. Of the smaller factories' output pieces made by Nymphenburg are particularly refined and correspondingly expensive, but all factories tend to have a keen following among porcelain collectors.

LUDWIGSBURG	NYMPHENBURG	
A group of dancers by J. Nees, c.1765. £4,000–6,000	Figure of a female dancer, c.1760. £8,000–12,000	
• Stiffly posed figures; faces have long bridgeless noses and doll-like features • Notched deeply modelled base • Subdued pastel palette • Smoky, greyish glaze	• Sophisticated handling of porcelain • Subtle Rococo pose with upwardly spiralling movement • Elongated bodies, delicate faces with slightly exaggerated expressions • Creamy coloured lightweight slip-cast body	CHARACTERISTICS
• Joseph Nees • Johann Beyer • J. H. von Damiecher	• F. A. Bustelli • J. P. Melchior • D. J. Auliczek (famous for hunting scenes)	KEY MODELLERS
• Interlaced Cs with coronet or a staghorn mark: three in a shield or one separate	• An impressed or incised shield mark often prominently placed at front of base	MARKS

BERLIN & LATER GERMAN FACTORIES

Berlin's first factory, founded in 1752 by W. K. Wegely, closed only five years later. In 1761 King Frederick the Great, together with a financier, established a second factory which the King later took sole charge of, succeeded by the State; this factory survives to the present day. Berlin porcelain of the 18th century is extremely rare and you are far more likely to come across 19th-century pieces, exquisitely decorated with detailed panels on solid coloured grounds, which are easily confused with porcelain made around the same time in Vienna and Paris. The Berlin factory also produced quantities of porcelain plaques, many of which were independently decorated with copies of well-known paintings. During the 19th and early 20th centuries the area near Meissen surrounding the city of Dresden became a major German centre of porcelain production with numerous factories producing ornamental and tea and dinner services in the style of Meissen and Sèvres.

◀ **VALUE**
Berlin and Dresden porcelain is widely available at a range of prices depending on the usual criteria of size, subject and decorative appeal. Pieces such as this Berlin *solitaire* (c.1775), beautifully painted in puce with Watteauesque figures in landscape vignettes, reflect the Rococo taste for monochrome decoration and no frames. £7,000–9,000

▶ **TOPOGRAPHICAL & CITY VIEWS**
Topographical and city views were a speciality of Berlin and Viennese porcelain decorators from the early 19th century. Scenes such as the one of the Gross Schloss decorating this egg are often named and among the most finely detailed of those produced anywhere in Europe. Earlier examples have either rectilinear or formal borders. The neo-Rococo gilt borders here indicate a date c.1880. Small decorative items like the egg shown here are rather unusual; plates and cabinet cups and saucers are much more readily available to collectors at lower prices. £1,000–2,000

FIGURES

A range of figures based on Meissen models was made by Wegely. The second factory made a series called "The Cries of Berlin", modelled by the Meyer brothers.

MARKS
Berlin marks include a sceptre on early pieces; from 1832 an orb appears; sometimes the letters "KPM" either in underglaze blue or red or impressed are also found.

KPM

► **PORCELAIN PLAQUES**
Popular paintings by artists such as Watteau, Murillo and Richter provided the subjects for huge numbers of porcelain plaques manufactured in Berlin. Ladies were favourite subjects too; this plaque shows a copy of a portrait of Beatrice Cenci by Guido Reni (c.1880). £2,000–3,000

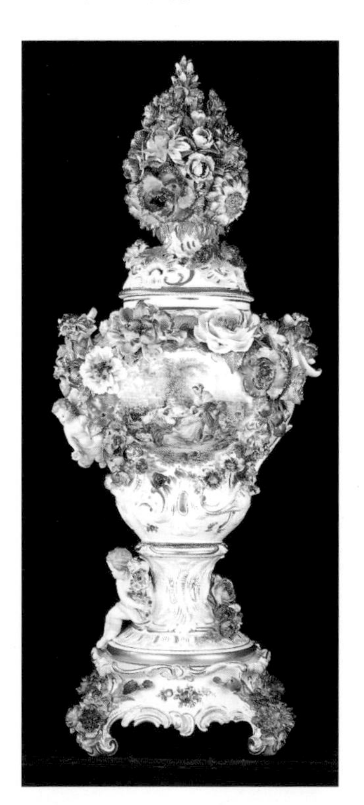

► **DECORATION**
Snowball vases or *Schneeballen*, decorated with May blossom, were first made by Meissen in the 18th century, but several 19th-century factories continued the tradition by covering the entire surface of objects with tiny florets. This large unmarked vase could be made by a number of factories in the Dresden area. £1,000–1,200

◄ **CARL THIEME**
A brilliant example of the 19th-century love of excessive ornament, this baluster vase was made by Carl Thieme, one of the major manufacturers in the Dresden area. Pieces such as this are obviously extremely vulnerable to damage and small chips are acceptable. £2,000–3,000

VIENNA

Vienna was the second European factory to begin producing hard-paste porcelain. The factory was founded by Claudius Innocentius du Pacquier, who bribed and cajoled the secret from various disgruntled Meissen employees. Until the 1780s Vienna produced a similar range of wares to those made by Meissen, but figures were rarer and more stiffly modelled. In 1744 the company was taken over by the State and the hitherto limited production increased in quantity. In 1784 a new director was appointed and the factory began producing wares in the Neo-classical style, with elaborate painted scenes and heavy gilding of very similar appearance to wares from centres such as Berlin and St Petersburg. The Vienna factory closed in 1864 but many imitations of its more elaborate wares were produced in the late 19th century.

Porcelain from the du Pacquier period is rare, and consequently prices start at around £1,500. Also highly desirable are pieces from later periods decorated by sought-after painters such as Joseph Nigg, a famous flower painter. Prices for other late pieces are chiefly dependent on their decorative appeal.

◀ **DECORATION**
The use of solid gilt grounds became more prevalent at the turn of the 18th century and was used by all leading factories from Paris to St Petersburg. This coffee cup is decorated with a combination of Classical key border and informal Rococo scattered flowers.
£400–600

▲ **WARES**
Although tablewares made up the bulk of the factory's output it also made some enchanting accessories such as this exquisite travelling casket (c.1785), decorated in French style with a dotted green ground.
£3,000–5,000

DU PACQUIER PORCELAIN

Porcelain from the du Pacquier period (1719–44) is rarer than Meissen and few pieces have survived. It is similar in composition to early Meissen and often has a greenish tone. Wares are usually decorated with chinoiserie and very formal Baroque strapwork borders. Dominant colours are puce, iron red and monochromes. Compared with Meissen the manner of painting is naive. Figures are rare and idiosyncratic.

MARKS

Vienna was marked with a distinctive shield mark – much copied on later pieces "in the style of" Vienna. From the 1780s the factory also used three impressed numerals denoting the last three numbers of the year of manufacture: for example, the mark 817 would indicate a piece made in 1817.

▲ **LATER "VIENNA-STYLE" WARES**
After 1864 other European factories continued to produce wares in similar style. This vase and cover, one of a pair decorated with scenes of Galatea, and Europa and the Bull within seeded gilt lines, are late 19th-century pastiches of Vienna wares of the early 19th century. Less sensitively painted than the genuine article, they are worth a fraction of the price. £600–700 for the pair

▲ **TOPOGRAPHICAL SCENES**
Topographical scenes were one of the specialities of Viennese decorators and the scenes were often named on the base. This cabinet cup and saucer (c.1830) is decorated with a named view of Schönbrunn palace. £1,200–1,500

▼ **STYLES**
The landscape subject within formal borders shows the transition between the decorative Rococo and the nascent Neo-classical styles. *Tête à têtes* (small teasets or cabarets) containing enough pieces for two became fashionable in the mid-18th century as part of the trend for boudoir intimacy. £12,000–18,000 for 8 pieces

ITALY

Italy was the first European country to try to produce porcelain and began experimenting in the 16th century. In 1575 Duke Francesco de Medici opened a short-lived factory in Florence but the first successful porcelain factory, Vezzi, was not established until 1720, when a former Meissen employee, Christoph Hunger, stole the formula and took it to Venice. Other leading factories at Doccia and Cozzi soon followed suit and the most famous Italian porcelain of all was made from 1743 at Capodimonte, Naples. When Naples fell under Spanish rule the Capodimonte factory moved to Buen Retiro in Spain, but returned to Naples in the late 18th century. Some Italian porcelain is characterized by its greyish granular appearance and is easily confused with pieces from minor German factories. Prices tend to be comparatively high and there is a strong home market. The most readily available pieces tend to be from Doccia, mostly in the quasi-Oriental or Neo-classical style.

MARKS

- The most common mark is the red or gold anchor of Cozzi – this is much bigger than the similar Chelsea mark.
- Capodimonte use a fleur-de-lis mark, usually in underglaze blue or gold.
- After 1757 Doccia marked with this star; a red, blue or gold "F" is used 1792–1815.

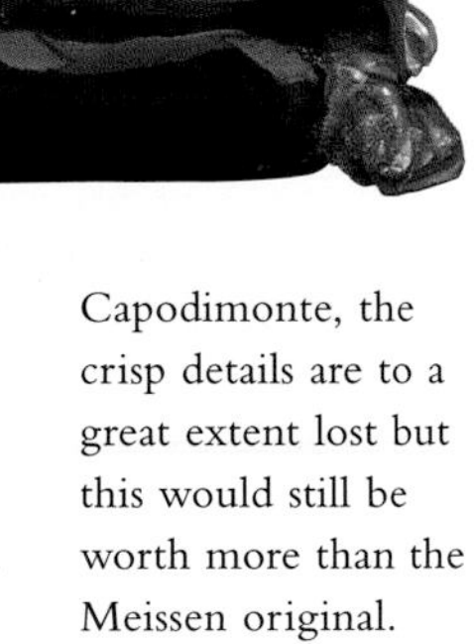

▲ **CAPODIMONTE**
This figure of a pug (c.1750) is a copy of a Meissen original by J. J. Kaendler of c.1740. Reinterpreted in the soft-paste porcelain of Capodimonte, the crisp details are to a great extent lost but this would still be worth more than the Meissen original. £4,000–6,000

▼ **IDENTIFICATION**
The rather delicate but primitive brushwork decorating this teacup and saucer is similar to Kakiemon styles of the 1740s. The angular scrolled handle is characteristic of the Capodimonte factory and can be found on coffee pots and other similar wares. £1,500–2,000

FIGURES

Italian factories made a range of figures dressed in a variety of ways. A classical subject is less desirable than a group in contemporary dress, and pieces should always be in good condition.

◀ **SHAPES**
The forms used by Italian porcelain-makers are often idiosyncratic and distinctive from those used by factories elsewhere in Europe. This Doccia coffee pot (made c.1775) combines an early baluster form, an oddly shaped spout and a shallow domed cover with Neo-classical decoration. The tentative, fussy gilding is also characteristic of much Italian porcelain. £5,000–8,000

► **DECORATION**
One feature of late 18th- and 19th-century decoration is the use of relief-moulded decoration. This small covered beaker, made by the Cozzi factory c.1780, is decorated with the low-relief lion of Venice. £800–1,200

◀ **NAPLES**
Neo-classical decoration was very popular on Neapolitan porcelain of the late 18th century and this coffee can and saucer (c.1790) are no exception. Fashionably decorated with Classical figures on a black ground panel, they boast borders of palmettes and swags which are also derived from antiquity. These ornaments are most probably inspired by the local archeological excavations at Herculaneum and Pompeii on the Bay of Naples.
£2,500–3,500

EARLY FRENCH PORCELAIN

The first French factory to produce porcelain successfully was at St Cloud, near Paris, in the 1690s, although some was probably made at Rouen earlier. Among the other early factories who followed suit in the 18th century, the most important were Chantilly and Mennecy. French factories produced tablewares and figures as well as a range of small decorative objects such as snuff boxes, *étuis* and *nécessaires.* Decorative styles and good proportion were influenced by Chinese and Japanese ceramics, although some slightly later pieces also reflect fashions introduced by Meissen. Until 1769 all porcelain produced in France was soft paste (see p26) and compared with German porcelain of the same date it looks softer and creamier. Pieces are often unmarked and can be difficult to attribute as factories often changed the pastes they used. In general, 18th-century French porcelain is less widely available than that made by leading German factories. While prices tend to reflect this relative rarity the naive quality of the French product (except Sèvres) tends to make it slightly less expensive than more sophisticated pieces of the same date made in Germany.

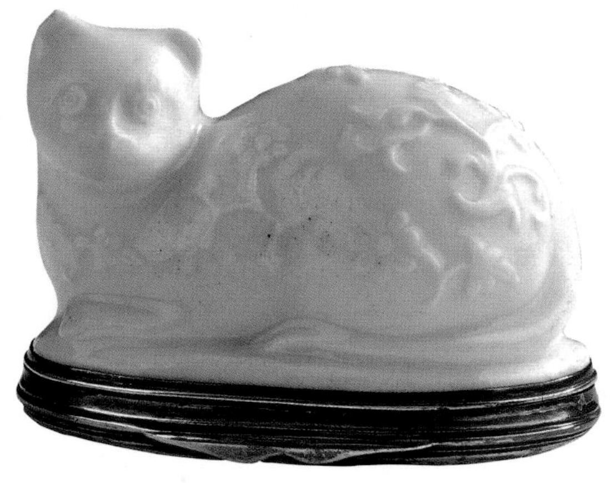

▲ ST CLOUD
This snuff box is typical of the range of small decorative items produced in France in the 18th century. Many of these objects have metal mounts which may, as in the case of this one, bear marks that can help with dating. The white cat shows the influence of *blanc de Chine* – uncoloured, thickly glazed Chinese porcelain produced from the Ming dynasty onwards. St Cloud paste is greyish-white and often peppered with black flecks. £2,000–3,000

BEWARE
Chantilly's distinctive style was imitated in the 19th century by Samson of Paris – and sometimes the original Samson marks have been removed. Samson however used hard-paste porcelain, and pieces look harder and shinier compared to true soft paste.

▼ CHANTILLY
The patron of the Chantilly factory was the Prince de Condé, a passionate collector of Kakiemon, and most of the factory's wares reflect his taste for the exotic. The flowers decorating this *bonbonnière* reflect this favourite style and indicate that it was probably made at Chantilly. £3,000–5,000

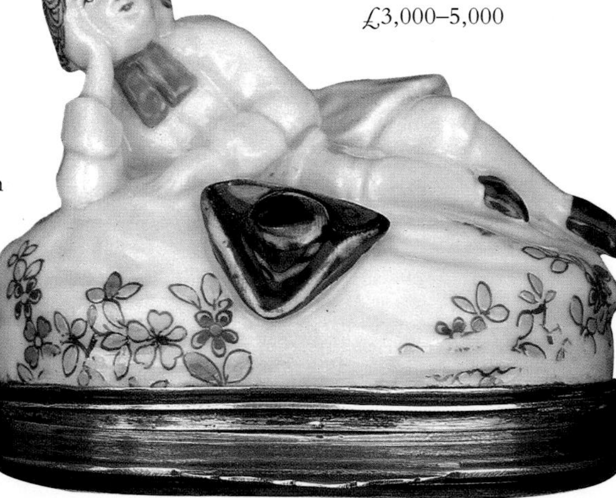

► **GILDING**
This Chantilly cachepot (c.1750) is decorated with a rustic European landscape inspired perhaps by Meissen. In common with most early French porcelain (and unlike 18th-century German porcelain) it is not gilded. Flower pots and cachepots were especially popular at Chantilly and Sèvres. The pleated contours and the handles applied with flowers are characteristic, as is the subtle delicate brushwork. This factory used tin glaze, perhaps to hide defects in the clay and to improve the surface, and consequently wares may often have a somewhat opaque appearance. £2,000–3,000

MARKS

- **St Cloud** variations of this mark are found in blue and red or impressed.

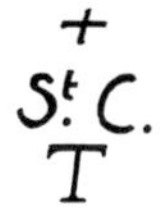

- **Chantilly** the hunting horn mark is used in red, black and blue. Rarely the word "Chantilly" appears.

- **Mennecy** the initials "DV" for Duc de Villeroy, patron of Mennecy, were used, sometimes with a crown above.

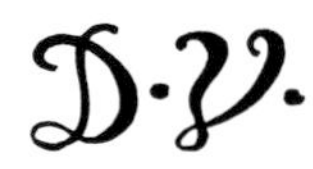

◄ **MENNECY**
This (c.1755–60) rather stiff and puppet-like pair of Mennecy figures are typically naively modelled and are painted in soft, rather washed-out colours, with the predominant use of pastel blue, irregularly streaked pink and yellow. Faces typically lack detail. The factory produced an extensive range of tea and coffee services, decorative wares such as vases, and figures. Figures of peasants and traders are available from £2,000 and upwards, but exotic subjects such as these are usually much more valuable. £5,000–8,000 for the pair

VINCENNES & SÈVRES

France's most famous porcelain factory, Sèvres, started life at Vincennes near Fontainebleau, where in 1738 a new factory began to produce soft-paste porcelain. In 1756 the factory moved to Sèvres, near the home of one of its keenest patrons, Louis XV's mistress Madame de Pompadour; soon after, the King became the owner of the factory. After 1769 Sèvres began making hard-paste porcelain (as well as soft paste) and after c.1803 stopped making soft paste completely.

Sèvres porcelain is famed for its lavish gilding and brilliantly coloured grounds, which formed a framework for panels decorated with flowers, figure subjects or landscapes. Its illustrious reputation has however attracted numerous imitators. Marks and styles were copied throughout Europe and confusion also arises because at the time of the French Revolution the factory was taken over by the state and numerous pieces were sold in the white to independent decorators in France and England. Coloured grounds are prone to faking; probably about 90 percent of all *bleu céleste* pieces, in the delicate sky blue for which Sèvres is known, are imitations.

MARKS

The interlaced Sèvres Ls were centred with a date letter from 1753. This is one of the most commonly faked of all porcelain marks. Look for strong definition on genuine pieces – a weak and attenuated mark is usually suspect.

▶ VINCENNES

All Vincennes and Sèvres forms are known by special names; this double-handled vase, made c.1755, is called a vase *Duplessis à fleurs* – after the celebrated sculptor employed by the factory to create interesting shapes. This colour, called *bleu céleste*, was the most popular of all coloured grounds – and also the most commonly faked. Pieces from the Vincennes period are rare and therefore command a premium. £5,000–8,000

▲ COLOURS

The brilliant shade of green used for this tea caddy (c.1780) was another of Sèvres' most distinctive colours. The range of coloured grounds was gradually introduced in the 18th century. Knowledge of these colours can help with dating and spotting fakes (see p24), since some pastiches combine late colours with early date marks. Green was introduced in 1756. Other Sèvres colours include: *bleu lapis* (lapis blue), 1749; *bleu céleste* (sky blue), 1752; *jaune jonquille* (pastel yellow), 1753; *violette* (violet), 1757; *rose* (pink), 1758; *bleu royal* (royal blue), 1763. £2,000– 3,000

◀ **GILDING**
Sèvres gilding is invariably soft and richly applied and the factory developed a wide range of distinctive techniques. This coffee can and saucer (made 1768) are decorated with the popular *œil-de-perdrix* (partridge eye) gilded decoration; *caillouté* (pebble) decoration was another favourite pattern; while on some pieces gilding was raised and tooled to create detailed bouquets of flowers and foliage. £800–1,200

▼ **DECORATION & FORM**
Monochrome decoration as on this tea service was less commonly used at Sèvres than in German factories but the shapes of the pear-shaped jug with tripod feet and ovoid teapot with ear-shaped handle are absolutely typical of 18th-century Sèvres forms. £5,000–7,000

BISCUIT FIGURES
The Vincennes and Sèvres factories also produced a range of unglazed porcelain figures modelled by E. M. Falconet, often after the designs of François Boucher. This type of porcelain is known as biscuit. Until recently these figures remained relatively inexpensive although they have risen considerably in value in the last few years.

◀ **LATER SÈVRES**
Like other leading European factories in the early 19th century Sèvres began producing more formally decorated pieces using semi-mechanical methods of gilding. This pale blue (*bleu agate*) tea service is made from hard-paste porcelain and looks much harder and sharper than the earlier pieces illustrated on these pages. £1,500–2,000

LATER FRENCH PORCELAIN

From the late 18th century and throughout the first half of the 19th Paris blossomed as a centre of porcelain manufacture as dozens of factories and enamelling shops began producing dinner services, teawares and ornamental pieces for the newly affluent population. As the 19th century progressed, the rising costs of running a commercial factory in the capital encouraged many manufacturers to move to the countryside near Limoges, where labour was cheaper and raw materials were readily available. Most French factories of this time have little individual style, relying heavily on earlier designs from Sèvres and elsewhere for inspiration. All factories produced only hard-paste porcelain (although some also made pottery) and marked their wares erratically. Many unmarked pieces and pieces with spurious Sèvres marks also survive; these are often catalogued either as "Paris" or as "Sèvres style". Prices depend heavily on the size and decorative appeal of the subject. Small plates are available for under £300, while large decorative vases can cost £1,000 or more.

◀ **DECORATIVE WARES**
Metal-mounted vases and covers are typical of the decorative items produced from the mid-19th century. The figurative panel, probably copied from a painting by Boucher or Watteau, is set against a dark blue ground – probably the most popular of all coloured grounds at this time. £2,000–3,000

IMPORTANT 19TH-CENTURY FRENCH FACTORIES

Sèvres continued to be France's best-known porcelain factory, but other leading names included:

- **Paris** Jacob Petit, Dagoty & Honoré, Darte Frères, Discry, Talmour, Pouyat, Nast, Clignancourt
- **Limoges** Denuelle, Michel & Valin, Alluaud, Ardant, Gibus & Cie, Jouhanneaud & Dubois, Haviland, Ruaud

MARKS

As with many pieces of this type the vase above is marked only with imitation Sèvres marks; few pieces are marked by their manufacturer, and some are not even porcelain. If genuine marks do appear they are usually of the overglaze, stencilled or printed varieties.

▲ **PICTORIAL SUBJECTS**
Scale and ostentation were typical of many of the decorative items produced at this time and pictorial plates featuring Louis XVI and Marie Antoinette were especially popular after c.1850. Many of the portraits on plates such as these were transfer-printed and then enamelled over. Their value lies chiefly in their decorative appeal. £300–500

◀ **THEMES**
The fashion for Egyptian subjects, a hallmark of French Empire style in the early decades of the 19th century, was fuelled by Napoleon's campaigns in North Africa. This elaborate centrepiece is adorned with characteristic sphinxes supporting a basket bordered with lotus petals. Both stylish and large (measuring 18in [46cm] high) this piece would command a high price in the market. Smaller pieces would cost a fraction of the price. £4,000–5,000

▶ **TEA SETS**
Among the wide range of tea sets produced at this time were cabarets – small tea sets usually for two also known as *tête à têtes* (see below) and *solitaires* for one. This late 19th-century cabaret is painted with 18th-century court beauties and includes a teapot, large sugar bowl, jug, cup and saucer and matching tray. £600–800

◀ **VALUE**
Quality and an illustrious provenance can dramatically affect the value of porcelain. Painted in a tasteful *grisaille* (grey monochrome), the decoration of this *tête à tête* made c.1785 by the Paris factory of the Duc d'Angoulême is highly refined, and the set has a royal history: it was a gift from Marie Antoinette to a Swedish count. £2,500–3,000

SWITZERLAND

In Switzerland a porcelain factory founded in 1763 at Scoren near Zurich by the painter Salomon Gessner continued to produce wares until the late 19th century. The factory began by producing soft-paste porcelain but by 1765 this was replaced by hard paste. Wares include tea and coffee services as well as a wide range of figures which are often recognizable by their rather static poses. Painted decoration tends to follow fashions established elsewhere in Europe although the style in which flowers and landscapes are painted is often sparse and stiff. The hard paste used by the factory tends to be greyish in colour and sugary in texture, and has a glassy greyish glaze. Colours before 1780 are typically soft though after this date a strong russet and yellow were introduced. Although Swiss porcelain is less widely available than that made by major German and French factories, it is not too scarce and remains popular with Swiss and other European collectors. Values are similar to those for smaller German factories, with the highest prices being paid for early rarities.

▶ WARES

This pair of Zurich dinner plates (c.1770) have obvious similarities with pieces made in Germany, but the figures are more stiffly painted and the three scattered landscapes on the rims would be unusual on German porcelain. The gilt dentil (indented) rims are also highly characteristic of Swiss porcelain. £3,500–4,500

▼ VALUES

Teabowls and saucers, such as this made c.1770, are among the most affordable pieces of 18th-century Swiss porcelain. The estuary scenes which decorate these examples are populated with Oriental figures – a reflection of the taste for chinoiserie. £300–500

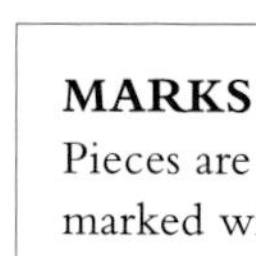

MARKS

Pieces are usually marked with a "Z" incised or in underglaze blue. Dots and incised letters and numbers are also sometimes found.

► **MINIATURES**
The Zurich factory also produced a range of miniature figures which are usually more affordably priced than larger subjects. These figures of a song seller and a fisherwoman (c.1770) measure about 4in (10cm); most figures are about twice this size.
£800–1,200 each

FIGURE SUBJECTS

Among the range of subjects produced by the Zurich factory, those you might come across include:

- street vendors
- fisherfolk
- musicians
- craftsmen
- fashionable ladies and gentlemen
- classical figures
- hunters
- shepherds and shepherdesses

FIGURES

THE ZURICH FACTORY PRODUCED A WIDE RANGE OF FIGURES THROUGHOUT THE 18TH CENTURY. THIS GIRL WITH A BROKEN POT, MODELLED BY VALENTIN SONNENSCHEIN C.1775, HAS MANY OF THE FEATURES YOU WOULD LOOK FOR ON A SWISS FIGURE. £2,500–3,000

Lack of detail in the face is a typical feature.

Soft autumnal or pastel shades are invariably found.

The modelling is so loose that the hard paste almost looks like soft paste.

Modelling of details such as hands and costume is very rudimentary.

Simple bases with little decoration are found on most Zurich figures.

The piece is marked with a blue "Z" and two dots.

OTHER PORCELAIN CENTRES – DENMARK & LOW COUNTRIES

The first porcelain factory was established in Copenhagen in 1760 with the help of German craftsmen from Meissen and Fürstenberg. The factory produced soft-paste porcelain in its early years (these pieces are very rare) and began making hard paste after 1783. Royal Copenhagen is perhaps most famous for the massive service made for the Russian Empress Catherine the Great c.1780–1805. Known as Flora Danica, each piece was adorned with a different botanical study, and the pattern is still produced today. Among Royal Copenhagen's other successes are Art Nouveau-style pieces produced in pale grey shades, small figures in regional costumes and animals. The second most famous Danish factory is that of Bing & Grondahl; founded in 1853, this company produced high-quality wares in similar styles to Royal Copenhagen.

Porcelain was also produced in various centres in the Low Countries. In Belgium the most famous centre was at Tournai; in the Netherlands Amstel, Weesp and Oude Loosdrecht all produced porcelain from 1757–1820. In general although wares from these factories are easily available in their native countries comparatively few examples are seen for sale on the international market.

▲ OTHER PATTERNS
Royal Copenhagen also produced a wide range of tablewares in styles derived from Meissen and other European centres, and its more recent products are still available for modest sums. This conventional service, made in the early 20th century and based on French Rococo patterns, contains over 125 pieces including: tureens; platters; sauceboats; leaf-shaped dishes; serving, dinner, dessert and hors d'oeuvres plates; salt, pepper and mustard pots; butter dishes; bowls; cups and saucers; jugs and many more items. Perhaps surprisingly it would be cheaper to buy than a modern equivalent. £3,000–5,000

▲ **FLORA DANICA**
Versions of this exquisitely painted service have been produced ever since the first one was made (late 18th century) for Catherine the Great. Modern services are less valuable than those made pre-1900. Early painted decoration looks lively and robust; modern decoration may seem flatter. The porcelain on modern versions is finer and more glassy in appearance. This service made c.1900 contains 55 pieces. £14,000–16,000

▶ **FIGURES**
Royal Copenhagen figures have enjoyed huge popularity with collectors and are among the most popular pieces. Groups such as this in historical dress provided a favourite theme, as did figures in regional dress. These 20th-century groups of elegant couples are typically well modelled and painted in detail with subtle shades.
£200–300 each

◀ **TOURNAI**
Figures were one of the specialities of the Belgian Tournai factory, which was the largest manufacturer of porcelain in the Low Countries. Many figures were unglazed and these, made c.1765 and part of a group depicting the Four Seasons, are also typical in their idyllic pastoral theme. They are probably based on a Boucher subject and are easily confused with Sèvres biscuit groups and later Derby wares. £1,500–2,500

OTHER PORCELAIN CENTRES – RUSSIA

Russia's foremost factory was established in St Petersburg c.1748 and became the Imperial Porcelain Factory in 1763. Early wares reflected the style of Meissen but under imperial patronage the factory began making more ambitious pieces. Wares of the early 19th century are particularly distinctive, and characteristically have detailed pictorial decoration and extravagant gilding. After the 1917 Revolution the factory was taken over by the State and went on to produce some interesting Constructivist designs which are much sought after. The factory is still in production. Russia's second most famous factory was established by Francis Gardner in 1776 and flourished until the 1880s when it was taken over by the rival Kuznetsov factory. Gardner was famed for figures in regional costumes and often used matt unglazed surfaces in sombre colours, especially Prussian blue.

▶ KUZNETSOV PORCELAIN
The Kuznetsov factory became Russia's biggest porcelain manufacturer at the turn of the 19th century, producing a wide range of wares for the home and export market. This Kuznetsov beaker (c.1900) is decorated with a portrait of the Tsar's son surrounded by typical Russian designs. Similar commemorative beakers were also produced in enamel. £400–600

▲ ST PETERSBURG
Painted with a scene showing a military encampment set against a solid blue ground with a formal ribbon border this sucrier (c.1780) reflects the early Neo-classical fashions popular in England, France and Austria. £700–900

▶ TOPOGRAPHICAL SUBJECTS
The scene of St Petersburg which decorates the centre of this Imperial Porcelain Factory plate (c.1820) is typically detailed and pictorial. The elaborately moulded rim is however unusual at this time and the gilded wreath decoration is unique to St Petersburg. £1,000–1,500

VALUE

THIS PLATE REFLECTS THE EARLY 19TH-CENTURY FASHION FOR LAVISH GILDING AND DETAILED DECORATION, AND IT HAS SEVERAL ADDITIONAL DESIRABLE FEATURES WHICH ADD SIGNIFICANTLY TO ITS VALUE.
£3,000–5,000

A signature by a leading artist – S. Daladugin.

The soldiers' regiment, rank and the date are also inscribed on the reverse of the plate.

Exceptionally high-quality painting. Similar pieces were made in Meissen and Berlin and often identification relies on marks and differences in the palette.

MARKS

St Petersburg porcelain is marked with a crown and also with the initials of the emperor or empress of that particular period in Roman letters. Other factories often used Cyrillic letters and marks can be the cause of confusion.

◄ POST-REVOLUTIONARY DESIGNS
Porcelain made after the 1917 Revolution was decorated with interesting and innovative angular designs which were inspired by the new styles developed by Russia's Constructivist designers. Subject matter typically reflected the country's new political ideals, and this cup and saucer designed by Maria V. Lebedeva in 1923 is decorated with factory scenes and workers in a typical Constructivist palette of various reds, black and grey.
£1,500–2,000

British Porcelain

England did not begin to produce porcelain until the 1740s when the earliest factories were opened at Chelsea, Bow and Bristol. Other important factories that were founded in the 18th century include Derby, Worcester, Longton Hall and Lowestoft. With few exceptions all these early factories produced soft-paste porcelain and most made luxury items as well as domestic wares. Porcelain making was, however, an expensive and precarious business and by the end of the 18th century most of the famous early factories had closed or had been amalgamated with rivals. Among the most successful and long-lived centres of production were Worcester and Derby where factories continued to produce porcelain throughout the 19th and 20th centuries. Industrial development in England in the 19th century led to the establishment of large factories producing a wide variety of goods, ranging from luxury bone china to domestic pottery.

So diverse are the wares produced by major factories such as Worcester, Minton, Spode and Derby that collectors often concentrate on buying ceramics of a particular type rather than choosing from the whole spectrum of products. English porcelain in general offers a rich and varied hunting ground for collectors. The most valuable pieces tend to be those from the 18th century, or larger elaborately decorated pieces made in the 19th century. However, it is still possible to buy some of the smaller 18th-century pieces which are in less than perfect condition for relatively low sums.

BOW

Arguably the earliest of the English porcelain factories Bow, founded in 1744, was called "New Canton" and in its heyday employed hundreds of workmen producing wares for a wide market. Until its closure in 1775, Bow continued to produce huge quantities of blue and white porcelain in the Chinese style, Japanese-inspired coloured wares and a range of figures based on those of Meissen. Slight imperfections in the glaze are characteristic of many pieces. Bow's extensive output means that there is plenty available for collectors to purchase and values tend to be lower than for Chelsea. Prices range from around £150 for tablewares to £5,000 or more for elaborate pieces. Generally most pieces fall within the £1,000–3,000 price range.

▲ **FIGURES**
This model of the actress Kitty Clive in a Garrick farce (c.1750) is simply modelled – a feature associated with early Bow figures. The warm glaze and slight discoloration are also typical. Theatrical subjects have a keen following, and this is early, hence the high price. £4,000–6,000

► DECORATION
This pair of bottles made c.1755 juxtaposes a Chinese shape with decoration in red, blue and yellow in the style of Japanese Kakiemon, although the flowers and birds are rather more stiffly painted. The peppering on the neck is often seen on Bow porcelain. Identical patterns with exotic long-tailed birds were also found on Chelsea and Arita porcelain (from Japan). £1,500–2,000

◄► IDENTIFICATION
One of the many domestic blue and white wares produced by Bow, this tankard (c.1760) has a handle finished with a heart-shaped terminal – a charming feature unique to Bow. The rusty discoloration on the base is found on both Bow and Lowestoft porcelain. £800–1,000

► LATER FIGURES
Later Bow figures are identifiable by the tall raised Rococo scroll bases and extensive *bocage* (leafy encrustations). Other factories made similar figures to this but Bow bases are very distinctive. Figures such as this made c.1770 are often marked on the reverse of the tree support. £600–900

MARKS
Various incised marks were used but wares are often unmarked. The most common mark on figures is a red anchor and dagger.

CHELSEA

One of the earliest and most important porcelain factories of its day, the Chelsea factory was founded in c.1745 by Nicholas Sprimont, a French Huguenot silversmith. Chelsea catered for the luxury London market and many of its inventive pieces reflect fashionable silver shapes. Chelsea porcelain is dated and classified according to the various marks used, and among the most sought-after pieces are those of the Red Anchor period, such as the "Hans Sloane" wares decorated with botanical specimens. The factory was taken over by the Derby Porcelain Works in 1769 and finally closed in 1784. Chelsea has a huge following and pieces copied from Meissen at Chelsea can fetch more than the originals.

CHELSEA PORCELAIN

Marks	Characteristics	Dates
Incised triangle	Small, often white pieces, silver shapes	1745–49
Raised anchor	Small, Meissen style, slightly opaque glaze	1749–52
Red anchor	Highly refined, Meissen and Kakiemon styles	1752–57
Gold anchor	Sumptuous decoration, Sèvres style	1759–69

GIRL IN A SWING

A second factory in London made very similar porcelain, especially tiny novelties. Known after its most famous figure, the "Girl in a Swing" it has now been identified as Charles Gouyn's factory in St. James (c.1749–59).

▲ RAISED ANCHOR

Jugs have more cachet than teabowls and hence always command a premium. This peach-shaped cream jug from the Raised Anchor period reveals the influence of Meissen in the floral decoration, which is based on the *deutsche Blumen* (German flower) designs but the painting is softer and the handle very naive. £5,000–7,000

▶ FIGURES

The British Ambassador in Saxony borrowed numerous Meissen figure models for Chelsea to copy and many figures were copied but the limitations of soft paste meant that the dramatic Meissen poses were impossible to achieve. This Oriental figure (c.1754) is typically softly modelled but columnar in form. £2,500–3,500

► **RED ANCHOR**
The body of Chelsea porcelain is rarely without imperfections and some of the insects and floral sprays decorating the surface of this lobed dish made c.1750–52 probably serve to disguise small blemishes. Chelsea Red Anchor porcelain is typically greenish in hue. The underside of dishes such as this will have three small raised spur marks – caused by the way the plates were stacked in the kiln during firing. £4,000–6,000

◄ **LATER FIGURES**
Chelsea figures of the later 18th century became more elaborate and usually stood on bases which were richly applied with floral *bocages* (leafy encrustations). The costumes of figures are usually heavily decorated too and pastel colours are favoured. £2,000–3,000

► **DECORATION**
In the mid-18th century Chelsea turned to the more florid Sèvres style and this pair of vases (c.1765) are typically decorated with a deep blue ground. Claret grounds were also popular at this time, and the glaze is often patchy, a fault camouflaged by the gilding. £5,000–6,000

DERBY

The Derby factory was founded in 1750 and was known in its early days as "new Dresden" because its particular speciality was producing figures in the style of Meissen. Judging by what has survived, Derby always aimed for the upper echelons of the ceramics market and few early domestic teawares survive; most pieces are elaborate and would have been expensive even when made. The factory is also well known for its decorative vases and other ornamental pieces. Early wares are of very high quality and tend to attract the highest prices. Later pieces are often less detailed but still highly decorative and tend to be more modestly priced. Derby's huge variety of figures reflect different styles, ranging from elaborately sculpted Rococo subjects to simply modelled Neo-classical figures.

▲ EARLY FIGURES
Derby's earliest figures are known as "dry-edge" figures because of the characteristic dry appearance of the edge of the base; many of these pieces were made in the white. Crisp, fine modelling, as demonstrated by this pair of sheep (c.1755), is also a feature of most early figures. White pieces are nearly always less valuable than similar enamelled ones so these would be worth £1,200–1,800.

▼ WARES
Few wares survive from Derby's early years because the paste used had a tendency to split, so pieces such as this coffee pot (c.1760) are rare. The slightly heavy shape has a distinctive lobed base – a typical Derby feature. £1,200–1,500

MARKS
Various marks were used in the early 1750–70 period. After 1770, the factory merged with Chelsea, and marking became more consistent – usually crowned Ds or crossed batons.

◀ **CLASSICAL FIGURES**
This figure of Lord Camden is in the Neo-classical style that by c.1775 had become the height of fashion. The figure is one of a series – Milton, Shakespeare and other national figures were also made. The plain square base is in keeping with the taste for classic simplicity. £1,000–1,500

DERBY & CHELSEA

Derby took over the Chelsea concern in 1769 and for a period it is very difficult to distinguish the products of the two firms since similar models were used and the pastes can be very similar.

▶ **FRILL VASES**
This so-called "frill vase" (c.1765) has pierced sides and was meant for holding pot pourri. The form was also made by other factories, and Samson produced copies in the 19th century. The large applied flowers are typical of Derby – Chelsea flowers would be more delicate. On pieces of this elaborate nature always check carefully for signs of damage. £700–1,000

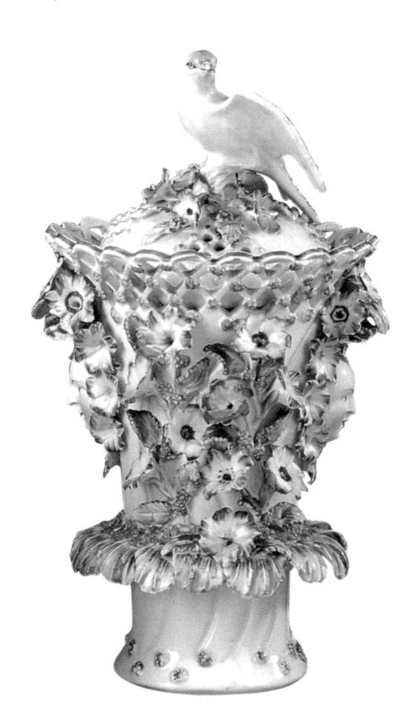

◀ **LATER FIGURES**
In contrast to the stiffer style of Classical figures, these late 18th-century Classical subjects are treated in Rococo style, with elaborate pierced bases. They were meant for vitrines (display cases) so unlike earlier examples which were made to be viewed in the round, the backs would be far less detailed or left entirely plain. £800–1,200 (for the pair)

WORCESTER

The Worcester factory was one of the most successful porcelain manufacturers of the 18th century and is famous, primarily, for the vast amount of tea and coffee wares it produced. Together with Chelsea, Worcester remains perhaps the most popular of the 18th-century factories with collectors. The porcelain made by Worcester contained soapstone which made it very stable and able to withstand boiling water. As a result vast quantities of hollowwares have survived and so daunting is the enormous range available that collectors often concentrate their efforts on a particular area such as early wares, transfer printed designs, teapots or coffee cups. Prices for Worcester pieces are generally lower than for Chelsea ones. Among the most readily available and collectable printed patterns are the Three Flowers pattern, Fence Pattern and Fisherman and Cormorant pattern.

CLASSIFICATION

Worcester is classified according to the factory's owners:

- Dr Wall or First period 1751–74
- Davis period 1774–83
- Flight period 1783–92
- Flight & Barr period 1792–1804
- Barr, Flight & Barr period 1804–13
- Flight, Barr & Barr period 1813–40
- Chamberlain & Co. 1840–52
- Kerr & Binns period 1852–62
- Worcester Royal Porcelain Co Ltd from 1862 (see p124)

MARKS

The usual Worcester mark is an open crescent, either painted or printed. The factory also marked with a square Chinese fret with cross hatching and used pseudo-Chinese or alchemical marks.

▲ **EARLY TRANSFER PRINTING**
Worcester started using underglaze transfer printing from c.1757. Printed wares, such as this double-handled pierced basket (c.1770), tend to be less expensive than hand-painted pieces of a similar date but are still keenly collected. £400–600

▼ **CHINOISERIE**
Early wares were heavily inspired by Chinese decoration and this sauce boat (c.1755) is typically painted with a fanciful Chinese landscape. The body has a bluish tone that is a feature of much early Worcester. £800–1,200

◄ DECORATION
Floral patterns inspired by Oriental designs were the most popular form of hand-painted design on 18th-century Worcester. This fluted coffee cup and saucer are typically painted in *gros bleu* panels separated by bunches of flowers. Elsewhere in England the Japanese style had long faded from favour but Worcester produced Japanese-inspired designs well into the 1770s when this was made. £400–600

► COLOURED GROUNDS
Coloured grounds, as seen on this covered vase, were popular in the 1760s–80s but had a tendency to blotchiness. This problem was often overcome by using a blue ground printed with an overlapping fish scale pattern. Yellow, green and claret grounds were also used sometimes. £2,500–3,500

▼ CHAMBERLAINS WORCESTER
Chamberlains Worcester was a rival factory which produced wares in similar styles. The factory eventually amalgamated with Flight, Barr & Barr to become Chamberlain & Co in 1840. Pieces are usually marked "Chamberlains Worcester".This Chamberlain vase (c.1815) is painted with feathers, a much sought after subject.
£2,500–3,500

◄ FLIGHT, BARR & BARR
Lavish gilding and exquisitely detailed painted decoration are hallmarks of Worcester porcelain of the early 19th century. This campana-shaped vase (c.1820), one of a pair, reflects the prevailing fashion for Neo-classical forms and is decorated with minutely observed botanical subjects.
£2,000–3,000 (for pair)

OTHER ENGLISH FACTORIES

In general most minor 18th-century English porcelain factories produced useful wares, such as tea and coffee wares and dinner services, rather than purely decorative items. Blue and white was the most popular form of decoration, but some factories also used a polychrome palette based loosely on the *famille rose* of Chinese porcelain. Of the minor factories, Limehouse was the first to be established in 1745 (until 1748), followed by Vauxhall in 1751 (until 1764), Lowestoft 1757 (until 1799) and Caughley (c.1775–99). In Liverpool several factories were established; among the most important were Richard Chaffers (1754/5–65), Samuel Gilbody (1754–61), Phillip Christian (1765–78) and Seth Pennington (1778–99). Although wares by these small manufacturers are less readily available than those by larger establishments such as Chelsea and Worcester, all are popular with collectors. Recently discovered factories such as Vauxhall and Limehouse have a particularly keen following and command a premium. Caughley, the most prolific of the minor English factories, tends to be less sought after and the factory also produced printed wares which are less valuable than some of its hand-decorated pieces.

▼ **LONGTON HALL**
Moulded patterns using floral and vegetable motifs were a particular speciality of the Longton Hall factory and this strawberry leaf plate, made c.1755, is typical of the factory's output. The body is similar to Chelsea Red Anchor porcelain and there are often flaws in the glaze. £600–800

▲ **CAUGHLEY**
A dense chinoiserie pattern is printed on this part-tea set (c.1790), a typical decorative technique for this factory. The use of gilding is also characteristic of Caughley. This is part of a set which originally would have had tea bowls, a pot stand and possibly a spoon tray too. Caughley porcelain is made using soaprock and may have a greyish tinge. £150–200 (for 6 pieces)

LONGTON HALL WARES

Blue and white tablewares, copies of Chinese *famille rose* designs, tureens and pot pourri vases are among the varied output.

▶ LIVERPOOL

This helmet-shaped sauce boat (1765–76) is attributed to Phillip Christian's factory, which produced cream jugs, sauce boats and a variety of tea and coffee wares. The decoration here is slightly awkward and the painting has overshot the moulded cartouche. The ribbed foot is seen on wares from other Liverpool factories. £500–700

◀ LOWESTOFT FIGURES

In general the Lowestoft factory tended to produce very few figures but dogs were more common than other subjects. These examples (1760–70) are inspired by Meissen originals. Lowestoft used a phosphatic body for its figures which often has a slightly greenish tinge. Over time, unfortunately, it suffers from brown rust-like discoloration. £2,000–3,000 each

MARKS

- **Caughley** marked with a "C" (similar to Worcester), or the word "Salopian" or "S"
- **Longton Hall** no marks used
- **Vauxhall** spurious Chinese marks are very occasionally found
- **Lowestoft** crossed swords; a crescent similar to Worcester is also used sometimes

◀ VAUXHALL

Vauxhall is a recently discovered factory. These bottle vases (1755) are very similar to Worcester wares but are more vibrantly coloured. Vases of this shape were particularly popular in London and were also made by the Bow factory. The shape has sagged slightly in firing, a defect characteristic of this factory. £8,000–12,000 for a pair

ROYAL CROWN DERBY

Two different factories flourished in Derby in the later part of the 19th century. The most important was the Derby Crown Porcelain Co Ltd, later to become Royal Crown Derby; second came the Derby King Street factory, headed by Samson Hancock. The two factories were rivals and made similar pieces mainly based on patterns and shapes made in Derby earlier in the century, which in turn were based on designs by Sèvres with coloured grounds and raised gilding. Imari wares were another particular Derby speciality and reached a peak of perfection in the 1890s–1915. After World War I production deteriorated and later pieces from the two factories are far less sought after today. Traditionally considered the poor relation of Worcester and Minton, Derby produced wares which were undervalued until relatively recently. Prices have escalated dramatically over the last few years, particularly for fine cabinet pieces with signed decoration.

◀ DECORATION
Derby porcelain is famed for its high-quality decoration and the finest Derby pieces made after 1895 were signed by the artist. This plate, made for the Duke of York's marriage in 1893, is by Derby's most famous decorator Désiré Leroy. He was a French decorator who had worked at Sèvres and was famed for his painting and gilding. £2,500–3,000

DECORATORS

Pieces by leading decorators are much sought after. Among the names to look out for are:

- **William Dean** a specialist in yachts and ships
- **Désiré Leroy** the best-known of all Derby painters, who trained at Sèvres
- **Cuthbert Gresley** and **Albert Gregory** well-known flower painters (see p123).

IMARI

One of Royal Crown Derby's most successful specialities was a range of Imari-style patterns. Each pattern has its own name; this toy saucepan (c.1910) is decorated in one of the most popular patterns, the Old Witches'. Cigar pattern and King's pattern are also popular. £350–400 Miniature pieces, such as this saucepan, are among the most valuable of all Imari wares. Rare shapes, such as milk churns, flat irons and casseroles, can fetch £500–800 each.

▲ FLOWER PAINTERS
This deep blue ground vase made in 1913 is typically based on 18th-century Sèvres porcelain styles; the central panel is filled with a bouquet of roses and other flowers, skillfully painted by Albert Gregory, one of the most famous flower painters at Derby. Cuthbert Gresley is also famed for his floral subjects. £800–1,000

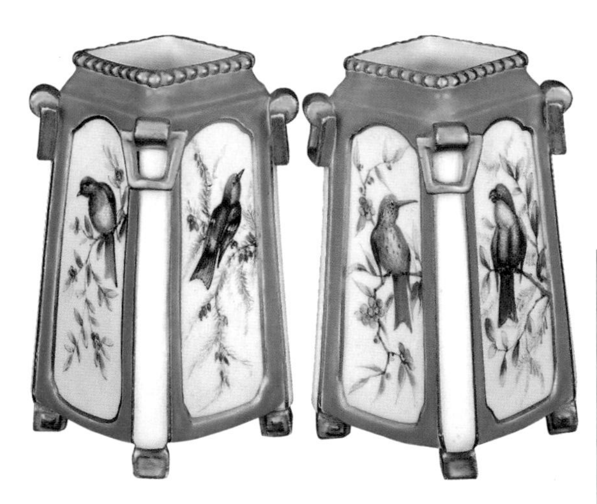

▲ CONDITION
One of these vases (made c.1880 by the Derby Crown Porcelain Co) was cracked and the gilding was slightly worn. In better condition they might fetch twice as much. £250–350

▼ FIGURES
A wide range of popular figures, many of which were originally made in the 18th century, were produced by the King Street factory. This traditional model of the Tailor on a Goat was reissued c.1900, and is marked "SH" for Samson Hancock. £150–200

MARKS

Derby pieces are nearly always clearly marked, usually in red, with a printed crown and cipher and normally a year code; these marks are rarely faked. The King Street factory used the original Derby painted mark with the initials "SH" on each side. Sometimes these initials have been ground away in an attempt to make the piece look older than it really is.

ROYAL WORCESTER

The Worcester Royal Porcelain Company was formed in 1862; the kaleidoscopic range of wares produced by this factory spanned from humble domestic goods to the finest cabinet pieces exquisitely decorated by leading painters. Royal Worcester's extraordinary commercial success was largely due to the inspired direction of the factory's artistic director, R. W. Binns, who kept his finger on the pulse of changing fashion and was able to cater to public demand. Worcester excelled in designs of Eastern inspiration; as Japanese art infiltrated the Western world the factory began producing patterns inspired by Japanese ceramics and Indian ivories, Persian ceramics and Oriental metalwork. Worcester is also famed for unusual finishes such as "blush ivory", an effect which was much copied in Germany and Austria. Among the most sought-after pieces of Royal Worcester are signed pieces by leading names such as the Stinton family, Charley Baldwyn and Harry Davis. Figures by Worcester's leading modeller James Hadley are also keenly collected, as are those made in the 20th century by Freda Doughty. Worcester also produced a series of limited-edition figures; these have not held their value as well as figures by Doulton.

◀ DECORATION
Japanese-inspired design was a popular form of decoration on many Royal Worcester pieces; this 1872 vase with frogs, one of a pair, is an amusing example copied exactly from an Oriental prototype. £2,000–2,500 pair

▶ JAMES HADLEY
Hadley was Royal Worcester's most famous modeller. Working in the 19th century, he was responsible for a series of blush ivory figures such as this "piping boy" (1889) which remain particularly popular with collectors today. Value depends on the rarity of the figure; this example would be worth £600–700.

▲ THE STINTONS
Harry and his father John Stinton specialized in painting Highland cattle in mountains. James, John's brother, painted game birds rather than cattle. Pieces such as this vase by Harry Stinton, made in 1919, are very popular but value varies greatly according to size. This measures 8in (21cm) high and would be worth £1,000–1,400.

▼ **FREDA DOUGHTY**
Freda Doughty, one of the most popular Worcester modellers working in the 1930s, produced a series of children at play, such as this figure known as "Dancing Waves". Royal Worcester also produced animal models by Doris Lindner. £170–200

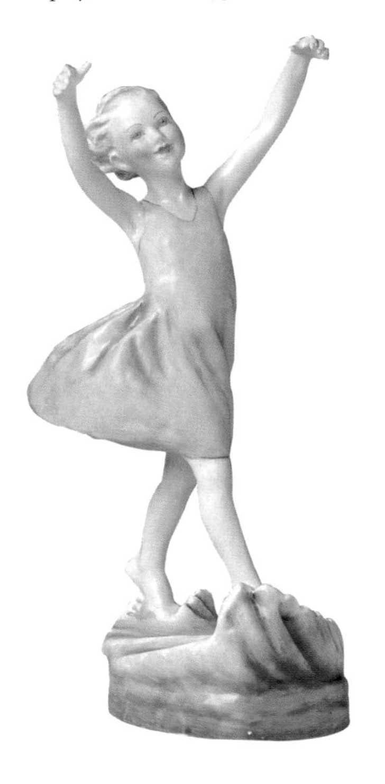

▲ **IVORY DECORATION**
The "blush ivory" finish provided a background for formal floral decoration which proved to be very popular in the late 19th/early 20th century. It is seen to great effect on this *pot pourri* vase. Many other makers both in England and abroad copied this unusal Royal Worcester colouring. £250–350

MARKS

Standard Royal Worcester marks are date-coded with a series of letters and dots. By looking in a specialist mark book you can work out the year of manufacture.

- From 1862 "Worcester Royal Porcelain Works", within a circle surmounted by a crown enclosing a cipher and crescent with the date code beneath were used.
- After 1891 "Royal Worcester England" was used (as above).
- From 1938 the words "Bone China" appear in addition to the standard mark.

▶ **VALUE**
Most blush ivory vases such as this one (c.1910) were decorated by hand, applying colour to a black printed etched outline of the design. The best designs are by Edward Raby who sometimes signed his work with his initials somewhere in the printed design. The blush ivory ground was sometimes painted entirely by hand and this adds to value. £250–300

NANTGARW & SWANSEA

The fortunes of Wales' two most famous porcelain factories were inextricably linked. William Billingsley, England's foremost flower painter on porcelain, opened the Nantgarw works in 1813, and shortly after moved to the Cambrian pottery, Swansea, where true porcelain had not hitherto been produced. A few years later he returned to Nantgarw, but such large quantities of porcelain were lost in firing that the company was soon in financial distress and closed down four years later. Billingsley is famed for producing fine white highly translucent porcelain of unrivalled quality and products were aimed at the upper echelons of the London market. Many pieces were sold in the white and decorated in London. Not surprisingly, considering its short-lived existence, Welsh porcelain is scarcer than that from other factories – and for this reason commands a premium. Many unmarked pieces are misascribed and identification is a matter of expert opinion. Pieces painted in Wales are more sought after than those decorated elsewhere – the most famous decorators were Thomas Baxter, Thomas Pardoe, David Evans and William Pollard as well as Billingsley himself.

▶ DECORATION
Decoration on much Welsh porcelain was inspired by French taste of the late Regency period, with neo-Rococo flowers on a white ground a typical form. This Swansea plate (c.1815) is typically decorated with delicately painted flowers on a white ground. £500–700

◀ LONDON DECORATED WARES
Although as a general rule pieces decorated in London tend to be less valuable than those decorated in Wales, decorative examples such as this made c.1820 with its detailed bird border are still popular with collectors. £700–1,000

BODY

On Swansea porcelain the body may be one of three different types and this can affect value. Early pieces were made from "glassy" paste; the most sought-after pieces are made from "duck egg" paste – with a slight greyish tinge; later pieces were made from what is known as "trident" paste and are considered somewhat less desirable.

▼ WARES

Nantgarw mainly produced plates, cups, saucers and a range of small ornaments. Swansea was famed for tea and dessert services, cabinet pieces and decorative objects such as taper sticks and inkwells. This tureen, made c.1815, is unmarked but the roses and shape are characteristic of Swansea. £500–600

◀ NANTGARW

Beautifully painted floral subjects were produced at both Nantgarw and Swansea and the marks are often the only way to tell the two apart. This Nantgarw plate (c.1817–20), stylistically very similar to the one opposite, is typically painted with a bouquet of irises, tulips and other flowers. £500–700

MARKS

NANT-GARW
C.W.

Nantgarw wares were usually marked with an impressed "NANTGARW CW" (for china works). Many Swansea pieces were marked in red with a painted, stencilled or impressed "SWANSEA" .

BEWARE

Many French porcelain blanks, especially vases, were also decorated in London with identical decoration to that used on Welsh plates. Fake Swansea marks are also known to have been added to English porcelain to make it more valuable.

MINTON

The Minton factory, founded by Thomas Minton in 1798, was one of the 19th century's biggest and most varied producers of every type of ceramic, ranging from architectural fittings such as tiles and majolica, to humble earthenware and fine bone china. Minton's greatest period spans from the mid- to late 19th century, although the company is still in production today. Among its finest wares were pieces made using the *pâte sur pâte* technique, which had been developed on the Continent. *Pâte sur pâte* was introduced to Minton, and the English market, by Marc-Louis Solon, a former employee of Sèvres. Japanese styles also influenced the factory c.1870–80; many patterns were based on Japanese porcelain, including cloisonné, using brilliant turquoise enamels. The Art Nouveau style was reflected in Secessionist wares designed for Minton by John Wadsworth and Leon Solon. Prices for Minton not surprisingly reflect the variety of objects made by the factory. Top examples of the sophisticated *pâte sur pâte* technique were expensive when made and similarly rank among the factory's most valuable products today; even plates cost around £1,000. Printed tiles designed by leading artists such as John Moyr Smith can be bought for £25, however, and many other less elaborate pieces can be found for similarly modest sums.

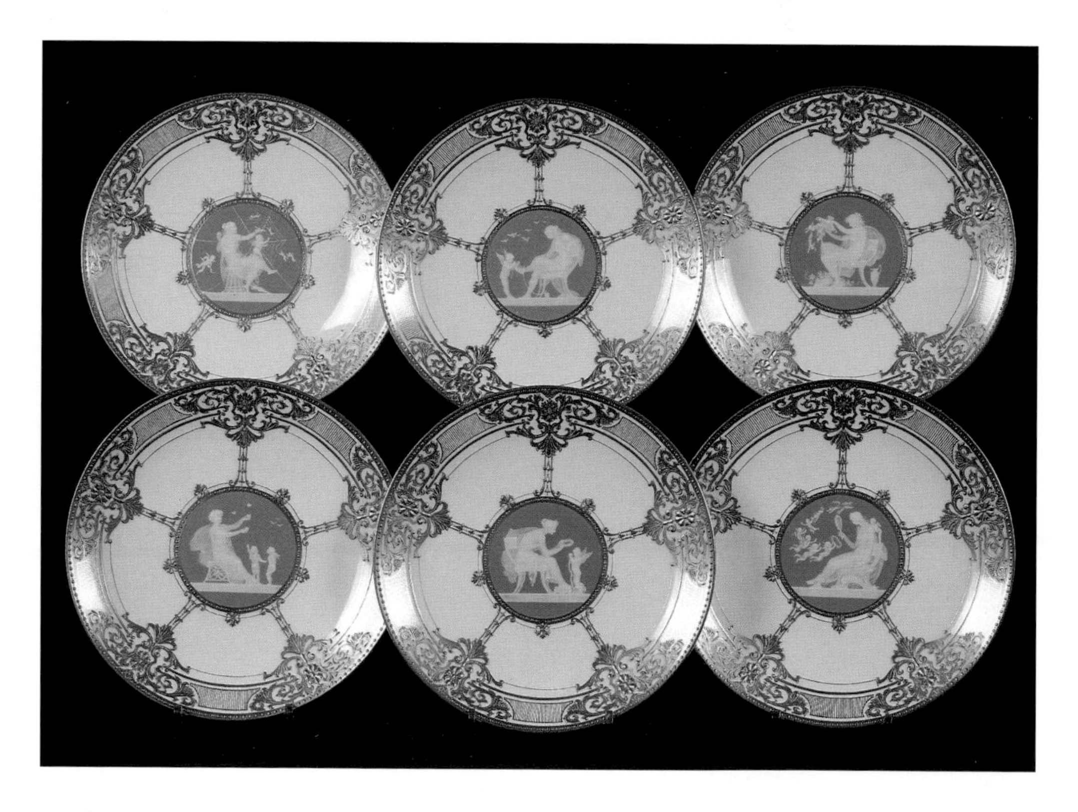

▲ ***PÂTE SUR PÂTE***

Pâte sur pâte was one of Minton's most sophisticated and expensive products. The technique involved building up decorative cameos by applying layers of white slip to a darker base. Many of these pieces were also adorned with lavish raised gilding. These plates (1911) are by Alboin Birks, one of the finest exponents of the technique. Vases from the Minton factory decorated using the *pâte sur pâte* technique are also highly sought after by collectors. £1,000–1,400 each

▲ **SERVICES**
As well as ornamental pieces Minton also produced dinner, dessert and tea services, many of which are still available at very reasonable prices. These plates, from an eight-piece part-dessert service date-marked 1870, are decorated with rich turquoise borders and sprays of wild flowers. £150–200 (for the part-set)

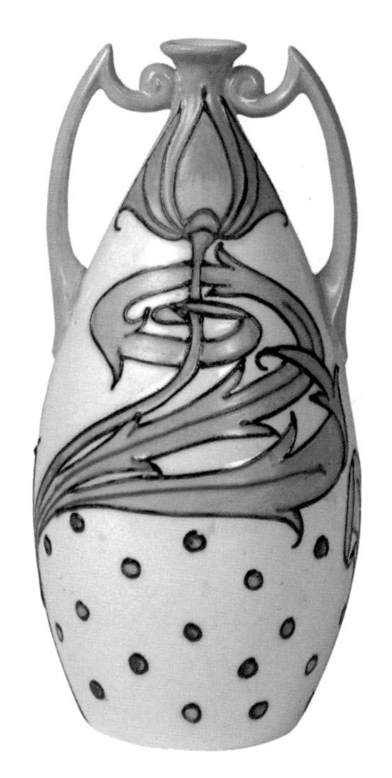

► **SECESSIONIST STYLE**
Minton was highly original in its use of innovative design in the late 19th century. Its "Secessionist" wares are identifiable by their unusual shapes and simple decoration. This vase is probably by John Wadsworth, a leading designer of the day; it is decorated using a slip-trailed technique and reflects the influence of continental Art Nouveau. Another leading designer in this style was Marc-Louis Solon. £200–300

▲ **ORIENTAL DECORATION**
Japanese cloisonné provided the inspiration behind a range of brilliantly decorated ornamental pieces such as this vase (c.1876) in which gilding was used to simulate the effect of the copper wires used in true cloisonné to divide the panels of differently coloured enamels. Pieces similar to this were designed by Dr Christopher Dresser but this example is unsigned. £350–500

MARKS
Nearly all Minton is marked, either with the name "MINTON" or "MINTONS". Pieces made before c.1914 usually have an additional date code mark.

COALPORT

Founded in the last decade of the 18th century the Coalport factory made fine ornamental wares alongside domestic pieces such as tea and dinner services, and the factory is still in production today. The varied tastes of the day are mirrored in the changing styles of the factory's output. Coalport mimicked styles first made by Meissen, Dresden and Sèvres with coloured grounds, eccentric shapes and generous helpings of decoration. The factory is also famous for flower-encrusted pieces made in the 1830s; these were known as "Coalbrookdale" but were in fact produced by the same company. Confusingly for collectors Coalport was also a major supplier of white porcelain to independent china painters and many wares carry decoration which was not done at the Coalport factory.

◀ **"COALBROOKDALE"**
Elaborate floral decoration and scrolls were hallmarks of the "English Dresden style" fashionable in the 1830s. Pieces in this style made at Coalport are popularly known as "Coalbrookdale"; this typical clock case is lavishly decorated in the Rococo style. £800–1,200

▼ **JEWELLED COALPORT**
At the turn of the century until the 1920s the Coalport works achieved huge success with jewelled porcelain, a distinctive decorative technique made by placing beads of turquoise enamel on a gilded or coloured ground. These pieces are much collected and even tiny cups and saucers can cost £400–500 or more.

▲ **DECORATORS**
Coalport allowed artists to sign pieces in the late 19th and 20th centuries. Among the best-known decorators are Edward Ball and Percy Simpson who painted landscapes. John Randall, a specialist in exotic bird subjects, and Frederick Chivers, a specialist in still life, decorated this elaborate vase (c.1905), one of a pair. £1,000–1,500 for the pair

ROCKINGHAM

Perhaps better known than it deserves to be, this short-lived factory is famous for its rather unconventional output. Founded in Yorkshire in 1826 by the Brameld family, the company took its name and mark from its financial backer, the Earl Fitzwilliam, Marquis of Rockingham. The company concentrated on ambitious decoration and in 1830 began making a service for King William IV, which ultimately led to its financial ruin; 12 years later, in 1842, the firm finally closed. Despite the limited output Rockingham is extremely popular with collectors and pieces command a premium compared with other more prolific factories. Simple cups and saucers fetch from £100 and more; small vases start at around £500–800. Most Rockingham was marked but many unmarked pieces by Minton and Coalport have been wrongly attributed over the years to Rockingham.

◀ **QUALITY**
Lavish gilding and accomplished painted decoration are typical of Rockingham. This plate is of particularly fine quality since it was a specimen design for the celebrated service produced by the factory for the coronation of William IV and first used for Queen Victoria's coronation.
£3,000–5,000

▲ **PLAQUES**
Plaques are rare among Rockingham's output and this example is particularly unusual and valuable because it is signed and dated by Thomas Steel, a leading artist who is known to have worked at the firm in 1830.
£4,000–5,000

▶ **ANIMALS**
Rockingham produced an assortment of animals including cats, dogs and rabbits. Rockingham animals, are invariably marked unlike those of other factories. This cat was made c.1830.
£800–1,000

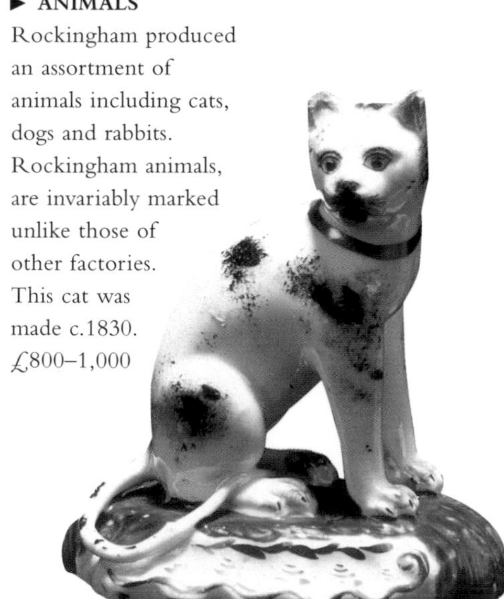

MARKS

Rockingham Works
Brameld

A griffin taken from Earl Fitzwilliam's arms was first printed in red, then purple after c.1830 when "Manufacturer to the King" was added.

SPODE

This large Staffordshire factory was founded by Josiah Spode in 1776, and from the end of the 18th century throughout the 19th century produced a vast range of domestic earthenwares and Parian, as well as grander bone china services and ornamental pieces reflecting the prevailing taste for Classical or Rococo style (see p78 and p134). The company was bought out by Copeland & Garrett in 1833 and is still in production. Spode's most distinctive products include the Imari patterns popular c.1815–25; pieces with coloured grounds; and pieces with hand-painted decoration by leading decorators such as C. F. Hürten. Lavish gilding is also a feature of many of the best pieces. This is a vast and diverse collecting field and although less distinctive than those of some other factories, products range from highly elaborate ornamental wares to humbler bat-printed wares.

◄ PRINTED WARES
Among the least expensive Spode pieces are bat-printed wares such as this milk jug of c.1815, which is decorated with a picturesque landscape. Bat printing was a method that relied on soft glue rather than paper to transfer printed designs.
£90–120

MARKS

Before 1830 Spode usually hand-painted this name on porcelain (or impressed on pottery). After 1833 the names "Copeland & Garrett" usually appear with "Late Spode" in the middle. The word "Copeland" is used alone after 1847.

► PATTERNS
Patterns are identified by numbers and this vase (c.1825), decorated in a floral design on a blue and gilt scale ground, is an example of pattern number 1166, one of the most popular of all Spode patterns with today's collectors.
£500–800

◄ DECORATORS
Decorators at Spode did not usually sign their work; their names are not generally known but their high-quality decoration can attract high prices none the less. This straight-sided chocolate cup and stand (c.1815) is skilfully painted with loose sprays of flowers within rectangular gilt panels but the low value reflects the damage it has suffered.
£200–300

DAVENPORT

Another successful Staffordshire factory, Davenport was established in 1794 by John Davenport and produced earthenware, basalt, creamware, bone china and even glass throughout the 19th century, until its closure in 1887. Davenport has no particularly distinctive style but is recognized for decoration of high quality. Pieces were usually marked with name and pattern number which, along with their style, are helpful clues for dating. Davenport patented a technique for making thinly potted porcelain plaques, many of which were sold to independent decorators. Although popular with collectors, Davenport tends to fetch lower prices than pieces by Worcester, Swansea or Rockingham, even though it is often of equally fine quality and can provide excellent value. Among the most affordable pieces to look out for are blue and white printed plates from £30–70.

◀ DECORATION
This is an early Davenport plate (c.1810–15), heavily decorated with chinoiserie inspired by Meissen. The highly detailed form of decoration seen here is unique to the Davenport factory. £140–180

MARKS
Davenport was marked with a crown between c.1870 and 1886; the words "Davenport Longport" also appear in some early and later marks.

SERVICES
Davenport produced some very extensive dinner services using porcelain and various other less expensive materials. This part-dessert service (c.1815) is made from pearlware; other large services were made from ironstone (see p75). £1,000–1,500 (for 17 pieces, allowing for some damage)

PARIAN

Named after Roman marble quarries at Paros, Parian was a hugely popular 19th-century material used to make inexpensive imitations of marble sculpture. Copeland and Minton both claim to have invented Parian in the 1840s and by the mid-century many other factories were also producing it. Copeland called the substance "statuary porcelain" in the early days, while Wedgwood termed its version "Carrara ware" after another famous Italian marble quarry. Parian's advantage was that it contained a high proportion of glass crystals and its vitreous nature meant that dirt did not adhere to it. Many of the leading 19th-century sculptors designed models to be reproduced in Parian; copies of famous antique works were also made, as well as busts of literary and historical personalities.

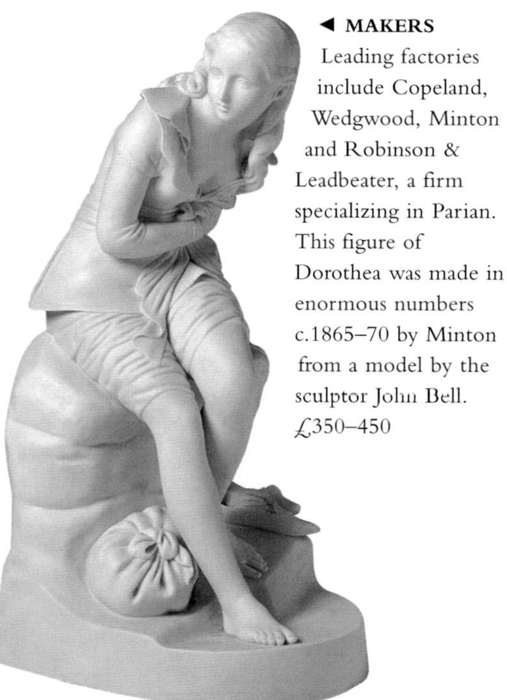

◀ **MAKERS**
Leading factories include Copeland, Wedgwood, Minton and Robinson & Leadbeater, a firm specializing in Parian. This figure of Dorothea was made in enormous numbers c.1865–70 by Minton from a model by the sculptor John Bell. £350–450

MARKS

Parian figures are usually marked by their maker on the base or on the back of the model. Some pieces may also be titled and have a date code.

VALUE

Value depends on a piece's maker, its decorative appeal and its size. Prices range from under £40 for small busts to over £1,000 for larger subjects.

◀ **SUBJECTS**
Parian figures depicting famous royal, political and literary figures were also popular. This figure of the Duke of Wellington is by Samuel Alcock & Co after a sculpture by G. Abbott, and dates from 1852. £350–500

ART UNIONS

Parian figures are often stamped with names such as "The Crystal Palace Union" or "Art Union of London". To raise funds for the Arts, "art unions" ran lotteries where art lovers could buy tickets in the hope that they might win an original work such as a specially commissioned Parian figure. Copeland was one of numerous firms who made figures for art unions.

BELLEEK

Established in 1863 by David McBirney and Robert Williams Armstrong in County Fermanagh, Northern Ireland, Belleek specialized in producing an incredibly thin high-quality white porcelain using a glazed Parian body. Warm and creamy in appearance, Belleek resembled the texture and translucence of sea shells and before long the factory began making wares in the forms of shells decorated with iridescent glaze, as well as finely woven baskets. Belleek has always been popular in the United States, and in fact workers from the Irish factory emigrated and set up rival establishments there; their products are termed "American Belleek". The factory is still in production and the same designs have been produced throughout its existence. Small pieces of Belleek cost from £30 while larger more elaborate pieces such as baskets can fetch between £4,000 and 5,000.

▲ **DATING**
Baskets can be dated from the number of strands from which the centres are woven. Baskets such as this from the 19th century are three-stranded; after c.1900, as potting became even finer, four strands were used. £250–350 (damaged)

▲ **DESIGNS**
The fine lustrous body perfected at Belleek looked like mother-of-pearl and shells provide a recurring theme for many ornamental pieces. This pair of oval shell jelly dishes (c.1870–90) are moulded with shells among coral branches enriched with pink and turquoise. £150–200

MARKS

The first mark with the word "BELLEEK" in black was used only until 1891 after which the second mark was used, adding the word "IRELAND" to the name. More modern marks are in green or gold.

▲ **TEAWARES**
Belleek's unusual teawares are also popular with collectors. The value of teasets depends on the number of pieces they have. This delicately green-tinged jug, made in the late 19th early 20th century in the Neptune pattern, is worth £70–80.

CONDITION

The fineness of Belleek makes it vulnerable to damage and even hairline cracks can devastate value. The basket top left has had some restoration; in perfect condition it might be worth £1,000–2,000.

Twentieth-century pottery and porcelain is without doubt the most rapidly growing collecting area in ceramics. As top examples from the traditional collecting fields of 18th- and 19th-century ceramics have become increasingly rare and expensive, collectors have turned their attentions to more recent manufacturers, and makers such as Carlton, Susie Cooper and Poole now have a very well-established following.

Twentieth-century ceramics encompasses on the one hand pieces by leading commercial designers and illustrious studio potters, and on the other a plethora of commercially produced wares which reflect the prevailing taste of the day. In general large pieces designed or made by well-established designers tend to be visually more striking and therefore attract the highest prices. But for collectors on a limited budget there is a rich hunting ground of smaller designer-made pieces as well as stylish wares by lesser manufacturers which can still be found very affordably. The modern geometric forms that typify Art Deco styles of the 1920s and 1930s are particularly popular with today's collectors, and pieces in characteristically dramatic shapes or with interesting designs will often be worth far more than less stylish pieces by the same designer.

Ornamental figurines, which have been made since the earliest days of ceramics, have continued to be an extremely popular field with collectors and among the most eagerly sought are figures by Doulton, Wade and Beswick which are still readily available for a huge range of prices.

CLARICE CLIFF

Acclaimed as "a poet in pottery" Clarice Cliff was one of the most prolific and innovative of potters working in the 1920s; her brightly coloured wares with streamlined modern shapes have become synonymous with Art Deco style and enjoy huge popularity with collectors. Clarice Cliff was born in Tunstall in 1899 and worked for A. J. Wilkinson, a Staffordshire pottery where she was given free rein to introduce a new range of hand-painted pottery called Hand Painted Bizarre by Clarice Cliff. By 1930 the success of Cliff's designs led to her appointment as art director of the firm, overseeing a hundred decorators and producing a range of designs.

▲ **DESIGNS**
Clarice Cliff drew on a range of sources as inspiration for her patterns. Floral and foliate patterns include Crocus, Autumn, Alpine, Capri and Latona. Landscape provides the theme for Secrets and textile design inspired Blue Chintz and Appliqué. This lotus-shaped jug features the rare Comet pattern. £3,000–4,000

DESIGNERS TO LOOK OUT FOR

During the 1930s Clarice Cliff commissioned designs from leading artists such as Paul Nash, Duncan Grant, Frank Brangwyn, Dame Laura Knight and Graham Sutherland.

◀ **VALUE**
Value depends on the rarity of the design and the appeal of the form. Modern streamlined shapes are especially desirable, as are unusual patterns and colour variants. This Tennis pattern tea set combines an unusual avant-garde pattern with highly innovative form; hence its high value. £6,000–7,000

▶ **FIGURATIVE PIECES**
Clarice Cliff produced a limited range of cut-out silhouette figures painted on both sides entitled The Age of Jazz series. This group is formed as a plaque showing two couples dancing the tango. £6,000–7,000

SETS
It is possible to build up sets of Clarice Cliff by buying individual pieces if you choose one of the more common designs, but pieces from rarer styles are more difficult to track down.

◀ **FAKES**
Fakes with photographically copied marks have appeared on the market in the last decade. These are often identifiable by the inferior quality of the painting and by the fact that the marks look fuzzy - the ones on the left are fake, those on the right genuine. £200–600 (for the genuine ones)

MOORCROFT

One of the most successful English ceramicists in the Art Nouveau style, William Moorcroft (1872–1945) was a Staffordshire-born and trained potter who began designing for the firm of Macintyre & Co in 1898. Moorcroft is best known for his use of slip-trailing decorations called Florian ware. Most of his early designs combine Art Nouveau naturalism, in which identifiable flowers, mushrooms or landscapes often form part of the design, with shapes influenced by Eastern and later Classical ceramics. Moorcroft supplied wares to Liberty's who often sold pieces with pewter mounts, and many wares were also sent abroad. In 1913 Moorcroft set up his own factory and new designs were gradually introduced. Moorcroft's son Walter joined the firm in 1935 and also produced mainly floral designs in a distinctive style. Value depends on the rarity of the design and the form. Prices range from £200 to £5,000 or more for unusual colour variants. The factory is still in production today and continues to turn out a range of innovative designs.

◀ FLORIAN WARE
Florian decoration was made by "drawing" on to the pot with thin lines of clay, rather like icing a cake, and then applying colours in between the lines. This oviform vase (c.1903) is decorated in autumnal shades with an Art Nouveau tulip pattern. £5,000–7,000

MARKS

Moorcroft marks vary; "W Moorcroft Des" or "WM des" is often found on Florian from the Macintyre period. Pieces produced in his own factory were usually marked "MOORCROFT BURSLEM", impressed with either "WM" or "W. Moorcroft".

▼ MODERN MOORCROFT
The Moorcroft factory is still in production and some of its limited edition products are set to become collectables of the future. This modern vase was designed in 1988 by Sally Tuffin, in a limited edition of 250 for the Canadian market, and shows a design with polar bears in an Arctic setting created by using the traditional slip-trailing techniques for which Moorcroft is known. £250–350

▶ PERSIAN PATTERN
Inspired by Iznik and Persian ceramics Moorcroft's Persian pattern is slightly more finely detailed than Persian ware by de Morgan (see p85), and is highly decorative and popular with collectors. This two-handled vase (c.1915) is typically decorated with stylized flowers and leaves. £3,000–4,000

WEMYSS WARE

Folksy and charmingly naive, Wemyss ware was first produced in the Fife pottery in south-east Scotland in 1880. The simple wares were decorated with highly distinctive hand-painted motifs of cabbage roses, fruit, birds and farmyard animals and sold through Thomas Goode in London. In 1930 the Scottish factory closed and the moulds were sold to the Bovey Tracey Pottery in Devon, where the same designs continued to be produced until the 1950s and beyond. Prices for Wemyss vary according to the age and rarity of the particular piece. A rare Wemyss cat has recently fetched £11,000, but most pieces sell for between £300 and 600, and you can find small items for under £100.

▶ DECORATION
Wemyss' distinctively colourful style of decoration was created using underglaze enamelling. Flowers were a recurring theme and this early 20th-century teapot is adorned with a cheerful buttercup pattern typical of the bold designs for which the Scottish factory is famed. £250–350

◀ PIGS
Ornamental pigs, often covered with cabbage roses, are among the best loved of Wemyss wares. Prices vary according to size; this pig made in 1920 is fairly small and would be worth £300–400, but larger pigs can fetch a great deal more.

▼ COCKEREL PATTERNS
Many Wemyss designs were by Karel Nekola, a Bohemian artist who worked for Wemyss for many years. This vase (c.1900) is decorated with cockerels and hens – one of Nekola's most popular designs. £400–500

MARKS

- Early Wemyss ware has an impressed mark and the words "Wemyss Ware R H & S" in a semi-circle.
- The single word "Wemyss" was used as a mark throughout the 19th and 20th centuries.
- The words "Thos Goode" usually indicate an early production.
- In 1930 the factory moved to Bovey Tracey in Devon, after this time the words "Made in England" can appear with the Wemyss mark.

COMMEMORATIVE CERAMICS

Commemorative ceramics are a vast collecting area encompassing anything made to mark a particular historical event. Royal weddings and deaths, coronations, jubilees, wars, elections and strikes have all been immortalized in ceramic form and collectors tend to concentrate on a particular theme or character rather than on the subject in general. Any ceramic commemoratives that predate the coronation of Queen Victoria are rare and correspondingly valuable. Most commemoratives date from the mid-19th century onwards, when improved methods of transfer printing allowed potters to make souvenirs inexpensively.

◄ WILLIAM IV CORONATION
Transfer prints on early commemorative wares can contain fascinating historical details. This Staffordshire pottery jug (1830) is unusual in showing William at the Coronation ceremony in Westminster Abbey. £300–400

▼ MAKERS
In the 20th century vast quantities of royal souvenirs were made. Values remain low unless the piece is unusual or by top manufacturers. This Wedgwood tankard by Richard Guyatt marks the marriage of Princess Anne and Captain Mark Phillips in 1973. £30–50

▲ VALUE
Commemoratives of obscure events are less sought after than those celebrating more significant events. Huge numbers of mugs marked the Diamond Jubilee of Queen Victoria in 1897; this mug has more details than most. £60–100

► RECENT COMMEMORATIVES
Most commemoratives from recent decades are still available for under £10 but cartoon mugs from the wedding of Prince Charles and Lady Diana Spencer are already increasing in value. This mug was designed by Mark Boxer to commemorate the royal wedding in 1981 and has a comical handle shaped as an ear. £30–45

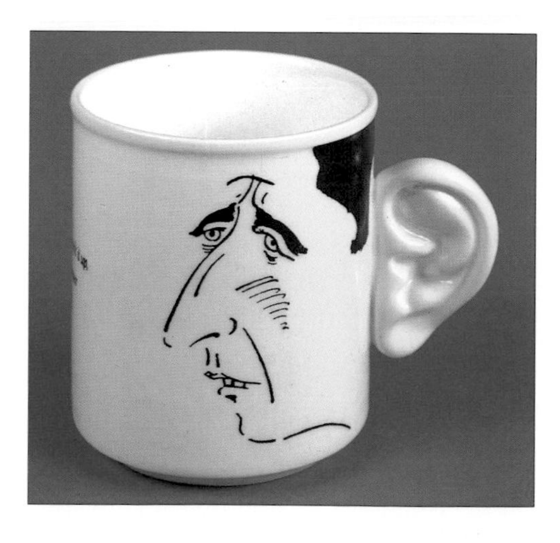

GOSS & CRESTED CHINA

Crested china was made as seaside souvenirs during the late 19th and early 20th centuries when the principal maker was W. H. Goss. Decorated with the coats of arms of the most popular holiday resorts, these pieces provided visitors with souvenirs of their stay and were made in vast quantities until c.1930 when the fashion for Goss abated. Goss pieces were invariably well made and although the Arcadian and Carlton works in England and other factories in Germany made similar crested pieces none could rival Goss for quality. The most sought-after pieces are "Goss cottages" which can fetch upwards of £200.

► COTTAGES
Goss buildings, representing the homes of national heroes such as Anne Hathaway's cottage (seen here), Lloyd George's home or Charles Dickens' house or other landmarks, can form the theme of a fascinating collection and have become extremely popular. £30–40

◄ FIGURES
Apart from heraldic ceramics and model buildings Goss also produced a keenly collected range of decorative figures, some based on famous sculptures such as this late 19th-century bust depicting "The Veiled Bride". £500–700

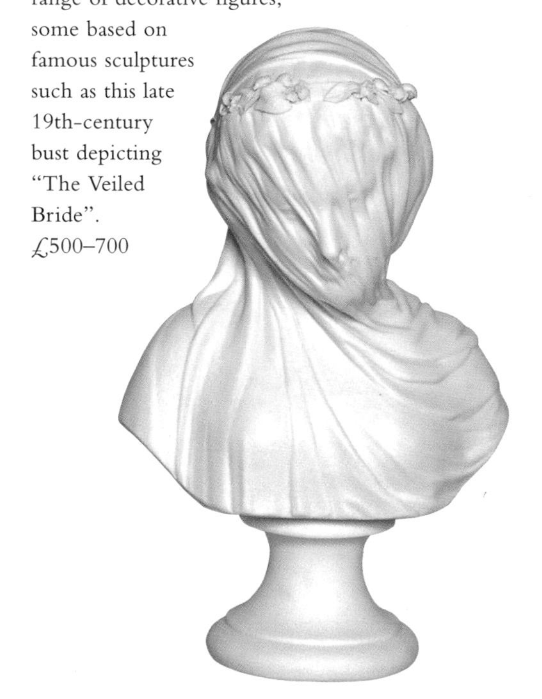

MARKS

- Genuine Goss china should have a printed mark with a hawk.
- Some crested pieces which were produced in Germany are marked "GEMMA".
- Beware of fake cottages produced with spurious hand-painted Goss marks.

WHAT TO LOOK FOR

World War I pieces from Barmouth, decorated with the flags of Britain's allies, are among the more desirable examples of crested china wares you are likely to come across. Other sought-after pieces include:

- figures
- lighthouses (and other buildings)
- animals.

SUSIE COOPER

Born in 1902, Susan Vera Cooper became one of the most influential ceramic designers of her generation, producing a wide range of commercial wares in a distinctively subtle style. After working for a while for A. E. Gray, Susie Cooper set up her own company in 1929. Early output mainly concentrated on earthenwares, often hand-painted; fine bone china began to be produced after the Second World War. In 1961 the firm of Susie Cooper merged with R. H. & S. L. Plant which in turn was taken over by Wedgwood in 1966. Wedgwood reissued some of Susie Cooper's early designs such as Pink Fern, Polka Dot and Yellow Daisy. The most sought-after pieces tend to be early hand-painted wares and lustrewares; services have also risen enormously in popularity in recent years.

◀ **SHAPES**
Wares are categorized according to their shape and the pattern with which they are decorated. This part-coffee set (c.1935) is in the Kestrel shape. Other popular forms include Curlew, Jay and Wren. (£350–450 for the pieces shown)

▶ **DECORATION**
Decoration on Susie Cooper is remarkably wide-ranging: floral designs, transfer prints, incised decoration and geometric patterns were all variously used. This pottery jug (c.1935) is carved with a charging goat, and covered in a moss green glaze. £120–180

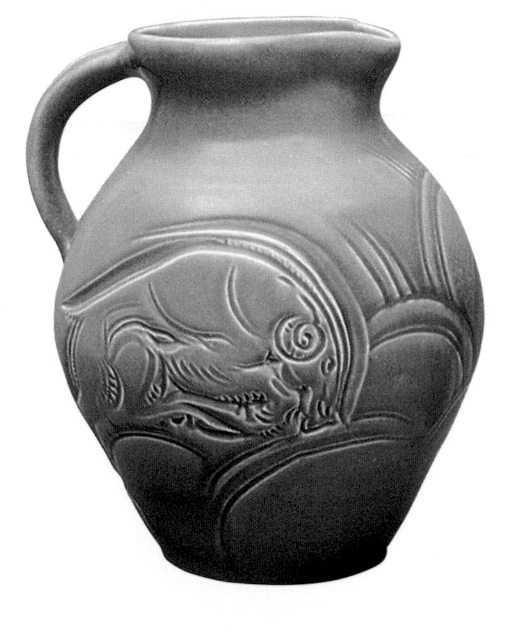

MARKS

- Most Susie Cooper pieces are marked with a facsimile signature.
- Pieces can be dated both by mark and by serial number.
- Pieces made while Susie Cooper was working at A. E. Gray's may have both the company name and the designer's own initials.

POOLE POTTERY

Pottery has been made in the region of Poole in Dorset for centuries but the company now known as Poole Pottery was established by Jesse Carter in 1873. Among the earliest pieces it turned out were terracotta jardinières, some designed by Archibald Knox for retail through Liberty & Co, and lustre-wares. Other designers who worked for Carter include James Radley Young who designed wares in an ethnic style, and Roger Fry whose designs were retailed through the Omega Workshops. In 1921 the company took on new partners Harold and Phoebe Stabler and John Adams to form a company known as Carter, Stabler & Adams. Other important designers at Poole are Truda Sharp (later Truda Carter), who produced colourful bold geometric and floral designs, and Olive Bourne, who produced plates decorated with stylized female faces. During the 1960s and 1970s the company continued to produce highly innovative studio wares.

◀ CARTER, STABLER & ADAMS
Under the influence of the new partners a range of sculptural pieces was introduced. Figures include "The Bull" and "Picardy Peasants". This roundel of the "Piping Faun" (1914) was designed by Phoebe Stabler. £1,500–2,000

MARKS
Poole pottery is invariably marked – a wide range of marks were used. They usually include the name of the company together with the decorator's initials or pattern code.

▶ ORNAMENTAL WARES
Among the most sought-after pieces of Poole pottery are the large hand-potted ornamental wares produced in the 1920s and 1930s. This large boldly painted jar and cover is signed by Truda Carter, one of Poole's leading designers. £3,000–4,000.

▶ ARCHITECTURAL CERAMICS
Architectural ceramics are among Poole's large and varied output. This tile advertising panel from the 1920s was made for the bookseller W. H. Smith, with lettering designed by the well-known illustrator Eric Gill. £500–600

DOULTON FIGURES

Doulton figures follow a long tradition of figurative ceramics and, although a few were made at the end of the 19th century by C. J. Noke, the vast majority were produced in the Burslem factory at some time between the 1920s and the present day. Extremely popular with collectors in both Britain and the United States, Doulton figures were produced in an enormously varied range of subjects. Figures are easily identifiable thanks to the marks and names on their bases. Each bears a series number and the prefix "HN" and the factory records are also helpful as they can tell you for how long every design was produced. Pretty ladies in elegant dresses became one of the firm's specialities, as did bathing belles, figures in historic dress, animals, dancers, jesters and street vendors. Value depends on the rarity of the model and on the subject, and rare colour variations also command a premium.

▶ THE PIED PIPER
One of the more elaborately decorated Doulton figures is this "Modern Piper" (1925–38). Based on the Pied Piper of Hamelin, the piece is marked "HN 756". It commands a relatively high value which reflects the decorative appeal of the subject and the fact that the design was made for a relatively short time. £1,000–1,500

▼ RARITY
Lesley Harradine designed this rare and sought-after figure entitled "Scotties" and marked "HN 1281". It was introduced in 1928 and withdrawn in 1938. The Scottish terriers were a fashionable pet at the time; hence the subject's popularity. £1,500–2,000

◀ SUBJECTS
This naked figure entitled "The Bather" was produced in the 1920s and 1930s in several versions. In the 1930s a similar figure was also introduced, dressed in a swimming costume. The clothed figure is rare and therefore slightly more valuable than the naked one. Naked £600–800; clothed £800–1200

CHARACTER JUGS

ANOTHER OF DOULTON'S MOST SUCCESSFUL 20TH-CENTURY PRODUCTS, CHARACTER JUGS CONTINUE THE TRADITION OF THE TOBY JUGS OF THE 18TH AND 19TH CENTURIES, AND WERE PRODUCED IN HUGE QUANTITIES FROM THE LATE 1930S ONWARDS. CHARACTER JUGS WERE OFTEN MADE IN FOUR DIFFERENT SIZES RANGING FROM LARGE, SMALL, MINIATURE AND TINY. SUBJECTS INCLUDE FAMOUS PERSONALITIES OF THE PAST AND PRESENT. DICK TURPIN, OLD KING COLE, WINSTON CHURCHILL AND FRANCIS DRAKE HAVE BEEN REPRESENTED IN THIS WAY. VALUES ARE HIGHEST FOR RARE VARIATIONS WHICH CAN FETCH £3,000 OR MORE, ALTHOUGH THE MAJORITY FALL INTO THE £100–400 PRICE RANGE AND YOU CAN FIND RECENT EXAMPLES FOR UNDER £50.

MARKS
As with figures, jugs are marked and well documented. They are invariably named on the piece and bear the full factory mark, a date showing when the design was introduced and a "D" series number.

◀ HANDLES
The handles of Doulton character jugs are usually modelled to reflect the subject. This "Gondolier" (1964–69) has a handle appropriately modelled as a Venetian gondola.
£200–300

▶ VALUE
Rarity rather than date has the biggest impact on value. This Alfred Hitchcock character jug is fairly recent in date (1995), but is a rare variation, because the shower curtain handle (a feature from the film *Psycho*) is pink; later versions, which are far more common, have a blue curtain. £600–800

WEDGWOOD

The tradition established by Josiah Wedgwood & Company of commissioning leading artists and illustrators to design ceramics for them was one established from the company's earliest days (see p70) and has continued unabated throughout the 20th century. As a result the company is responsible for some of the most distinctive of all British ceramic designs of the modern period. Designers include John Skeaping, Richard Guyatt, Keith Murray, Eric Ravilious, and Daisy Makeig-Jones. Twentieth-century Wedgwood has little uniformity of style and collectors tend to concentrate on the work of particular designers rather than the pottery as a whole. Value depends on the size and elaborateness of the decoration as well as on the reputation of the designer concerned.

◀ **DAISY MAKEIG- JONES**
One of the chief designers working for Wedgwood in the 1920s, Daisy Makeig-Jones designed the ornamental range known as "Fairyland lustre" – a variegated dark-coloured ground printed with colours and gilding depicting imaginary landscapes and fantastic figures. Although these pieces are in part mechanically produced, some of the rarer designs can attract very high prices. This vase is decorated with Woodland Bridge and Woodland Elves II patterns. Daisy Makeig-Jones also produced less elaborate, more modestly priced ranges of lustreware including "Dragon lustre" and "Butterfly lustre". £3,500–4,500

▶ **ERIC RAVILIOUS**
The illustrator Eric Ravilious produced a number of designs for Wedgwood in the 1930s including alphabet nursery ware, a zodiac set, a boat race cup and bowl, and a design for a commemorative tankard for the coronation of Edward VIII, which was adapted for the coronations of George VI and Elizabeth II. This nursery plate designed in 1937 is typical of his simple but effective designs. Ravilious himself was killed while working as a war artist in 1942, before many of the wares printed with his designs could be produced. £80–100

◀ **KEITH MURRAY**
Keith Murray was a prominent architect whom Wedgwood commissioned to create a range of more modern designs. Bold geometric shapes in muted shades and simple patterns are typical of his distinctive style. This bowl (c.1935) is a typically strong form with a plain off-white surface, made more stylish by the addition of a silver lustre glazed band inside the rim and on the foot. £250–300

MARKS

Works by designers are usually marked with a printed signature or the words "designed by..." plus the name and the Wedgwood mark. Letter date marks were also used; after 1929 the last two years of the date appear in full.

▶ **OTHER DESIGNERS**
A wide range of designers were employed by Wedgwood in the 20th century, and the value of their work depends on the appeal of the form. This jug (c.1910–30), designed by Alfred and Louise Powell, was made for commercial production and painted by hand. Rarely were ceramics of this type signed by the Powells. £100–120

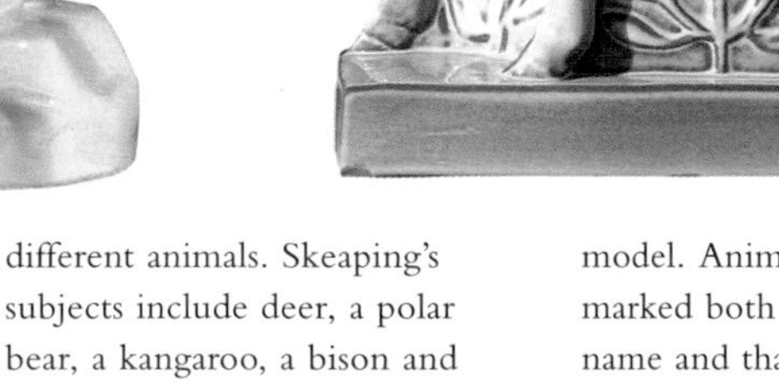

▲ **JOHN SKEAPING**
Animals in matt glazes were the speciality of John Skeaping who produced designs in 1927 for Wedgwood for a group of ten different animals. Skeaping's subjects include deer, a polar bear, a kangaroo, a bison and monkeys in various glazes and he was paid £10 for each model. Animals are typically marked both with the designer's name and that of Wedgwood. Examples such as this sealion and calf fetch £300–400 each.

CARLTON

Carltonware was produced by the Carlton Works, based in Stoke-on-Trent. The company was founded in 1890 and renamed Carlton Ware Ltd in 1957. The pottery enjoyed a heyday of popularity during the 1920s and 1930s when it produced a huge range of decorative items and tablewares; geometric designs, moulded tablewares, novelty items and lustre decoration are particular specialities. The most desirable pieces tend to be those painted with bold abstract designs or lustre pieces decorated in Oriental style. Tablewares tend to be far less sought after than Carlton's decorative items. Prices range from £50–300 for small vessels to £1,000 and upwards for larger decorative pieces.

► GUINESS ADVERTISING
Objects advertising Guinness have a cult following among collectors. Although estimated at £200–400 for the set, these 1930s toucan wall plaques sold for over £1,000. Beware there are modern fakes of these.

▼ DESIGNS
Decorative items painted in distinctive abstract designs with bright colours are among the most highly sought after of all Carlton Ware items. This vase typifies the innovative patterns found on pieces of the 1930s; later colours tended to be more subdued. £300–400

► ORIENTAL DESIGNS
Lustrewares were produced in large quantities by the Carlton factory. Pieces such as this hexagonal covered vase made c.1930 reflect the influence of Oriental-style design both in their decoration and choice of shape. £200–300

MARKS

- Some early marks have the initials of Wiltshaw & Robinson as well as the name Carlton Ware.
- Designers are not generally named.

SHELLEY

First known as Wileman & Co, then as Foley, the Shelley factory is best known for the highly original Art Deco-style tablewares produced during the 1920s along with nursery wares and pieces based on Mabel Lucie Attwell figures. The most valuable pieces are Art Deco tablewares boldly painted with futuristic designs in very distinctive colour combinations. Less striking designs with floral decoration were produced during the 1930s and 1940s and these tend to be far less sought after. Shelley also produced commemorative pieces and their value depends mainly on the rarity of the design (see p140).

◀ VALUE
The streamlined geometric shape, futuristic design and bold yellow/black colour scheme combine to make this Sunray pattern part-coffee set (c.1930) extremely sought after. Wares are also categorized according to their shape. This form is known as Mode; other shapes are Eve, Vogue, Queen Anne and Regent. £300–500 (for the pieces shown)

▶ FIGURES
Shelley also produced a popular range of figurines and nursery wares, which are keenly sought after, based on the illustrations of Mabel Lucie Attwell. This group (c.1937) is known as "Our Pets" and features children with their pet rabbits. It is rare and therefore more valuable than more common subjects.
£800–1,200

MARKS

Pieces are usually marked with a signature in a cartouche and serial number. Marks which include the words "Fine Bone China", date from after 1945. A serial number beginning with a 2 shows the piece was a second.

▶ LATER DESIGNS
The more elaborate designs of the 1930s and later tend to fetch considerably less than those in the Art Deco style and are still available for modest sums.
This part-tea service (c.1928), attractively decorated with a landscape design, contains more than 20 individual pieces and would be worth £220–280.

WADE FIGURES

Famous for its range of endearing porcelain and cellulose nursery figures and animals, the Wade factory was founded in 1922 in Burslem, Staffordshire, by George Wade. The company also established a factory in Ireland where it made porcelain for the tourist industry and export market. Price generally depends on the rarity and popularity of the subject. The most sought-after early Disney figures of the 1930s can fetch £1,500. Pieces modelled by Faust Lang, a leading Wade modeller, are also keenly sought after and command a premium, but you will still find animals and other figures for under £1.

MARKS

Figures are not always marked and do not have serial numbers. To help identify and date figures, refer to the specialist collectors' guides on the subject listed on pp168–9.

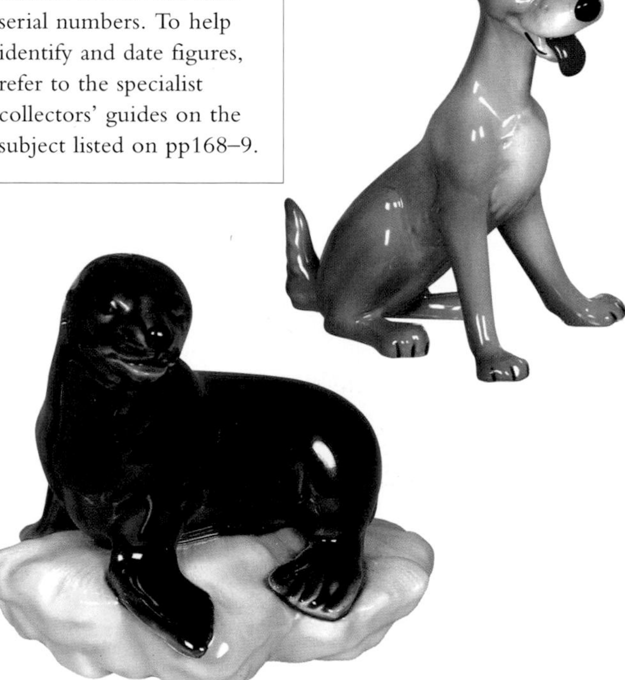

◀ DISNEY FIGURES
Disney figures were produced to coincide with new films and are still being produced today. This figure depicts Tramp from the film *Lady and the Tramp* and was produced between 1961 and 1965. £75–100

▲ ANIMALS
A vast range of endearing animals such as this seal were produced by Wade in the 1950s. Condition does affect value, however, so look out for small chips and imperfections which will reduce the price quite significantly. £50–100

▼ OTHER SUBJECTS
Apart from animal figures Wade produced a wide range of other small decorative objects including models of Pearly Kings and Queens, the latter (c.1959) shown here, advertising wares, novelty egg cups, vases, money boxes, jugs and trays. Most of these are less valuable than the figures. £40–60

CELLULOSE FIGURES

A range of inexpensive cellulose figures were produced by Wade in the 1930s. These are especially vulnerable to flaking paint and prices are lower than for ceramic figures from other factories.

BESWICK FIGURES

The pottery-making company founded in the late 19th century by James Beswick was taken over by his son in 1920 and sold to Royal Doulton in 1969. Since the 1920s it has established a reputation for its popular range of decorative china animals. Subjects include figurines depicting animals from Beatrix Potter and Winnie the Pooh. The firm has also produced a range of figures of animals and birds as well as low-relief wall plaques in graduated sets. Horses are a particular favourite with collectors and some figures portray named jockeys with well-known racehorses. The highest prices tend to be paid for rare popular subjects – a series of Beatrix Potter wall plaques made in 1967–69 can fetch over £1,200 each, and even figures produced in the 1980s can make over £100.

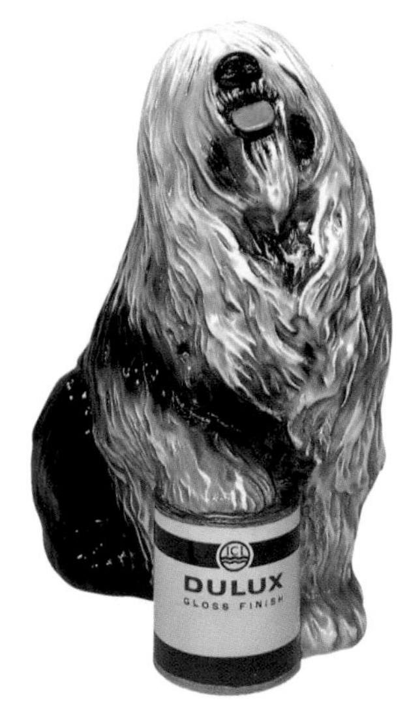

◀ **DATING**
Pieces can be dated from the mark and model by referring to the specialist guides on the subject listed on pp168–9. This Old English sheepdog, "The Dulux dog", was produced between 1964 and 1970. £120–150

▲ **HORSES**
Model horses are extremely popular with collectors and from the 1950s onwards the firm employed some highly skilled modellers to produce their designs. The Lippizaner stallion and rider from the Spanish Riding School, Vienna, above, modelled by Graham Tongue, was produced between 1973 and 1980. £150–160.

▼ **VALUE**
Among the most sought-after Beswick figures are Beatrix Potter subjects, Alice in Wonderland figures and this set representing Snow White and the Seven Dwarfs, produced 1954–67. £1,000–1,500

STUDIO CERAMICS

One of the most dramatic recent developments in the world of ceramics is the rise of interest in post-war studio pottery. Individually hand-crafted pieces by certain leading potters have enjoyed a rapid increase in popularity and price in recent decades, but value depends on quality rather than upon the name of the potter concerned and only a handful command high prices.

Bernard Leach is regarded by many as the founder of the revival in British studio ceramics. After a visit to Japan where he met the potter Shoji Hamada and became fascinated by Oriental ceramics, Leach with Hamada established a factory at St Ives in 1920 which influenced and inspired a generation of student potters. Lucie Rie is perhaps the best known of all the studio potters; she trained in Vienna in her native Austria and came to London in 1938 where she initially earned her living making buttons. Rie along with Hans Coper with whom she worked closely from 1946, are renowned for their strong innovative forms and unusual glazes.

◀ BERNARD LEACH Made during the early years of Leach's St Ives pottery, this slipware dish reflects the influence of traditional slipware techniques in the vein of Thomas Toft (see p62) and was made by dipping the earthenware body in a creamy slip and trailing designs in darker slip on top. £3,000–5,000

▶ DESIGN The influence of Oriental design can be seen on this Bernard Leach stoneware flask which is decorated with calligraphic motifs against a soft grey and blue quartered background. Leach's work is always thickly potted and simple in form. £1,000–1,500

◀ HANS COPER Unusual shapes with little additional decoration are a hallmark of many of Hans Coper's vases. This stoneware "Thistle" vase made c.1950 is characteristically finished with a pale finely textured surface and a contrasting brownish black interior. £8,000–10,000

◄ **VALUE**
Size, rarity and decorative appeal all play a part in determining value. This fine green porcelain bowl (c.1983) with a bronze glazed band by Lucie Rie is a highly decorative object, and even though it is only 6 ¾in (17cm) in diameter it sold for £6,325.

MARKS

- **Bernard Leach** usually marked with "BL" either painted, impressed monogram or cipher.
- **Lucie Rie** an impressed seal of "LR" monogram.
- **Hans Coper** an impressed "HC" cipher resembling a potter's wheel on its side.

► **LUCIE RIE**
This tapering stoneware vase (c.1974) by Lucie Rie is decorated with mottled glazes, reflecting her fascination with different glaze techniques, which she often used to give texture to the surface of her pots. £800–1,200

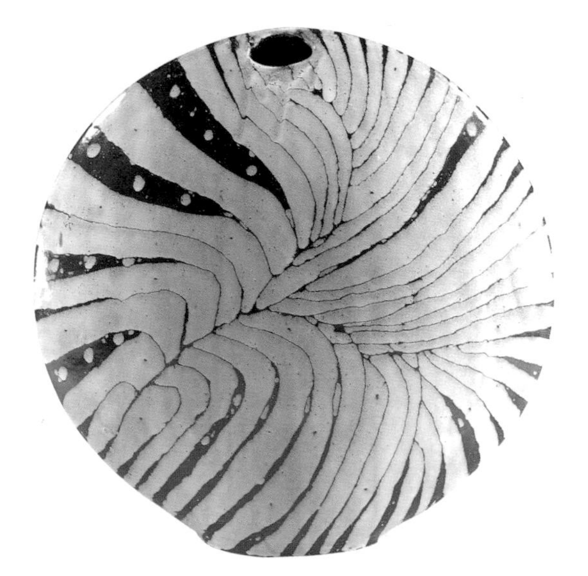

◄ **OTHER POTTERS**
This circular earthenware vessel (1983) is by James Tower, one of a new generation of studio potters whose work is increasingly collectable. Other potters whose work is also achieving notable success include Elizabeth Fritsch, Sue Mundy, Alison Britton, Gabriele Koch and Abdo Nagi. £2,500–3,500

The development of the American ceramic industry was hampered by the fashion for imported foreign wares and a shortage of skilled manpower. Foreign wares had always been considered superior and provided the inspiration for most early American ceramics. The first American porcelains are hard to distinguish from the imported wares they copied and are rarely marked. Wares made by Bonnin & Morris in Philadelphia from 1770–72 copied Bow; Tucker porcelains made in Philadelphia from 1826–38 were based on French 18th-century designs; D. & H. Henderson's Jersey City stonewares are mistaken for those made by Ridgway; brown-glazed wares from Bennington, Vermont, are similar to Derby wares; while Belleek made in Trenton rivalled Irish Belleek and Royal Worcester.

Despite the gradual burgeoning of potteries in the 19th century, American ceramics shown at the great Centennial Exposition in Philadelphia in 1876 received little notice. Since the turn of the century, however, interest in American ceramics has grown, and collectors now prefer these rarer American versions, and pay a premium for them. Among the most popular are 18th-century slip-decorated redwares, produced from c.1750 onwards in New England, and salt-glazed stonewares developed by immigrant French and German potters from c.1750. English imported pottery such as spatterware is also becoming increasingly popular. Innovative forms and decorations are hallmarks of art pottery made from the late 19th century and this is avidly collected too.

EARLY AMERICAN CERAMICS

America's first porcelain factory, Bonnin & Morris, was founded by the British-born Gousse Bonnin and the Philadelphian Antony Morris in Philadelphia in 1769. The factory produced dinner- and teawares in underglaze blue and polychrome decoration for only two years. After that the Non-Importation Act restricting British imports was repealed. Wares are extremely rare and expensive. The second American porcelain manufacturer, William Ellis Tucker, was in business from 1826–38. Much is unmarked and hard to distinguish from the French and Italian prototypes on which it is based. The mid-19th century English fashion for Parian figures quickly crossed the Atlantic and Parian was produced in Bennington, Vermont, the Greenpoint section of Brooklyn, and Trenton, New Jersey.

▲ **BONNIN & MORRIS**
This pickle stand (1770–72) was the company's most expensive form. Based on similar pieces made at Plymouth and Bow, London it reflects the great rarity of all Bonnin & Morris porcelain. Only two pickle stands have been sold in recent decades; one bought at a jumble sale for £1 sold for over £44,000; this one sold for £55,000 in 1989.

▶ TUCKER PORCELAIN
Tucker produced a wide range of tablewares and decorative objects. The undecorated coffee pot is a rare form. The three-vase garniture is decorated with floral decoration on one side and sepia scenes on the other (1828–38). Coffee pot £3,500–5,500; pitchers £1,000–3,500 each; garniture £13,000–16,500

▼ VALUE
The vase-shaped pitcher is a form unique to Tucker and was produced with a variety of decoration; prices depend on the painting – Philadelphia views are particularly sought after. This example (c.1830), painted on either side with sprays of flowers, is identified as No. 7 in the Tucker pattern book and it sold for £4,216.

▶ PARIAN
This marked bust of William Shakespeare (c.1876) was produced by Ott & Brewer, Parian manufacturers of Trenton, New Jersey, who produced a range of busts including George Washington, Abraham Lincoln, Ulysses Grant, Benjamin Franklin and Cleopatra, all by the sculptor Isaac Broome. £4,000–7,000 (less if small or unmarked)

IDENTIFICATION & MARKS

- Bonnin & Morris pieces are unmarked or simply marked with a letter "P".
- Most Tucker is unmarked; some pieces are marked "Tucker and Hulme" or "Tucker and Hemphill". Early examples can be identified by their greenish cast when held up to the light. Later pieces seem orange or straw-coloured.

AMERICAN ART POTTERY I

The nationwide fashion for painted ceramics in the years following the Centennial Exhibition gave rise to the American Art Pottery movement, which flourished from c.1875–1920. Two prominent Cincinnati women were largely responsible for its development. Mary Louise McLoughlin created Cincinnati faience, based on the slip-painted wares of Limoges and on Japanese wares, and founded the Cincinnati Pottery Club. Maria Longworth Nichols established the Rookwood Pottery in 1880 and by 1890 was offering "standard" wares slip-cast and decorated in dark brown, red, orange and yellow under a yellow-tinted high-gloss glaze; Iris and Sea Green glazes in 1894; and matte glazes by 1902. The most desirable pieces are those signed by artists; over 100 decorators marked with their initials or a cipher. Other prominent American Art potteries include Grueby of Boston, the Dedham pottery, Massachusetts, and the Newcomb pottery, founded in 1895 at Sophie Newcomb Memorial College for Women in New Orleans by Mary G. Sheerer.

► GRUEBY

William Grueby of Boston specialized in organic forms decorated with matte glazes in green, yellow, ochre and brown. His most popular works have crisp applied decorations in the form of leaves and buds. The Grueby pottery was patronized by Gustav Stickley, a leader of the Arts and Crafts movement. These vases were made c.1900. From left to right: £5,500; £3,500; £5,000

ROOKWOOD MARKS

Rookwood is impressed with the "RP" mark. A flame over the mark was introduced in 1886, and another flame added each year until 1900 when a Roman numeral was added to indicate the year. Many pieces are also signed by their artists, and these are much sought after.

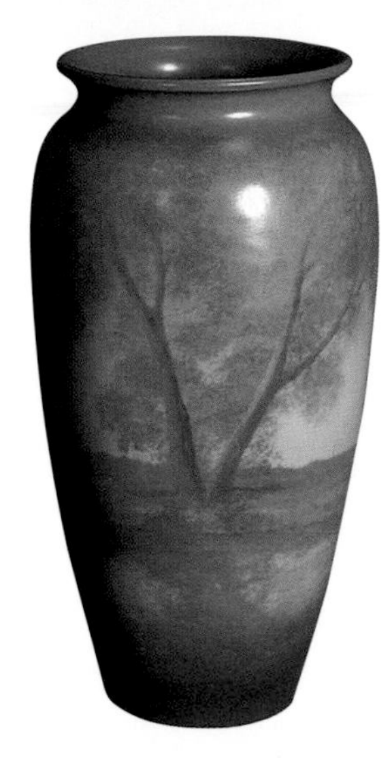

◄ ROOKWOOD

The influence of Japanese design on Rookwood pottery became more prevalent after the arrival of the young Japanese artist Kataro Shirayamadani in 1887. This ovoid vellum landscape vase (c.1920) was decorated by Edward Diers, another prominent Rookwood artist. Vellum, a matte glaze transparent enough to be used over coloured slip decorations, was introduced in 1904. £1,400–£2,000

◄ **NEWCOMB**
These Newcomb vases (c.1900-1908) are typically decorated with designs based on the flora and fauna of the surrounding bayou country. The decoration is created by drawing into the wet clay and painting with oxides after the first firing. Wares were then given a transparent high-gloss glaze in a distinctive palette of blue, green, yellow and black. Designs were rarely repeated. From left to right £10,000; £25,500; £12,000; £13,500 (damaged £3,500)

► **DEDHAM**
Oriental ceramics and French art pottery provided the inspiration for the Dedham pottery. Hugh Robertson, the pottery's founder, developed a wide variety of decorative effects including flambé and volcanic glazes and a dragon's blood finish, as well as crackle glazes decorated with blue and white borders of various animals marching clockwise. This crackle-glazed teapot (1896–1928) has a charming rabbit pattern and is marked with a blue stamp reading "Dedham Pottery" above a rabbit enclosed in a square frame. £530–800

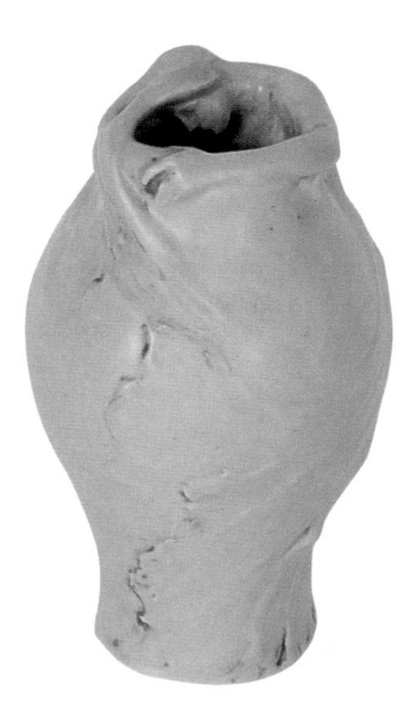

◄ **VAN BRIGGLE**
Artus van Briggle, a decorator at Rookwood, set up his own pottery in Colorado Springs in 1901. Influenced by Art Nouveau his designs combine figural, floral and plant forms. This Lorelei vase reflects the influence of Rodin and was produced at Rookwood by van Briggle in 1898. £10,000–15,000

BEWARE
The van Briggle pottery still produces Lorelei vases. New ones sell for around £50; those from the late 1920s, called "dirty bottom ware" fetch around £500. Always buy from a reputable source or check against a known original if you are at all in doubt.

AMERICAN ART POTTERY II

Other leading American art potters include Louis Comfort Tiffany, Teco, Fulper and George Ohr. Pottery became an interest for Tiffany in 1904, long after his reputation as an interior designer was established. Pieces were typically cast in Art Nouveau style but Tiffany's pottery proved to be less successful than his glass and was discontinued after 1917. One of the United States' most prolific potters was George Ohr, "the Mad Potter from Biloxi", who made more than 10,000 pots of unparalleled virtuosity and crammed them into barrels in the family attic. Regarded today as the father of the studio pottery movement, Ohr produced innovative forms that anticipate many of the developments of contemporary ceramics. The American Terracotta and Ceramic Company, known as Teco or as the Gates Pottery, registered a trademark in 1895 and is especially known for its architectural ceramics, some designed by Frank Lloyd Wright. The Fulper Pottery Co in Flemington, New Jersey, was established in 1815 and produced stoneware and utilitarian earthenware before turning to art pottery. William Hill Fulper made moulded or hand-thrown stoneware in Classical and Oriental forms and used a variety of glazes.

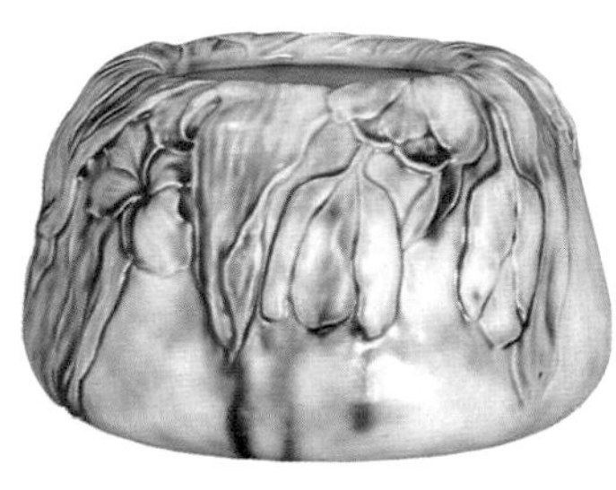

◀ ▲ TIFFANY
Tiffany pottery was mostly made by casting into moulds. Organic forms and decoration reflect the Art Nouveau style. Light yellow-green tones were favoured on early pieces. The baluster vase left (c.1912) has a textured semi-gloss glaze, while the vase above (c.1910) with embossed foliage is matte-glazed. Left £800–1,000; above 1,500–2,000

MARKS
Most Tiffany pottery, such as the tall baluster flambé vase seen on the left, is marked with the "LCT" cipher; the words "L. C. Tiffany", "Favrile Pottery" or "Bronze Pottery" are sometimes etched on the base.

▶ TECO
The silvery matte green glaze seen on this floor vase (c.1905) is typical of Teco pottery and was inspired by matte glazes on French ceramics shown at the 1893 World's Columbian Exposition in Chicago. Teco mass-produced 500 shapes in various matte colours including brown, grey, blue, rose, purple and yellow. The innovative architectural form is characteristic of this company's ceramics. Teco is marked with a long-stemmed "T" with the "e", "c' and "o" arranged under each other. £6,500–10,000

▶ FULPER
Fulper pottery lamps were patented in 1912. This example (c.1915) has a ceramic base and shade with yellow and green slag glass inserts mounted with copper foil. The lamp is decorated with a crystalline dark leopard skin glaze in brown, cream and grey. £6,000–10,000

MARKS
Fulper is usually marked with an impressed name but is rarely dated. Vertical letters are earlier than horizontal ones. A circular mark with the date of 1805 is sometimes seen.

▼ GEORGE OHR
Unctuous glazes richly mottled in metallic colours, bizarre forms and thin walls are distinctive features of George Ohr's pottery. Pieces such as these vases (c.1905–1910) are impressed "GEO E OHR/BILOXI, MISS." Left to right £1,000–1,500; £5,000–7,000; £3,000–3,500

EARTHENWARE & STONEWARE

Red earthenware pottery was made by the earliest settlers at Jamestown, Virginia, and Plymouth, Massachusetts. In the South, as the plantation owners prospered, slave-made pottery was relegated to the kitchen while imported wares were used on the table by the masters. In New England and the Middle States there was also demand for storage and kitchen wares made by local potters. Before the end of the 17th century potters decorated their wares with simple slip-trailed designs. Inspired by the folk pottery of the Rhine Valley and Switzerland, sgraffito and slip-decorated redware reached its fullest flowering on ornamental pie plates. Stoneware was also produced in the United States by such potters as Remmey and Crolius in New York and Norton in Bennington, Vermont. Among a wide range of stoneware objects made were crocks, jugs, jars, chamber pots, colanders, butter tubs and pitchers, often decorated with incised or painted cobalt blue designs. Various new types of stoneware were developed in the 19th century including brown-glazed Rockingham ware and flint enamel, a process developed by C. W. Fenton in Bennington.

▶ REDWARE

This redware jar made in Pennsylvania in the late 18th century is typically decorated with a slip design of tulips. This form of decoration is identified with Conrad Mumbouer of Haycock Township, Bucks County (1794–1844). Decorated hollow wares such as this command a premium with collectors; hence the high value. £10,000–13,000

▼ STONEWARE

This 19th-century salt-glazed stoneware pitcher is decorated in cobalt blue and is stamped by John Bell of the Bell Pottery, Waynesboro, Shenandoah Valley, Pennsylvania (1833–89). Decoration determines value; jugs and crocks with cobalt floral decoration are less valuable than those with figures, birds and animals. Incised designs bring a premium. £1,000–1,600

◀ SGRAFFITO

This redware plate (c.1805) has been covered with white slip and the design created by scratching through to the red clay, a technique known as sgraffito. The figure on horseback is surrounded by an inscription in a German dialect which translates as: "I have ridden over hill and dale and found girls everywhere". Plates decorated with figures are rare and expensive; this plate is worth over £50,000. Floral decorated plates are more readily available and cost from £1,000.

ANGLO-AMERICAN WARES

Quantities of hand-painted and transfer-printed ceramics were made in England in the 19th century for export to the United States. Not long after transfer printing was discovered, potteries in Liverpool made jugs, mugs, bowls and plates decorated with sailing ships and patriotic subjects that would appeal to visiting seamen. Staffordshire potters also produced pitchers with transfer-printed naval battles and portraits of heroes. After the war of 1812, hundreds of different American views were transfer-printed in underglaze blue on various domestic items (see p78). English potters also produced hand-decorated pearlware for export, imitating Imari patterns made at Derby and Worcester. Popular with Pennsylvanian Germans, these colourful wares are known as "Gaudy Dutch".

◀ LIVERPOOL POTTERY
Made at the Herculaneum pottery in Liverpool c.1769–1800, this jug is decorated with a transfer print of *Peace, Plenty and Independence.* On the reverse side there is a print of a ship flying an American flag. Jugs with patriotic designs range from £1,000–14,000. This one is particularly valuable because it is hand-coloured and in mint condition; it sold for £12,000.

▲ "GAUDY DUTCH"
"Gaudy Dutch" came in several patterns including Butterfly, Grape, War Bonnet, Single Rose, Oyster, Zinnia, Carnation and Sunflower. This cup (1830–50) is an example of the Butterfly pattern. All "Gaudy Dutch" designs are hand-painted in a distinctive palette of orange, yellow, green and red with underglaze blue. £600–800

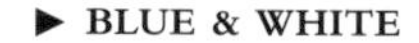

▶ BLUE & WHITE TRANSFER-PRINTED WARES
Blue and white wares decorated with American subjects were produced in Staffordshire for export. This soup tureen (c.1830) featuring Pennsylvania Hospital would be worth £5,000–6,000. Plates start at around £130.

◀ SPATTERWARE
Produced by Staffordshire potters from 1820–1860, spatterware was another popular type of export pottery. It came in a wide range of colours often with an image and a spatter border. Pieces such as these are rare and costly. The rainbow spatter teapot and bowl sold for £10,700; the pitcher for £8,000.

KINGS 2

PART 5

INFORMATION

ABOVE A VALUABLE SOURCE ON PRICES.

LEFT VIEWING AND EXAMINING THE CERAMICS ON OFFER PRIOR TO A SALE.

WHERE TO BUY

MAJOR AUCTION HOUSES

The following leading London auction houses also have regional offices. Look in the local press for information on local auctions.

Bonhams (Knightsbridge)
Montpelier Street
London SW7 1HH

Bonhams (Chelsea)
65–9 Lots Road
London SW10 0RN

Christie's
8 King Street
St James's
London SW1Y 6QT

Christie's South Kensington
85 Old Brompton Road
London SW7 3LD

Phillips
101 New Bond Street
London W1Y OAS

Sotheby's
34-35 New Bond Street
London W1A 2AA

MAJOR ANTIQUES FAIRS

Antiques for Everyone
National Exhibition Centre
Birmingham
Annual event held in August.

British International Antiques Fair
National Exhibition Centre
Birmingham
Annual event held in April.

The Fine Art & Antiques Fair
Olympia, London
Thrice yearly event held in March, June and November, organised by P & O Events, Exhibition Centre, Warwick Road, London SW5 9TA.

International Ceramics Fair & Seminar
31 Old Burlington Street
London W1X 1LB
Specialist fair covering a wide range of ceramics held in June.

The Grosvenor House Antiques Fair
Grosvenor House Hotel
Park Lane
London
Annual event in June for top of the range works of art.

Northern Antiques & Fine Art Fair
Royal Assembly Rooms
Harrogate
North Yorkshire
Annual event held in September.

Thames Valley Antiques
Dealers Association Fair
School Hall Eton college
Eton
Berkshire
Annual event held in April.

The South of England Show
Selsfield Road
Ardingly
Sussex
One of the largest outdoor antiques events held in July.

Alexandra Palace Antique & Collector's Fairs
The Great Hall
Alexandra Palace
London
Huge antiques event held five times a year.

Contact the following trade associations for information on annual antiques fairs, specialist dealers, valuations and insurance:

British Antique Dealers' Association (BADA)
20 Rutland Gate
London SW7 1BD

London and Provincial Antique Dealers' Association (LAPADA)
535 King's Road
London SW10 0SZ

Council for the Protection of Art Theft (COPAT)
17 Whitcomb Street
London WC2 7PL

ANTIQUES CENTRES AND MARKETS

A more comprehensive list of names and addresses is available in *Miller's Collectables Price Guide*

Alfie's Antique Market
13-25 Church Street
London NW8 8DT

Antiquarius Antiques Market
131-141 King's Road
London SW3 8DT

Bath Antiques Market
Guinea Lane
(off Landsdowne Road)
Bath
Avon BA1 5NB

Bermondsey Antiques Warehouse
173 Bermondsey Street
London SE1 3UW

City of Birmingham Antiques Market
St Martins Market
Edgbaston Street
Birmingham B5 4QL

Bristol Antique Market
The Exchange
Corn Street
Bristol BS1 1HQ

The Lanes
Brighton
East Sussex

Camden Antiques Market
Camden Town
London NW10

Campden Passage Antique Centre
12 Campden Passage
Islington
London N1 8ED

Chelsea Antiques Market
245–53 King's Road
London SW3 5EL

Cirencester Antiques Market
Market Place
Cirencester
Gloucestershire GL7 2PY

Crystal Palace Collectors' Market
Jasper Road
Westow Hill
London SE19

Dorking Antiques Centre and Victoria & Edward Antiques Centre
17–18 West Street
Dorking
Surrey RH4 1DD

Grays Antique Market
58 Davies Street
London W1Y 1LL

Ironbridge Antique Centre
Dale End
Ironbridge
Shropshire TF8 7DW

Oxford Street Antiques Centre
16–26 Oxford Street
Leicester LE1 5XU

Leominster Antiques Market
14 Broad Street
Leominster
Hereford HR6 8BS

Newark Antiques Centre
Regent House
Lombard Street
Newark NG24 1XP

St. Mary's Antiques & Collectors Centre
Duke Street
Norwich NR3 3AF

Oxford Antiques Market
Gloucester Green
Oxford OX1 2DF

Petworth Antiques Market
East Street
Petworth
Sussex GU28 0AB

Portobello Road Antiques Market
Portobello Road
London W11

Stoke on Trent Antique & Collectors Centre
The Potteries Antique Centre
271 Waterloo Road
Cobridge
Stoke on Trent ST6 3HR

Antiques and Bric-a-Brac Market
Market Square
Wellingborough NN8 1AR

SPECIALIST COLLECTORS' CLUBS

More useful collectors' information can be obtained through specialist clubs.

Goss & Crested China Club
62 Murray Road
Horndean
Hants P08 9JL

The Beswick Collectors' Circle
Corner Cottage
Hedgerley Lane
Gerrards Cross
Bucks SL9 7NS

Clarice Cliff Collectors Club
Fantasque House
Tennis Drive
The Park
Nottingham NG7 1AE

Commemorative Collector's Society
25 Farndale Close
Long Eaton
Nottingham NG10 3PA

Oriental Ceramic Society
30b Torrington Square
London WC1E 7JL

Royal Doulton International Collectors' Society
Minton House
London Road
Stoke on Trent
Staffs ST4 7QD

Shelley Group
12 Lilleshall Road
Clayton
Newcastle under Lyme
Staffs ST5 3BX

Susie Cooper Collectors' Group
PO Box 48
Beeston
Nottingham NG9 2RN

Torquay Pottery Collectors' Society
Avenue Road
Torre Abbey
Torquay TQ2 5JX
Devon

Wade Collector's Club
14 Windsor Road
Selston
Nottingham NG16 6JJ

WHERE TO SEE

Visiting major collections in museums and houses open to the public is a valuable way of learning about ceramics. Listed below are some of the the most noteworthy places.

The Balston Collection
Bantock House Museum
Bantock Park
Wolverhampton
West Midlands WV3 7DL

Belton House
Grantham
Lincs NG32 2LS

Birmingham City Museum and Art Gallery
Chamberlain Square
Birmingham
West Midlands B3 3DH

Bristol City Art Gallery
Queen's Road
Bristol BS8 1RL

British Museum
Great Russell Street
London WC1B 3DG

Cecil Higgins Art Gallery and Museum
Castle Close
Bedford MK40 3NY

City Museum and Art Gallery
Bethesda Street
Hanley
Stoke on Trent
Staffs ST1 3DE

Clandon Park
West Clandon
Guildford
Surrey GU4 7RQ

Derby Museum and Art Gallery
The Strand
Derby DE1 1BS

Fenton House
Windmill HIll
London NW3 6RP

Fitzwilliam Museum
Trumpington Street
Cambridge CB 1RB

Keddeston Hall
Derby DE22 5JH

Harewood House
Harewood
Leeds LS17 9LQ

Leeds City Art Gallery
The Headrow
Leeds LS1 3AA

Luton Hoo
Luton
Bedfordshire LU1 3TQ

Museum of Worcester Porcelain
Severn Street
Worcester WR1 2NE

Norwich Castle Museum
Castle Meadow
Norwich
Norfolk NR1 3JU

Nottingham Castle Museum and Art Gallery
The Castle
Nottingham NG1 6EL

Polesdon Lacey
Great Bookham
Dorking
Surrey RH5 6BD

Royal Crown Derby Museum
Royal Crown Derby Porcelain Co
194 Osmaston Road
Derby DE23 8JZ

The Royal Pavilion
Brighton
East Sussex BN1 1UE

Royal Museum of Scotland
Chambers Street
Edinburgh EH1 1JF

Saltram
Plymton
Plymouth
Devon
Cornwall PL7 3UH

Shugborough Hall
Milford
Stafford ST1 70XA

Standen
East Grinstead
West Sussex RH19 4NE

Tatton Park
Knutsford
Cheshire WA16 6QN

Temple Newsam House
Leeds
West Yorkshire
LS15 0AE

Upton House
Banbury
Oxfordshire OX15 6HT

Victoria & Albert Museum
Cromwell Road
London SW7 2RL

Waddesdon Manor
Waddesdon
Aylesbury
Bucks HP18 0JH

The Wallace Collection
Manchester Square
London W1

Wedgwood Museum
Josiah Wedgwood & Sons Ltd
Barlaston
Stoke on Trent
Staffs ST12 9EF

Windsor Castle
Windsor
Berkshire SL4 1NT

Woburn Abbey
Woburn
Bedfordshire MK43 0TP

Yorkshire Museum
Museum Gardens
York YO1 2DR

WHAT TO READ

GENERAL

Atterbury, Paul, *History of Porcelain*, 1982

Battie, David, Ed., *Sotheby's Concise Encyclopedia of Porcelain*, 1990

Berges, Ruth, *From Gold to Porcelain; the Art of Porcelain and Faience*, 1963

Cushion, John, *Porcelain*, 1973

Charleston, Robert, Ed., *World Ceramics*, 1981

Fleming, John and Honour, Hugh, *The Penguin Dictionary of the Decorative Arts*, 1977

Honey, W. B., *The Art of the Potter*, 1980 and *Ceramic Art from the End of the Middle Ages to about 1815*, 1952

Knowles, Eric, *Miller's Antiques Checklists: Victorianna* (1991), *Art Nouveau* (1992), *Art Deco* (1991)

Lang, Gordon, *Miller's Antiques Checklists: Porcelain* (1991), *Pottery* (1995), *Pottery & Porcelain Marks* (1995)

Morley Fletcher, Hugo, Ed., *Techniques of the World's Great Masters of Pottery and Ceramics*, 1984

Savage, George, *Porcelain through the Ages*, 1954

Savage, George & Newman, Harold, *An Illustrated Dictionary of Ceramics*, 1958

EUROPEAN, ISLAMIC & AMERICAN CERAMICS

Atterbury, Paul, *Moorcroft Pottery*, 1987 and *Dictionary of Minton*, 1989

Austwick, J. and B., *The Decorated Tile*, 1980

Barber, Edwin Atlee, *The Pottery and Porcelain of the United States* 1909 (3rd edition); *Marks of American Potters*, 1904 and *Tulip ware of the Pennsylvania German Potters*, 1903

Barnard, J., *Victorian Ceramic Tiles*, 1972

Batkin, Maureen, *Wedgwood Ceramics 1846–1959*, 1982

Bergesen, V., *Majolica – British, Continental and American Wares*, 1989

Birks, Tony, *Lucie Rie*, 1987

Caiger-Smith, A., *Tin Glaze Pottery in Europe and the Islamic World*, 1973

Bivins Jr., John, *The Moravian Potters in North Carolina*, 1972

Clark, Garth and Hughto, Margie, *A Century of Ceramics in the United States 1878–1978*, 1979

Clark, Garth, Ellison Jr., Robert A. and Hecht, Eugene, *The Mad Potter of Biloxi; the Art & Life of George E. Ohr*, 1989

Comstock, H.E., *The Pottery of the Shenandoah Valley Region*, 1994

Cox and Cox, *Rockingham Pottery and Porcelain*, 1983

Cushion, J. B. *British Ceramic Marks*, 1988

Danckert, Ludwig, *Directory of European Porcelain*, 1981

Dennis, Richard, *Doulton Stoneware and Terracotta 1870–1925*; *Doulton Pottery Lambeth to Burslem 1873–1939*, 1975; *Royal Doulton 1815–1965*, 1965 and *The Parian Phenomenon*, 1989

Eidelberg, Martin, Editor, *From Our Native Clay, Art pottery from the Collections of the American Ceramics Arts Society,* 1987

Evans, Paul, *Art Pottery of the United States*, 1987

Eyles and Dennis, *Royal Doulton Figures*, 1978

Frelinghuysen, Alice Cooney, *American Porcelain 1770–1920*, 1989

Gilbert, Alfred, *English Art Pottery 1865*–1915, 1975

Godden, G. A., *Encyclopedia of British Pottery and Porcelain Marks*, 1994;
An Illustrated Encyclopedia of British Pottery and Porcelain, 1968;
The Illustrated Guide to Mason's Ironstone China, 1971;
Staffordshire Porcelain, 1983 and
Victorian Porcelain, 1961

Haggar, R. G., *Staffordshire Chimney Ornaments*, 1955

Hall, J., *Staffordshire Portrait Figures*, 1972

Haslam, Malcolm, *English Art Pottery 1865–1915*, 1975

Henzke, L. *American Art Pottery*, 1970

Hughes, Bernard G., *English and Scottish Earthenware; Victorian Pottery and Porcelain*, 1959

Lane, A., *Early Islamic Pottery*1947;
Later Islamic Pottery, 1957

Lloyd, Thomas E., *Victorian Art Pottery*, 1974

Oliver, A., *The Victorian Staffordshire Figure*, 1971

Rago, David, *The Fulper Book*, 1986

Rice, Paul and Gowing, Christopher, *British Studio Ceramics in the 20th Century*, 1989

Sandon, H., *The Illustrated Guide to Worcester Porcelain*, 1970

Sandon, H. and Grainger, J., *Worcester Porcelain*, 1989

Snyder, Jeffrey B., *Historical Staffordshire, American Patriots and Views*, 1995

Watson, Howard, *Collecting Clarice Cliff*, 1988

Wills, Geoffrey, *Wedgwood*, 1989

Wilson, T., *Ceramic Art of the Italian Renaissance*, 1987

Winstone, Victor, *Royal Copenhagen*, 1984

ORIENTAL

Beurdeley, M., and Raindre, G., *Qing Porcelain; Famille Verte, Famille Rose*, 1987

Carswell, John, *Chinese Blue and White and its Impact on the Western World*, 1985

Garner, Henry, *Oriental Blue and White*, 1970

Godden, G. A., *Oriental Export Market Porcelains*, 1979

Jenyns, Soame, *Japanese Pottery*, 1971 and *Japanese Porcelain*, 1985

Medley, Margaret, *The Art of the Chinese Potter*, 1981

Munsterberg, Hugo, *The Ceramic Art of Japan*, 1964

Reichel, Friedrich, *Early Japanese Porcelain*, 1981

Sato, M., *Chinese Ceramics, A Short History*, 1981

Vainker, S. J., *Chinese Pottery & Porcelain*, 1991

GLOSSARY

Agateware Pottery resembling the marbled effect of agate made in Staffordshire in the 18th century by mixing together clays of various colours.
Albarello Drug jar usually of waisted cylindrical form.
Applied decoration Ornamental motifs made separately from the main body and attached to it to provide decoration.
Arita Region in the West of Japan where the early porcelain industry was centred: Imari, blue«e and white, Kakiemon and Nabeshima were all produced in the area.
Armorial wares Pieces embellished with heraldic coats of arms, or crests.
Baluster Bulbous vase-shaped form.
Basalt ware (or Basaltes) Type of black stoneware developed by Josiah Wedgwood in the 1760s.
Basket weave Low relief motif simulating woven twigs or reeds.
Bat printing Printing decoration made using glutinous bats (sheets) popular in the early 19th century in Staffordshire.
Bellarmine Bulbous, narrow-necked German stoneware jug with a moulded mark, named after Cardinal Roberto Bellarmino.
Birnkrug Pear-shaped vessels originally from the Netherlands or Germany.
Biscuit Pottery or porcelain fired once but unglazed.
Blackware Iron rich type of ancient Chinese pottery.
Blanc de Chine Uncoloured, thickly glazed Chinese porcelain produced from the Ming dynasty.
Bleu celeste A sky blue enamel ground colour developed in the Vincennes factory in 1752 and particularly associated with Sevres and Vincennes porcelain.
Blue and white Ceramics decorated with cobalt blue enamel.
Bocage Encrustations of flowers, grass and moss generally used to decorate the supporting plinths of ceramic figures.
Body The material from which a piece of pottery or porcelain is made (the word paste is also used for porcelain).
Bonbonniere Small, lidded novelty box or bowl for sweets.
Bone ash Powder derived from burnt, ground animal bone that was added to porcelain in England in the mid-18th century.
Bone china Type of porcelain incorporating dried bone developed by Josiah Spode and others c.1794, that became the most popular type of English porcelain.
Bracket lobes Flat-topped projecting motif used as a moulding on plates and other tablewares.
Cachepot Holder for a flower pot, similar but smaller than a Jardinière.
Cailloùté Term derived from the French word for "pebble", meaning gilding applied in a series of dotted patterns.
Camaieu Monochrome painted decoration resembling ancient shell cameos.
Campana vase Vase shape derived from ancient prototypes resembling an inverted bell.
Caneware Stoneware of pale buff colour fashionable in the late 18th and early 19th century.
Canton porcelain Elaborately decorated and gilded chinese wares produced in the Canton (Guangzhou) district for export to the West.
Cartouche Decorative frame usually surrounding an inscription or pictorial decoration.
Caudle cup Vessel used for drinking caudle – a type of spiced porridge.
Celadon Semi-opaque glaze usually of greenish colour applied to Chinese stoneware.
Charger A large circular or oval dish or plate, usually richly decorated.
China clay A fine white clay also known as kaolin, which is mixed with petuntse to form true hard paste porcelain.
Chinoiserie European interpretation of oriental style decoration, which became particularly fashionable in the 17th and 18th century.
Cobalt blue Mineral used to create blue and white decoration.
Crabstock Spouts and handles modelled as knotted trees and branches, a popular form in the 18th century.
Crackle/crazing Fine network of cracks seen in certain types of glazed surfaces, sometimes used as a deliberate decorative effect, or as a result of ageing.
Creamware Refined lead glazed pottery of a pale colour popular in the 18th century.
Delftware Tin glazed earthenware made in England (delftware) or Holland (Delftware).
Deutsche Blumen Naturalistically painted flowers derived from prints used as a popular decorative motif on 18th century pottery and porcelain.
Dingyao Creamy white type of Chinese porcelain made in the Song dynasty.
Enamel Colours made from glass which fuse with the glaze during firing.
Enghalskrug Type of narrow necked jug with bulbous body popular in Germany and Holland from the mid-17th to the mid-18th century.
Faience Tin-glazed earthenware produced in France or Germany (similar to maiolica).
Famille rose Chinese porcelain decorated with a distinctive palette of opaque enamels.
Famille verte Chinese porcelain decorated with a

distinctive palette of predominantly green translucent enamels.
Flat backs Figurative pottery groups with undetailed almost flat backs, intended to stand on a mantelpiece, they were produced mainly in Staffordshire from c.1840.
Fritware Islamic pottery wares made from silica-rich clay that created a semi translucent body.
Garniture A set of three or more vases of matching and complimentary form intended to stand on a mantelpiece or on the top of a cabinet.
Gilding The use of gold to create decorative effect. Gilding can be applied in a variety of ways: mixed with honey or mercury and fired or applied after firing.
Glaze Glassy coating applied onto the biscuit body of pottery and porcelain and fired to make it non- porous.
Ground Background area of a single colour onto which further enamelled decoration or gilding may be applied.
Hard-paste porcelain Type of porcelain body made using kaolin and petuntse, first developed in China.
Hausmaler Term used to describe an independent painter of German porcelain blanks (most often with reference to Meissen).
Hookah Middle eastern smoking pipe, also known as a Narghili or Kendi.
Imari Japanese porcelain made in the Arita region and exported through the port of Imari. The term refers to pieces decorated with designs probably inspired by brocade textiles, predominately with underglaze blue, iron red and lavish gilding.
Impressed Term generally used to describe a mark or decoration that is stamped or indented into the surface of pottery.
Incised Term generally used to describe a mark or decoration that is cut or scratched into the surface.
Ironstone Hard white earthenware popular from the early 19th century for dinner services and other ornamental and architectural pieces.
Istoriato Mythological, historical or biblical subjects used to decorate the surface of plates, dishes and other maiolica wares from 16thC onward in Italy.
Jasper ware Type of unglazed stoneware body popularized by Josiah Wedgwood.
Jewelled decoration Method of decoration created by drops of enamel over gilding, resembling jewels, popular in France in the 18th century and at Royal Worcester.
Kakiemon Sparse asymmetrical style of decoration on a white ground introduced by Japanese potters in the 17th century.
Kendi Term used for a bulbous drinking vessel, used originally to hold holy water, but also as a hubble-bubble (see Hookah).
Knop Term used to describe the decorative finial on teapot and vase lids.
Kraak-porselein Late Ming Chinese blue and white porcelain exported by Dutch traders to Europe.
Lustre ware Pottery decorated with metallic pigments, popular in the Middle East, and Europe.
Majolica Colourful type of earthenware loosely based on Italian maiolica, popular in Europe and America from c.1850.
Manganese Mineral used to make purple brown pigments.
Overglaze Decoration applied on top of the glaze.
Palette Selection of colours used in decoration.
Parian Unglazed glassy type of porcelain popular for making imitation marble sculpted figures in the 19th century.
Pâte sur pâte Decorative technique made by building up layers of slip on a contrasting ground to simulate the effect of a cameo.
Press moulding Method of producing ceramic objects by forcing clay into a mould.
Salt glaze Glaze used on some stoneware made by throwing salt into the kiln during firing first developed in the Rhineland.
Satsuma Type of Japanese pottery with elaborate decoration, crackle glaze and heavy gilding named after the Japanese port where it was made.
Soft-paste porcelain Type of body developed in Europe in the 16th century made from various materials which may include kaolin, glass, soapstone or bone ash, in an attempt to create true hard-paste porcelain.
Slip Mixture of clay and water used as a form of decoration on some types of pottery.
Spur marks Indentations left in the glaze caused by small stilts used to support an object during firing.
Stoneware Non porous ceramic body made from clay and sand or flint, fired at high temperature.
Tin glaze Opaque white glaze made from tin oxide, used on maiolica, faience and delftware pottery.
Terracotta Red-coloured earthenware pottery, usually unglazed.
Toby jug Tankard or jug in the form of a toper.
Transfer printing Printed decoration made by transferring a design created on a copper plate via transfer paper onto a pottery or porcelain body.
Underglaze Decoration applied to a body before glazing and final firing.
Yue ware Distinctive type of Chinese stoneware coated in green glaze made from the Han dynasty to 10th century.

INDEX

Page numbers in **bold** refer to main entries; those in *italic* refer to illustrations.

ACKNOWLEDGMENTS

The publishers would like to thank all those who provided pictures for this book and in particular Christie's Images, Letitia Roberts, Director of the Porcelain Department at Sotheby's, for the pictures on pp154 and 155bl, and John Hays, Director of the Americana Department at Christie's for the pictures on pp160c and 160br.

Front cover: **tl** CI; **tr** RIB/CH/P; **cr** SL; bl RIB/CH/P; **br** SL. Back cover CI. Front flap RIB/SL Back flap RIB

2tl&tr RIB/SL; **2cl,c&cr** SPLL; **2bl&br** CI; **3tl** RIB/SL; **3tr** CI; **3c** RIB/SL; **3bl** P; **3br** RIB/SL; **11** SPLL; **15** DS; **16** RIB/JH; **17** RIB/JH; **18** RIB/JM/DS; **19** CL; **20** RIB/JH; **22** AM; **25** SZ; **26t** Be; **26b** CL; **27** SL; **28** B; **29t** RIB/CH/P; **29b** SL; **30** SL; **31** RIB/CH/P; **32** RIB/JM; **33** P; **34** RIB/CH/P; **35b** RIB/CH/P; **35tl** SL; **35tr** RIB/CH/P; **36l** SPLL; **36r** RIB/CH/P; **37t&c** SL; **37b** SPLL; **38t&b** SL; **39t,bl&br** SL; **40l&r** SL; **41t** SPLL; **41bl** SL; **41br** SPLL; **42l** SL; **42r** RIB/CH/P; **43l&r** RIB/CH/P; **43b** SNY; **44l&r** SL; **45t,bl&br** SL; **46t** Be; **46b** CI; **47l&tr** SPLL; **47br** RIB/PB; **48** SPLL; **49tr** RIB/CH/P; **49cr** SL; **49tl** SPLL; **49bl** RIB/CH/P; **50** RIB/CNY; **51tl** RIB/SL; **51tr** CI; **51bl** RIB/SL; **51br** RIB/SL/IB; **52l** RIB/CH/P; **52tr** CI; **52br** RIB/CH/P; **53&b** CI; **54l** CI; **54r** SPLL; **55l** CI; **55tr** SPLL; **55br** CI; **56l** CI; **56r** SL; **57tl&bl** CI; **57r** SL; **58tl,tr&b** RIB/CH/P; **59tl,tr,c&b** CI; **60** RIB/CH/P; **61tl&tr** CI; **61b** SL; **62l** SL; **62tr** CI; **62br** SL; **63t&b** SL; **64t&b** CI; **65l** P; **65tr,bc&br** CI; **66l&r** CI; **67t** RIB/CH/P; **67b** CI; **68t** RIB/CH/P; **68bl** CI; **68br** CI; **69t,bl** CI&br CI; **70l** P; **70b** CI; **71c** RIB/SL; **71b** CI; **72** P; **73tl** P; **73bl&r** RIB/SL; **74l** CI; **74tr** JJM; **74b** CI; **75l** RIB/CH/P; **75tc** P; **75br** CI; **76l** CI; **76r** JHA; **77t&r** Be; **77bl** CI; **78t** P; **78b** RIB/CH/P; **79t** P; **79b** RIB/CH/P; **80l** RIB/CH/P; **80br** P; **81l,tr&br** RIB/CH/P; **82tl** Be; **82tr&b** RIB/CH/P); **83l&r** P; **84l,tr&br** P; **85tl,tr&b** P; **86l** P; **86b** RIB/CH/P; **87l** Be; **87tr&br** RIB/CH/P; **88** CI; **89l&tr** RIB/CH/P; **90l,tr,&br** CI; **91tl** SL; **91bl** SPLL; **91r** CI; **92l&r** CI; **93l** CI; **93r** RIB/CH/P; **94t&b** CI; **95t** SPLL; **95bl&br** CI; **96l&r** CI; **97tl** SPLL; **97tr&b** CI; **98t&b** CI; **99tl,bl&r** CI; **100l&r** CI; **101t&l** CI; **102l** CI; **102r** SPLL; **103bl** CI; **103tl** SPLL; **103r** CI; **104l&r** CI; **105tl** CI; **105bl** SPLL; **105r** CI; **106t&b** CI; **107t&b** CI; **108** C; **109t** CI; **109l** SL; **109r** Be; **110tl,tr&br** CI; **111t&b** CI; **112** CI; **113l** RIB/CH/P; **113tr** CI; **113rc** RIB/CH/P; **113rb** RIB/CH/P; **113b** CI; **114l&r** CI; **115l,tr&br** CI; **116t&b** RIB/CH/P; **117tl** RIB/CH/P; **117bl** CI; **117r** CI; **118l&r** RIB/CH/P; **119tl** Be; **119c,bl&br** CI; **120l&r** CI; **121tl** CI; **121bl** SL; **121r** SPLL; **122t** P; **122b** RIB/CH/P; **123tl** P; **123bl** CI; **123r** RIB/CH/P; **124l** RIB/CH/P; **124c** P; **124r** RIB/CH/P; **125l&tr** RIB/CH/P; **125br** CI; **126l** RIB/CH/P; **126r** P; **127l** P; **127t** RIB/CH/P; **128** P; **129t** CI; **129l** RIB/CH/P; **129r** RIB/CH/P; **130tl** P; **130tr** Be; **130b** CI; **131tl** P; **131bl&r** CI; **132tl** RIB/CH/P; **132bl** CI; **132r** RIB/CH/P; **133t** RIB/CH/P; **133b** CI; **134l** RIB/CH/P; **134r** CI; **135tl,bl&r** CI; **136** P; **137t,cl&bl** P; **137r** RIB/CH/P; **138tl** P; **138bl** RIB/CH/P; **138r** P; **139l** RIB/DL; **139tr** RIB/CH/P; **139br** P; **140tl** RIB/B; **140tr** RIB/CH/P; **140bl** RIB/IB; **140br** RIB/B; **141t&b** GCC; **142t** SL; **142b** Be; **143t,bl&r** P; **144l,c&r** P; **145l&r** RIB/CH/P; **146l** SL; **146r** RIB/CH/P; **147tl** RIB/CH/P; **147rc** B; **147r** RDP; **148t&bl** CSK; **148br** RIB/CH/P; **149t&c** RIB/CH/P; **149b** Be; **150l,tc&r** P; **151tl** MJ; **151tr** RIB/CH/P; **151b** P; **152r** RIB/CH/P; **152tl&bl** P; **153tl&bl** B; **153r** RIB/CH/P; **154** SL; **155tl** GDS; **155bl** SL; **155br** GDS; **156t&b** DR; **157t** DR; **157c** CL; **157b** DR; **158l,c&r** DR; **159t&b** DR; 1 **160c** CL; **160bl** GDS; **160br** CL; **161tl** RT; **161tr,c&b** WTK

KEY
b bottom, **c** centre, **l** left, **r** right, **t** top

AM Allyson McDermott
B Bonhams, London
Be Bearnes, Torquay
CI Christie's Images, London
CL Christie's London
CNY Christie's New York
CSK Christie's South Kensington
DL Dr Laird
DR David Rago
DS D Stratton
DSe Dennis Severs
GCC Goss & Crested China Ltd
GDS Gary & Diana Stradling
IB Ian Booth
JH Jacqui Hurst
JHA Jonathan Home Antiques
JJM J & J May/John May
JM James Merrell
MJ Malcolm Jennings
P Phillips, London
PB Patricia Bayer
RDP Richard Dennis Publications (Plate XLVIII "Wedgwood Ceramics" by Maureen Batkin published by Richard Dennis
RIB Octopus Publishing Group Ltd
RIB/CH/P Octopus Publishing Group Ltd./Chris Halton/Phillips
RT Robert Teitelman
SL Sotheby's London
SPLL Sotheby's Picture Library, London
SNY Sotheby's New York
SZ Sotheby's Zurich
WTK William and Teresa Kurau